PRINCIPLES OF
ELECTRICITY AND MAGNETISM

SECOND EDITION

EMERSON M. PUGH,
Carnegie-Mellon University

EMERSON W. PUGH,
IBM Corporate Headquarters
Armonk, New York

PRINCIPLES OF
ELECTRICITY AND MAGNETISM

SECOND EDITION

 ADDISON-WESLEY PUBLISHING COMPANY
Reading, Massachusetts · Menlo Park, California
London · Amsterdam · Don Mills, Ontario · Sydney

ISBN 0-201-06014-0
CDEFGHIJKL-MA-79876

PREFACE

The primary purpose of this book is to provide an advanced treatment of electricity and magnetism with as much clarity, simplicity, and rigor as the vector notation permits. In order to assure adequate familiarity of the student with vectors, we have used Chapters 2 and 3 for a complete course in vector notation and a review of basic electrostatic concepts. These chapters, essentially unchanged from the first edition, have been found to be useful not only as a first course in vectors but also as a valuable review for students who have previously had introductory courses either in vectors, in electrostatics, or in both. Other advanced mathematical concepts such as complex variables and line integrals are used where necessary to make the treatment simpler or more rigorous. However, these concepts are fully developed when introduced so that no prior knowledge of mathematics beyond algebra and calculus is necessary for effective use of the book.

The first edition was written primarily for physicists, but it has been used so much in electrical engineering and applied physics courses that we have made a number of additions and alterations to the revised edition to increase its effectiveness for these courses. Where physicists and engineers have different terminology for the same thing both terms are introduced. Illustrative examples for basic physics concepts have been taken from important areas of modern engineering, for example high-frequency transmission lines for electronic circuitry, wave guides, non-linear magnetic devices (including ferrite core memory elements), and nonohmic semiconducting devices. The interrelations between electricity and magnetism and solid state physics which were highlighted in the first edition have been retained and updated in this new edition. We have retained the heavier emphasis on magnetic, rather than semiconductor, materials and devices because most students will have one or more independent courses dealing with semiconductors, while this may be their only introduction to magnetics.

Magnetic moments in materials are introduced using the fundamental Amperian current approach. This is followed by a rigorous derivation of the magnetic pole concept starting with Amperian currents. The two concepts are then used for solving some illustrative problems. Because the pole concept leads to scalar mathematics whereas the Amperian current concept leads to vector mathematics, most practicing engineers think in terms of poles while physicists cling to the more basic current concept. We believe this book now offers the clearest available treatment of the relationship between these two points of view.

Two new chapters have been added to the text: *The Poynting Vector* and *Relativity*. The latter is entirely new while the former utilizes material previously distributed elsewhere in the book combined with some new material which we believe will provide the reader with additional insight into this important subject. Particular attention has been given to providing a clear and correct treatment of the following subjects which are so often carelessly, if not incorrectly, treated: Faraday's law, Poynting vector momentum, static fields in dc circuits, dipole radiation near the dipole, magnetic poles and Amperian currents, and sources of emf. Finally, a large number of new problems have been added to the book and most of these have been tested in the classroom.

While the text can be used effectively for students with no mathematics beyond algebra and calculus, students with minimal preparation should not be expected to cover all the material in the book in a normal two semester course. The following sections consisting of one hundred pages can be omitted without loss of continuity: 4–11; 6–7, 10; 7–8; 8–10; 9–6, 12, 13, 14, 15, 16, 17; 10–4, 7, 8; 11–11, 12, 14; 12–5, 9, 10, 11; and 13–2, 5, 8. The remaining material provides at least skeletal coverage of all major topics in the book.

Some teachers have used the first edition in one semester courses for students with substantially better than average preparation by selectively omitting some additional material in the book. This should be equally possible with the revised edition. For example, chapters 6 and 11 on direct and alternating currents may be omitted entirely without breaking the continuity of the book. It is in fact anticipated that part or all of these chapters will be omitted by many one and two semester courses since these subjects are often fully treated in concurrent or forthcoming courses.

Many changes have been made in the original text in order to clarify the presentation, add important new material, and to take account of recent developments in the understanding and practical utilization of the principles of electricity and magnetism. We believe these changes will make the text more valuable to the student and teacher alike. To the many users of the text whose advice on revising was "Just don't change anything" we hope we have maintained enough of the original to keep your support and that you will find merit in the numerous modifications and additions.

The authors are indebted to the students and teachers who have made constructive suggestions based on their experience with the first edition. We also take pleasure in acknowledging the contributions of the following persons who have read and commented critically on all or parts of this revised edition: W. R. Beam, H. Chang, F. Culp, G. C. Feth, M. J. Freiser, G. J. Lasher, G. E. Pugh, and H. D. Young.

September 1969   E.M.P.
                                                                     E.W.P.

CONTENTS

Basic Relations of Electricity and Magnetism

	Gaussian	Heaviside-Lorentz	Rationalized mks
Lorentz force equation	$F = q\left(E + \dfrac{1}{c}v \times B\right)$	$F = q\left(E + \dfrac{1}{c}v \times B\right)$	$F = q(E + v \times B)$
Maxwell's field relations	$D = E + 4\pi P$ $B = H + 4\pi M$	$D = E + P$ $B = H + M$	$D = \epsilon_0 E + P$ $B = \mu_0 H + \mathcal{M}$
First	$\nabla \cdot D = 4\pi\rho$	$\nabla \cdot D = \rho$	$\nabla \cdot D = \rho$
Second	$\nabla \cdot B = 0$	$\nabla \cdot B = 0$	$\nabla \cdot B = 0$
Third	$\nabla \times E = -\dfrac{1}{c}\dot{B}$	$\nabla \times E = -\dfrac{1}{c}\dot{B}$	$\nabla \times E = -\dot{B}$
Fourth	$\nabla \times H = \dfrac{1}{c}(4\pi J + \dot{D})$	$\nabla \times H = \dfrac{1}{c}(J + \dot{D})$	$\nabla \times H = J + \dot{D}$
Poynting vector	$S = \dfrac{c}{4\pi}E \times H$	$S = cE \times H$	$S = E \times H$

Chapter 1 **SYSTEMS OF UNITS**

Electrostatic forces were observed as early as the time of Homer and probably had been observed throughout the ages. The magnetic forces due to natural permanent magnets also must have been observed in prehistoric times. Nevertheless, it remained for Gilbert, personal physician to Queen Elizabeth in the 16th century, to make the first scientific study of both of these forces.

It is now well known that all matter is made up of elementary particles, most of which carry either a positive or a negative charge. Compared with gravitational forces, the electrostatic forces between these particles are colossal.

With such powerful forces around them, how could the early philosophers have remained so ignorant of their existence? Probably because nature had hidden these forces so effectively by generally providing just as many negative charges (electrons) as there are positive charges of equal magnitude (protons) in each object. Only when this delicate balance is upset can these forces be observed as electrostatic forces.

Coulomb, in 1789, was the first to determine experimentally the fundamental law of attraction and repulsion between localized electrostatic charges. Coulomb expressed this law, which is named after him, in terms of the force F between two point charges of magnitude q and q' respectively. It states that

$$F = k \frac{qq'}{r^2} ,$$

where r is the distance between the point charges, and k is a constant which in a vacuum depends only on the units used in measuring the q's, r, and F. A positive sign for the product qq' means that the charges are repelled by each other, while a negative sign means they are attracted.

1

Chemical, mechanical, electric, and magnetic properties of all materials are strongly dependent upon the coulomb forces. However, when fundamental particles like the electron and proton are separated by much less than 10^{-9} m, as they are within most materials, the laws of interaction are too complex to be easily understood. An understanding of the role played by these coulomb forces in producing the physical and chemical properties of materials has been achieved only since 1927, with the development of quantum mechanics.

While the theoretical understanding of electricity and magnetism progressed so rapidly in the 19th century that Maxwell had predicted the existence of electromagnetic waves by 1864, the subject was largely a laboratory curiosity until the latter part of that century. It is true that Morse had invented the telegraph, which was put into operation by several companies as early as 1851, but the general public did not become familiar with the phenomena of electricity until the telephone and electric light brought electric wires into private homes. The first telephone switchboard was established by Bell in New Haven, Connecticut, in 1878 with 21 subscribers, and the first central power station was established by Edison to light 12 blocks of New York City in 1882. Someone in this latter group of customers installed an electric motor to grind coffee, and the electric power industry was born. To be able to charge customers equitably, the electric industry had to adopt a standard system of units for measurement. The system they adopted was called "practical" because the magnitudes for the various units were selected to be more useful in the practical problems of the day than were the cgs units used by scientists. The names of these practical units—ampere, volt, ohm, farad, coulomb, watt, joule, henry, etc.—are now familiar to most people.

In searching for a single system of units that could properly be called "the absolute system," scientists had developed a number of systems all based upon cgs units. Having established a single cgs system for use in mechanics, they had set about extending this system to the more recently discovered laws of electrostatics and magnetostatics. Soon it was discovered that at least two different systems were emerging, each of which had equally good claims to the term "absolute." Most of the textbooks in use between the two World Wars used one cgs system for electrostatics, another cgs system for magnetostatics, and the "practical" system for direct and alternating currents. The multiplicity of systems caused confusion, especially for beginning students. Two additional cgs systems, each attempting to unify the original systems, were developed and used in advanced theoretical texts. It was eventually recognized that there is no single system that properly can be called "absolute."

The arbitrary nature of systems of units can be seen from the development of units in mechanics. All mechanical systems of units have started from Newton's second law, stated as $F = ma$ rather than $F = kma$. In using $F = ma$, the law for gravitational attraction has had to be written as

$F = Gmm'/r^2$, where G is neither dimensionless nor unity. There is no reason to consider either of these laws to be more fundamental than the other. If the system had been started from the gravitational law, the laws would have been written as $F = mm'/r^2$ and $F = kma$. Force would then have the dimensions of (mass/length)^2, with units of (g/cm)^2. For consistency, the k in Newton's second law then must have the units g s^2/cm^3 (or g s^2 cm^{-3}). This system is just as consistent as is the one that has been adopted for mechanics. Fortunately, the system that has been adopted provides simpler calculations in most cases.

The variety of units in use in electromagnetic theory was especially confusing to students. Studies revealed, for example, that students had little carry-over from elementary physics courses to electrical engineering courses, largely because of the use of different units and definitions. The American Association of Physics Teachers organized a committee to study the problem.* This committee recommended adopting the mks system, proposed by Giorgi in 1901, for all teaching and experimental work in physics, and it proposed the name Newton (N) for the previously unnamed unit of force in the mks system. It also recommended the adoption of the Gaussian system, one of the numerous cgs systems, for advanced theoretical work. A sizable minority preferred to drop the Gaussian system as well and retain only the mks system, in the belief that it was as easy to use and that it provided much simpler comparisons with experimental results. The fact that the majority of articles in physics journals used the Gaussian system was a strong argument against dropping it; and this trend has continued, although some articles using the mks units have appeared in physics journals.

Shortly after the appearance of the recommendations by the committee of the AAPT, books using the mks system began to appear. Most of them adopted the rationalized mks units recommended by the committee. Unfortunately, a few of them adopted unrationalized units. Others called their systems either mksa or mksq, adding the a for ampere or the q for coulomb respectively, as an additional fundamental quantity. It is generally recognized that even in classical theory there is no need for more than three arbitrarily defined units, since the ampere and coulomb can be defined in terms of length, mass, and time, as shown in Section 1–5. In this text we use the rationalized mks units recommended by the committee, in accordance with common practice. This system is identical to the mksa system.

Although the mks units are based on length, mass, and time, it should be stressed that the number of fundamental units is arbitrary. In quantum field theory, for example, one often uses a set of units in which c and $\hbar$ are equated to unity. This, together with the value of the fine-structure constant $e^2/\hbar c$, reduces the number of fundamental units to one, usually taken to be length.

---

* *Am. Phys. Teacher* **3**, 90 (1935) and **6**, 144 (1938).

Table 1–1

STANDARD ABBREVIATIONS

ampere	A	meter	m
coulomb	C	kilogram	kg
volt	V	second	s
ohm	Ω	newton	N
farad	F	joule	J
henry	H	watt	W
weber	Wb	radian	rad
tesla	T or Wb m^{-2}	steradian	sr

An "International System of Units" was adopted by the General Conference on Weights and Measures in Paris during October, 1960. At that time 36 nations agreed to adhere to the treaty. This system was based entirely on the mks system, although the Kelvin degree, candela, and lumen were added for units of temperature, luminous intensity, and luminous flux respectively. The National Bureau of Standards announced* that they were adopting this system of units together with the abbreviations proposed for them and for other related units. These abbreviations are now used quite generally and Table 1-1 lists those required for this text.

We believe that considerable understanding of electricity and magnetism can be obtained by studying the historical development of systems of units. Therefore, we include here a brief account of that development.

1–1 The original cgs systems of units

Historically, electrostatics and magnetostatics were studied before current electricity. Knowledge of these phenomena increased rapidly after the demonstration by Coulomb that the forces between static charges and the forces between magnetic poles both followed the inverse-square law. This similarity between the two phenomena made possible the parallel development of the laws of electrostatics and magnetostatics. It also resulted in the simultaneous development of two independent systems of units. Both systems were based upon the centimeter-gram-second system that had become so useful in mechanics. In the Coulomb's law equations for the force between two concentrated charges and between two isolated magnetic poles:

$$F = k \frac{qq'}{r^2}, \tag{1–1}$$

$$F = k_m \frac{q_m q'_m}{r^2}, \tag{1–2}$$

---

* "NBS Adopts International System of Units." *NBS Tech. News Bull.*, **48**, 4, 61 (April, 1964).

the k and k_m were left out, which amounted to assuming that $k = k_m = 1$ in a vacuum and that both constants were dimensionless. With F in dynes and r in centimeters, Eqs. (1–1) and (1–2) provided the means for defining and measuring "unit charges" and "unit poles," respectively.

Analyses of indirect measurements* have shown that when experiments are in air or any other isotropic medium instead of in a vacuum, the results depend upon this medium. To allow for the effect of the medium, ϵ, the dielectric constant, and μ, the permeability, were placed in the denominators of (1–1) and (1–2) respectively. These quantities were arbitrarily chosen to be dimensionless and to be $\epsilon_v = \mu_v = 1$ in a vacuum. Thus, using $F = qq'/r^2$ as the defining equation, a charge that would repel an equal and like charge with a force of 1 dyne at a distance of 1 cm in a vacuum was defined as a unit charge. The unit magnetic pole was defined from $F = q_m q'_m/r^2$ in the same fashion.

Next, unit electrostatic and unit magnetostatic fields were defined as the electric field that would exert one dyne of force per unit charge, and the magnetic field that would exert one dyne of force per unit pole, respectively. The defining equations were $E = F/q$ and $H = F/q_m$.

When it was discovered that magnetic fields could be produced by electric currents or by the motion of electric charges, the two systems of units could no longer be considered as independent. Biot and Savart showed that magnetic fields produced by current circuits could be calculated by integrating

$$dH = K_I \frac{I \, dl \sin \theta}{r^2}$$

around the complete electric circuit. It was then realized that this equation made it possible to define a unit current in a fundamental manner by letting K_I be unity and dimensionless. The equation for defining a unit current then became

$$dH = \frac{I \, dl \sin \theta}{r^2}.$$

This unit became the cgs-electromagnetic unit (cgs-emu) of current.

However, the electrostatic system (cgs-esu) provides another very straightforward method of defining a unit current. Since current is merely the flow of electric charges, a unit current can be defined as the current that transports a unit charge past any point in the circuit in one second, $I = dq/dt$. Using the electrostatic unit for the charge, a unit of current was defined and called the cgs-electrostatic unit (cgs-esu) of current. These two units for electric current proved to differ from each other, both in magnitude and in

---

* No one has succeeded in making direct measurements of force between charges or poles with sufficient accuracy to jusify these equations. Their proofs rest upon analyses of very accurate, though indirect, measurements. See Plimpton and Lawton, *Phys. Rev.* **50**, p. 1066 (1936).

dimensions. In fact, it is quite obvious that one complete set of units can be obtained by starting with Eq. (1–1) and another complete set can be obtained by starting with Eq. (1–2).

Originally, no names were assigned for units either in the cgs-esu or in the cgs-emu, probably because they were presumed to be *the* fundamental units, for which no particular names were needed. This attitude now seems strange since there were, in fact, two complete and different sets of so-called "fundamental" units. Since both systems were cgs units, this caused considerable confusion. In the 1930's before the adoption of the mks system, some of the confusion was alleviated by assigning names to the units in each of the cgs systems. Because the practical units were obtained by multiplying the corresponding units in the cgs-emu system by powers of ten, the cgs-emu system was considered to be the absolute system, whose units should be the absolute ampere, the absolute ohm, etc. The names actually chosen and used for these quantities in the cgs-emu system were the practical units with the prefix *ab*, i.e., abampere, abvolt, abcoulomb, abohm, abfarad, etc. In a similar manner, the cgs-esu units were named statampere, statvolt, statcoulomb, statohm, statfarad, etc.

Long before these names were coined the surprising discovery had been made that the ratio obtained by dividing the quantity measuring a given charge in statcoulombs by the quantity measuring the same charge in abcoulombs was 2.9979×10^{10} cm s^{-1}, which equaled c, the velocity of light, both numerically and dimensionally, within the limits of error of the most accurate measurements. This discovery gave Maxwell one of the best clues for his remarkable theory of electromagnetic fields which led to his electromagnetic theory of light and his prediction of the existence of electromagnetic waves of lower frequencies.

1–2 The gaussian system of units

Gauss had pointed out that the two cgs systems could be made into one system by placing c (the velocity of light) in cm/sec in the denominator of the Biot-Savart law; that is,

$$dH = \frac{I \, dl \, \sin \theta}{cr^2}.$$

The gaussian system thus developed is the system that is now most commonly used by physicists in their advanced theoretical work. The five elementary equations required for defining this system are listed in the first column of Table 1–2. The first two of these equations define unit pole and unit magnetic field as they had been defined in the cgs-emu system, and the last two define unit current and unit charge as they had been defined in the cgs-esu system. The middle equation, from Biot and Savart, accomplishes a unification of the first two with the last two; that is, by the introduction of c into the denominator of this relation dH is expressed in oersteds from the

Table 1–2

SOME ELEMENTARY DEFINING EQUATIONS

Gaussian	Heaviside-Lorentz	Rationalized mks
$F = \dfrac{q_m q'_m}{r^2}$	$F = \dfrac{q_m q'_m}{4\pi r^2}$	$F = \dfrac{q_m q'_m}{4\pi\mu_0 r^2}$
$H = \dfrac{F}{q_m}$	$H = \dfrac{F}{q_m}$	$H = \dfrac{F}{q_m}$
$dH = \dfrac{I\,dl\sin\theta}{cr^2}$	$dH = \dfrac{I\,dl\sin\theta}{c4\pi r^2}$	$dH = \dfrac{I\,dl\sin\theta}{4\pi r^2}$
$q = \displaystyle\int I\,dt$	$q = \displaystyle\int I\,dt$	$q = \displaystyle\int I\,dt$
$F = \dfrac{qq'}{r^2}$	$F = \dfrac{qq'}{4\pi r^2}$	$F = \dfrac{qq'}{4\pi\epsilon_0 r^2}$

emu system, when I is in statamperes from the esu system. Most of the units in the gaussian system have the prefix *stat*, as statcoulomb, statampere, and statvolt; however, the units of magnetic field are the same as in the cgs-emu; namely, oersteds for H, and gauss for B. It is interesting that these two names are in very common use even though it has often been pointed out that, in these systems, B and H must have the same units, since $B = H + 4\pi M$.

1–3 Rationalization

One further refinement of the gaussian system was introduced independently by Heaviside and Lorentz. This refinement, which is called rationalization, is illustrated by the second column of elementary equations in Table 1–2. The resulting system, called the Heaviside-Lorentz system, is an excellent one. Since no names have been chosen for its units, its adoption at this late date would seem to upset all of the well-established units. Although the advantages to be gained are not sufficient to justify rationalization of the gaussian system, the development was important, for it led to the adoption of the rationalized rather than the unrationalized system of mks units.

The process of rationalization consists in substituting the expression for the area of a sphere, $4\pi r^2$, for the r^2 in the denominator of the first, third, and fifth of the defining equations in Table 1–2. While, at first sight, this appears to make some equations more complex, it does remove the 4π's from equations that are very frequently used in calculations. The net result of rationalization is to simplify the most useful equations. A complete understanding of the advantages and disadvantages of the various systems can be obtained only by studying the subject with each system in mind.

1–4 The rationalized mks system of units

The so-called "absolute" system (cgs-emu) was developed from the first four elementary relations in the first column of Table 1–2, but with the c left out of the denominator of the third relation. This defined I in abamperes and q in abcoulombs. The "practical" units ampere and coulomb were each arbitrarily defined as 1/10 of the corresponding "absolute" units. The abvolt then was defined by the energy relation 1 abvolt × 1 abcoulomb = 1 erg, and the "practical" unit, the volt, was arbitrarily defined as 10^8 abvolts. The combined effect of the arbitrary definitions for the coulomb and volt fortunately produced the relation 1 volt × 1 coulomb = 10^7 erg = 1 joule. Giorgi pointed out that a fundamental system, based upon the standard meter and the standard kilogram, could be devised to include the "practical" units of electricity and magnetism. While the defining equation, $F = ma$, defines the dyne as the unit of force, with m in grams and a in cm s^{-2}, it defines a unit of force equal to 10^5 dynes with m in kilograms and a in m s^{-2}. In the 1930's this new unit of force was appropriately christened the *newton*. The unit of energy in the mks system then became the newton·meter = 10^5 dynes × 10^2 cm = 1 joule. Since the "practical" units were defined from the "absolute" units, using arbitrarily chosen powers of ten, a fundamental system including the "practical" units can be developed from the elementary relations in the third column of Table 1–2 by giving the arbitrary constant μ_0 the correct value. It is easily demonstrated that $\mu_0 = 4\pi 10^{-7}$ is the correct numerical value, where the 4π stems from the decision to rationalize the equations. While μ_0 could be chosen dimensionless, it is much more convenient to assign it the value

$$\mu_0 = 4\pi 10^{-7} \text{ H m}^{-1}$$

The necessity for introducing μ_0 into the first mks relation and ϵ_0 into the fifth does increase the complexity of the system. However, these constants do take care of all conversions of units. Furthermore, they are related by $\mu_0 \epsilon_0 c^2 = 1$, where c is 2.9979×10^8 m s^{-1}, the velocity of light in a vacuum. Hence,

$$\epsilon_0 = \frac{1}{\mu_0 c^2} \cong 8.854 \times 10^{-12} \text{ F m}^{-1}$$

The constants μ_0 and ϵ_0 have been arbitrarily chosen to set up the mks system of units. They are sometimes called the *permeability* of free space and the *permittivity* of free space, respectively.

The advantages of the mks system over the other two systems can be seen by examining the frontispiece and Table 1–2. The gaussian system is simplest only in the two Coulomb laws and the Biot-Savart law. Although these equations are useful for teaching elementary principles, they are less frequently used in calculations of real problems. The Lorentz force equation, Maxwell's field relations, and the Poynting vector are included in the frontispiece. Although these do require the vector notation, introduced in the next two chapters, and the general principles developed later in the text, it is not necessary to understand these relations to compare the simplicity of their forms in the three systems.

1–5 Legal definitions for the mks units

The five elementary equations of the rationalized mks system that are listed in Table 1–2 could provide the basis for establishing all of the mks units. With μ_0 chosen to be exactly $4\pi10^{-7}$ H m^{-1}, all units would have their present legal value. However, the Coulomb law for forces between isolated magnetic poles cannot be used for precise measurements or precise definitions. Magnetic poles are always poorly defined and, in fact, are undoubtedly fictitious. Far more precise measurements can be made upon the forces between current-carrying wires. *Since the forces between two circuits carrying the same current can be precisely calculated and precisely measured, these forces are now used to establish a standard ampere as the basic unit for the mks system.* The principles involved in calculating the forces between current circuits are discussed in more detail in Chapter 8. It is sufficient to note here that the magnetic fields calculated from the third equation in the last column of Table 1–2 are proportional to I, but the Lorentz equation for the force on an element of wire must be written $dF = \mu_0 HI\,dl \sin \phi$. These resulting forces are proportional to I^2 but also depend upon the value of μ_0. Thus the magnitudes of all the electric and magnetic units are determined by the arbitrarily chosen constant μ_0.

In 1908 an International Conference on Electrical Units and Standards met in London and adopted independent definitions for an international ohm, an international ampere, and an international volt. Later, when it became possible to make measurements to five and six significant figures, inconsistencies appeared. In fact, the magnitudes of all these international units differed slightly from those obtained from fundamental principles beginning with the units of mass, length, and time. It was soon realized that the electrical units should be legally defined in terms of the well-established units of mass, length, and time. Careful experiments by the national laboratories of major nations were required. These were scheduled for completion by 1940 but were delayed by World War II.

In October 1946 the International Committee on Weights and Measures recommended the following definitions,* which are substantially the same as the definitions that were passed by the Congress of the United States prior to March 1, 1949:

1. *The ampere* (unit of electric current). The ampere is the constant current which, if maintained in two straight parallel conductors of infinite length, of negligible circular sections, and placed 1 meter apart in a vacuum, will

---

* National Bureau of Standards Circular 475, p. 36, June 1949.

produce between these conductors a force equal to 2×10^{-7} newton per meter of length.

2. *The volt* (unit of difference of potential and of electromotive force). The volt is the difference of electric potential between two points of a conducting wire carrying a constant current of 1 ampere, when the power dissipated between these points is equal to 1 watt.

3. *The ohm* (unit of electric resistance). The ohm is the electric resistance between two points of a conductor when a constant difference of potential of 1 volt, applied between these two points, produces in this conductor a current of 1 ampere, this conductor not being the seat of any electromotive force.

4. *The coulomb* (unit of quantity of electricity). The coulomb is the quantity of electricity transported in 1 second by a current of 1 ampere.

5. *The farad* (unit of electric capacitance). The farad is the capacitance of a capacitor between the plates of which there appears a difference of potential of 1 volt when it is charged by a quantity of electricity equal to 1 coulomb.

6. *The henry* (unit of electric inductance). The henry is the inductance of a closed circuit in which an electromotive force of 1 volt is produced when the electric current in the circuit varies uniformly at a rate of 1 ampere per second.

7. *The weber* (unit of magnetic flux). The weber is the magnetic flux which, linking a circuit of 1 turn, produces in it an electromotive force of 1 volt as it is reduced to zero at a uniform rate in one second.

Object of these definitions. These definitions are intended solely to fix the magnitude of the units and not the methods to be followed for their practical realization.* This realization is effected in accord with the well-known laws of electromagnetism. For example, the definition of the ampere represents only a particular case of the general formula expressing the forces which are developed between conductors carrying electric currents, chosen for the simplicity of its verbal expression. It serves to fix the constants in the general formula which has to be used for the realization of the unit.

1–6 Conversion of units

Table 1-3 is designed to make the conversion from one system of units to another as simple and automatic as possible. It contains four columns of

---

* For the methods used by the U.S. Bureau of Standards to establish the ampere, see Section 8–10.

units, i.e., mks, gaussian, cgs-esu, and cgs-emu, since practically all published papers on the subject of electricity and magnetism have employed units from one of these four systems.

The only units listed in these columns are those that fit into completely consistent systems of equations. This means that only one unit is mentioned in any square, although in a few instances alternate names for the same unit are listed.

Each row in the table contains equal quantities. For example, 1 coulomb $= 2.998 \times 10^9$ statcoulomb $= 10^{-1}$ abcoulomb. The automatic method for converting a quantity expressed in one system into a proper value expressed in another system has been described by the statement "all conversion factors are unity" when the names of units are treated as algebraic quantities. When it is desired to convert a quantity expressed in one set of units into the same quantity expressed in another set of units, the original quantity is multiplied by ratios, whose values are unity, arranged so that names of units in the denominator cancel the same names in the numerator and leave only the desired names of units. For example, to express the length 20,000 inches terms of miles,

$$l = 20{,}000 \, \cancel{in.} \times \frac{1 \, \cancel{ft}}{12 \, \cancel{in.}} \times \frac{1 \, mi}{5280 \, \cancel{ft}} = \frac{20{,}000}{12 \times 5280} \, mi = 0.317 \, mi.$$

Suppose it is desired to find in farads the capacitance C of a capacitor whose capacitance in the cgs-esu system of units is 1000 statfarads:

$$C = 1000 \, \cancel{\text{statfarads}} \frac{1 \text{ farad}}{2.998^2 \times 10^{11} \, \cancel{\text{statfarads}}},$$

$$C = 111.2 \times 10^{-11} \text{ farad}.$$

Since 1 farad $= 2.998^2 \times 10^{11}$ statfarads, the ratio

$$\frac{1 \text{ farad}}{2.998^2 \times 10^{11} \text{ statfarads}}$$

is unity. Note that the statfarad in the denominator of the ratio cancels algebraically with the statfarad in the numerator of the given quantity.

Actually, it is possible to check many of the numbers in this table by the above-mentioned system. For example, knowing that the electric field in the cgs-emu system of units must be in abvolt per centimeter, one can obtain the number given for this electric field by taking an electric field of one volt per meter and writing it as

$$\frac{1 \text{ volt}}{m} \times \frac{10^8 \text{ abvolts}}{\text{volt}} \times \frac{1 \text{ m}}{100 \text{ cm}} = 10^6 \frac{\text{abvolts}}{\text{cm}}.$$

Hence, 1 volt/m $= 10^6$ abvolts/cm.

Table 1–3 TABLE FOR CONVERSION OF UNITS

Unit	mks	gaussian	cgs-esu	cgs-emu
Length	1 meter	100 centimeters	100 centimeters	100 centimeters
Mass	1 kilogram	1000 grams	1000 grams	1000 grams
Force	1 newton	10^5 dynes	10^5 dynes	10^5 dynes
Work, Energy	1 joule	10^7 ergs	10^7 ergs	10^7 ergs
Power	1 watt	10^7 ergs/sec	10^7 ergs/sec	10^7 ergs/sec
Charge	1 coulomb	2.998×10^9 statcoulombs	2.998×10^9 statcoulombs	10^{-1} abcoulomb
Current	1 ampere	2.998×10^9 statamperes	2.998×10^9 statamperes	10^{-1} abampere
Potential difference	1 volt	$\dfrac{1}{299.8}$ statvolt	$\dfrac{1}{299.8}$ statvolt	10^8 abvolts
Electric field, E	$1 \dfrac{\text{newton}}{\text{coulomb}}$ or $1 \dfrac{\text{volt}}{\text{meter}}$	$\dfrac{1}{2.998 \times 10^4} \dfrac{\text{dyne}}{\text{statcoul}}$ or $\dfrac{\text{statvolt}}{\text{cm}}$	$\dfrac{1}{2.998 \times 10^4} \dfrac{\text{dyne}}{\text{statcoul}}$ or $\dfrac{\text{statvolt}}{\text{cm}}$	$10^6 \dfrac{\text{dynes}}{\text{abcoul}}$ or $\dfrac{\text{abvolts}}{\text{cm}}$
Resistance	1 ohm	$\dfrac{1}{(2.998)^2 \, 10^{11}}$ statohm	$\dfrac{1}{(2.998)^2 \, 10^{11}}$ statohm	10^9 abohms
Resistivity	1 ohm·meter	$\dfrac{1}{(2.998)^2 \, 10^9}$ statohm·cm	$\dfrac{1}{(2.998)^2 \, 10^9}$ statohm·cm	10^{11} abohms·cm

Capacitance	1 farad	$(2.998)^2\,10^{11}$ statfarads	$(2.998)^2\,10^{11}$ statfarads	$\dfrac{1}{10^9}$ abfarad
Inductance	1 henry	$\dfrac{1}{(2.998)^2\,10^{11}}$ stathenry	$\dfrac{1}{(2.998)^2\,10^{11}}$ stathenry	10^9 abhenries
Magnetic flux density, B	$1\,\dfrac{\text{weber}}{\text{m}^2}$	10^4 gauss	†	10^4 gauss
Magnetic flux, ϕ	1 weber	10^8 maxwells or gauss·cm²	†	10^8 maxwells or gauss·cm²
Magnetic field, H	$1\,\dfrac{\text{ampere}}{\text{meter}}$	$4\pi10^{-3}$ gauss or oersted	†	$4\pi10^{-3}$ gauss or oersted*
Permeability of free space, μ_0	$4\pi \times 10^{-7}\,\dfrac{\text{webers}}{\text{ampere·meter}}$ or $\dfrac{\text{henry}}{\text{meter}}$	1	1	1
Permittivity of free space, ϵ_0	$8.854 \times 10^{-12}\,\dfrac{\text{coulomb}}{\text{volt·meter}}$ or $\dfrac{\text{farad}}{\text{meter}}$	1	1	1
Velocity of light in free space, c	$2.9979 \times 10^8\,\dfrac{\text{meter}}{\text{second}}$	$2.9979 \times 10^{10}\,\dfrac{\text{centimeter}}{\text{second}}$	$2.9979 \times 10^{10}\,\dfrac{\text{centimeter}}{\text{second}}$	$2.9979 \times 10^{10}\,\dfrac{\text{centimeter}}{\text{second}}$

* Originally the term gauss was used for both B and H in the gaussian system of units. Some felt, however, that the two quantities should be distinguished from each other and the word oersted was adopted for H only. Since in the gaussian system one always uses the expression $B = H + 4\pi M$, it makes little sense to use different names for the units of B and H.

† These spaces are left blank because, although units for these quantities could be devised in the cgs-esu system, they have not been used.

In the mks system there are two constants that occur frequently, neither of which is dimensionless. They are

$$\epsilon_0 \cong \frac{10^{-9}}{36\pi} \text{ farad/m} \cong 8.85 \times 10^{-12} \text{ farad/m}$$

and

$$\mu_0 = 4\pi \times 10^{-7} \text{ henry/m},$$

where μ_0 is an exact number by definition. The value of ϵ_0 is obtained from μ_0 and the velocity of light, i.e.,

$$\epsilon_0 = \frac{1}{\mu_0 c^2}, \qquad \text{where } c \text{ is } 2.99792 \pm 0.00002 \times 10^8 \text{ m/s.*}$$

The experimentally determined value of c occurs several places in the table, where it has been rounded off to 2.998×10^8 m/s.

PROBLEMS

1–1 The following problem is very artificial but it does illustrate the nature and intensity of electrostatic forces. Two copper spheres, each 0.2 mm in diameter, are 40 cm apart in a vacuum. They do not exert electrostatic forces upon each other if both are electrically neutral. However, this lack of force can be considered as due to the balancing of two large repulsive forces by two equally large attractive forces.

a) Calculate the repulsive force exerted by the copper nuclei of one sphere on the copper nuclei of the other, ignoring the effects of the electrons.

b) Convert this repulsive force into tons.

c) Consider the number of electrons to be exactly equal to the number of protons in each of these neutral copper spheres. Could electrostatic force measurements be used to determine whether the magnitudes of the proton and electron charges are exactly equal? Suppose the proton charge is given by $e(1 + \delta)$, where e is the magnitude of the electronic charge. Then suppose an electrostatic force of 0.0001 g could be but is not detected when the spheres are 10 cm apart. What is the largest value δ can have under these circumstances?

[*Note.* Under these conditions each sphere would have to be charged to about 3 million volts to detect a repulsion between them. Such charging would remove about 200 billion electrons which is 2×10^{-8} of the total electrons.]

1–2 Compute the electrostatic force of repulsion between two α-particles at a separation of 10^{-11} cm, and compare with the force of gravitational attraction between them. Each α-particle has a charge of $+2e$, or $2 \times 1.60 \times 10^{-19}$ C and a mass (2 protons + 2 neutrons) of $4 \times 1.67 \times 10^{-24}$ g.

1–3 6.02×10^{23} atoms of monatomic hydrogen have a mass of one gram. How far would the electron of a hydrogen atom have to be removed from the nucleus for the force of attraction to equal the weight of the atom?

---

* Rank, Bennett, and Bennett, *Phys. Rev.* **100**, 993, 1955.

1–4 Radium decomposes radioactively to form radon (atomic number 86) by emitting an α-particle from its nucleus.

a) What is the force of repulsion between the radon nucleus and the α-particle when the distance between them is 5×10^{-11} cm?

b) What is the acceleration of the α-particle at this distance?

1–5 In the Bohr model of the hydrogen atom, an electron revolves in a circular orbit around a nucleus consisting of a single proton. If the radius of the orbit is 5.28×10^{-9} cm, find the number of revolutions of the electron per second. The force of electrostatic attraction between proton and electron provides the centripetal force.

1–6 Two charges of $+10^{-9}$ C each are 8 cm apart in air. Find the magnitude and direction of the force exerted by these charges on a third charge of $+5 \times 10^{-11}$ C that is 5 cm distant from each of the first two charges.

1–7 Two equal positive point charges are a distance $2a$ apart. Midway between them and normal to the line joining them is a plane. The locus of points where the force on a point charge placed in the plane is a maximum is, by symmetry, a circle. Calculate the radius of this circle.

1–8 How many excess electrons must be placed on each of two small spheres spaced 3 cm apart if the force of repulsion between the spheres is to be 10^{-19} N?

1–9 Each of two small spheres is positively charged, the combined charge totaling 26 statcoul. What is the charge on each sphere if they are repelled with a force of 3 dynes when placed 4 cm apart?

1–10 Point charges of 2×10^{-9} C are situated at each of three corners of a square whose side is 0.20 m. What would be the magnitude and direction of the resultant force on a point charge of -1×10^{-9} C if it were placed (a) at the center of the square? (b) at the vacant corner of the square?

1–11 Calculate the force exerted on 1 μC of positive charge, concentrated at a point, by another equal charge 1 m from it. Make your calculations in cgs-esu and in mks units, leaving ϵ_0 as an unknown quantity in the latter case. Calculate the magnitude of ϵ_0 from these two results.

1–12 The quantity of an electric charge can be expressed as q_s in statcoulombs or as q_a in abcoulombs. Since these two were defined by different equations they have different dimensions in terms of mass, length, and time (M, L, and T). The first four gaussian equations in Table 1–2, with the c omitted from the Biot-Savart relation, establish the dimensions of q_a in abcoulombs. Remembering that force has the dimensions (mass·length/time^2) or the units (g·cm/s^2):

a) Determine the units of q_a in g, cm, and s.

b) Determine the units of q_s in g, cm, and s (from the fifth gaussian equation).

c) Show that the ratio of the two quantities involves a velocity in cm s^{-1}.

d) Show that the introduction of a velocity into the denominator of the Biot-Savart equation makes the five gaussian equations consistent.

e) Could this c have been introduced into one of the Coulomb's law equations, instead of into the Biot-Savart relation, to make the five equations consistent dimensionally? Demonstrate your statement.

1–13 In both of the cgs systems the unit of energy is the erg = dyne·cm and in all of the systems the work required to move a charge dq through a potential difference of V is $dW = V\,dq$. From the fact that there are 2.998×10^{10} statamp in 1 abamp, calculate the relation between statvolts and abvolts. Would you expect to find portable instruments in a well-equipped laboratory capable of measuring 1 statvolt and 1 abvolt?

1-14 The expression *electron-volt* is often found in modern magazines. Of what physical quantity is this a unit? Express the electron-volt in terms of the proper mks unit.

Chapter 2 VECTOR MATHEMATICS FOR ELECTROSTATIC FIELDS

Coulomb's law for the attraction or repulsion between two point charges was discussed briefly in Chapter 1. When the force on any charge q is due to the nearness of a number of other charges such as q_1, q_2, q_3, etc., the force exerted on q by each of these charges must be calculated separately, since each of these forces usually points in a different direction. The total force on q is the resultant of all these forces and can be obtained by graphical construction, but unless the forces are all in the same plane graphical solutions will be difficult.

Fortunately, rather simple vector algebra and calculus have been invented that are ideally suited for the analytical solutions of the problems involving such forces. Since most of the subject of Electricity and Magnetism deals with forces of one kind or another, in the aggregate time will be saved by spending a little time now to learn these vector methods.

In this and the next chapter the vector rules and operations that will be needed are included. Only the essential operations are presented, and these will be used over and over again in the chapters that follow. The presentation of these vector operations provides an opportunity for reviewing some of the important principles covered in more elementary texts.

2–1 The vector notation

Physical quantities of the simplest type are completely defined by the assignment of a single number combined with the unit of measure. Examples are a mass of 25 lb, a temperature of 53°C, and a pressure of 110 lb in.$^{-2}$. These are called *scalar* quantities.

Some other physical quantities can be completely defined only by specifying a direction in addition to the magnitude. These are called *vector* quantities. Examples of vector quantities are a velocity of 25 mi h^{-1} in the northeast

direction and a current density of 29 A m⁻² to the right and parallel to the
surface of a conductor. Such statements are cumbersome. It is much simpler
to draw straight lines with lengths proportional to the magnitudes of the
quantities and with arrowheads to indicate their directions. For vectors not
in a single plane, a three-dimensional coordinate system can be used for
drawing these lines in perspective.

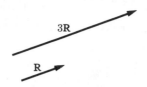

Fig. 2–1 Two vectors, **R** and **3R**.

Ordinary algebra was invented to deal primarily with scalar quantities.
A somewhat similar algebra, vector algebra, is used to deal with vector
quantities. Like ordinary algebra, vector algebra makes use of addition,
subtraction, and multiplication. A notation is needed to distinguish vectors
from scalars. By rather general agreement, vectors are printed in boldface
type, while italic type is used for scalars. For example, the symbol **R**
designates a vector and R designates its scalar magnitude.

The following axioms define the properties of these vectors:

1. Two vectors are equal to each other if they have the same magnitude and
   point in the same direction. The relative location of the vectors in space
   does not affect this equality.

2. The multiplication of a vector by a scalar results in a new vector, pointing
   in the same direction as the original vector and having a magnitude equal
   to the product of the individual magnitudes. The vector **3R** is three times
   as long as the vector **R**, as shown in Fig. 2–1.

2–2 Vector addition and subtraction

Suppose a wide river is flowing northward at 3 mi h⁻¹ and an outboard
motorboat on this river is heading due east at 4 mi h⁻¹ with respect to the
water. With respect to the land, this boat moves with a velocity equal to the
resultant of these two velocities; namely, at 5 mi h⁻¹ in the direction 36°52′
north of east. In vector language, the velocity **v** of the boat with respect to
the land is the vector sum of the velocity of the river, $\mathbf{v}_R$, and the velocity of
the boat with respect to the water, $\mathbf{v}_B$. The vector equation

$$\mathbf{v} = \mathbf{v}_R + \mathbf{v}_B$$

designates that **v** is the resultant or vector sum of $\mathbf{v}_R$ and $\mathbf{v}_B$. *It does not mean that the magnitude v of the vector* **v** *is equal to the algebraic sum of the magnitudes of the other two vectors;* i.e.,

$$v \neq v_R + v_B.$$

The general definition for the sum of any two vectors, for example **A** and **B** is shown in Fig. 2–2, where the process of vector addition is illustrated. If **A** and **B** are any two vectors representing physical quantities having the same units and drawn to the same scale, the sum

$$\mathbf{R} = \mathbf{A} + \mathbf{B}$$

is obtained by drawing **B**, using its correct length and direction, with its tail at the head of **A**. The vector sum **R** is the single vector that can be drawn from the tail of **A** to the head of **B**. Obviously,

$$\mathbf{A} + \mathbf{B} = \mathbf{B} + \mathbf{A},$$

since the vector **R**, drawn on the left as a solid line in Fig. 2–2, represents the sum of the left side of the equation, while the vector **R** drawn on the right as a dashed line represents the sum on the right side of this equation. These two **R**'s are identical, since they point in the same direction and have the same length.

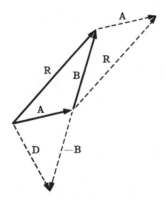

Fig. 2–2 Vector addition and subtraction. The vectors illustrated are related by $\mathbf{R} = \mathbf{A} + \mathbf{B} = \mathbf{B} + \mathbf{A}$ and $\mathbf{D} = \mathbf{A} - \mathbf{B}$.

Figure 2–2 also illustrates the subtraction of vectors. To subtract **B** from **A**, merely add to vector **A** the vector −**B**, which is the same length as **B** but points in the opposite direction. In Fig. 2–2 the vector

$$\mathbf{D} = \mathbf{A} - \mathbf{B}$$

is the difference between vectors **A** and **B**.

The vector addition of two vectors is easily extended to the addition of three or any number of vectors. If **C** is a third vector in Fig. 2–2, the resulting relation,

$$\mathbf{A} + \mathbf{B} + \mathbf{C} = \mathbf{R} + \mathbf{C},$$

contains the sum of just two vectors on the right side of the equation. Since the sum of two vectors has been defined, this equation defines the sum of three vectors. Note that **C** does not need to be in the plane determined by **A** and **B**. To obtain the sum

$$\mathbf{S} = \mathbf{A} + \mathbf{B} + \mathbf{C} + \mathbf{D} + \mathbf{E},$$

draw **B** starting with its tail at the head of **A**, then **C** with its tail at the head of **B**, then **D** with its tail at the head of **C**, etc. The vector sum **S** is that single vector that can be drawn from the tail of **A** to the head of **E**. No matter what order is used in adding the five vectors, vector **S** will be found to have the same length and direction.

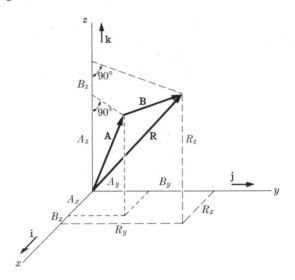

Fig. 2–3 The vectors **R** = **A** + **B** are shown with their x-, y-, and z-components. Note that $A_x + B_x = R_x$, $A_y + B_y = R_y$, and $A_z + B_z = R_z$.

2–3 Unit vectors and rectangular components

In the previous section vector additions and subtractions were performed graphically. While graphical methods are excellent in learning to visualize the processes, some analytical methods will be needed for obtaining numerical results. This is especially true when the vectors cannot be drawn in the same plane.

Set up a rectangular coordinate system with the x-, y-, and z-axes having their origin at the tail of a vector **A**, as shown in Fig. 2–3. The vector **A** can be thought of as the sum of three vectors: one parallel to the x-axis of length A_x, a second parallel to the y-axis of length A_y, and a third parallel to the z-axis of length A_z, where A_x, A_y, and A_z are the respective rectangular components of **A** along the three axes. It is not correct, however, to write A as the sum of A_x, A_y, and A_z, since these are all scalar quantities for which the correct relation is

$$A^2 = A_x^2 + A_y^2 + A_z^2.$$

To write a vector equation for **A**, three unit vectors **i**, **j**, and **k** are introduced, which respectively point in the positive x-, the positive y-, and the positive z-directions. The equation

$$\mathbf{A} = \mathbf{i}A_x + \mathbf{j}A_y + \mathbf{k}A_z$$

then correctly expresses the magnitude and direction of **A**. The unit vectors **i**, **j**, and **k** have magnitudes of one, with no dimensions of their own. The rectangular components A_x, A_y, and A_z have the same units or dimensions as A. For example, the velocity of the boat on the river, mentioned at the beginning of Section 2–2, would be written

$$\mathbf{v} = \mathbf{i}(4 \text{ mi h}^{-1}) + \mathbf{j}(3 \text{ mi h}^{-1})$$

if the coordinate x- and y-axes are drawn to point east and north, respectively.

Figure 2–3 also illustrates in rectangular coordinates the vector sum

$$\mathbf{R} = \mathbf{A} + \mathbf{B}.$$

From the diagram it can be seen that the x-component of the resultant vector **R** equals the algebraic sum of the x-components of **A** and **B**. The same is true for the other components:

$$R_x = A_x + B_x,$$
$$R_y = A_y + B_y,$$
$$R_z = A_z + B_z.$$

When all the vectors involved are written in terms of the unit vectors **i**, **j**, and **k**, the operations involving addition or subtraction become simple algebraic operations. For example, if

$$\mathbf{R} = \mathbf{A} + \mathbf{B} - \mathbf{C}$$

then

$$\mathbf{R} = \mathbf{i}(A_x + B_x - C_x) + \mathbf{j}(A_y + B_y - C_y) + \mathbf{k}(A_z + B_z - C_z).$$

2–4　Coordinate systems

There are many coordinate systems besides the rectangular x, y, z system just treated. These systems have been invented to simplify the solution of particular problems and, in general, the solution of any problem is simplified by choosing the best coordinates for that problem. The systems most com-

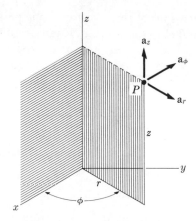

Fig. 2–4 Cylindrical coordinates r, ϕ, and z.

monly used are the rectangular, the cylindrical, and the spherical coordinate systems, which have many common properties. All the problems in this book can be solved in one of these three systems.

Cylindrical and spherical coordinates. Each of the three coordinate systems employs three coordinates to locate a point in space. In the cylindrical system the coordinates are r, ϕ, and z, which locate the point P as shown in Fig. 2–4. The x-, y-, and z-axes are given to show the relationships between the two systems. Notice that r and ϕ are measured in the xy-plane. The angle ϕ may be measured from any convenient plane through the z-axis, though in general it is measured from the xz-plane as in Fig. 2–4. In this system three unit vectors, $\mathbf{a}_r$, $\mathbf{a}_\phi$, and $\mathbf{a}_z$ (or $\mathbf{k}$), are defined as follows:

$\mathbf{a}_r$ points in the direction in which P moves when r is increased with ϕ and z constant,

$\mathbf{a}_\phi$ points in the direction in which P moves when ϕ is increased with r and z constant,

$\mathbf{a}_z$ points in the direction in which P moves when z is increased with r and ϕ constant.

Note that in rectangular coordinates the respective directions of $\mathbf{i}$, $\mathbf{j}$, and $\mathbf{k}$ can also be defined, respectively, as the directions in which P moves when only x increases, when only y increases, and when only z increases. Note also that $\mathbf{i}$, $\mathbf{j}$, and $\mathbf{k}$ are mutually perpendicular and that $\mathbf{a}_r$, $\mathbf{a}_\phi$, and $\mathbf{a}_z$ are also mutually perpendicular. *Coordinate systems in which the unit vectors are mutually perpendicular are called orthogonal coordinate systems.*

In Fig. 2–5 the point P is located in a spherical coordinate system. In this system the unit vectors are $\mathbf{a}_R$, $\mathbf{a}_\theta$, and $\mathbf{a}_\phi$ which point, respectively, in the

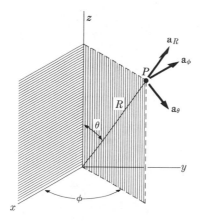

Fig. 2–5 Spherical coordinates R, θ, and ϕ.

directions in which P moves on increasing only R, only θ, and only ϕ. These unit vectors are mutually perpendicular, and therefore the system is orthogonal. When there is no likelihood of confusing the unit vectors $\mathbf{a}_R$ in spherical coordinates and $\mathbf{a}_r$ in cylindrical coordinates, it is common to write the unit-radial vector of spherical coordinates as $\mathbf{a}_r$, also. This then reserves a unit vector $\mathbf{a}_R$ which can point in any direction, not necessarily radially.

While the three systems have many similarities, the rectangular system differs markedly from the other two in one respect; namely, its unit vectors do not depend on the coordinates, but always point in the same direction. For example, a point may be located uniquely in rectangular coordinates by giving its distance from the origin as

$$\mathbf{R} = \mathbf{i}3 \text{ m} + \mathbf{j}4 \text{ m} + \mathbf{k}12 \text{ m},$$

whereas the corresponding expression in spherical coordinates

$$\mathbf{R} = \mathbf{a}_R 13 \text{ m},$$

shows only that the point is 13 m from the origin and does not locate the point uniquely. The uncertainty arises from the fact that the direction of $\mathbf{a}_R$ depends on both θ and ϕ. This uncertainty can hardly be considered a disadvantage, however, since it is a distinct advantage in some problems (for example, when an expression is needed for a general point arbitrarily located on the surface of a sphere).

Vector examples. In Fig. 2–6 a position vector $\mathbf{R}$ and a velocity vector $\mathbf{v}$ are plotted. Since $\mathbf{R}$ is in meters, while $\mathbf{v}$ is in m s^{-1}, these two cannot be added or subtracted. However, diagrams containing vectors representing different kinds of quantities are often very useful. For example, Fig. 2–6 could represent a mass point located in space at P but having a velocity $\mathbf{v}$. The

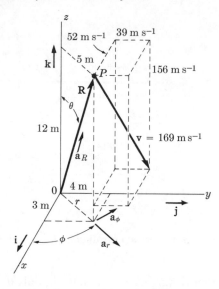

Fig. 2–6 Vector diagram showing a position vector **R** and a velocity vector **v**. The relations among the three coordinate systems are illustrated.

magnitudes and directions of **R** and **v** have been chosen so that each of them can be expressed in all three coordinate systems with rather simple numbers. The position of P can be expressed by

$$\mathbf{R} = \mathbf{i}3 + \mathbf{j}4 + \mathbf{k}12$$

or

$$\mathbf{R} = \mathbf{a}_r5 + \mathbf{a}_\phi0 + \mathbf{a}_z12$$

or

$$\mathbf{R} = \mathbf{a}_R13.$$

The last expression does not mean that P can be located with only one quantity. To determine its location, the spherical coordinates θ and ϕ for $\mathbf{a}_R$ must be known. Likewise, the velocity can be expressed by

$$\mathbf{v} = -\mathbf{i}52 + \mathbf{j}39 - \mathbf{k}156$$

or

$$\mathbf{v} = \mathbf{a}_r0 + \mathbf{a}_\phi65 - \mathbf{a}_z156$$

or

$$\mathbf{v} = -\mathbf{a}_R144 + \mathbf{a}_\theta60 + \mathbf{a}_\phi65.$$

The reader should study these examples carefully to help him in understanding vector expressions. These examples will be used in our study of other vector principles. The study of the electrostatic fields that are produced by distributions of charges can be used to illustrate the usefulness of vector algebra.

2–5 Electrostatic fields

Electrostatic fields are found in the neighborhood of any distribution of electric charges. *The electrostatic field* **E** *at any point is defined by the relation* **E** $= \lim_{q \to 0}$ **f**$/q$, *where* **f** *is the force that would act on a test charge q if it were placed at that point.* It is necessary to define the field as the limit of the force/charge as the test charge approaches zero only because a finite test charge may change the distribution of neighboring charges, especially if these reside on conductors. The simple definition **E** $=$ **f**$/q$ is valid for any size of test charge, *provided the charges producing the field are not displaced by the presence of the test charge.* The definition states that the electric field **E** is a vector that points in the same direction as the force vector **f** when the test charge q is positive.

If a point charge Q is located at the origin of coordinates in Fig. 2–5 and a test charge q is located at P, Coulomb's law for the force **f** on q can be written as

$$\mathbf{F} = \mathbf{a}_r \frac{Qq}{\epsilon_0 4\pi r^2} = \mathbf{a}_r k \frac{Qq}{r^2},$$

where $k = 1/4\pi\epsilon_0 \cong 9 \times 10^9$ N m^2 C^{-2} and $\mathbf{a}_r$ is the unit radial vector in spherical coordinates. The field **E** produced at P by the charge Q at the origin is then

$$\mathbf{E} = \mathbf{a}_r k \frac{Q}{r^2}. \tag{2–1}$$

If in addition to Q there are other point charges, the field at P will be the vector sum of the individual vector fields due to each of the charges. To write a general expression for such a field, it is desirable first to write the field at P due to a charge that is not at the origin but is at some general point in space. When the charge Q is not at the origin, the expression for the field at P due to Q is somewhat more easily visualized when it is written in rectangular rather than in spherical coordinates. Rectangular coordinates can be placed so that Q is located at (a, b, c) and the general point P is located at (x, y, z). The distance R between Q and P is then given by

$$R^2 = (x - a)^2 + (y - b)^2 + (z - c)^2.$$

A vector drawn from Q to P is expressed by

$$\mathbf{R} = \mathbf{i}(x - a) + \mathbf{j}(y - b) + \mathbf{k}(z - c).$$

A unit vector $\mathbf{a}_R$ parallel to **R** can be obtained by dividing **R** by its magnitude; i.e.,

$$\mathbf{a}_R = \frac{\mathbf{R}}{R} = \mathbf{i}\frac{x - a}{R} + \mathbf{j}\frac{y - b}{R} + \mathbf{k}\frac{z - c}{R},$$

and the field at P due to the charge Q is then

$$\mathbf{E} = \mathbf{a}_R \frac{kQ}{R^2}.$$

Note the distinction between the unit radial vector $\mathbf{a}_r$ and the more general unit vector $\mathbf{a}_R$.

If a series of n point charges are located with q_1 at (x_1, y_1, z_1), q_2 at (x_2, y_2, z_2), etc., with, finally, q_n at (x_n, y_n, z_n), the electric field $\mathbf{E}$ at P will be given by the vector sum of the electric fields due to the individual charges. Thus

$$\mathbf{E} = k \sum_{i=1}^{i=n} \frac{q_i}{r_i^2} \left[\mathbf{i} \frac{x - x_i}{r_i} + \mathbf{j} \frac{y - y_i}{r_i} + \mathbf{k} \frac{z - z_i}{r_i} \right], \qquad (2\text{–}2)$$

where

$$r_i = \sqrt{(x - x_i)^2 + (y - y_i)^2 + (z - z_i)^2}.$$

Field calculations. The electric field at all points in space due to any particular distribution of point charges can be calculated by direct substitution into Eq. (2–2). To obtain such a general expression, the point P at which the field is calculated is given the general coordinates (x, y, z). Each of the charges in the distribution contributes one term to the summation; its particular magnitude and its particular coordinates are substituted into that term. If many charges are involved, the expression is too long to be practical. Fortunately, simpler methods can be developed for calculating fields in most actual problems. However, since the simpler methods are based upon the fundamental law expressed by Eq. (2–2), its implications should be mastered. The calculation of the field at some particular point due to two point charges should help to clarify this law.

As an illustration of the use of the general expression, a charge $q_1 = -16 \times 10^{-10}$ C can be placed at the origin, where $x_1 = y_1 = z_1 = 0$, and another charge $q_2 = 12 \times 10^{-10}$ C can be placed at the point in the yz-plane where $x_2 = 0$, $y_2 = 3$ m, and $z_2 = 4$ m. The field can then be found at any other point, for example on the z-axis where $x = 0$, $y = 0$, and $z = 4$ m.

$$\mathbf{E} = \frac{9 \times 10^9 (-16 \times 10^{-10})}{4^2} \left[\mathbf{k} \frac{4}{4} \right]$$
$$+ \frac{9 \times 10^9 (12 \times 10^{-10})}{3^2} \left[\mathbf{j} \frac{0 - 3}{3} + \mathbf{k} \frac{4 - 4}{3} \right],$$
$$\mathbf{E} = -1.20\mathbf{j} - 0.90\mathbf{k}, \qquad \text{all in N C}^{-1}.$$

The field at a point 4 m above the origin on the z-axis is then 1.50 N C^{-1} pointing in such a direction that its components along the x-, y-, and z-axes are respectively 0, -1.20 N C^{-1}, and -0.90 N C^{-1}. This calculation could have been simplified by choosing either cylindrical or spherical coordinates with the axis of symmetry passing through the two charges.

2–6 Dipole fields

In nature electric charges frequently occur in pairs having equal magnitudes but opposite signs. When a dielectric is brought into an electric field, its electrons do not drift as they do in metals because they are too tightly bound to the nuclei. While they do not drift, they do become displaced so that the center of the cloud of electrons surrounding each nucleus no longer coincides with the center of that nucleus. While the external electric field produced by each atom is zero when these centers coincide, it is not zero when the centers are displaced.

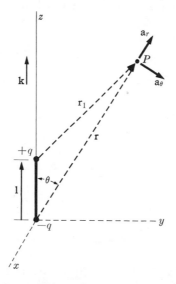

Fig. 2–7 An electric dipole consisting of two charges $+q$ and $-q$ separated by a distance l. The field at P due to the dipole is calculated in the text.

Any pair of equal and opposite charges whose centers are separated constitute a dipole. The most important dipoles have very small separations between two relatively large charges; these will be called *ideal* dipoles. Since ideal dipoles will be encountered frequently, it is desirable to obtain a general expression for the electric fields they produce.

Consider charges $-q$ at the origin and $+q$ on the z-axis at $z = l$, where $l \ll r$, as shown in Fig. 2–7. With complete symmetry about the z-axis, the ϕ coordinate of the spherical coordinates can be ignored. The vector distance to P from $+q$ is $\mathbf{r}_1$ and from $-q$ is $\mathbf{r}$. The electric field at P is then given by

$$\mathbf{E} = \frac{kq\mathbf{r}_1}{r_1^3} - \frac{kq\mathbf{r}}{r^3} = \frac{kq}{r^3}\left[\frac{\mathbf{r}_1 r^3}{r_1^3} - \mathbf{r}\right]. \qquad (2\text{–}3)$$

The first term in the brackets should be converted to the spherical coordinates to obtain a more understandable expression. From Fig. 2–7,

$$\mathbf{r}_1 = \mathbf{r} - 1 = \mathbf{r} - \mathbf{k}l$$

and

$$r_1^2 = r^2 + l^2 - 2rl \cos \theta.$$

Hence

$$r_1^{-3} = r^{-3}[1 + (l/r)^2 - 2(l/r) \cos \theta]^{-3/2}.$$

Since $(l/r) \ll 1$, the binomial theorem may be used to expand the expression in brackets into a power series in l/r, in which powers higher than the first can be neglected. Then

$$\frac{r^3}{r_1^3} = 1 + 3(l/r) \cos \theta.$$

Multiplying this by $\mathbf{r}_1 = \mathbf{r} - \mathbf{k}l$ and again discarding higher terms in l/r gives

$$\frac{\mathbf{r}_1 r^3}{r_1^3} = \mathbf{r}[1 + 3(l/r) \cos \theta] - \mathbf{k}l.$$

But

$$\mathbf{k} = \mathbf{a}_r \cos \theta - \mathbf{a}_\theta \sin \theta$$

and

$$\mathbf{r} = \mathbf{a}_r r;$$

hence

$$\frac{\mathbf{r}_1 r^3}{r_1^3} - \mathbf{r} = \mathbf{a}_r 2l \cos \theta + \mathbf{a}_\theta l \sin \theta.$$

Substituting this into Eq. (2–3) gives the electric field of the ideal dipole:

$$\mathbf{E} = \frac{kql}{r^3} (\mathbf{a}_r 2 \cos \theta + \mathbf{a}_\theta \sin \theta).$$

Note that the field produced is proportional to the product of the magnitude of one of the charges and the distance they are separated. *This product ql is called the moment of the dipole*, which will be designated by p. The *ideal dipole field* is then given by

$$\mathbf{E} = \frac{p}{4\pi\epsilon_0 r^3} (\mathbf{a}_r 2 \cos \theta + \mathbf{a}_\theta \sin \theta), \tag{2–4}$$

since $k = 1/4\pi\epsilon_0 = 8.988 \times 10^{-9}$ or 9×10^{-9} within 0.13 percent.

Fields of ideal dipoles vary inversely as the cube of the distance of the observer from the center of the dipole. They are twice as strong on the extension of the dipole axis as on the normal plane that bisects this axis. It will be found that the electric field of an ideal dipole given in Eq. (2–4) can be obtained more easily by first calculating the electrostatic potentials around the dipole. In order to calculate electrostatic potentials, a vector method

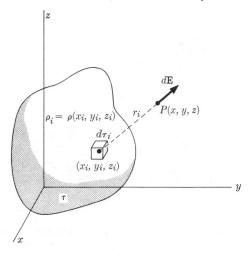

Fig. 2–8 Showing the direction and magnitude of $d\mathbf{E}$, the electric field at P due to the charge in an infinitesimal element of the volume τ.

must be introduced for determining the work required to move a charge in an electrostatic field. This method will utilize the scalar or dot product of vectors.

2–7 Fields by direct integration

While Eq. (2–2) is useful only for calculating fields due to distributions of point charges, it can be expressed in the form of a vector integral for use in calculating fields due to continuous distributions. Just as a scalar integral can be considered to constitute an algebraic summation of infinitesimal scalar elements, *a vector integral constitutes a vector summation of infinitesimal vector elements*. In Fig. 2–8 charges are smoothly distributed within the irregular volume τ, and it is desired to calculate the electric field at the point P whose coordinates are (x, y, z). A small element of volume $d\tau_i$ located at (x_i, y_i, z_i) contains the charge $\rho_i \, d\tau_i$, where ρ_i is the charge density at (x_i, y_i, z_i). If $d\tau_i$ is sufficiently small the field it produces at P can be calculated from Coulomb's law:

$$d\mathbf{E}_i = \mathbf{a}_i k \frac{\rho_i \, d\tau_i}{r_i^2}, \qquad (2\text{–}5)$$

and the total field at P can be obtained by adding vectorially the contributions of all such elements within the volume

$$\mathbf{E} = \int_\tau k \mathbf{a}_i \frac{\rho_i \, d\tau_i}{r_i^2}, \qquad (2\text{–}6)$$

where $\mathbf{a}_i$ is a unit vector directed from (x_i, y_i, z_i) toward (x, y, z), r_i is the

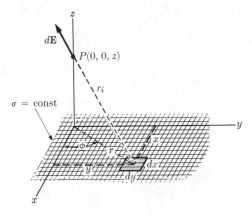

Fig. 2–9 The direction and magnitude of dE, the electric field at P due to the surface charge in an infinitesimal element of area in the xy-plane.

distance between these two points, and $d\tau_i$ is the element of volume $dx_i \, dy_i \, dz_i$. Note that during this integration the point P is fixed, so that x, y, z must be treated as constants. Frequently the process of integration can be simplified by writing the integrand in terms of the unit vectors $\mathbf{i}, \mathbf{j}$, and $\mathbf{k}$, which provides three scalar integrals, each multiplied by one of these unit vectors.

This process can be illustrated by calculating the electric field near an infinite plane that has a uniform surface density of charge σ, whose units in the mks system are C m^{-2}. Coordinates can be chosen so that the xy-plane coincides with the charged plane. Since the charged plane extends to infinity in all directions, the field point P can be placed at z on the z axis, without loss of generality. The field at P due to the charge enclosed in the area $dx \, dy$ at the coordinates $(x, y, 0)$, as shown in Fig. 2–9, is

$$d\mathbf{E} = \frac{k\sigma \, dx \, dy}{r_i^2}\left(-\mathbf{i}\,\frac{x}{r_i} - \mathbf{j}\,\frac{y}{r_i} + \mathbf{k}\,\frac{z}{r_i}\right),$$

where

$$r_i^2 = x^2 + y^2 + z^2.$$

A double integration of $d\mathbf{E}$, with x and y each varying from $-\infty$ to ∞, is required to determine $\mathbf{E}$. Since $\mathbf{i}, \mathbf{j}$, and $\mathbf{k}$ are constants, $\mathbf{E}$ can be expressed by three scalar double integrals, each multiplied by a unit vector. That is,

$$\mathbf{E} = k\sigma\left(-\mathbf{i}\int_{-\infty}^{\infty}\int_{-\infty}^{\infty}\frac{x \, dx \, dy}{r_i^3} - \mathbf{j}\int_{-\infty}^{\infty}\int_{-\infty}^{\infty}\frac{y \, dx \, dy}{r_i^3} + \mathbf{k}\int_{-\infty}^{\infty}\int_{-\infty}^{\infty}\frac{z \, dx \, dy}{r_i^3}\right).$$

(2–7)

The first two double integrals are each zero, as should be expected from symmetry considerations, but the third gives

$$\mathbf{E} = \mathbf{k}\frac{\sigma}{2\epsilon_0}, \tag{2-8}$$

a field which is independent of the coordinates at all points above the charged plane. Below the plane the field has this same magnitude but is in the $-\mathbf{k}$ direction.

The foregoing calculations may appear artificial, since infinite planes do not exist. However, Eq. (2–8) is an excellent approximation to the fields at points that are relatively close to a large flat sheet having a uniform surface density of charge. For points less than 0.005 cm from its center, a flat disk of 1-cm diameter can be assumed infinite, unless inaccuracies must be kept below 1%.

Rectangular coordinates are not suitable for the solution of all problems involving vector integrations. When other coordinates are used, the possibility that unit vectors may not be constant during the integration must be carefully considered. The symmetry of the problem of Fig. 2–9 suggests that cylindrical coordinates should be more suitable than the rectangular coordinates. The element of area on the charged plane would then be $ds = r \, d\phi \, dr$ and

$$dE = \frac{k\sigma r \, d\phi \, dr}{r_i^2}\left(-\mathbf{a}_r\frac{r}{r_i} + \mathbf{k}\frac{z}{r_i}\right),$$

where

$$r_i^2 = r^2 + z^2.$$

The field at P is then given by

$$\mathbf{E} = k\sigma\left(-\int_0^\infty\int_0^{2\pi}\mathbf{a}_r\frac{r^2 \, dr \, d\phi}{r_i^3} + \mathbf{k}z\int_0^\infty\int_0^{2\pi}\frac{r \, dr \, d\phi}{r_i^3}\right),$$

where $\mathbf{a}_r$ must remain in the integrand because its direction varies with ϕ, whereas $\mathbf{k}$ can be removed from its integrand, since it is independent of both r and ϕ. The first double integral can be written

$$-\int_0^\infty\frac{r^2 \, dr}{r_i^3}\int_0^{2\pi}\mathbf{a}_r \, d\phi,$$

and since

$$\int_0^{2\pi}\mathbf{a}_r \, d\phi = \int_0^{2\pi}(\mathbf{i}\cos\phi + \mathbf{j}\sin\phi) \, d\phi = 0,$$

the first double integral vanishes. The field is then given by the second double integral alone, whose value is the same as the third double integral of Eq. (2–7), namely, $\mathbf{E} = \mathbf{k}(\sigma/2\epsilon_0)$.

It should be noted that $\int_0^{2\pi}\mathbf{a}_r \, d\phi = 0$, whereas $\mathbf{a}_r\int_0^{2\pi}d\phi = 2\pi\mathbf{a}_r$, and is not zero. The vanishing of $\int_0^{2\pi}\mathbf{a}_r \, d\phi$ is easily demonstrated without formal integration by noting that it constitutes a vector sum of coplanar vectors of equal magnitude and, since they are directed uniformly in all radial directions, this sum must be zero. An important corollary is that $\oint d\mathbf{l} = 0$ for any closed path of integration. This can be understood geometrically from the fact that $\int d\mathbf{l}$ over any path represents the vector sum of all displacements included in traversing that path; that is, this integral gives the vector that can be drawn from the point of beginning to the point of ending. When the path is a closed one this vector must be zero. This statement can be proved analytically as follows:

$$\oint d\mathbf{l} = \oint (\mathbf{i}\, dx + \mathbf{j}\, dy + \mathbf{k}\, dz)$$

$$= \mathbf{i}\int_{x_1}^{x_1} dx + \mathbf{j}\int_{y_1}^{y_1} dy + \mathbf{k}\int_{z_1}^{z_1} dz = 0.$$

2–8 Scalar or dot product of vectors

It has been assumed that scalar and vector quantities can be multiplied together to provide new vectors of different lengths. The question arises whether or not an operation involving the multiplication of one vector by another can be defined to yield useful and consistent results. It will be shown that at least two such operations can be defined.

One of the most common operations in physics is to calculate the work done when a given force $\mathbf{F}$ moves a body a given displacement $\mathbf{d}$. If the force is not in the direction of motion, the work done is given by $Fd \cos \widehat{Fd}$. Thus a useful definition for the product of two vectors $\mathbf{A}$ and $\mathbf{B}$ is

$$\mathbf{A} \cdot \mathbf{B} = AB \cos \widehat{AB}. \tag{2–9}$$

This equation defines the operation known either as the dot product or the scalar product, since it gives a scalar quantity and the multiplication is indicated by a dot. The dot distinguishes this from another type of vector multiplication which is indicated by a cross. To determine the usefulness of the dot product it is desirable to investigate (1) the algebraic laws governing dot products of vectors and vector sums, and (2) the geometric and physical properties of these products.

The algebra of dot products. From the definition, $\mathbf{A} \cdot \mathbf{B} = \mathbf{B} \cdot \mathbf{A}$. While this vector operation does commute, the cross product does not. By simple geometry it can be shown that if $\mathbf{S} = \mathbf{D} + \mathbf{E}$, then $\mathbf{R} \cdot \mathbf{S} = \mathbf{R} \cdot \mathbf{D} + \mathbf{R} \cdot \mathbf{E}$. Furthermore, if $\mathbf{R} = \mathbf{A} + \mathbf{B}$,

$$\mathbf{R} \cdot \mathbf{S} = \mathbf{A} \cdot \mathbf{D} + \mathbf{B} \cdot \mathbf{D} + \mathbf{A} \cdot \mathbf{E} + \mathbf{B} \cdot \mathbf{E}, \tag{2–10}$$

which is easily extended to products of the sums of any number of vectors.

Properties of dot products. The definition of the dot product given in Eq. (2–9) indicates that $\mathbf{A} \cdot \mathbf{B} = 0$ if $\mathbf{A}$ and $\mathbf{B}$ are perpendicular, and that $\mathbf{A} \cdot \mathbf{B} = AB$ if they are parallel. These results can be utilized to obtain some useful relations between unit vectors in orthogonal coordinate systems. Since such unit vectors are mutually perpendicular, one unit vector dotted into another is zero unless it is the same unit vector, and then the product is unity. The relations between unit vectors in rectangular coordinates are

$$\mathbf{i} \cdot \mathbf{i} = \mathbf{j} \cdot \mathbf{j} = \mathbf{k} \cdot \mathbf{k} = 1,$$
$$\mathbf{i} \cdot \mathbf{j} = \mathbf{i} \cdot \mathbf{k} = \mathbf{j} \cdot \mathbf{k} = 0;$$

in cylindrical coordinates are

$$\mathbf{a}_r \cdot \mathbf{a}_r = \mathbf{a}_\phi \cdot \mathbf{a}_\phi = \mathbf{k} \cdot \mathbf{k} = 1,$$
$$\mathbf{a}_r \cdot \mathbf{a}_\phi = \mathbf{a}_r \cdot \mathbf{k} = \mathbf{a}_\phi \cdot \mathbf{k} = 0;$$

and in spherical coordinates are

$$\mathbf{a}_r \cdot \mathbf{a}_r = \mathbf{a}_\theta \cdot \mathbf{a}_\theta = \mathbf{a}_\phi \cdot \mathbf{a}_\phi = 1,$$
$$\mathbf{a}_r \cdot \mathbf{a}_\theta = \mathbf{a}_r \cdot \mathbf{a}_\phi = \mathbf{a}_\theta \cdot \mathbf{a}_\phi = 0.$$

The dot product provides a convenient method for calculating the angle between two vectors or any two directed lines in space. From Eq. (2–9) the angle $\widehat{AB}$ between the positive directions of the vectors $\mathbf{A}$ and $\mathbf{B}$ can be obtained by using

$$\cos \widehat{AB} = \frac{\mathbf{A} \cdot \mathbf{B}}{AB}. \tag{2–11}$$

If two vectors $\mathbf{A}$ and $\mathbf{B}$ are each written with unit vectors in terms of their components, as illustrated in Section 2–4, their dot product is easily calculated from Eq. (2–10). Nine terms are obtained, but because of the relations between the orthogonal unit vectors all but three of these are zero, and therefore

$$\mathbf{A} \cdot \mathbf{B} = A_x B_x + A_y B_y + A_z B_z. \tag{2–12}$$

The scalar magnitudes for Eq. (2–11) may be calculated from $\mathbf{A} \cdot \mathbf{A} = A^2$ and $\mathbf{B} \cdot \mathbf{B} = B^2$. An important application of this method of calculating angles is the testing of two vectors to determine if they are perpendicular. If neither $\mathbf{A}$ nor $\mathbf{B}$ is zero, but $\mathbf{A} \cdot \mathbf{B} = 0$, then $\mathbf{A}$ and $\mathbf{B}$ are perpendicular.

Expressions almost identical to Eq. (2–12) may be obtained for cylindrical and spherical coordinates. They are

$$\mathbf{A} \cdot \mathbf{B} = A_r B_r + A_\phi B_\phi + A_z B_z$$

and

$$\mathbf{A} \cdot \mathbf{B} = A_r B_r + A_\theta B_\theta + A_\phi B_\phi.$$

To demonstrate that the dot product between two vectors is independent of the coordinate system, the $\mathbf{R} \cdot \mathbf{v}$ in Fig. 2–6 may be calculated in all three coordinate systems.

In rectangular coordinates,

$$\mathbf{R} \cdot \mathbf{v} = -3 \cdot 52 + 4 \cdot 39 - 12 \cdot 156 = -1872;$$

in cylindrical coordinates,

$$\mathbf{R} \cdot \mathbf{v} = 5 \cdot 0 + 0 \cdot 65 - 12 \cdot 156 = -1872;$$

and in spherical coordinates,

$$\mathbf{R} \cdot \mathbf{v} = -13 \cdot 144 = -1872.$$

The product of the absolute magnitudes of the two vectors is $Rv = 13 \cdot 169 = 2197$, and hence the angle between positive directions of $\mathbf{R}$ and $\mathbf{v}$ is

$$\cos^{-1}(-1872/2197) = 148°34'.$$

Many geometric relations that are in common use can be so easily derived with vector mathematics that they need not be remembered. For example, the expression for r_1 in Fig. 2–7, used in calculating the dipole field, is easily derived. From the diagram,

$$\mathbf{r}_1 = \mathbf{r} - \mathbf{l} \quad \text{and} \quad \mathbf{r}_1 \cdot \mathbf{r}_1 = (\mathbf{r} - \mathbf{l}) \cdot (\mathbf{r} - \mathbf{l});$$

therefore

$$r_1^2 = r^2 + l^2 - 2\mathbf{r} \cdot \mathbf{l} = r^2 + l^2 - 2rl \cos \theta,$$

which is the relation used in Section 2–6.

2–9 Line element

Frequently in physics it is necessary to calculate the work done when a force moves some object along a path in space. In Chapter 1 the potential difference between two points along a wire is defined by the power dissipated by a steady rate of flow of charges between these two points. The wire was not essential to this definition. In general, either positive or negative work is done by an electric field any time a charge moves through it. To calculate this work it is desirable to use a line element $d\mathbf{l}$, which expresses the direction of the displacement of the center of a charge as it moves a short distance through space. In rectangular coordinates this line element is

$$d\mathbf{l} = \mathbf{i}\, dx + \mathbf{j}\, dy + \mathbf{k}\, dz, \tag{2–13}$$

where dx, dy, and dz are the rectangular components of the small displacement. In cylindrical coordinates the line element is

$$d\mathbf{l} = \mathbf{a}_r\, dr + \mathbf{a}_\phi r\, d\phi + \mathbf{k}\, dz.$$

The unit vector $\mathbf{a}_\phi$ is multiplied by $r\, d\phi$ instead of just $d\phi$, because the latter does not give the displacement of the object in length units. From Fig. 2–4 it can be seen that when ϕ increases by $d\phi$ the point P moves the distance $r \, d\phi$. Likewise, from Fig. 2–5 it can be seen that when θ increases by $d\theta$, P moves a distance $r \, d\theta$. Furthermore, when ϕ increases by $d\phi$ in this figure, P moves a distance $r \sin \theta \, d\phi$. In spherical coordinates the line element is then

$$d\mathbf{l} = \mathbf{a}_r \, dr + \mathbf{a}_\theta r \, d\theta + \mathbf{a}_\phi r \sin \theta \, d\phi.$$

2–10 Line integral

When a force $\mathbf{f}$ moves an object a distance $d\mathbf{l}$, the work it does on that object is given by $dW = f \, dl \cos \gamma$, where γ is the angle between $\mathbf{f}$ and $d\mathbf{l}$. The work may be expressed vectorially by $dW = \mathbf{f} \cdot d\mathbf{l}$. Using Eq. (2–12), this can be written $dW = f_x \, dx + f_y \, dy + f_z \, dz$, which contains nothing but scalar quantities. If the force $\mathbf{f}$ moves the object from a point with coordinates (x_1, y_1, z_1) to one with coordinates (x_2, y_2, z_2), the total work done is

$$W = \int_1^2 \mathbf{f} \cdot d\mathbf{l} = \int_{x_1}^{x_2} f_x \, dx + \int_{y_1}^{y_2} f_y \, dy + \int_{z_1}^{z_2} f_z \, dz.$$

In performing the integration with respect to x, for example, the value substituted for f_x must express its value for every point on the path the object travels. This defines the line integral. An example, using f in newtons and l in meters, should help to clarify the process.

Suppose an object moves in the yz-plane from a point P_1 on the z-axis at $(0, 0, 3 \text{ m})$ to a point P_2 whose coordinates are $(0, 4m, 5m)$. The work done on this object by a force $\mathbf{f}$ whose value varies from point to point in space in accordance with the relation $\mathbf{f} = \mathbf{i}x + \mathbf{j}z + \mathbf{k}z$ is

$$W = \int_0^0 x \, dx + \int_0^4 z \, dy + \int_3^5 z \, dz.$$

The values for these integrals will depend upon the path traveled from P_1 to P_2, and although there is an infinite number of these paths that could be traveled, calculations for the two paths shown in Fig. 2–10 should clarify the method. The straight line joining the two points directly is expressed analytically by $z = 3 + y/2$. For this path,

$$W = \int_0^4 \frac{6 + y}{2} \, dy + \int_3^5 z \, dz = 24 \text{ J},$$

since $dx = 0$ throughout this path. When the dashed path is followed, the integration must be done in two steps, first over the horizontal and then over the vertical portion. On the horizontal path $z = 3$, independently of y, while on the vertical portion $y = 4$ independently of z. The integral then becomes

$$W = \left(\int_0^4 3 \, dy + \int_3^3 3 \, dz \right) + \left(\int_4^4 z \, dy + \int_3^5 z \, dz \right) = (12 + 8 = 20) \text{ J}.$$

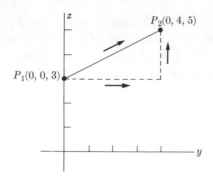

Fig. 2–10 Two of the many possible paths of integration between points P_1 and P_2 are illustrated, one by a dashed line and the other by a solid line.

The work done on the object by **f** depends upon the path traveled in Fig. 2–10, namely, 24 J over the solid line and only 20 J over the dashed line. Since the work done by **f** is not independent of the path, it is called a *nonconservative force. By definition, the work done by conservative forces must be independent of the path traveled.*

2–11 Electrostatic potentials

It was pointed out in Section 2–5 that whenever the distribution of charges throughout space is known, the fields can in principle be calculated directly, although the mathematical difficulties are often very great. Usually one does not know the location and value of all charges in space, and therefore more powerful methods for solving electrostatic problems are needed. Even when the charge distributions are known, the fields usually can be computed more easily with potential theory than by the direct summations or integrations of Sections 2–5, 2–6, and 2–7.

In Chapter 4 a general method for solving electrostatic problems is developed. This method is based upon differential equations given by Poisson and Laplace which express the variations of electrostatic potentials with coordinates in space. Two fundamental laws that must be obtained before these equations can be derived are (1) the law stating that in any system of static charges the work done in carrying a test charge around any closed path is zero, and (2) Gauss' law that the electric field **E** can be expressed consistently in terms of the density of electric flux lines. The first of these will now be developed; the second is postponed to Chapter 3.

The work done by the field in moving a test charge q through a distance $d\mathbf{l}$ in an electrostatic field **E** is $dW = q\mathbf{E} \cdot d\mathbf{l}$. The difference in potential between the two ends of the element $d\mathbf{l}$ is defined by $-dV = dW/q = \mathbf{E} \cdot d\mathbf{l}$. The negative sign indicates that the field **E** points in the direction of decreasing potential. The difference of potential between any two points P_a and P_b

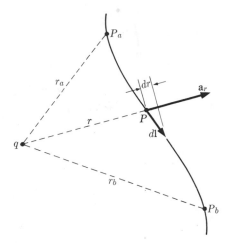

Fig. 2–11 The electric field at point P is given by $\mathbf{E} = k(q/r^2)\mathbf{a}_r$. The difference in potential between points P_a and P_b is calculated in the text by integrating $\mathbf{E} \cdot d\mathbf{l}$ from P_a to P_b.

is then given by

$$V_a - V_b = \int_{P_a}^{P_b} \mathbf{E} \cdot d\mathbf{l}.$$

Considering only the field due to a single point charge q, the potential difference between points P_a and P_b in Fig. 2–11 can be calculated. The field at any point such as P along the path of integration is given by $(kq/r^2)\mathbf{a}_r$, where $\mathbf{a}_r$ is a unit vector parallel to $\mathbf{r}$ and directed away from q. From the diagram, $dr = \mathbf{a}_r \cdot d\mathbf{l}$ and hence the integral is

$$V_a - V_b = \int_{r_a}^{r_b} \frac{kq}{r^2}\, dr = kq\left(\frac{1}{r_a} - \frac{1}{r_b}\right).$$

Note that it was not necessary to choose a specific path to be able to evaluate this integral. The same value is obtained for every path that can be drawn connecting P_a with P_b. Furthermore, if the integral is evaluated over some path in the opposite direction, from P_b to P_a, an equal but opposite expression is obtained. Obviously the integral of $\mathbf{E} \cdot d\mathbf{l}$ around any closed circuit must be equal to zero; i.e.,

$$\oint \mathbf{E} \cdot d\mathbf{l} = 0.$$

The difference of potential between P_a and P_b, calculated for the case shown in Fig. 2–12, where there are two point charges q_1 and q_2, is given by

$$V_a - V_b = \int_{P_a}^{P_b} \mathbf{E} \cdot d\mathbf{l} = \int_{P_a}^{P_b} \mathbf{E}_1 \cdot d\mathbf{l} + \int_{P_a}^{P_b} \mathbf{E}_2 \cdot d\mathbf{l}$$

$$= kq_1\left(\frac{1}{r_{a1}} - \frac{1}{r_{b1}}\right) + kq_2\left(\frac{1}{r_{a2}} - \frac{1}{r_{b2}}\right), \qquad (2\text{–}14)$$

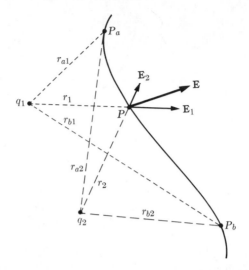

Fig. 2–12 The electric field $\mathbf{E}$ at P is the vector sum of $\mathbf{E}_1$ and $\mathbf{E}_2$, the fields at P due respectively to q_1 only and q_2 only. The difference in potential between points P_a and P_b is calculated by integrating $\mathbf{E} \cdot d\mathbf{l}$ from P_a to P_b.

which is also independent of the path. Obviously, this can be extended to any number of charges, and the work done in moving a test charge around any closed path in a field produced by a system of point charges will be zero. A field is said to be *conservative* when the total work done by it on a small test charge traversing any closed loop is zero. Electrostatic fields are always conservative. In any electrostatic problem it is possible to assign a unique value of potential for every point in space, since the difference of potential between any two points is independent of the path of integration.

Only differences in potentials have been treated thus far, so that some point of reference must be chosen if unambiguous potentials are to be assigned to each point in space. Actually, the point of reference can be chosen quite arbitrarily, although it is common practice to assign zero potential to points at infinity. With this choice, the potential at P_a can be calculated from Eq. (2–14) by placing P_b at infinity, which gives $V_a = kq/r_a$ for Fig. 2–11 and

$$V_a = \frac{kq_1}{r_{a1}} + \frac{kq_2}{r_{a2}}$$

for Fig. 2–12. Since the potentials at a point due to any number of point charges are additive, from Eq. (2–14), the potential at point P due to n point charges $q_1, q_2, \ldots, q_n$ can be expressed as

$$V = \sum_{i=1}^{i=n} \frac{kq_i}{r_i}, \tag{2–15}$$

where r_i is the distance of P from the point charge q_i.

Equation (2–15) is useful only for calculating potentials due to distributions of point charges, but, like Eq. (2–2), it can be replaced with an integral for use in calculating potentials due to continuous distributions of charge. Since Eq. (2–15) is a scalar summation, the potential due to continuous distributions of charges can be obtained from the scalar integral

$$V = k \int_\tau \frac{\rho \, d\tau}{r}. \qquad (2\text{--}16)$$

Mathematically, potentials and fields become infinite at each point charge. However, point charges do not exist in nature and all real fields and potentials are finite. Nevertheless, the concept of point charges is a very convenient mathematical fiction, since any collection of charges in a small volume can be treated as a point charge for calculating fields at points at a considerable distance from the center of these charges.

Since potentials are scalar quantities, while the fields are vector quantities, potentials are usually easier to calculate than fields. In Section 2–13 it is shown that once the potential distribution is known, the electric fields can be easily determined. For example, if an electric field **E** can be considered uniform and two points are chosen a short distance δd apart on a line parallel to the direction of **E**, the difference of potential is just $\delta V = -E\delta d$, that is, $E = -\delta V/\delta d$. In fact, for any distribution of potentials, the electric field is just equal to the negative rate of change of potential and its direction is that in which the potential reduces most rapidly.

2–12 Partial differentiation

It is frequently desirable to determine the rate of change of potential in some particular direction. The coordinates should be oriented so that the particular direction is parallel to one coordinate axis. Suppose the potential V is known as a function of the rectangular coordinates. The rate of change of $V = V(x, y, z)$ in the x-direction can be obtained by differentiating V with respect to x, considering both y and z as constants. Such differentiation is called *partial differentiation*, and is designated by $\partial V/\partial x$. As an example, if $V = Axyz + Bz^2$, then, $\partial V/\partial x = Ayz$.

The concept of partial differentiation is so generally useful in physics, mathematics, and other fields that many theorems have been developed for its use. Detailed discussions of partial differentiation and some of these theorems may be found in several texts.* One of these theorems on the *total differential* is presented in this section. An attempt is made to show the plausibility of this relationship, but no rigorous proof is intended.

---

* W. Kaplan, *Advanced Calculus*, Addison-Wesley, Reading, Mass., 1952; W. T. Martin and E. Reissner, *Elementary Differential Equations*, Chap. 8, Addison-Wesley, Reading, Mass., 1956; M. Morris and O. E. Brown, *Differential Equations*, 2nd ed., Prentice-Hall, Englewood Cliffs, N.J., 1942.

When V is a function of the coordinates, its value may change from one point in space to another. If the distance between the two points is infinitesimal, the increase in V is given by the total differential. The total differential is defined in rectangular coordinates by

$$dV = \frac{\partial V}{\partial x}\, dx + \frac{\partial V}{\partial y}\, dy + \frac{\partial V}{\partial z}\, dz; \qquad (2\text{--}17)$$

in cylindrical coordinates by

$$dV = \frac{\partial V}{\partial r}\, dr + \frac{\partial V}{\partial \phi}\, d\phi + \frac{\partial V}{\partial z}\, dz; \qquad (2\text{--}18)$$

and in spherical coordinates by

$$dV = \frac{\partial V}{\partial r}\, dr + \frac{\partial V}{\partial \theta}\, d\theta + \frac{\partial V}{\partial \phi}\, d\phi. \qquad (2\text{--}19)$$

The similarity between these equations is obvious, and in fact, the total differential may be expressed by

$$dV = \frac{\partial V}{\partial u}\, du + \frac{\partial V}{\partial v}\, dv + \frac{\partial V}{\partial w}\, dw$$

in any orthogonal system of coordinates in which the coordinates are u, v, and w. These total differential expressions for dV exactly represent the increase in V only when the displacement from the first to the second point is so small that dx, dy, dz; dr, $d\phi$, dz; and dr, $d\theta$, $d\phi$ are infinitesimals. Future discussions will be limited to just these situations. That the dV in these expressions does represent the increase in V under these circumstances is quite plausible. The increase in V if only x changes by dx must be the rate of change with x multiplied by dx; that is, $(\partial V/\partial x)\, dx$. Likewise, the increase in V with a small change in y only is $(\partial V/\partial y)\, dy$, and with a small change in z only is $(\partial V/\partial z)\, dz$. When all three change by dx, dy, and dz, respectively, the total increase is the sum of the separate increases provided these differentials are infinitesimal. Whenever the change in V, due to any term, is a decrease rather than an increase, that term will be negative. The arguments for the other coordinate systems are similar.

2–13 Gradient of potential

It was pointed out in Section 2–11 that the electric field **E** is given by the negative rate of change or the negative gradient of potential and that it points in the direction in which the potential reduces most rapidly. This statement is abbreviated to the expression $\mathbf{E} = -\text{grad } V$ or $\mathbf{E} = -\nabla V$. The latter abbreviation suggests that there is an operation that can be performed on the potential function to obtain an expression for the field, and that the nature of this operation should be determined.

To determine the nature of this operation an expression for **E** in terms of V and the coordinates is needed. If V increases by dV with an infinitesimal displacement $d\mathbf{l}$, then $\mathbf{E} \cdot d\mathbf{l} = -dV$. This relation can be used to obtain a definition for the gradient of V (∇V) that is valid for any system of coordinates. Such a definition of ∇V, which is invariant to the coordinates, is found in the relation

$$\nabla V \cdot d\mathbf{l} = dV. \tag{2-20}$$

The gradient of V, ∇V, is then defined as that vector operation on V that gives the total differential dV when it is dotted into the displacement $d\mathbf{l}$. Obviously, the ∇V thus defined is a vector quantity. To find the expression ∇V in a particular system of coordinates it is only necessary to find the vector expression which, when dotted into the correct expression for $d\mathbf{l}$, gives dV in those coordinates. In rectangular coordinates,

$$d\mathbf{l} = \mathbf{i}\, dx + \mathbf{j}\, dy + \mathbf{k}\, dz$$

and

$$dV = \frac{\partial V}{\partial x}\, dx + \frac{\partial V}{\partial y}\, dy + \frac{\partial V}{\partial z}\, dz.$$

Referring to Eq. (2–8) for dot products in general, it appears that this expression for dV is the dot product of two vectors whose rectangular components are $\partial V/\partial x$, $\partial V/\partial y$, $\partial V/\partial z$ and dx, dy, dz, respectively. Thus the partial derivatives here represent the rectangular components of the vector ∇V. In rectangular components, then,

$$\nabla V = \mathbf{i}\frac{\partial V}{\partial x} + \mathbf{j}\frac{\partial V}{\partial y} + \mathbf{k}\frac{\partial V}{\partial z}. \tag{2-21}$$

In cylindrical coordinates, when Eq. (2–20) is compared with

$$d\mathbf{l} = \mathbf{a}_r\, dr + \mathbf{a}_\phi r\, d\phi + \mathbf{k}\, dz$$

and

$$dV = \frac{\partial V}{\partial r}\, dr + \frac{\partial V}{\partial \phi}\, d\phi + \frac{\partial V}{\partial z}\, dz,$$

the resulting expression for ∇V is

$$\nabla V = \mathbf{a}_r\frac{\partial V}{\partial r} + \mathbf{a}_\phi\frac{1}{r}\frac{\partial V}{\partial \phi} + \mathbf{k}\frac{\partial V}{\partial z}. \tag{2-22}$$

In spherical coordinates,

$$d\mathbf{l} = \mathbf{a}_r\, dr + \mathbf{a}_\theta r\, d\theta + \mathbf{a}_\phi r \sin\theta\, d\phi$$

and

$$dV = \frac{\partial V}{\partial r}\, dr + \frac{\partial V}{\partial \theta}\, d\theta + \frac{\partial V}{\partial \phi}\, d\phi.$$

The vector that can be dotted into $d\mathbf{l}$ to produce dV is

$$\nabla V = \mathbf{a}_r \frac{\partial V}{\partial r} + \mathbf{a}_\theta \frac{1}{r} \frac{\partial V}{\partial \theta} + \mathbf{a}_\phi \frac{1}{r \sin \theta} \frac{\partial V}{\partial \phi} . \qquad (2\text{–}23)$$

Similarities between these three orthogonal systems should now be fairly obvious. To point out these similarities and to provide general relations valid for any orthogonal* system, a general system using the three coordinates u, v, and w can be introduced and the line element can be defined as

$$d\mathbf{l} = \mathbf{a}_u U \, du + \mathbf{a}_v V \, dv + \mathbf{a}_w W \, dw.$$

Here U, V, and W are chosen as in the three familiar systems, so that $U \, du$, $V \, dv$, $W \, dw$ all represent elements of length. To make the result more general, a new differentiable function of coordinates, $\Phi = \Phi(u, v, w)$ can be defined that is not necessarily an electrostatic potential function. Since the total differential is

$$d\Phi = \frac{\partial \Phi}{\partial u} \, du + \frac{\partial \Phi}{\partial v} \, dv + \frac{\partial \Phi}{\partial w} \, dw,$$

the vector that can be dotted into $d\mathbf{l}$ to produce $d\Phi$ is

$$\boldsymbol{\nabla}\Phi = \mathbf{a}_u \frac{1}{U} \frac{\partial \Phi}{\partial u} + \mathbf{a}_v \frac{1}{V} \frac{\partial \Phi}{\partial v} + \mathbf{a}_w \frac{1}{W} \frac{\partial \Phi}{\partial w} .$$

This is the general expression for the gradient of Φ in any orthogonal system of coordinates.

Dipole potential and field. The ideal dipole provides an excellent demonstration of the usefulness of potentials in calculating fields. The potential at P in Fig. 2–7 is, from Eq. (2–15),

$$V = kq \left(\frac{1}{r_1} - \frac{1}{r} \right).$$

Since $r_1^2 = r^2 + l^2 - 2rl \cos \theta$ and, when the approximations of Section 2–6 are used, $1/r_1 = (1 + l \cos \theta / r)/r$, the potential of the dipole becomes

$$V = \frac{kql \cos \theta}{r^2} .$$

Since $k = 1/4\pi\epsilon_0$ and $ql = p$,

$$V = \frac{p \cos \theta}{\epsilon_0 4 \pi r^2} \qquad (2\text{–}24)$$

is the potential in spherical coordinates of a dipole whose moment is p.

---

* Several orthogonal and some nonorthogonal systems are clearly treated in H. Margenau and G. M. Murphy, *Mathematics of Physics and Chemistry*, Chap. 5, 2nd ed., D. Van Nostrand, Princeton, N.J., 1956.

The field is given by

$$\mathbf{E} = -\nabla V = -\left(\mathbf{a}_r \frac{\partial V}{\partial r} + \mathbf{a}_\theta \frac{1}{r} \frac{\partial V}{\partial \theta}\right).$$

Now

$$\frac{\partial V}{\partial r} = -\frac{2kp \cos \theta}{r^3}$$

and

$$\frac{\partial V}{\partial \theta} = -\frac{kp \sin \theta}{r^2},$$

and therefore the dipole field is

$$\mathbf{E} = \frac{p}{\epsilon_0 4\pi r^3}(\mathbf{a}_r 2 \cos \theta + \mathbf{a}_\theta \sin \theta), \tag{2–25}$$

which is the expression developed in Section 2–6. The expression for the field of an ideal dipole occurs so frequently that it is desirable to define a vector $\mathbf{d}_f$ that might be called the dipole-field vector:

$$\mathbf{d}_f \equiv \mathbf{a}_r 2 \cos \theta + \mathbf{a}_\theta \sin \theta. \tag{2–25a}$$

This dipole-field vector could also be defined by

$$\mathbf{d}_f = -r^3 \nabla \left(\frac{\cos \theta}{r^2}\right).$$

Properties of the gradient. The gradient ∇V is a vector operation on V (a function of the coordinates) that provides information concerning V in the neighborhood very close to the general point at which it is evaluated. Defining ∇V by the relation $\nabla V \cdot d\mathbf{l} = dV$ of Eq. (2–20) makes it easy to demonstrate its important properties, namely:

a) ∇V points in the direction in which V increases most rapidly,
b) ∇V is everywhere perpendicular to the imaginary surfaces (the equipotentials) on which V is constant, and
c) the magnitude $|\nabla V|$ is equal to the maximum of the space rate of increase of V.

Now ∇V is a vector that has definite magnitude and direction at P, as shown in Fig. 2–13. On a sphere of infinitesimal radius dl drawn about P the value of V differs from its value at P by an amount dV. According to Eq. (2–20), dV is maximum when $d\mathbf{l}$ is parallel to ∇V and zero when $d\mathbf{l}$ is perpendicular. Therefore V increases most rapidly with displacements parallel to ∇V and stays constant with displacements perpendicular to ∇V, demonstrating properties (a) and (b). When $d\mathbf{l}$ is parallel to ∇V,

$$|\nabla V| = \frac{dV_{max}}{dl},$$

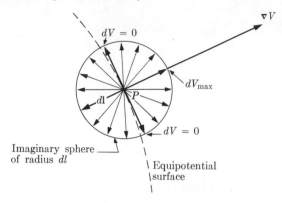

Fig. 2–13 The gradient of V, ∇V, which is defined by $\nabla V \cdot d\mathbf{l} = dV$, is perpendicular to equipotential surfaces and gives the direction and magnitude of the maximum space rate of increase of V.

so that the magnitude of ∇V is the maximum space rate of change of V, demonstrating property (c).

2–14 Vector or cross product of vectors

The dot product of two vectors, which is a scalar quantity, has been defined in Section 2–8, and its usefulness has been demonstrated in Sections 2–9 through 2–13. The defining equation (2–9) uses the cosine of the angle between the vectors. There is also a need, especially in electrodynamics, for a product defined with the sine instead of the cosine of this angle. Usually, where the product, including the sine, is needed, the result should have both magnitude and direction; i.e., it should be a vector. Consequently, a vector product has been defined that is distinguished from the scalar product by the use of a cross in place of the dot. The definition is

$$\mathbf{A} \times \mathbf{B} = \mathbf{a}_n AB \sin \widehat{AB},$$

where the unit vector $\mathbf{a}_n$ is normal to the plane defined by the vectors $\mathbf{A}$ and $\mathbf{B}$, as shown in Fig. 2–14. The direction of $\mathbf{a}_n$ is the direction in which a right-handed screw would move if it were forced through the plane by rotating it in the direction in which $\mathbf{A}$ would have to rotate to become parallel to $\mathbf{B}$ and point in the same direction. In Fig. 2–14 $\mathbf{a}_n$ points upward, since the screw proceeds in that direction. However, for the product $\mathbf{B} \times \mathbf{A}$, the screw must rotate in the opposite direction and $\mathbf{a}_n$ would then point downward. Obviously,

$$\mathbf{B} \times \mathbf{A} = -\mathbf{A} \times \mathbf{B} \tag{2–26}$$

and the order in which the vectors are written is very important. This

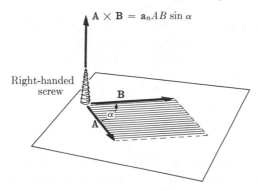

Fig. 2–14 Definition of the vector cross product $A \times B$, where a_n is a unit vector normal to the plane of A and B.

constitutes one of the important differences between the cross and the dot products, since with the dot product the vectors may be written in any order.

If A and B in Fig. 2–14 have units of length, $A \times B$ is numerically equal to the area of the parallelogram (shaded in the figure) that is defined by the vectors A and B. This product then suggests that an area sometimes may be considered to be a vector quantity. An excellent example is that of a tank filled with gas at high pressure. The force acting on a small area δs of the tank surface is $\delta f = p\,\delta s$, where the vector direction is assigned to δs, since the pressure p is a scalar quantity. While the direction of the area vector represented by $A \times B$ is well defined, ordinarily when areas are used as vectors, rules must be invented for determining their directions. The rule to be used for the direction of the δs of the surface of the gas tank is one that is quite generally used when closed surfaces are involved, namely, δs is parallel to the outward drawn normal.

Simple geometry shows that if $S = D + E$, then $R \times S = R \times D + R \times E$. Furthermore, if $R = A + B$, then

$$R \times S = A \times D + B \times D + A \times E + B \times E, \qquad (2\text{–}27)$$

which is easily extended to the cross products of the sums of any number of vectors. *It is very important here to maintain the correct order for writing the vectors.* Equation (2–27) makes it possible to obtain relations for the analytical calculation of cross products when the vectors are written with components and unit vectors. The cross products of unit vectors will be needed. In rectangular coordinates,

$$i \times i = j \times j = k \times k = 0; \qquad (2\text{–}28)$$

$$i \times j = k, \qquad j \times k = i, \qquad k \times i = j,$$
$$j \times i = -k, \qquad k \times j = -i, \qquad i \times k = -j. \qquad (2\text{–}29)$$

An analytical expression for $\mathbf{A} \times \mathbf{B}$ can be obtained from Eq. (2–27). There will be nine terms, but three of these are zero because of Eq. (2–28). The remaining terms are conveniently written in the form of a determinant:

$$\mathbf{A} \times \mathbf{B} = \begin{vmatrix} \mathbf{i} & \mathbf{j} & \mathbf{k} \\ A_x & A_y & A_z \\ B_x & B_y & B_z \end{vmatrix}. \qquad (2\text{–}30)$$

The relations between unit vectors shown in Eq. (2–29) resulted from choosing a right-handed coordinate system. If the x- and y-axes were interchanged in Fig. 2–3, these relations would become $\mathbf{i} \times \mathbf{j} = -\mathbf{k}$, etc., and the system would be called left-handed. The other orthogonal systems have also been chosen to be right-handed. Hence, in cylindrical coordinates, $\mathbf{a}_r \times \mathbf{a}_\phi = \mathbf{k}$, $\mathbf{a}_\phi \times \mathbf{k} = \mathbf{a}_r$, etc., and in spherical coordinates $\mathbf{a}_r \times \mathbf{a}_\theta = \mathbf{a}_\phi$, $\mathbf{a}_\theta \times \mathbf{a}_\phi = \mathbf{a}_r$, etc.

The usefulness of the cross product can best be shown by examples. Several of the following examples are also useful for reviewing elementary principles.

Angular momentum and torque. In Fig. 2–6 an example was given of a mass point moving near the origin of coordinates. Its angular momentum about the origin is $m\mathbf{R} \times \mathbf{v}$, if m is the mass of the point. The vector $\mathbf{R} \times \mathbf{v}$ is parallel to the instantaneous axis of rotation and in the direction a right-handed screw would move with this rotation. Since the numerical magnitudes and directions of both vectors are given, the cross product can be obtained numerically. In fact, this could be obtained by using any of the three coordinate systems in which these vectors are given. The cylindrical coordinates seem the simplest here. The cross product in cylindrical coordinates is given by a determinant similar to that of Eq. (2–30). To simplify this determinant, factor 13 m s^{-1} out of $\mathbf{v}$ and 1 m out of $\mathbf{R}$:

$$\mathbf{R} \times \mathbf{v} = \begin{vmatrix} \mathbf{a}_r & \mathbf{a}_\phi & \mathbf{k} \\ 5 & 0 & 12 \\ 0 & 5 & -12 \end{vmatrix} 13 \text{ m}^2 \text{ s}^{-1},$$

or $\mathbf{R} \times \mathbf{v} = (-60\mathbf{a}_r + 60\mathbf{a}_\phi + 25\mathbf{k})\, 13 \text{ m}^2 \text{ s}^{-1}$, which represents a vector of magnitude 1150 m^2 s^{-1}, pointing 6°48′ above the xy-plane and $135° + 53°8′ = 188°8′$ from the positive x-axis. When this is computed in rectangular coordinates, the result is

$$\mathbf{R} \times \mathbf{v} = (-84\mathbf{i} - 12\mathbf{j} + 25\mathbf{k})13 \text{ m}^2 \text{ s}^{-1},$$

which is a vector having the same magnitude and direction as the result using cylindrical coordinates.

Force on current in a magnetic field. In elementary texts the force acting on an element of wire of length dl carrying a current is usually given by $df = BI\, dl \sin \theta$, where θ and the direction df must be separately specified. When this is written in vector notation as $d\mathbf{f} = I\, d\mathbf{l} \times \mathbf{B}$, where $d\mathbf{l}$ points in the direction of the current, no further specification is required. With I, $d\mathbf{l}$, and $\mathbf{B}$ in amperes, meters, and T respectively, $d\mathbf{f}$ is in newtons. This law is treated in more detail in Chapter 7.

Biot-Savart law. The law for the magnetic field dB produced by an electric current I in an element of wire dl is often written as

$$dB = \mu_0 \frac{I\, dl \sin \theta}{4\pi r^2},$$

where r is the distance measured from the element to the point where the field is being calculated. The angle θ and the direction of dB must be separately specified. In vector notation this is written as

$$d\mathbf{B} = \mu_0 \frac{I\, d\mathbf{l} \times \mathbf{a}_r}{4\pi r^2}$$

and no additional specification is needed. The writing of such equations in vector notation usually eliminates the need for diagrams and written explanations. A more detailed discussion of this law will be found in Chapter 8.

2–15 Identities in vector algebra

Dot and cross products between two vectors have been discussed. Often in the course of some derivation or other mathematical exercise, products between more than two vectors arise. There are just two possible types of products involving three vectors, namely, $(\mathbf{A} \times \mathbf{B}) \cdot \mathbf{C}$ and $(\mathbf{A} \times \mathbf{B}) \times \mathbf{C}$.

Triple scalar product. The first of these triple products is usually written $\mathbf{A} \times \mathbf{B} \cdot \mathbf{C}$ without the enclosures, since the expression would be meaningless with parentheses around $\mathbf{B} \cdot \mathbf{C}$. From Fig. 2–14, $\mathbf{A} \times \mathbf{B}$ has a magnitude equal to the shaded area and is perpendicular to it (vertical in the diagram). Dotting this into $\mathbf{C}$ multiples this area by the vertical component of $\mathbf{C}$, which gives the volume of the parallelepiped shown in Fig. 2–15. Obviously, $\mathbf{B} \times \mathbf{C} \cdot \mathbf{A}$ and $\mathbf{B} \cdot \mathbf{C} \times \mathbf{A}$ give this same volume, and the dot and cross can be interchanged at will. However, the sign of the product is reversed by interchanging any two of the vectors, such as $\mathbf{B} \cdot \mathbf{C} \times \mathbf{A} = -\mathbf{B} \cdot \mathbf{A} \times \mathbf{C}$, but is not changed so long as the three vectors are kept in cyclic order. A convenient expression is

$$\mathbf{A} \times \mathbf{B} \cdot \mathbf{C} = \begin{vmatrix} A_x & A_y & A_z \\ B_x & B_y & B_z \\ C_x & C_y & C_z \end{vmatrix}.$$

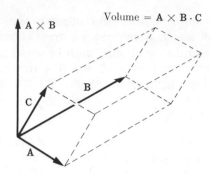

Fig. 2–15 The volume represented by the triple scalar product $\mathbf{A} \times \mathbf{B} \cdot \mathbf{C}$.

Triple vector product. The product $(\mathbf{A} \times \mathbf{B}) \times \mathbf{C}$ is a vector in the plane determined by $\mathbf{A}$ and $\mathbf{B}$, the vectors in the parentheses. This can be seen from Fig. 2–15. The $\mathbf{A} \times \mathbf{B}$ vector is perpendicular to the $\mathbf{AB}$ plane, but the cross product of $\mathbf{C}$ with this is perpendicular to both $\mathbf{A} \times \mathbf{B}$ and $\mathbf{C}$, which puts it right back into the $\mathbf{AB}$ plane. If $\mathbf{A}$ and $\mathbf{B}$ are finite and are not parallel, any vector in the $\mathbf{AB}$ plane can be expressed by $k_1\mathbf{A} + k_2\mathbf{B}$, where k_1 and k_2 are scalar constants, properly determined. The correct expression is

$$(\mathbf{A} \times \mathbf{B}) \times \mathbf{C} = \mathbf{B}(\mathbf{A} \cdot \mathbf{C}) - \mathbf{A}(\mathbf{B} \cdot \mathbf{C}),$$

as is easily verified by writing each vector in terms of its components and the unit vectors, and then carrying out the indicated operations. If the parentheses enclose $\mathbf{B} \times \mathbf{C}$, this operation is performed first, the final product is in the $\mathbf{BC}$ plane, and the expression becomes

$$\mathbf{A} \times (\mathbf{B} \times \mathbf{C}) = \mathbf{B}(\mathbf{A} \cdot \mathbf{C}) - \mathbf{C}(\mathbf{A} \cdot \mathbf{B}).$$

Both of these equations have two terms on the right, each consisting of one vector from the parentheses on the left multiplied by the dot product of the other two vectors. These two terms on the right have opposite signs, but the positive sign is associated with the vector from the middle of the original expression on the left.

PROBLEMS

2–1

a) By referring to Figs. 2–4 and 2–5, show that relations written with the unit vectors of rectangular coordinates can be transformed into cylindrical coordinates with the following equations:

$$\mathbf{i} = \mathbf{a}_r \cos \phi - \mathbf{a}_\phi \sin \phi,$$
$$\mathbf{j} = \mathbf{a}_r \sin \phi + \mathbf{a}_\phi \cos \phi,$$
$$\mathbf{k} = \mathbf{a}_z;$$

and into spherical coordinates with the following equations:

$$\mathbf{i} = \mathbf{a}_R \sin \theta \cos \phi + \mathbf{a}_\theta \cos \theta \cos \phi - \mathbf{a}_\phi \sin \phi,$$

$$\mathbf{j} = \mathbf{a}_R \sin \theta \sin \phi + \mathbf{a}_\theta \cos \theta \sin \phi + \mathbf{a}_\phi \cos \phi,$$

$$\mathbf{k} = \mathbf{a}_R \cos \theta - \mathbf{a}_\theta \sin \theta.$$

b) Show that the above equations can be used to give the last six equations in Section 2–4.

2–2 The general relations

$$\mathbf{a}_r = \mathbf{i} \cos \phi + \mathbf{j} \sin \phi, \qquad \mathbf{a}_z = \mathbf{k},$$

$$\mathbf{a}_R = \mathbf{i} \sin \theta \cos \phi + \mathbf{j} \sin \theta \sin \phi + \mathbf{k} \cos \theta$$

are easily verified by studying Figs. 2–4 and 2–5, respectively. By means of vector products calculate the following unit vectors in terms of the unit vectors $\mathbf{i}$, $\mathbf{j}$, and $\mathbf{k}$:

a) $\mathbf{a}_\phi$, for both figures, and

b) $\mathbf{a}_\theta$, for Fig. 2-5.

2–3 Using the general relations obtained in Problem 2–1, calculate: (a) $\partial \mathbf{a}_r / \partial \phi$, (b) $\partial \mathbf{a}_R / \partial \theta$, (c) $\partial \mathbf{a}_R / \partial \phi$, and (d) $\partial \mathbf{a}_\phi / \partial \phi$. Express these partial derivatives in terms of unit vectors of the cylindrical or spherical coordinate systems.

2–4 Two vectors are defined as $\mathbf{A} = \mathbf{i}4 + \mathbf{j}3 + \mathbf{k}12$ and $\mathbf{B} = \mathbf{i}3 + \mathbf{j}12 + \mathbf{k}4$. Calculate the following:

a) $\mathbf{S} = \mathbf{A} + \mathbf{B}$ and $\mathbf{D} = \mathbf{A} - \mathbf{B}$.

b) The angle between $\mathbf{A}$ and $\mathbf{B}$.

c) The expression, in rectangular coordinates, for a unit vector perpendicular to the plane containing $\mathbf{A}$ and $\mathbf{B}$.

2–5

a) Use rectangular coordinates to obtain an expression for $\mathbf{A} \times (\mathbf{B} \times \mathbf{C})$.

b) Convert the expression obtained in (a) to two terms, each containing a dot product of two of the original vectors, which is multiplied by the other original vector. Compare this with the last equation in Chapter 2.

2–6 In a rectangular coordinate system a charge of 25×10^{-9} C is placed at the origin of coordinates, and a charge of -25×10^{-9} C is placed at the point $x = 6$ m, $y = 0$. What is the electric intensity at (a) $x = 3$ m, $y = 0$? (b) $x = 3$ m, $y = 4$ m?

2–7 A charge of 16×10^{-9} C is fixed at the origin of coordinates, a second charge of unknown magnitude is at $x = 3$ m, $y = 0$, and a third charge of 12×10^{-9} C is at $x = 6$ m, $y = 0$. What is the magnitude of the unknown charge if the resultant field at $x = 8$ m, $y = 0$ is 20.25 NC^{-1} directed to the right?

2–8 A ring of radius 10 cm has a positive charge q of 5×10^{-9} C.

a) Compute the electric intensity at points on the axis of the ring, at distances from its center of 0, 5, 8, 10, and 15 cm. Show the results in a graph.

b) In terms of the radius R of the ring, at what axial distance from the center of a charged ring is the electric intensity a maximum? Compare with the graph in part (a).

2–9 In rectangular coordinates point charges of -150×10^{-9} C and $+150 \times 10^{-9}$ C are at (0, 0, 0) and (0, 20 cm, 0), respectively. Locate field points A and B at (0, 0, 15 cm) and at (0, 20 cm, 15 cm), respectively.
a) Write the vector expression for E at B.
b) Find the potentials V_A and V_B at A and B, respectively.
c) Find the work required to move a charge of $+3 \times 10^{-9}$ C from A to B along the straight line joining them.

2–10 A thin ring of radius a is in the xy-plane with its center at the origin. This ring has the linear charge density $\lambda = A \sin \phi$. Find the electric field at a point on the z-axis where $z = h$.

2–11 A thin circular disk of insulating material has a radius b and is in the xy-plane with its center at the origin. This disk has a uniform surface charge σ.
a) Calculate the electric field on the z-axis at $z = z$.
b) Show that when $z \ll b$, E approximates its value near an infinite plane.
c) Show that when $b/z \ll 1$, E approximates the field that would be produced by a point charge of magnitude $\sigma \pi b^2$ at the origin.

2–12 An ideal dipole, having the dipole moment $p = 24\pi\epsilon_0$, is at the origin of a rectangular coordinate system and is pointing in the positive y-direction. Draw this in the first quadrant of the yz-plane. Draw arcs of two circles of radii 1 m and 2 m. Using an appropriate scale factor, draw vectors representing the direction and magnitude of the electric field due to the dipole at points on each of these two circles where $\theta = 0°$, 30°, 60°, and 90°.

2–13 A small object carrying a charge of -5×10^{-9} C experiences a downward force of 20×10^{-9} N when placed at a certain point in an electric field.
a) What is the electric intensity at the point?
b) What would be the magnitude and direction of the force acting on an α-particle placed at the point?

2–14 A point charge of -2μ C is at the origin and one of $+8\mu$ C is on the negative x-axis, 30 cm from the origin.
a) Write a vector expression for the field E at the point (0, 40 cm, 0).
b) At what point $P_0(x_0, y_0, z_0)$ in these coordinates is the electric field zero?
c) Would a small test charge be in stable equilibrium at point P_0; i.e., would it tend to return to P_0 if it was displaced a small distance in any direction? To answer this question find the direction of E at the two points $(x_0 - 1 \text{ cm}, y_0, z_0)$ and $(x_0, y_0 + 1 \text{ cm}, z_0)$.

2–15 Charges of $60\pi\epsilon_0$ C and $-60\pi\epsilon_0$ C are located in a rectangular coordinate system at (0, 0, 0) and at (0, 4 m, 0), respectively.
a) Find E at $P_1(0, 0, 3 \text{ m})$ and at $P_2(0, 4 \text{ m}, 3 \text{ m})$. Draw a diagram with arrows to represent these fields.
b) Find the difference of potential between P_1 and P_2.

2–16 An infinite nonconducting plane coincides with the xy-plane passing through the origin. This plane has a uniform surface charge σ, and a point charge q is located on the z-axis at $z = 5$ m. Determine the electric field at $z = z$ on the z-axis.

2–17 Two point charges whose magnitudes are $+20 \times 10^{-9}$ C and -12×10^{-9} C are separated by a distance of 5 cm. An electron is released from rest between the two charges, 1 cm from the negative charge, and moves along the line connecting the two charges. What is its velocity when it is 1 cm from the positive charge?

2–18 A thin metallic ring describes a circle in the xz-plane of a rectangular co-ordinate system. It has a radius of 10 cm, its axis coincides with the y-axis, and it is uniformly charged with 2.6×10^{-8} C of electric charge.
a) Calculate the difference between the potential at the origin and at the point on the y-axis where $y = 24$ cm.
b) Calculate and plot the potential along the y-axis as a function of y.
c) Calculate and plot the velocity of a proton that has been projected in the positive direction along the y-axis and has arrived at $y = 24$ cm with the velocity 0.55×10^6 m s^{-1}.

2–19 Most commerical capacitors consist of conducting sheets so close together that the electric fields between these sheets can be closely approximated by assuming the conductors to be infinite planes parallel to each other. Fields can be much more simply calculated near an infinite plane than near a finite plane. From the expression for the field of a charged plane of infinite extent, find the fields due to two such planes when they are parallel and uniformly charged with $+\sigma$ and $-\sigma$ surface densities, respectively, (a) between the planes, and (b) outside the space between the planes. Are these fields uniform?

2–20 Assume that the charged plane of Fig. 2–9 is not infinite but is a disk of radius a about the z-axis.
a) Calculate **E** at P.
b) Calculate the percent error involved in using Eq. (2–8) when $a \gg z$.

2–21 Two disks, each like that of Problem 2–16, are parallel and are separated a distance $d \cong a$. They are charged to uniform surface densities of σ on the upper and $-\sigma$ on the lower disk and the positively charged disk coincides with the xy-plane.
a) Calculate **E** at a point on the axis above both disks.
b) Simplify the expression from part (a) for $z \gg a$ and compare the result with the expression for the field of an ideal dipole.

2–22 An infinitely long wire coinciding with the z-axis has a uniform charge λ in C/m.
a) Find the expression for **E** in cylindrical coordinates at points in the xy-plane.
b) Find **E** at these same points due only to the charge along the positive z-axis, that is, assuming that there is no charge along the negative z-axis.

2–23 Demonstrate geometrically the correctness of the relation

$$\mathbf{R} \times \mathbf{S} = \mathbf{R} \times \mathbf{D} + \mathbf{R} \times \mathbf{E}$$

when

$$\mathbf{S} = \mathbf{D} + \mathbf{E}.$$

2–24 Show geometrically that if $\mathbf{S} = \mathbf{D} + \mathbf{E}$, then $\mathbf{R} \cdot \mathbf{S} = \mathbf{R} \cdot \mathbf{D} + \mathbf{R} \cdot \mathbf{E}$.

2–25 Two vectors are defined as follows: $\mathbf{A} = \mathbf{i}4 + \mathbf{j}3 + \mathbf{k}12$ and $\mathbf{B} = \mathbf{i}12 - \mathbf{j}4 - \mathbf{k}3$. Calculate (a) $\mathbf{S} = \mathbf{A} + \mathbf{B}$, (b) $\mathbf{D} = \mathbf{A} - \mathbf{B}$, (c) $P = \mathbf{A} \cdot \mathbf{B}$, and (d) $\mathbf{C} = \mathbf{A} \times \mathbf{B}$. (e) Find the angle between **A** and **B**.

2–26 Using vector operations:
a) Calculate the angles that the vector $\mathbf{A} = \mathbf{i}3 + \mathbf{j}4 + \mathbf{k}12$ makes with the x-, y-, and z-axes respectively.
b) Calculate the angle that **A** of part (a) makes with $\mathbf{B} = \mathbf{i}4 + \mathbf{j}12 + \mathbf{k}3$ and draw the diagram.

2–27 The direction of a vector in space can be expressed in terms of the cosines of the angles it makes with the three rectangular coordinate axes x, y, and z. These are called the "direction cosines" of the vector. Using vector operations, find the angle between vectors **A** and **B** whose respective direction cosines are l_1, m_1, n_1, and l_2, m_2, n_2.

2–28 The force on a small object in rectangular coordinates is given in N by $\mathbf{F} = \mathbf{i}(x + z) + \mathbf{j}(x + z) + \mathbf{k}(y + 2z)$. Find the work done by this force if it moves the object along a straight line for (0, 0, 3m) to (0, 4m, 5m).

2–29 Suppose $\mathbf{E} = \mathbf{a}_r r \sin \phi$ in cylindrical coordinates. Find $\mathbf{E} \cdot d\mathbf{l}$ along the straight line from (0, b, 0) to (a, b, 0) in rectangular coordinates.

2–30 Assume the component of the electric field along the x-axis varies with x as follows:
a) $E_x = 2\ \mathrm{N\,C^{-1}}$ for $0 < x < 1$ m,
b) $E_x = -1\ \mathrm{N\,C^{-1}}$ for $1\ \mathrm{m} < x < 2$ m,
c) $E_x = (2\ \mathrm{m} - x/2)\mathrm{N\,C^{-1}\,m^{-1}}$ for $2\ \mathrm{m} < x < 4$ m,
d) $E_x = 0$ for $x > 4$ m.
Assuming that the potential V is zero at $x = \infty$, draw graphs of E_x and V, each versus x.

2–31 A circular ring in the xy-plane, centered at the origin, has the radius a and carries a uniformly distributed charge totaling q in coulombs.
a) Calculate the potential at its center.
b) Calculate the potential on the z-axis at $z = z$.
c) Calculate the velocity v acquired by a proton of mass m and charge e, as it moves from the origin to $z = a\sqrt{3}$. What is the maximum velocity it can acquire from this charged ring?

2–32 Integrate $\oint \mathbf{A} \cdot d\mathbf{l}$ for $\mathbf{A} = \mathbf{i}y - \mathbf{j}x$ around a closed rectangle whose corners are at the following (x, y) coordinates: (0, 0), (a, 0), (a, b), and (0, b).

2–33 Expressed in cylindrical coordinates, the force on a very small object is $\mathbf{f} = A\mathbf{a}_\phi / r$. Find the work done in moving this object around the following closed paths:
a) A circle of radius a about the origin in the xy-plane.
b) A composite path of four parts, namely, (1) the arc of a circle of radius a with ϕ increasing from 0 to π, (2) the radial line at $\phi = \pi$ extending from $r = a$ to $r = b$ (with $b > a$), (3) the arc of a circle of radius b with ϕ increasing from π to 2π, and (4) the radial line at $\phi = 2\pi$ extending from $r = b$ to $r = a$.
c) The same path as in (b) for parts 1, 2, and 4, but with part 3 being that arc of a circle of radius b for which ϕ decreases from π to 0.

2–34 The force acting on a small object that is restrained to move in the xy-plane is $\mathbf{F} = \mathbf{i}y^2 + \mathbf{j}4x$. Calculate the work that must be done on this object to cause it to move through the origin from (4, −4) to (4, 4) along the parabola $y^2 = 4x$ and then back to the starting point along the straight line joining the points.

2–35 Two point charges, each of magnitude q, are placed at $y = b$ and $y = -b$.
a) Find the field **E** at ($x = x$, $y = 0$).
b) Find the work required to move a small charge q' from $x = 10b$ to $x = b$.

2–36 Calculate the electric fields in the two regions where the following are the potential distributions:

$$\text{(a)} \quad V = x^3 + 3x^2y + 2y^2z,$$
$$\text{(b)} \quad V = (x - 3)(y - 4) + 5.$$

In both of these potential fields, locate the points where $\mathbf{E} = 0$.

2–37 Write $\mathbf{A}$, $\mathbf{B}$, and $\mathbf{C}$ in rectangular components with unit vectors and demonstrate by direct multiplication that

a) $\mathbf{A} \times \mathbf{B} \cdot \mathbf{C}$ = the volume of the parallelepiped of Fig. 2–15.

b) $\mathbf{A} \times (\mathbf{B} \times \mathbf{C}) = \mathbf{B}(\mathbf{A} \cdot \mathbf{C}) - \mathbf{C}(\mathbf{A} \cdot \mathbf{B})$.

2–38 It is often mathematically convenient to deal with charged sheets of infinite extent or charged wires of infinite length, even though such devices cannot exist. When Eq. (2–16) is used to calculate potentials due to these devices, the results may be infinite. It is more realistic in such cases to give these devices constant though very large dimensions, so that the potentials remain finite. Assume that there is a uniform distribution of electrostatic charges, given by λ in C m^{-1}, along the z-axis from $z = -Z$ to $z = Z$, where the Z's are large constants.

a) By direct integration determine the electric field $\mathbf{E}$ at the point P, where in cylindrical coordinates $z = 0$ and $r = r$.

b) By direct integration determine the potential V at P.

c) Simplify the expressions for $\mathbf{E}$ and V by assuming that $Z \gg r$.

d) Show that the simplified expression for $\mathbf{E}$ can be obtained directly from the simplified expression for V.

2–39 Consider a charged filament like the one in the previous problem, which extends only from $z = 0$ to $z = Z$.

a) By direct integration determine $\mathbf{E}$ at P, where $z = 0$ and $r = r$.

b) By direct integration determine V at P.

c) Simplify the expressions for $\mathbf{E}$ and V by assuming that $Z \gg r$.

d) Show that the simplified expression for $\mathbf{E}$ *cannot be obtained* from the simplified expression for V. Explain why $\mathbf{E}$ cannot be obtained here from the V of part (b), although in the last problem $\mathbf{E}$ could be obtained from the V of part (b) there.

2–40 For the charged filament of Problem 2–39, which extends only along the positive z-axis, obtain a potential function from which the $\mathbf{E}$ at P can be obtained. Demonstrate that the $\mathbf{E}$ of part (c) of that problem can be obtained from your expression for V.

2–41 Two point charges Q and $-q$, where $Q > q > 0$, are located on a z-axis which is the axis of symmetry of a spherical coordinate system. Charge Q is at $z = d$ and charge $-q$ is at $z = b$, where $d/b = (Q/q)^2$. Find the equation for the closed surface where the potential is zero, in terms of r and θ.

2–42 Figure 2–16 shows a geometrical situation commonly encountered in electrostatics. A point charge, a dipole, or other collections of charges may be at P_1 with rectangular coordinates (a, b, c) so that $r_1^2 = a^2 + b^2 + c^2$. The general point at which potentials or fields are to be calculated has the coordinates (x, y, z) or (r, θ, ϕ), so that $r^2 = x^2 + y^2 + z^2$. The vector distance from P_1 to P is $\mathbf{R}$ and $R^2 = (x - a)^2 + (y - b)^2 + (z - c)^2$.

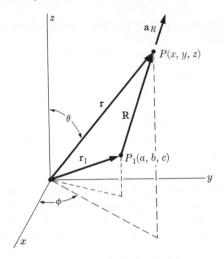

Fig. 2–16 Geometrical relations for calculating fields or potentials at a general point P due to charges located at P_1.

a) Using vector algebra, show that since $\mathbf{R} = \mathbf{r} - \mathbf{r}_1$ from the diagram, $R^2 = r^2 + r_1^2 - 2\mathbf{r} \cdot \mathbf{r}_1$.

b) Suppose that $V = R$ and calculate ∇V, first using rectangular coordinates, and second using spherical coordinates. Express these results in terms of the unit vector $\mathbf{a}_R$.

c) Show that if $V = 1/R$, $\nabla V = -\nabla R / R^2$.

2–43 Two point charges $+4q$ and $-q$ are separated a distance l.

a) Find the position of the equilibrium point where the electric field is zero.

b) Find the potential at the equilibrium point.

c) Show that the equipotential surface that passes through the equilibrium point consists of two completely closed surfaces, one inside the other and contacting it only at the equilibrium point. [*Hint:* With the origin at charge $-q$ spherical coordinates can be used to calculate where this equipotential cuts the axis. To determine where it cuts the basal plane that passes through $-q$, the origin may be shifted to the location of the $+4q$ charge to allow the solution to be obtained in terms of trigonometric functions.]

2–44 To help you to understand the potential field near the charges of Problem 2–43, draw additional equipotentials on the diagram. Of special interest are the equipotentials $V = 0$, $V = (1/4\pi\epsilon_0)(2q/3l)$, and the two closed but separated surfaces on which $V = (1/4\pi\epsilon_0)(3q/2l)$. These surfaces can be roughly drawn by calculating three or four well-chosen points. Probably the simplest procedure is to write an expression for potentials on the basal plane, which can be plotted to obtain graphical solutions.

Chapter 3 FUNDAMENTAL THEOREMS OF ELECTROSTATICS WITH DIFFERENTIAL OPERATORS

There are many scientific and practical applications for electrostatic phenomena relating to conductors and insulators shaped into a multitude of geometrical forms. These phenomena may involve charges, fields, potentials, forces, torques, and capacitances. A general method for calculating the relationships between these various quantities for all geometrical configurations is presented in Chapter 4. Briefly stated, the method is to find a solution of a differential equation, known as Laplace's equation, which satisfies the boundary conditions. The fundamental principles needed for this general method are presented in this chapter. All of the principles are here specifically related to electrostatic problems; however, the method is actually applicable to the solution of problems in several other fields of physics, for example in current electricity, in heat flow, and in hydrodynamics.

The general method is based upon a simple physical concept that is often used, without proof, in elementary texts on electric and magnetic fields. This concept is that *continuous flux lines can be drawn in empty space so that, at every point in this space, the density and the direction of these lines represent the intensity and the direction of the field.* The concept is usually extended with some modification to include spaces that are not empty but do contain materials.

It is easy to show that this concept is valid for describing the electrostatic field due to a single isolated point charge q. Imagine radial lines drawn uniformly in all directions from the point charge. If the total number of lines drawn is q/ϵ_0, the density of lines passing through an imaginary sphere surrounding the point is found by dividing the total number of lines by the area of the sphere. For a sphere of any radius r the area is $4\pi r^2$. Thus the density of lines at the surface is just $q/\epsilon_0 4\pi r^2$, which is numerically equal to the electric field at the surface of the sphere. In agreement with Coulomb's law, the density of lines varies with the inverse square of the distance from the point charge and thus the concept is valid for a single isolated point charge. To show that it is valid in general for multitudes of point charges with various distributions, it is desirable to make use of the concept of solid angles.

3–1 Solid angles

To define a plane angle, a circle of radius r is drawn about the apex as the center. The angle in radians is the ratio l/r, where l is the length of the arc subtending the angle. This definition is useful because l/r is independent of the radius of the circle. A similar procedure is used to define a solid angle. A typical solid angle is indicated by the irregular cone shown in Fig. 3–1. When two spheres of radii r_1 and r_2 are drawn with centers at the apex of this cone, areas s_1 and s_2 are inscribed by the cone on the surfaces of the respective spheres. Since these surfaces are proportional to the squares of the respective radii, the dimensionless ratio $s_1/r_1^2 = s_2/r_2^2 = \omega$ is independent of the size of the sphere and is used to define the solid angle of the cone.

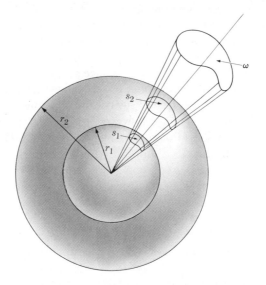

Fig. 3–1 The solid angle subtended by areas s_1, s_2 on two concentric spheres.

Steradian is the name given to one unit solid angle. *A steradian is the solid angle subtended at the center of a sphere of one meter radius by an area of one square meter on its surface.* An increase of the area inscribed on a given sphere indicates a proportionate increase in the solid angle, and a total solid angle inscribes the total surface area of $4\pi r^2$. The total solid angle is then 4π steradians, in the same sense as the total plane angle is 2π rad. However, it is often convenient to think of solid angles that exceed 4π steradians, just as it is often convenient to think of plane angles that exceed 2π rad.

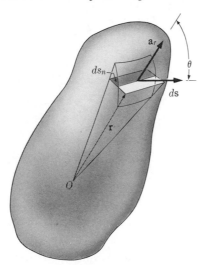

Fig. 3–2 An illustration of the elemental solid angle $d\omega = ds_n/r^2$ subtended by the area ds on the surface of a potato.

Elemental solid angles. In defining solid angles, areas inscribed on the surface of a specially located sphere were utilized. It is often desirable to express solid angles in terms of areas inscribed on very irregular surfaces. Since potatoes have such irregular shapes, they are ideal for visualizing general volumes and general surfaces. If, as in Fig. 3–2, a point O is chosen anywhere inside such a potato, an element of the skin area ds will subtend at O an element of solid angle $d\omega$. By definition, $d\omega = ds_n/r^2$, where r is the distance from O to the center of ds and ds_n is the area inscribed by this elemental solid angle on the surface of a sphere of radius r drawn about O as a center. When ds is sufficiently small $ds_n = ds \cos \theta$, which can be written as $ds_n = \mathbf{a}_r \cdot d\mathbf{s}$ by defining the element of area on this closed surface as a vector $d\mathbf{s}$ directed along the outward drawn normal. The expression for the elemental solid angle then becomes

$$d\omega = \mathbf{a}_r \cdot \frac{d\mathbf{s}}{r^2}, \tag{3–1}$$

Note that if this is integrated over the whole surface of the potato, the result must be 4π:

$$\int_s \mathbf{a}_r \cdot \frac{d\mathbf{s}}{r^2} = 4\pi. \tag{3–2}$$

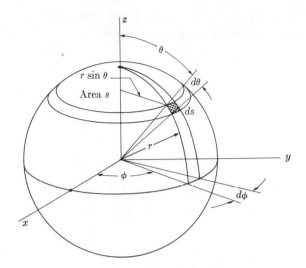

Fig. 3–3 An illustration showing how the solid angle of a right conical section is obtained from an integration of the elemental solid angle ds/r^2.

An expression for the solid angle of a right circular cone in terms of its apex angle is often needed. This can be obtained by integrating Eq. (3–1) over the circular cap of the sphere shown in Fig. 3–3. Using spherical co-ordinates, the surface element is $(r\, d\theta)(r \sin \theta\, d\phi) = \mathbf{a}_r \cdot d\mathbf{s}$. The elemental solid angle then is $d\omega = \mathbf{a}_r \cdot d\mathbf{s}/r^2 = \sin \theta\, d\theta\, d\phi$, and the solid angle subtended by the circular cap becomes

$$\omega = \int_0^\theta \sin \theta\, d\theta \int_0^{2\pi} d\phi = 2\pi(1 - \cos \theta). \tag{3–3}$$

3–2 Gauss' law

The basis for the mathematical solutions of electrostatic problems is Gauss' law. This law states that the total flux out of any closed surface is proportional to the total charge enclosed by that surface.

At the center of ds in Fig. 3–2 a point charge q at point O produces an electrostatic field $\mathbf{E} = \mathbf{a}_r q/\epsilon_0 4\pi r^2$, which is the average field over ds, when ds is infinitesimal. The electric flux through the element $d\Phi = E_n\, ds$, where E_n is the component of $\mathbf{E}$ normal to the surface, can be written as $d\Phi = E \cos \theta\, ds = \mathbf{E} \cdot d\mathbf{s} = \mathbf{a}_r \cdot d\mathbf{s}\, q/\epsilon_0 4\pi r^2$. From Eq. (3–2), the total flux out of the closed surface due to q is then

$$\Phi = \int_s \frac{q \mathbf{a}_r \cdot d\mathbf{s}}{\epsilon_0 4\pi r^2} = \frac{q}{\epsilon_0}. \tag{3–4}$$

While this proves the law for a single charge q at any point O inside a closed surface of any shape, more general situations must be investigated.

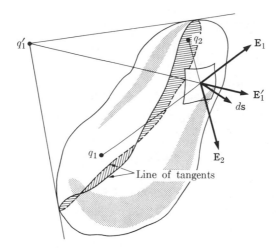

Fig. 3–4 An element of area ds is shown on the surface of a general volume shaped like a potato. Vectors $\mathbf{E}_1$, $\mathbf{E}_2$ and $\mathbf{E}_1'$ represent fields at the element due to q_1, q_2 and q_1', respectively. Since the solid angles subtended by the portions of the potato to the right and left of the line of tangents are equal, all flux entering the left side of the potato from q_1' leaves on the right side.

Many point charges in space. Consider the more general case shown in Fig. 3–4, where q_1 and q_2 are two charges of any magnitude at any two points inside the closed surface and q_1' is another such charge outside this surface. The field at the surface element due to these three charges is the vector sum of the separate fields due to each of the charges, $\mathbf{E} = \mathbf{E}_1 + \mathbf{E}_2 + \mathbf{E}_1'$. The total flux out of the closed surface is then

$$\Phi = \int_s \mathbf{E} \cdot d\mathbf{s} = \int_s \mathbf{E}_1 \cdot d\mathbf{s} + \int_s \mathbf{E}_2 \cdot d\mathbf{s} + \int_s \mathbf{E}_1' \cdot d\mathbf{s}.$$

From Eq. (3–4) the first two integrals give q_1/ϵ_0 and q_2/ϵ_0 respectively, but the third integral must be considered separately. Note that the solid angle subtended by the closed surface is 4π for both q_1 and q_2, while straight lines drawn from q_1' tangent to the closed surface describe a solid angle ω' that is less than 4π. The points of tangency make a line dividing the closed surface into two sheets. Over the sheet farther from q_1',

$$\int \mathbf{E}_1' \cdot d\mathbf{s} = \frac{q_1'\omega'}{4\pi\epsilon_0},$$

while over the sheet closer to q_1',

$$\int \mathbf{E}_1' \cdot d\mathbf{s} = -\frac{q_1'\omega'}{4\pi\epsilon_0},$$

which is negative because $\mathbf{E}_1'$ is directed inward at all points on this nearer sheet. Therefore,

$$\int_s \mathbf{E}_1' \cdot d\mathbf{s} = \frac{(\omega' - \omega')q_1'}{4\pi\epsilon_0} = 0.$$

No charge located outside a closed surface can contribute to the net flux emanating from that closed surface. Obviously, if there are charges q_1, $q_2, \ldots, q_n$ inside and charges $q_1', q_2', \ldots, q_m'$ outside any closed surface, the net flux out of that surface is

$$\Phi = \int_s \mathbf{E} \cdot d\mathbf{s} = \sum_1^n \frac{q_i}{\epsilon_0},$$

which is the mathematical statement of Gauss' law.

Usually, the presence of charges in space is evident only because some of the objects in the space have either more or less electrons than they have protons. Such conditions are most simply characterized by a density distribution of charges, symbolized by ρ, in units of C m^{-3}. The summation of charges is then expressed best by an integral. Gauss' law becomes

$$\int_s \epsilon_0 \mathbf{E} \cdot d\mathbf{s} = \int_\tau \rho \, d\tau, \tag{3–5}$$

where τ must be the total volume enclosed by the closed surface s. Charges outside of s do not contribute to the integral.

Actually, Eq. (3–5) is the more general expression, since the density ρ at any point is defined by

$$\rho = \lim_{\Delta\tau \to 0} \frac{\sum_i q_i}{\Delta\tau},$$

where the q_i's represent all positive and negative charges in $\Delta\tau$. As $\Delta\tau$ approaches zero, it is assumed that ρ will approach some fixed value that is dependent upon the location in space. In physical problems, it is convenient to think of the volume becoming small but still remaining finite. This avoids the violent fluctuations in ρ that would occur if the elemental volume decreased beyond the point where it contains only two or three atoms. The reduction of $\Delta\tau$ should be stopped before noticeable fluctuations are produced. This small but finite volume may be referred to as a *physical point*. A physical point might be 10^{-21} m^3 or smaller, since a volume this small would still contain over 10^8 atoms in a solid or 10^5 atoms in air at atmospheric pressure.

Because mathematical operations on continuous functions are simpler than on discontinuous quantities like point charges, charge distributions are represented by a density function wherever possible.

Flux lines and field intensities. The fact that electric field intensities at all points in space can be represented by the densities of continuous flux lines can now be proved. For this proof, groups of positive charges or individual positive charges can be surrounded by closed equipotential surfaces, so that no positive charge exists in the remaining space not enclosed by an equipotential surface. It was proven in Section 2–13 that such surfaces are everywhere perpendicular to the electric field **E**, and according to Gauss' law of Eq. (3–5), the electric field lines of flux must start from positive charges and must end on negative charges. An extremely large number of flux lines can be drawn parallel to the electric field throughout the intervening space in such a manner that, where they start from each element of an equipotential surface surrounding positive charges, the density of the lines is proportional to the field at that element. Starting flux lines in this manner guarantees that their densities are proportional to the fields on these particular equipotential surfaces, but it remains to be shown that this proportionality exists at all locations in space.

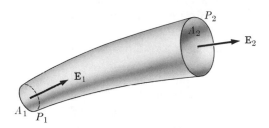

Fig. 3–5 Tube of flux lines between points P_1 and P_2. The flux through A_1 equals the flux through A_2, and the fields E_1 and E_2 are equal to the total flux lines divided by the areas A_1 and A_2, respectively.

As shown in Fig. 3–5, on one of these initial equipotential surfaces where the field is E_1 over a small area A_1, a point P_1 can be chosen for constructing a tube which is generated by the lines of flux passing through the periphery of the surface A_1 and progressing to a point P_2 where the field is E_2. Constructing a new surface of area A_2 perpendicular to the lines of flux at this point produces a closed surface containing no charge. According to Gauss' law, the net flux out of this enclosure must be zero. Since the flux lines are parallel to the curved surface between the two areas A_1 and A_2, no flux lines can enter or leave through the curved surface and the total flux entering at P_1 must equal the flux leaving at P_2. Mathematically, $\phi = E_1A_1 = E_2A_2$, the field at P_1 is $E_1 = \phi/A_1$, and at P_2 is $E_2 = \phi/A_2$, which shows that the density of the flux lines does represent the strength of the field at the points P_1 and P_2. Since all the intervening space can be filled with such tubes of flux, the same relationships must hold everywhere. This establishes the validity of the physical picture for all electrostatic fields.

Gauss' law calculations of symmetrical fields. There are a number of problems in electrostatics which can be solved easily by the direct application of

Gauss' law. These problems have symmetry such that imaginary closed surfaces can be drawn with shapes such that $\mathbf{E}$ is known to be constant and normal to some surfaces but parallel to all other surfaces. Over the former surfaces the integral can be written $\int \mathbf{E} \cdot d\mathbf{s} = E \int ds$, whereas over the latter surfaces $\int \mathbf{E} \cdot d\mathbf{s} = 0$, so that E can be evaluated. The problem concerning the tube of flux in Fig. 3–5 is of this type.

The tube of force closed at the ends by A_1 and A_2 constitutes the imaginary closed surface, and the vector fields at P_1 and P_2 are uniform and perpendicular to the surfaces A_1 and A_2. The integral $\int_s \mathbf{E} \cdot d\mathbf{s} = 0$ over the long curved portions between these two areas because $\mathbf{E}$ is perpendicular to $d\mathbf{s}$ everywhere on this part of the surface.

This method can be used to calculate the field near a large plane with a uniform density of surface charge. Consider that the charged plane coincides with the xy-plane, has dimensions great enough to be considered infinite in the x- and y-directions, and has a uniform areal density σ_t. The problem is to find the field due to this distribution of charges at a point P whose distance from the plane is z. Since the plane is infinite, any such point must have a field directed parallel to the z-axis. A right circular cylinder may be constructed extending from $-z$ to $+z$ with its axis parallel to the z-axis and with these ends closed by small circular disks of area A. This closed surface s, which is often called a gaussian surface or a pillbox, encloses a total charge of $\sigma_t A$, and from Gauss' law $\int_s \mathbf{E} \cdot d\mathbf{s} = \sigma_t A/\epsilon_0$. Considering σ_t to be positive, the field is upward at $+z$ and downward at $-z$ but the magnitudes are equal by symmetry and each is directed out of the closed cylindrical surface, so that over the two end disks $\int \mathbf{E} \cdot d\mathbf{s} = E \int ds = E2A$. Since the fields are parallel to the cylindrical surface, this surface contributes nothing to the integral and hence $2EA = \sigma_t A/\epsilon_0$ or $E = \sigma_t/2\epsilon_0$. Therefore the magnitude of the field at P is independent of its position, although this can be strictly true only so long as the plane is truly infinite. However, even when such a plane is finite, the expression $E = \sigma_t/2\epsilon_0$ is a very accurate approximation for points close to the plane.

Another important problem is that of calculating the field near the surface of a conductor in terms of the density σ of its surface charge. The important relation for conductors is

$$\sigma = \epsilon_0 E,$$

which can be demonstrated by means of Gauss' law. It is first necessary to realize that practically all real surfaces appear flat when the observation points are close enough. Even the sharp point of a needle is rounded under a microscope and would appear flat if it could be observed from a point within 10^{-8} m. A small cylindrical pillbox can be constructed with one disk-shaped end (of small area A) inside the conductor and the other outside, with both parallel to the surface of the conductor. The charge enclosed by this pillbox is σA. Since there is no field in the conductor, $\int \mathbf{E} \cdot d\mathbf{s} = 0$ over the surface of the pillbox inside the conductor. The surface of the conductor is an equipotential surface and the field just outside must, therefore, be normal to the surface, so that $\int \mathbf{E} \cdot d\mathbf{s} = 0$ over the cylindrical sides of the pillbox. Since the field E is normal to the outer disk and is uniform when A is taken very small, Gauss' law gives $\sigma A = \epsilon_0 EA$ or

$$\sigma = \epsilon_0 E = -\epsilon_0 \frac{\partial V}{\partial n}, \tag{3–6}$$

where n is the normal distance from the conducting surface.

The field near a thin charged sheet of infinite extent has been shown to be $E = \sigma_t/2\epsilon_0$, where σ_t is the total charge density. If the thin sheet is a conductor, one-half the charge would be on each side of the sheet and the surface charge density on either surface would then be $\sigma = \sigma_t/2$, which makes the last two field calculations consistent.

It is left as an exercise to show that the field around a very long straight wire, charged to a uniform density of charge λ, is $E = \lambda/2\pi\epsilon_0 r$ and that the field around a uniformly charged sphere is given by $E = q/4\pi\epsilon_0 r^2$. In other words, Gauss' law shows that with plane symmetry the field is independent of the distance, with cylindrical symmetry it varies inversely as the first power of the distance, and with spherical symmetry it varies inversely as the square of the distance.

Differential form of Gauss' law. In the integral form the usefulness of Gauss' law for solving electrostatic problems is strictly limited to very simple geometries. On the other hand, the number of problems in electrostatics that need to be solved is essentially infinite because the number of possible distributions of charges and possible shapes and locations of conductors or other materials is infinite. Gauss' law can be modified to solve many of these by exact mathematical methods and many others by approximate methods. This involves developing a differential equation by applying Gauss' law to a small elemental volume in space.

In rectangular coordinates the elemental volume is a rectangular parallelepiped at the point x, y, z, whose dimensions are Δx, Δy, and Δz. Let us apply Gauss' law to this closed surface as shown in Fig. 3–6. The closed surface consists of six rectangles, two of area $\Delta x \, \Delta y$, two of area $\Delta y \, \Delta z$, and two of area $\Delta z \, \Delta x$. Assume that the field in this volume element is a function of the coordinates x, y, and z. According to Gauss' law, the total flux out of this volume is $(\rho/\epsilon_0)(\Delta x \, \Delta y \, \Delta z)$, where ρ is the average density of charge within the volume. The flux out of the shaded rectangle on the left is $-E_y \, \Delta x \, \Delta z$ and the flux out of the shaded rectangle on the right is $E_y \, \Delta x \, \Delta z + (\partial/\partial y)(E_y \, \Delta x \, \Delta z) \, \Delta y$, provided Δx, Δy, and Δz are infinitesimal, and the net flux out of these two shaded rectangles is $(\partial E_y/\partial y)(\Delta x \, \Delta y \, \Delta z)$, since Δx and Δz are independent of y. By symmetry, the net flux out of the other

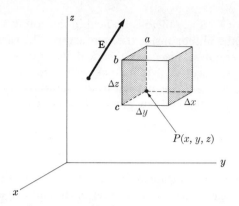

Fig. 3–6 Volume element in rectangular coordinates. Some readers may prefer to place the rectangular element with *P* at its center. This complicates the mathematics but yields the same result as Δx, Δy, and Δz approach zero.

pairs of surfaces is given by $(\partial E_x/\partial x)(\Delta x\,\Delta y\,\Delta z)$ and $(\partial E_z/\partial z)(\Delta x\,\Delta y\,\Delta z)$, respectively. The net flux out of this volume element is the sum of these three terms. Equating this sum to $\rho\,\Delta x\,\Delta y\,\Delta z/\epsilon_0$ and dividing both sides by the volume $(\Delta x\,\Delta y\,\Delta z)$ results in

$$\frac{\partial E_x}{\partial x} + \frac{\partial E_y}{\partial y} + \frac{\partial E_z}{\partial z} = \frac{\rho}{\epsilon_0}.$$

The terms on the left-hand side of this equation occur so frequently in physical and mathematical theories that they have been given the name *divergence* of **E**, which is usually written as $\mathbf{\nabla} \cdot \mathbf{E}$. Gauss' law for an elementary volume is then given by the differential relation

$$\mathbf{\nabla} \cdot \mathbf{E} = \frac{\rho}{\epsilon_0}. \tag{3–7}$$

Notice that $\mathbf{\nabla} \cdot \mathbf{E}$ expresses the flux out of the volume element per unit volume. Wherever there is a distribution of charges of density ρ, the divergence of **E** is proportional to ρ. Furthermore, when there are no charges in a given space, the right side becomes zero and the expression states that the flux out of the elemental volume is equal to the flux into that volume.

3–3 Divergence of vector fields

The divergence is a mathematical operation that can be applied to any vector function of coordinates. The general definition of the divergence of any vector such as **A** is

$$\text{div } \mathbf{A} = \mathbf{\nabla} \cdot \mathbf{A} = \lim_{\Delta \tau \to 0} \frac{\int_s \mathbf{A} \cdot d\mathbf{s}}{\Delta \tau}, \tag{3–8}$$

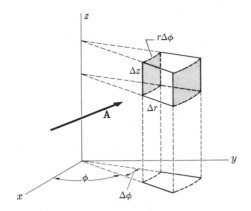

Fig. 3–7 Volume element in cylindrical coordinates.

where $\Delta\tau$ is the value of the elemental volume. This definition of divergence is independent of the system of coordinates used.

The general definition can be used to find the expression for divergence in cylindrical coordinates by constructing a volume element as shown in Fig. 3–7, using the system of coordinates introduced in Fig. 2–4. In cylindrical coordinates, the sides of the elemental volume are of length Δr, $r\,\Delta\phi$, and Δz. The integral $\int \mathbf{A} \cdot d\mathbf{s}$ taken over the shaded surface on the left is $-A_r r\,\Delta\phi\,\Delta z$ and the integral over the shaded surface on the right is $A_r r\,\Delta\phi\,\Delta z + (\partial/\partial r)(A_r r\,\Delta\phi\,\Delta z)\,\Delta r$. Hence the integral over these two shaded surfaces gives $(\partial(rA_r)/\partial r)\,\Delta r\,\Delta\phi\,\Delta z$ and by symmetry the integrals over the other pairs of surfaces yield

$$\frac{\partial}{\partial\phi}(A_\phi\,\Delta r\,\Delta z)\,\Delta\phi = \frac{\partial A_\phi}{\partial\phi}\,\Delta r\,\Delta\phi\,\Delta z$$

and

$$\frac{\partial}{\partial z}(A_z r\,\Delta\phi\,\Delta r)\,\Delta z = \frac{\partial A_z}{\partial z}\,r\,\Delta r\,\Delta\phi\,\Delta z$$

respectively. From Eq. (3–8),

$$\mathbf{\nabla} \cdot \mathbf{A} = \frac{1}{r\,\Delta r\,\Delta\phi\,\Delta z}$$

$$\times \left[\frac{\partial}{\partial r}(rA_r)\,\Delta r\,\Delta\phi\,\Delta z + \frac{\partial A_\phi}{\partial\phi}\,\Delta r\,\Delta\phi\,\Delta z + \frac{\partial A_z}{\partial z}\,r\,\Delta r\,\Delta\phi\,\Delta z\right],$$

which reduces to

$$\mathbf{\nabla} \cdot \mathbf{A} = \frac{1}{r}\frac{\partial(rA_r)}{\partial r} + \frac{1}{r}\frac{\partial A_\phi}{\partial\phi} + \frac{\partial A_z}{\partial z}, \tag{3–9}$$

Table 3–1

MULTIPLIERS AND UNIT VECTORS FOR
ORTHOGONAL COORDINATE SYSTEMS

	Coordinates			Multipliers			Unit vectors		
General	u	v	w	U	V	W	$\mathbf{a}_u$	$\mathbf{a}_v$	$\mathbf{a}_w$
Rectangular	x	y	z	1	1	1	$\mathbf{i}$	$\mathbf{j}$	$\mathbf{k}$
Cylindrical	r	ϕ	z	1	r	1	$\mathbf{a}_r$	$\mathbf{a}_\phi$	$\mathbf{a}_z$
Spherical	r	θ	ϕ	1	r	$r \sin \theta$	$\mathbf{a}_r$	$\mathbf{a}_\theta$	$\mathbf{a}_\phi$

which is the expression for the divergence of any vector $\mathbf{A}$ in cylindrical coordinates. This calculation of the divergence of $\mathbf{A}$ in cylindrical coordinates follows the pattern used for the calculation in rectangular coordinates. In each case the volume element has six surfaces over which $\int \mathbf{A} \cdot d\mathbf{s}$ must be evaluated. These are separated into three pairs, each pair of surfaces being perpendicular to one of the three unit vectors. The flux into the element through the first surface of any pair is almost equal to the flux out of the other surface. In any one pair of surfaces, the net flux out of the elemental volume through the two surfaces is just the increase in flux passing through the second surface over that passing through the first. In Fig. 3–7 the flux through the first surface at r and perpendicular to $\mathbf{a}_r$ is $\Phi_r = A_r r \, \Delta\phi \, \Delta z$. The net flux out of the element due to the two surfaces perpendicular to $\mathbf{a}_r$ is just the increase in the flux passing through the second surface, at $r + \Delta r$, over that through the first surface; that is,

$$\frac{\partial \Phi_r}{\partial r} \Delta r = \frac{\partial}{\partial r} (A_r r \, \Delta\phi \, \Delta z) \, \Delta r.$$

Repeating this procedure with the surface pairs perpendicular to $\mathbf{a}_\phi$ and $\mathbf{a}_z$ gives three expressions, and the divergence is obtained by adding these three and dividing by the volume of the element.

The procedure is best illustrated by calculating the divergence in general orthogonal coordinates. Table 3–1 can be used to convert any expression given in general orthogonal coordinates to the correct expression in any of the specific orthogonal coordinate systems. There are many other orthogonal systems that could be included in this table.

If the three orthogonal coordinates are u, v, and w, and the unit vectors in these three directions are $\mathbf{a}_u$, $\mathbf{a}_v$, and $\mathbf{a}_w$, any displacement in such a system of coordinates can be written as $d\mathbf{l} = \mathbf{a}_u U \, du + \mathbf{a}_v V \, dv + \mathbf{a}_w W \, dw$, where U, V, and W are the quantities which must multiply du, dv, and dw to determine linear distances in the directions $\mathbf{a}_u$, $\mathbf{a}_v$, and $\mathbf{a}_w$, respectively. For example, in rectangular coordinates, U, V, and W are unity and dimensionless while u, v, and w are x, y, and z. In the cylindrical coordinates just described

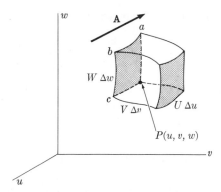

Fig. 3–8 Volume element in general orthogonal coordinates.

$U = 1$, $V = r$, and $W = 1$, and $u = r$, $v = \phi$, and $w = z$. The elemental volume for these general coordinates is shown in Fig. 3–8.

Now consider the pair of surfaces perpendicular to $\mathbf{a}_v$, shaded in the figure. The flux through the surface at v is $\Phi_v = A_v U\,\Delta u\,W\,\Delta w$, and the net flux out through these two surfaces is

$$\frac{\partial \phi_v}{\partial v}\,\Delta v = \frac{\partial}{\partial v}\,(A_v U\,\Delta u\,W\,\Delta w)\,\Delta v.$$

The net flux out through the pair of surfaces perpendicular to $\mathbf{a}_w$ can be obtained by cyclic substitution; that is, by changing u to v, v to w, w to u, U to V, V to W, W to U, etc. A second cyclic substitution gives the net flux out through the pair of surfaces perpendicular to $\mathbf{a}_u$. The divergence then is obtained by adding these three expressions and dividing by the volume $\Delta \tau = UVW\,\Delta u\,\Delta v\,\Delta w$. Since Δu, Δv, and Δw are independent of the coordinates, the final expression is

$$\nabla \cdot A = \frac{1}{UVW}\left[\frac{\partial}{\partial u}\,(VWA_u) + \frac{\partial}{\partial v}\,(WUA_v) + \frac{\partial}{\partial w}\,(UVA_w)\right].$$

This general expression for the divergence of **A** is easily translated into a specific expression for any orthogonal coordinate system. The problems in this text can be solved by use of rectangular, cylindrical, or spherical coordinates. For convenience, Table 3–1 contains definitions of the quantities needed for these three coordinate systems in terms of the symbols of the general coordinates. The spherical coordinates used in this text are shown in Fig. 2–5.

3–4 The Gauss mathematical theorem

Gauss' law in integral form, $\int_s \epsilon_0 \mathbf{E} \cdot ds = \int_s \rho\,d\tau$, is given in Eq. (3–5) and in differential form, $\nabla \cdot \mathbf{E} = \rho/\epsilon_0$, in Eq. (3–7). If the differential form is substituted into the integral form, we obtain

$$\int_s \mathbf{E} \cdot d\mathbf{s} = \int_\tau \mathbf{\nabla} \cdot \mathbf{E} \, d\tau.$$

This purely mathematical relation is called Gauss' theorem, and since it is true for any vector function of position in space provided that function is continuous and differentiable, it may be written as

$$\int_s \mathbf{A} \cdot d\mathbf{s} = \int_\tau \mathbf{\nabla} \cdot \mathbf{A} \, d\tau,$$

where $\mathbf{A}$ is any such function of the coordinates. Although the statements above do not constitute a rigorous proof of this theorem, the general definition for divergence given in Eq. (3–8) makes it possible to provide a simple direct proof.

A perfectly general volume, illustrated by the potato shape of Fig. 3–2, which is surrounded by a closed surface like the skin of the potato can be considered stationary in a vector field $\mathbf{A}$. A symbol ψ can then be defined by $\int_s \mathbf{A} \cdot d\mathbf{s} = \psi$ when the integral is taken over the entire closed surface. A very sharp knife cutting through the volume without disturbing the positions of the parts will produce two volumes with the two added surfaces k and k'. Whatever flux passes out of one volume through k passes into the other volume through k' and integration over all the surfaces of both volumes then gives

$$\int \mathbf{A} \cdot d\mathbf{s} = \int_s \mathbf{A} \cdot d\mathbf{s} + \int_k \mathbf{A} \cdot d\mathbf{s} + \int_{k'} \mathbf{A} \cdot d\mathbf{s} = \psi,$$

since the integral over k' cancels that over k.

More elements of volume and more surfaces can be created by slicing through the potato in various directions with this sharp knife. The new surfaces are always created in pairs such that the integral $\int \mathbf{A} \cdot d\mathbf{s}$ over one surface of the pair cancels this integral over the other. Specially, the volume may be divided up into a large number of very small volume elements. The defining Eq. (3–8) for divergence then may be applied separately to each of these elemental volumes and the results added. Each volume element inside the potato has six surfaces, but each of these six surfaces can be paired with an adjacent surface of some other volume element. Adding the right-hand-side terms of Eq. (3–8) for all of the volume elements amounts to integrating $\int \mathbf{A} \cdot d\mathbf{s}$ over all of the surfaces. Since the integrals over the pairs of surfaces cancel, the only surface integrals that contribute to the final result are those over the skin of the original potato. The sum of all terms on the right-hand side of Eq. (3–8) then gives $\int_s \mathbf{A} \cdot d\mathbf{s} = \psi$, which is equal to the sum of terms on the left-hand side of Eq. (3–8) for all of the volume elements. In the limit as $\Delta\tau$ approaches zero, this sum, $\sum_i \mathbf{\nabla} \cdot \mathbf{A}_i \, \Delta\tau_i = \psi$, becomes $\psi = \int_\tau \mathbf{\nabla} \cdot \mathbf{A} \, d\tau$.

Therefore Gauss' mathematical theorem

$$\int_\tau \mathbf{\nabla} \cdot \mathbf{A} \, d\tau = \int_s \mathbf{A} \cdot d\mathbf{s} \qquad (3\text{–}10)$$

is true for any volume τ enclosed by a surface s in a region where $\mathbf{A}$ is some continuous and differentiable vector function of the coordinates. This theorem is very useful for transforming certain volume integrals into surface integrals and conversely.

Poisson's and Laplace's equations. Gauss' law, developed directly from Coulomb's law of forces between static charges, has the integral form

$$\int_s \mathbf{E} \cdot d\mathbf{s} = \int_\tau \frac{\rho}{\epsilon_0} \, d\tau$$

of Eq. (3–5) and the differential form

$$\mathbf{\nabla} \cdot \mathbf{E} = \frac{\rho}{\epsilon_0}$$

of Eq. (3–7). The integral form states that the flux out of any closed volume equals the enclosed charges divided by ϵ_0. The differential form makes the same statement for a small element of volume, although here both sides of the equation are divided by the volume of the element. Equation (3–7) can be derived directly from Eq. (3–5) by using Gauss' mathematical theorem, which allows Eq. (3–5) to be written as

$$\int_s \mathbf{E} \cdot d\mathbf{s} = \int_\tau \mathbf{\nabla} \cdot \mathbf{E} \, d\tau = \int_\tau \frac{\rho}{\epsilon_0} \, d\tau.$$

Since this relation is true for any arbitrary volume τ, it is true when τ shrinks to the dimensions of a physical point, where the integrands of the volume integrals become constant and $\mathbf{\nabla} \cdot \mathbf{E} = \rho/\epsilon_0$.

It was shown in Sections 2–11 and 2–13 that any electrostatic field can be expressed as the gradient of a potential, that is, $\mathbf{E} = -\mathbf{\nabla}V$.

Substituting this into Eq. (3–7) gives the important relation, first developed by Poisson,

$$\mathbf{\nabla} \cdot \mathbf{\nabla}V = -\frac{\rho}{\epsilon_0}. \qquad (3\text{–}11)$$

The operator relation $\mathbf{\nabla} \cdot \mathbf{\nabla}V$, which is usually written as $\nabla^2 V$, is just the divergence of the gradient of V.

Laplace, investigating special situations in free space where $\rho = 0$, obtained the form of this equation,

$$\nabla^2 V = 0, \qquad (3\text{–}12)$$

that has been given his name, and provided many useful solutions for electrostatic problems.

3–5 Capacitors and capacitance

The useful devices produced by placing two conductors close to but insulated from each other, which were originally called condensers, are now designated by the more consistent name *capacitor*. In use, a charge is moved from one conductor of the capacitor to the other, to give a charge $-q$ to one and q to the other, and to produce a potential difference V between the two conductors. When these conductors are separated in a vacuum, in air, or by most of the insulations that are used in such devices, the ratio q/V, designated by C, is constant. Originally C was called the capacity of the condenser but recently C has been given the more consistent designation of the *capacitance* of the *capacitor*. While in commercial capacitors the two conductors are usually placed close together to produce a relatively large capacitance, any two conductors insulated from each other constitute a capacitor whose capacitance can be determined by measuring both q and V. In many cases, when the two conductors and their separation follow a regular geometric pattern, the capacitance can also be calculated. The simplest capacitor, which is very often used, consists of two parallel plates, each of area A, separated by a small distance d. With the charge q on one plate and $-q$ on the other, the electric field is uniform between the plates, except near the edges of the plates, and is given by Eq. (3-6) as

$$E = \frac{\sigma}{\epsilon_0} = \frac{q}{\epsilon_0 A}.$$

With this uniform field between the plates the potential difference is simply $V = Ed = qd/\epsilon_0 A$, and the capacitance is

$$C = \frac{q}{V} = \frac{\epsilon_0 A}{d},$$

provided the nonuniformity at the edges of the plates can be ignored. In this simple capacitor the capacitance is proportional to A and inversely proportional to d.

Capacitance of isolated conductors. It is sometimes desirable to determine the capacitance of a single isolated conductor. The electric flux lines that leave a positively charged conductor must, by Gauss' law, end on an equal negative charge. Usually the flux lines from a single charged conductor end on the walls of the laboratory or on nearby conductors, but if the charged conductor could be truly isolated, these flux lines would extend to infinity. The capacitance of an isolated conductor may be defined as the capacitance of a capacitor, which consists of the conductor surrounded by a hollow conducting sphere of infinite radius. Since the potential at infinity is zero by definition, the infinite conducting sphere is not essential in the definition. *The capacitance of an isolated conductor is the ratio of its charge divided by its potential.*

The potential of any isolated sphere of radius a and charge q is given by

$$V = \int_{r=a}^{r=\infty} \mathbf{E} \cdot d\mathbf{l} = \int_a^\infty \frac{q\,dr}{4\pi\epsilon_0 r^2} = \frac{q}{4\pi\epsilon_0 a}.$$

Therefore the capacitance of an isolated sphere is

$$C = \frac{q}{V} = 4\pi\epsilon_0 a,$$

where a is the radius of the sphere.

3–6 Curl of a vector field

In Section 2–11 it is shown that the line integral around any closed path is everywhere zero in an electrostatic field:

$$\oint \mathbf{E} \cdot d\mathbf{l} = 0,$$

which proves that electrostatic fields are conservative and that $\mathbf{E}$ in static fields can be written as the gradient of a potential function. There are many other fields that can be expressed as gradients of potential functions because they are conservative. A simpler method for determining whether or not a field is conservative is provided by a vector operator known as the *curl*. Whenever the curl of a vector field is zero, the field is conservative and can be expressed as the gradient of some potential function. The testing of vector fields to determine whether they are conservative is one of the very important uses of the curl operator, but it is by no means its only important use. For example, the curl is used later to obtain a very simple differential expression for the magnetic field in the vicinity of electric currents.

The curl of any vector function of coordinates such as $\mathbf{A}$ is symbolized by $\nabla \times \mathbf{A}$, which is also a vector. The definition, which is independent of the coordinate system chosen, involves line integration around an element of area Δs oriented arbitrarily in space. The curl of $\mathbf{A}$, $\nabla \times \mathbf{A}$, is then defined by the relation

$$\lim_{\Delta s \to 0} (\nabla \times \mathbf{A}) \cdot \Delta \mathbf{s} = \lim_{\Delta s \to 0} \oint \mathbf{A} \cdot d\mathbf{l}, \qquad (3\text{–}13)$$

where the positive direction of $\Delta \mathbf{s}$ is the direction the right-handed screw in Fig. 3–9 would progress if it were rotated in the positive direction of $d\mathbf{l}$.

To specify the direction and magnitude of any vector it is best to determine its components in three coordinate directions. For example, in rectangular coordinates the vector $\mathbf{A}$ is completely defined by $\mathbf{A} = \mathbf{i}A_x + \mathbf{j}A_y + \mathbf{k}A_z$. From the definition of the curl in Eq. (3–13), its components in any coordinate system can be calculated. In rectangular coordinates the

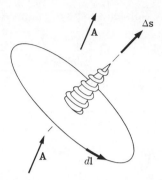

Fig. 3–9 In the defining equation for the curl, Eq. (3–13), the positive direction for Δs is chosen to be the direction in which a right-handed screw would move if rotated in direction $d\mathbf{l}$.

x-component of the curl is calculated by orienting the arbitrary surface Δs perpendicular to the x-axis to make the vector $\Delta \mathbf{s}$ parallel to the x-axis. The right-hand side of Eq. (3–13) always gives the component of $\nabla \times \mathbf{A}$ parallel to $\Delta \mathbf{s}$.

Curl in rectangular coordinates. To obtain the expression for $\nabla \times \mathbf{A}$ in rectangular coordinates, it is necessary to choose three elements of area, each perpendicular to one of the coordinate axes. The volume element shown in Fig. 3–6 has six faces from which three area elements can be chosen. The shaded surface on the left is perpendicular to the y-axis and can be used to obtain the y-component of $\nabla \times \mathbf{A}$. Using this surface, the left-hand side of Eq. (3–13) is

$$(\nabla \times \mathbf{A}) \cdot \Delta \mathbf{s} = (\nabla \times \mathbf{A})_y \, \Delta s_y = (\nabla \times \mathbf{A})_y \, \Delta x \, \Delta z,$$

and the right-hand side is obtained by integrating around the rectangle from P to a to b to c and back to P. Integrations along these four straight lines provide the four terms

$$A_z \, \Delta z, \qquad \left(A_x + \frac{\partial A_x}{\partial z} \Delta z \right) \Delta x, \qquad \left(A_z + \frac{\partial A_z}{\partial x} \Delta x \right)(-\Delta z),$$

and

$$A_x(-\Delta x).$$

When these are added, all but two of the terms cancel, leaving

$$(\nabla \times \mathbf{A})_y \, \Delta x \, \Delta z = \left(\frac{\partial A_x}{\partial z} - \frac{\partial A_z}{\partial x} \right) \Delta x \, \Delta z,$$

or

$$(\nabla \times \mathbf{A})_y = \frac{\partial A_x}{\partial z} - \frac{\partial A_z}{\partial x}.$$

Because of the symmetry of the coordinates, the z- and x-components can be obtained by the cyclic substitutions of z for y, x for z, and y for x, giving

$$(\nabla \times \mathbf{A})_z = \frac{\partial A_y}{\partial x} - \frac{\partial A_x}{\partial y},$$

$$(\nabla \times \mathbf{A})_x = \frac{\partial A_z}{\partial y} - \frac{\partial A_y}{\partial z},$$

and the curl of $\mathbf{A}$ in rectangular coordinates becomes

$$\nabla \times \mathbf{A} = \mathbf{i}\left(\frac{\partial A_z}{\partial y} - \frac{\partial A_y}{\partial z}\right) + \mathbf{j}\left(\frac{\partial A_x}{\partial z} - \frac{\partial A_z}{\partial x}\right) + \mathbf{k}\left(\frac{\partial A_y}{\partial x} - \frac{\partial A_x}{\partial y}\right). \quad (3\text{–}14)$$

It is convenient to write this as a determinant:

$$\nabla \times \mathbf{A} = \begin{vmatrix} \mathbf{i} & \mathbf{j} & \mathbf{k} \\ \dfrac{\partial}{\partial x} & \dfrac{\partial}{\partial y} & \dfrac{\partial}{\partial z} \\ A_x & A_y & A_z \end{vmatrix}. \quad (3\text{–}15)$$

Unlike the ordinary determinant, this is valid only when expanded in terms of the top row.

Curl in general orthogonal coordinates. An expression for the curl in each coordinate system will be needed. Table 3–1 makes it easy to convert mathematical expressions in general coordinates into any of the other three systems, making it necessary only to obtain the expression for the curl in general coordinates. The volume element in Fig. 3–8 has six surfaces, three of which can be used to obtain the three components that are needed. In fact, it is necessary to use only one surface to determine one of the components, since the other two can be obtained by cyclic substitution.

The shaded surface on the left can be chosen to obtain the v-component of the curl in general coordinates. Following the method used in calculating the y-component in rectangular coordinates, the line integral on the left in Eq. (3–13) should be taken from P to a to b to c and back to P. When this integral was evaluated around the rectangle in Fig. 3–6, the integral along ab almost cancelled that along cP, and the integral along bc almost cancelled that along Pa. The only terms remaining in rectangular coordinates, after the four line integrals were added, were the ones giving the change in magnitude of the integral from cP to ab, $(\partial/\partial z)(A_x \,\Delta x)\,\Delta z$, and from Pa to bc, $-(\partial/\partial x)(A_z \,\Delta z)\,\Delta x$. Because of the direction of the integration, the latter term has a negative sign. A similar type of cancellation takes place for general orthogonal coordinates when integrating along the path $Pabc$ around the shaded surface of Fig. 3–8. The integral along Pa is $A_w W \,\Delta w$, which is equal and opposite to that along *bc* except for the small variation given by

$$-\left(\frac{\partial}{\partial u}\right)(A_w W \, \Delta w) \, \Delta u.$$

In the same way, the integrals along *ab* and *cP* cancel except for the small variation

$$\left(\frac{\partial}{\partial w}\right)(A_u U \, \Delta u) \, \Delta w.$$

For the shaded surface of the general coordinates, *Pabc* in Fig. 3–8, the curl defining Eq. (3–13) becomes

$$(\mathbf{\nabla} \times \mathbf{A})_v U \, \Delta u \, W \, \Delta w = \left[\frac{\partial}{\partial w}(UA_u) - \frac{\partial}{\partial u}(WA_w)\right] \Delta u \, \Delta w,$$

or

$$(\mathbf{\nabla} \times \mathbf{A})_v = \frac{1}{UW}\left[\frac{\partial}{\partial w}(UA_u) - \frac{\partial}{\partial u}(WA_w)\right],$$

which is the *v*-component of the curl of **A**. The other two components can be obtained from this by the cyclic substitution of *v* for *u*, *w* for *v*, *u* for *w*, *V* for *U*, *W* for *V*, and *U* for *W*. The curl is then given by the vector sum of these components, each multiplied by its appropriate unit vector. It is conveniently given by the determinant form

$$\mathbf{\nabla} \times \mathbf{A} = \frac{1}{UVW}\begin{vmatrix} U\mathbf{a}_u & V\mathbf{a}_v & W\mathbf{a}_w \\ \dfrac{\partial}{\partial u} & \dfrac{\partial}{\partial v} & \dfrac{\partial}{\partial w} \\ UA_u & VA_v & WA_w \end{vmatrix}, \tag{3–16}$$

which again must be expanded in terms of the top row to give meaningful results.

The physical interpretation of the divergence is easy to understand, but the physical meaning of the curl is more difficult. Using an imaginary disk of small area Δs, the curl defining Eq. (3–13) can be integrated around its periphery to give the component of the curl of **A** perpendicular to the surface of the disk. The direction of the curl of **A** could be obtained by changing the orientation of the disk until this integral is maximum. When the integral is maximum Δs points in the direction of $\mathbf{\nabla} \times \mathbf{A}$. In other words, the curl of **A** points in that direction in space which maximizes the line integral of **A** around a small surface normal to it.

3–7 Stokes' theorem

In Section 3–3, by using an infinitesimal volume element, the divergence is defined as a point function. It is then shown in Section 3–4 that the two sides of the defining equation (3–8) could be integrated over a finite volume to obtain the mathematical theorem of Gauss that is applicable to any volume. A very similar procedure is possible with the equation that defines the curl. By using an infinitesimal surface element, the curl is defined as a point function in Eq. (3–13). Integration of the two sides of this equation over a finite surface of general shape gives Stokes' theorem. Just as Gauss' theorem is useful for transforming volume integrals into surface integrals and vice versa, Stokes' theorem is useful for transforming surface integrals into line integrals and vice versa.

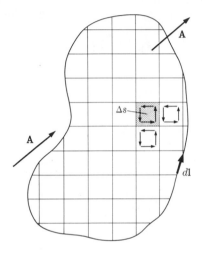

Fig. 3–10 Line integral of **A** is taken counterclockwise around element of area Δs. When this result is added to adjoining elements, the contributions along adjacent boundaries cancel, leaving only the contribution from the outside boundary.

To demonstrate Stokes' theorem, an irregular surface like that shown in Fig. 3–10 can be used to represent a general shape of surface that is not necessarily in one plane. This surface can be divided into a large number of small four-sided elements of surface somewhat like those shown as rectangles in Fig. 3–10. These surface elements can be made small enough so that Eq. (3–13), defining the curl, can be applied to each of them. If the right-hand side of Eq. (3–13) is applied to two adjacent elements, the common boundary will be traversed twice in opposite directions by the line integral. Thus the integrals along this common boundary cancel each other, and only the integral around the periphery of the combined surfaces remains. If another adjacent surface element is added, the integrals along the common boundary again cancel. The process is repeated as each adjacent element is added until only the integral around the periphery of the general surface remains. Therefore, when the two sides of Eq. (3–13) are integrated over the whole surface shown in Fig. 3–10, the right side becomes just $\oint \mathbf{A} \cdot d\mathbf{l}$, integrated all around the periphery. The integration of the left-hand side gives just $\int_s \nabla \times \mathbf{A} \cdot d\mathbf{s}$, which must be integrated over the entire surface s, and the final result is

$$\int_s \nabla \times \mathbf{A} \cdot d\mathbf{s} = \oint \mathbf{A} \cdot d\mathbf{l}, \qquad (3\text{–}17)$$

which is Stokes' mathematical theorem for transforming line to surface integrals and surface to line integrals.

When $\nabla \times \mathbf{A}$ is zero everywhere, both sides of Eq. (3–17) are zero. Since the line integral of $\mathbf{A} \cdot d\mathbf{l}$ around any closed path is zero, the vector field $\mathbf{A}$ is conservative and can be expressed as the gradient of some potential function. The condition $\nabla \times \mathbf{A} = 0$ is both necessary and sufficient for determining that $\mathbf{A}$ constitutes a conservative field.

3–8 The operator del

Three differential operations involving scalar and vector quantities have been defined and some of their uses have been explained. The definitions of these operations, called the gradient, the divergence, and the curl, were each made independent of the system of coordinates used. The brief symbols ∇V, $\nabla \cdot \mathbf{A}$, and $\nabla \times \mathbf{A}$ that were introduced to identify the operations appear, at first glance, to be completely unrelated to the defining relations. However, the operations suggested by the symbols can be used to obtain the correct expressions, provided the symbol ∇, called *del*, is treated as a differentiating vector operator,

$$\nabla = \mathbf{i}\frac{\partial}{\partial x} + \mathbf{j}\frac{\partial}{\partial y} + \mathbf{k}\frac{\partial}{\partial z},$$

in rectangular coordinates. The gradient of V is then

$$\text{del } V = \nabla V = \left(\mathbf{i}\frac{\partial}{\partial x} + \mathbf{j}\frac{\partial}{\partial y} + \mathbf{k}\frac{\partial}{\partial z}\right)V$$

or

$$\nabla V = \mathbf{i}\frac{\partial V}{\partial x} + \mathbf{j}\frac{\partial V}{\partial y} + \mathbf{k}\frac{\partial V}{\partial z}.$$

The dot product $\nabla \cdot \mathbf{A}$ would be

$$\frac{\partial A_x}{\partial x} + \frac{\partial A_y}{\partial y} + \frac{\partial A_z}{\partial z},$$

which is the divergence in rectangular coordinates. Likewise, the cross product of del with $\mathbf{A}$, $\nabla \times \mathbf{A}$, gives Eq. (3–14), the curl in rectangular coordinates.

This formal interpretation of the operator del provides expressions for gradient, divergence, and curl in rectangular coordinates as easily as they can be obtained from their general definitions. In most coordinate systems, however, these expressions can be obtained much more easily from the general definitions.

A more precise interpretation of the operator del must be used with cylindrical, spherical, or general coordinates than was required with the rectangular coordinates. Care must be taken to see that this differential operator differentiates everything to its right. In cylindrical coordinates,

$$\nabla = \mathbf{a}_r \frac{\partial}{\partial r} + \mathbf{a}_\phi \frac{1}{r}\frac{\partial}{\partial \phi} + \mathbf{k}\frac{\partial}{\partial z}, \tag{3–18}$$

and to obtain the divergence of **A**, a formal expansion of $\nabla \cdot \mathbf{A}$, can be made to obtain

$$
\begin{aligned}
\nabla \cdot \mathbf{A} = \ &\mathbf{a}_r \frac{\partial}{\partial r} \cdot \mathbf{a}_r A_r + \mathbf{a}_r \frac{\partial}{\partial r} \cdot \mathbf{a}_\phi A_\phi + \mathbf{a}_r \frac{\partial}{\partial r} \cdot \mathbf{k} A_z \\
&+ \mathbf{a}_\phi \frac{1}{r}\frac{\partial}{\partial \phi} \cdot \mathbf{a}_r A_r + \mathbf{a}_\phi \frac{1}{r}\frac{\partial}{\partial \phi} \cdot \mathbf{a}_\phi A_\phi + \mathbf{a}_\phi \frac{1}{r}\frac{\partial}{\partial \phi} \cdot \mathbf{k} A_z \\
&+ \mathbf{k}\frac{\partial}{\partial z} \cdot \mathbf{a}_r A_r + \mathbf{k}\frac{\partial}{\partial z} \cdot \mathbf{a}_\phi A_\phi + \mathbf{k}\frac{\partial}{\partial z} \cdot \mathbf{k} A_z.
\end{aligned}
$$

This expression is written with nine terms, but it actually contains eighteen, since each term may be expanded into two. For example,

$$\mathbf{a}_r \frac{\partial}{\partial r} \cdot \mathbf{a}_r A_r = \mathbf{a}_r \cdot \mathbf{a}_r \frac{\partial A_r}{\partial r} + \mathbf{a}_r A_r \cdot \frac{\partial \mathbf{a}_r}{\partial r}.$$

Simplification or elimination of these terms may be accomplished, using the relations

$$\mathbf{a}_r \cdot \mathbf{a}_r = \mathbf{a}_\phi \cdot \mathbf{a}_\phi = \mathbf{k} \cdot \mathbf{k} = 1,$$
$$\mathbf{a}_r \cdot \mathbf{a}_\phi = \mathbf{a}_r \cdot \mathbf{k} = \mathbf{a}_\phi \cdot \mathbf{k} = 0,$$

as well as the relations

$$\frac{\partial \mathbf{a}_r}{\partial r} = 0, \qquad \frac{\partial \mathbf{a}_\phi}{\partial r} = 0, \qquad \frac{\partial \mathbf{k}}{\partial r} = 0;$$

$$\frac{\partial \mathbf{a}_r}{\partial \phi} = \mathbf{a}_\phi, \qquad \frac{\partial \mathbf{a}_\phi}{\partial \phi} = -\mathbf{a}_r, \qquad \frac{\partial \mathbf{k}}{\partial \phi} = 0;$$

$$\frac{\partial \mathbf{a}_r}{\partial z} = 0, \qquad \frac{\partial \mathbf{a}_\phi}{\partial z} = 0, \qquad \frac{\partial \mathbf{k}}{\partial z} = 0.$$

The change in direction of $\mathbf{a}_r$ and $\mathbf{a}_\phi$ resulting from a change in ϕ may be seen from an examination of Fig. 3–11. For example, the change in $\mathbf{a}_\phi$ resulting from an incremental increase $\Delta\phi$ in ϕ is a vector pointing in the negative $\mathbf{a}_r$ direction, which is just $-\mathbf{a}_r \, \Delta\phi$. Application of these relations

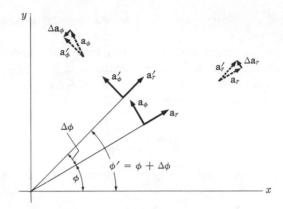

Fig. 3–11 Diagram illustrates why an increase of $\Delta\phi$ in the angle ϕ of cylindrical coordinates results in the following changes in the unit vectors:

$$\Delta\mathbf{a}_\phi = \frac{\partial \mathbf{a}_\phi}{\partial\phi}\Delta\phi = -\mathbf{a}_r\,\Delta\phi \quad\text{and}\quad \Delta\mathbf{a}_r = \frac{\partial \mathbf{a}_r}{\partial\phi}\Delta\phi = \mathbf{a}_\phi\,\Delta\phi.$$

reduces the 18-term expression to

$$\mathbf{\nabla}\cdot\mathbf{A} = \frac{\partial A_r}{\partial r} + 0 + 0$$

$$+\frac{1}{r}A_r + \frac{1}{r}\frac{\partial A_\phi}{\partial\phi} + 0$$

$$+ 0 + 0 + \frac{\partial A_z}{\partial z}.$$

This is frequently written as

$$\mathbf{\nabla}\cdot\mathbf{A} = \frac{1}{r}\frac{\partial}{\partial r}(rA_r) + \frac{1}{r}\frac{\partial A_\phi}{\partial\phi} + \frac{\partial A_z}{\partial z}, \tag{3–19}$$

in agreement with Eq. (3–9), which was obtained more simply from the fundamental definition of the divergence.

In a similar manner, the curl or Laplacian may be obtained in cylindrical or spherical coordinates. It must be remembered, of course, that the order of operators in an equation must be maintained. These operations can become involved, as the foregoing example illustrates. The operations were simple in rectangular coordinates because the unit vectors are constant in direction as well as in magnitude. The operations are complicated in cylindrical and spherical coordinates because the unit vectors are not constant in direction.

It is recommended that the del operator nomenclature be considered as a memory device and as an abbreviated notation for expressing the quantities formally defined in Sections 2–13, 3–3, and 3–6. Vector operations involving del are given below in various coordinate systems for convenient reference.

Rectangular coordinates

$$\nabla\psi = \mathbf{i}\frac{\partial\psi}{\partial x} + \mathbf{j}\frac{\partial\psi}{\partial y} + \mathbf{k}\frac{\partial\psi}{\partial z}.$$

$$\nabla\cdot\mathbf{A} = \frac{\partial A_x}{\partial x} + \frac{\partial A_y}{\partial y} + \frac{\partial A_z}{\partial z}.$$

$$\nabla\times\mathbf{A} = \mathbf{i}\left(\frac{\partial A_z}{\partial y} - \frac{\partial A_y}{\partial z}\right) + \mathbf{j}\left(\frac{\partial A_x}{\partial z} - \frac{\partial A_z}{\partial x}\right) + \mathbf{k}\left(\frac{\partial A_y}{\partial x} - \frac{\partial A_x}{\partial y}\right).$$

$$\nabla^2\psi = \frac{\partial^2\psi}{\partial x^2} + \frac{\partial^2\psi}{\partial y^2} + \frac{\partial^2\psi}{\partial z^2}.$$

Cylindrical coordinates

$$\nabla\psi = \mathbf{a}_r\frac{\partial\psi}{\partial r} + \mathbf{a}_\phi\frac{1}{r}\frac{\partial\psi}{\partial\phi} + \mathbf{k}\frac{\partial\psi}{\partial z}.$$

$$\nabla\cdot\mathbf{A} = \frac{1}{r}\frac{\partial}{\partial r}(rA_r) + \frac{1}{r}\frac{\partial A_\phi}{\partial\phi} + \frac{\partial A_z}{\partial z}.$$

$$\nabla\times\mathbf{A} = \mathbf{a}_r\left(\frac{1}{r}\frac{\partial A_z}{\partial\phi} - \frac{\partial A_\phi}{\partial z}\right) + \mathbf{a}_\phi\left(\frac{\partial A_r}{\partial z} - \frac{\partial A_z}{\partial r}\right)$$
$$+ \mathbf{k}\left(\frac{1}{r}\frac{\partial}{\partial r}(rA_\phi) - \frac{1}{r}\frac{\partial A_r}{\partial\phi}\right).$$

$$\nabla^2\psi = \frac{1}{r}\frac{\partial}{\partial r}\left(r\frac{\partial\psi}{\partial r}\right) + \frac{1}{r^2}\frac{\partial^2\psi}{\partial\phi^2} + \frac{\partial^2\psi}{\partial z^2}.$$

Spherical coordinates

$$\nabla\psi = \mathbf{a}_r\frac{\partial\psi}{\partial r} + \mathbf{a}_\theta\frac{1}{r}\frac{\partial\psi}{\partial\theta} + \mathbf{a}_\phi\frac{1}{r\sin\theta}\frac{\partial\psi}{\partial\phi}.$$

$$\nabla\cdot\mathbf{A} = \frac{1}{r^2}\frac{\partial}{\partial r}(r^2A_r) + \frac{1}{r\sin\theta}\frac{\partial}{\partial\theta}(A_\theta\sin\theta) + \frac{1}{r\sin\theta}\frac{\partial A_\phi}{\partial\phi}.$$

$$\nabla\times\mathbf{A} = \mathbf{a}_r\frac{1}{r\sin\theta}\left[\frac{\partial}{\partial\theta}(A_\phi\sin\theta) - \frac{\partial A_\theta}{\partial\phi}\right]$$
$$+ \mathbf{a}_\theta\frac{1}{r}\left[\frac{1}{\sin\theta}\frac{\partial A_r}{\partial\phi} - \frac{\partial}{\partial r}(rA_\phi)\right] + \mathbf{a}_\phi\frac{1}{r}\left[\frac{\partial}{\partial r}(rA_\theta) - \frac{\partial A_r}{\partial\theta}\right].$$

$$\nabla^2\psi = \frac{1}{r^2}\frac{\partial}{\partial r}\left(r^2\frac{\partial\psi}{\partial r}\right) + \frac{1}{r^2\sin\theta}\frac{\partial}{\partial\theta}\left(\sin\theta\frac{\partial\psi}{\partial\theta}\right) + \frac{1}{r^2\sin^2\theta}\frac{\partial^2\psi}{\partial\phi^2}.$$

General orthogonal coordinates

$$\nabla\psi = \mathbf{a}_u \frac{1}{U}\frac{\partial\psi}{\partial u} + \mathbf{a}_v \frac{1}{V}\frac{\partial\psi}{\partial v} + \mathbf{a}_w \frac{1}{W}\frac{\partial\psi}{\partial w}.$$

$$\nabla\cdot\mathbf{A} = \frac{1}{UVW}\left[\frac{\partial}{\partial u}(VWA_u) + \frac{\partial}{\partial v}(WUA_v) + \frac{\partial}{\partial w}(UVA_w)\right].$$

$$\nabla\times\mathbf{A} = \mathbf{a}_u \frac{1}{VW}\left[\frac{\partial}{\partial v}(WA_w) - \frac{\partial}{\partial w}(VA_v)\right]$$

$$+ \mathbf{a}_v \frac{1}{WU}\left[\frac{\partial}{\partial w}(UA_u) - \frac{\partial}{\partial u}(WA_w)\right]$$

$$+ \mathbf{a}_w \frac{1}{UV}\left[\frac{\partial}{\partial u}(VA_v) - \frac{\partial}{\partial v}(UA_u)\right].$$

$$\nabla^2\psi = \frac{1}{UVW}\left[\frac{\partial}{\partial u}\left(\frac{VW}{U}\frac{\partial\psi}{\partial u}\right) + \frac{\partial}{\partial v}\left(\frac{WU}{V}\frac{\partial\psi}{\partial v}\right) + \frac{\partial}{\partial w}\left(\frac{UV}{W}\frac{\partial\psi}{\partial w}\right)\right].$$

Operator identities. There are a number of vector identities that are useful in electromagnetic theories. For convenience, the most useful of those involving differentiation are listed here. No derivation is given for any of these equations because they are all easily proved by expanding them into components in any of the coordinate systems. Since vector equations are independent of the coordinate system in which they are expressed, the equations can be verified in rectangular coordinates. The proof of any of the following relations, by expanding them into their rectangular components, constitutes a revealing exercise. The reader should perform each proof sometime before he has occasion to use the relation.

$$\nabla\cdot(\psi\mathbf{A}) = \psi\nabla\cdot\mathbf{A} + \mathbf{A}\cdot\nabla\psi, \tag{3–20}$$

$$\nabla\times(\psi\mathbf{A}) = \psi\nabla\times\mathbf{A} - \mathbf{A}\times\nabla\psi, \tag{3–21}$$

$$\nabla\cdot(\mathbf{A}\times\mathbf{B}) = \mathbf{B}\cdot\nabla\times\mathbf{A} - \mathbf{A}\cdot\nabla\times\mathbf{B}, \tag{3–22}$$

$$\nabla\cdot(\nabla\times\mathbf{A}) = 0, \tag{3–23}$$

$$\nabla\times(\nabla\times\mathbf{A}) = \nabla(\nabla\cdot\mathbf{A}) - \nabla^2\mathbf{A}, \tag{3–24}$$

where $\nabla^2\mathbf{A} \equiv (\nabla\cdot\nabla)\mathbf{A}$,* which in rectangular coordinates is

$$\nabla^2\mathbf{A} = \mathbf{i}\,\nabla^2 A_x + \mathbf{j}\,\nabla^2 A_y + \mathbf{k}\,\nabla^2 A_z,$$

$$\nabla\cdot(\psi\nabla\psi) = (\nabla\psi)^2 + \psi\nabla^2\psi, \tag{3–25}$$

$$\nabla\times(\nabla\psi) = 0. \tag{3–26}$$

---

* Since $\nabla\cdot\nabla$ is a scalar it can operate on a vector. Unit vectors are differentiated as in Section 3-8; H. Margenau and G. M. Murphy, *Mathematics of Physics and Chemistry*, Vol. 1, 2nd ed., D. Van Nostrand, Princeton, N.J., 1956, p. 154. It is usually simpler and safer to work out the components of $\nabla\times(\nabla\times\mathbf{A})$.

For these equations **A** and **B** must be vector functions of the coordinates and ψ must be a scalar function of the coordinates, with finite first and second derivatives.

PROBLEMS

3–1 Verify the statements in the note of Problem 1–1.

3–2 A spherical shell of radius a has a charge q uniformly distributed over its surface. Find the potential due to q, (a) at $r > a$, (b) at $r = a$, and (c) at $r < a$.

3–3 A small sphere of mass 0.2 g hangs by a thread between two parallel vertical plates 5 cm apart. The sphere holds a charge of 6×10^{-9} C. What difference of potential between the plates will cause the thread to assume an angle of 30° with the vertical?

3–4 Two hollow concentric spherical conductors have radii of 2 cm and 4 cm, respectively. The inner sphere carries a charge of 12×10^{-9} C, the outer 20×10^{-9} C. Determine the potential at the following distances from the center: 5 cm, 4 cm, 3 cm, 2 cm, and 1 cm.

3–5 The two spheres of Problem 3–4 are connected by a conductor, which is then removed. What is the potential at the same points?

3–6 The plates of a parallel-plate capacitor are 5 mm apart and 2 m² in area. The plates are in vacuum. A potential difference of 10,000 V is applied across the capacitor. Compute (a) the capacitance, (b) the charge on each plate, and (c) the electric intensity in the space between them.

3–7 Three capacitors have the capacitances C_1, C_2, and C_3 respectively. Prove that (a) when they are connected in parallel the combined capacitance is $C_p = C_1 + C_2 + C_3$, and (b) when they are connected in series the combined capacitance C_s is given by $1/C_s = 1/C_1 + 1/C_2 + 1/C_3$.

3–8 A 1-μF capacitor and a 2-μF capacitor are connected in series across a 1200-V supply line.

a) Find the charge on each capacitor and the voltage across each.

b) The charged capacitors are disconnected from the line and from each other, and reconnected with terminals of like sign together. Find the final charge on each and the voltage across each.

3–9 A 1-μF capacitor and a 2-μF capacitor are connected in parallel across a 1200-V supply line.

a) Find the charge on each capacitor and the voltage across each.

b) The charged capacitors are then disconnected from the line and from each other, and reconnected with terminals of unlike sign together. Find the final charge on each and the voltage across each.

3–10 In Fig. 3–12, each capacitance $C_3 =$ 3 μF and each capacitance $C_2 = 2$ μF.

a) Compute the equivalent capacitance of the network between points a and b.

b) Compute the charge on each of the capacitors nearest a and b, when $V_{ab} = 900$ V.

c) With 900 V across a and b, compute V_{cd}.

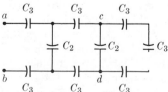

Figure 3–12

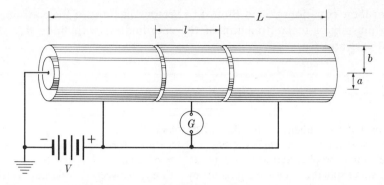

Figure 3–13

3–11 Two conductors that form a capacitor consist of concentric spherical shells of radii a and b, where $b > a$ and the space between the spheres is evacuated. With the outer sphere grounded and with a charge Q on the inner sphere, calculate

a) the electric field between the spheres, using Gauss' law,

b) the potential difference between the spheres, and

c) the capacitance of the system.

d) Show that if $d = b - a$, where $d \ll b$, the parallel-plate relation for capacitance, $C = \epsilon_0 A/d$, is valid.

3–12 A solid sphere of radius a has a volume charge of uniform density ρ. Obtain expressions for both **E** and V where $r < a$ and where $r > a$. Plot these expressions as a function of r.

3–13 Two conductors that form a capacitor consist of concentric cylindrical shells of radii a and b and of length L, where $L \gg b > a$. As shown in Fig. 3–13, the outer cylinder consists of three sections insulated from each other. While each of the three sections is charged to the potential V, the charge Q measured as it flows through the instrument G resides only on the central section of length l. The two outer sections, called "guard rings," serve to keep the field inside the central section radial, as it would be if the cylinders were all of infinite length. With the space between cylinders evacuated, calculate (a) both **E** and V between cylinders in the center, and (b) the capacitance of the central cylinder with respect to the inner cylinder.

3–14 A solid cylinder of radius a and of infinite length has a volume charge of uniform density ρ. Obtain expressions for both **E** and V as functions of r where $r < a$ and where $r > a$. Plot these expressions.

3–15 Two point charges $+4q$ and $-q$ are separated a distance l. Due to these charges the field is zero at an equilibrium point on the extension of a straight line through the two charges, the axis of symmetry. Many lines of flux leave the $+4q$ charge but one group of these follows curved paths to this equilibrium point. Draw the diagram. Choose one of this latter group of lines and calculate the angle it makes with the axis of symmetry just as it leaves the $+4q$ charge.

3–16 A slab of nonconducting material has faces parallel to the xz-plane. It extends from $-\infty$ to $+\infty$ in both the x- and z-directions and extends from 0 to a in the y-direction. The entire slab contains an electric charge whose volume-density is ρ, a constant. Find **E** at $y = 3a/4$, $y = a$, and $y > a$.

3-17 In the geometry of Problem 3–16 assume $\rho = Gy$ and find **E** at $y < a$, $y = a$, and $y > a$.

3-18 An infinitely long cylindrical shell of nonconducting material has inner and outer radii equal to a and b, respectively. Real charges produce a uniform volume density ρ throughout this shell. Find **E** at r, where $a < r < b$.

3-19 Consider an infinitely long solid cylinder of nonconducting material surrounded by the shell of Problem 3–18. The radius of the solid cylinder is p, where $p < a < b$, and it has the same uniform charge density ρ as the outer cylinder. Find **E** at $r < p$, $p < r < a$, and $r > b$.

3-20 A vector **A** has the value $\mathbf{i}(x^2 - y^2) - \mathbf{j}2xy + \mathbf{k}3z$ throughout space. Calculate $\int \mathbf{A} \cdot d\mathbf{s}$ over the surface of an imaginary sphere of radius a, centered at the origin.

3-21 Two thin spherical shells are centered on the z-axis. Both are nonconductors. The first has a radius a, is centered at the origin, and has the charge q_a uniformly distributed on its surface. The second has a radius b, is centered at $z = c$, and has the charge q_b uniformly distributed on its surface. Here $a + b > c > a > b$, so the spheres overlap. Write the potentials in spherical coordinates at the following three general points:

a) P_1 outside both spheres.

b) P_2 inside both spheres.

c) P_3 inside the sphere of radius a but outside the sphere of radius b.

3-22 A certain electronic tube has two electrodes which consist of coaxial cylinders of radii a and b, where $b > a$. They have equal lengths $l \gg b$. A charge distribution between these electrodes is found to be given by $\rho = A/r$, where A is constant. Assume that the inner cylinder is grounded at zero potential.

a) Find **E** between the cylinders.

b) Find V of the outer cylinder.

3-23 Calculate the angle that the flux line chosen in Problem 3–15 makes with the axis of symmetry as it passes through the plane that is perpendicular to the axis through the charge $-q$.

3-24 Charges are symmetrically distributed about the origin of spherical coordinates so that $\rho = \rho_0 e^{-ar^2}/r$. Find the electric field.

3-25 A thick spherical shell between $r = a$ and $r = b$ contains a body charge given by $\rho = \rho_0 e^{zr^3}$. The remaining space $r < a$ and $r > b$ is a vacuum. Calculate **E** within this shell.

3-26

a) A spherical shell (that need not be metallic) has a uniform surface density of charge σ but has no charge inside it. Use Gauss' law to prove that the field inside is everywhere zero.

b) A hollow cube that is not metallic has a uniform surface density of charge σ but has no charge inside it. Can you use Gauss' law as in part (a) to prove that the field inside is everywhere zero? Discuss any important differences between parts (a) and (b).

3-27 Determine the charge distribution required in regions where the potentials are given by the following expressions:

In cylindrical coordinates:

a) r^{-1} b) $\ln r$ c) $r^n \cos n\phi$ d) $r^2(3 \cos^2 \phi - 1)$

In spherical coordinates:

e) r^{-1} f) $\ln r$ g) $r^n \cos n\theta$ h) $r^2(3\cos^2\theta - 1)$

3–28 Find the electric fields in spaces where there are the following potential distributions:

a) $V = \ln \dfrac{r + A}{r - A}$ (cyl. coord.)

b) $V = \dfrac{\cos\theta \sin\phi}{r^2}$ (sph. coord.)

c) $V = \dfrac{qe^{-r/a}}{4\pi\epsilon_0 r}$ (sph. coord.)

3–29 The following are expressions for fields (not necessarily electrostatic) in rectangular coordinates. Determine whether or not each of these can be expressed as a gradient of a potential function. For each field that can be so expressed, determine whether or not the potential function would satisfy Laplace's equation, $\nabla^2 V = 0$:

a) $\mathbf{A} = \mathbf{i}x + \mathbf{j}y + \mathbf{k}z$,   b) $\mathbf{A} = \mathbf{i}y + \mathbf{j}z + \mathbf{k}x$,
c) $\mathbf{A} = \mathbf{i}yz + \mathbf{j}zx + \mathbf{k}xy$,   d) $\mathbf{A} = \mathbf{i}xy + \mathbf{j}yz + \mathbf{k}zx$.

3–30 Electrostatic fields are always conservative.
a) Determine which, if any, of the following $\mathbf{A}$ fields could represent an electrostatic field, $\mathbf{E} = \mathbf{A}$:
   (i) $\mathbf{A} = \mathbf{i}x^2 + \mathbf{j}y^2 + \mathbf{k}z^2$,
   (ii) $\mathbf{A} = \mathbf{i}xy + \mathbf{j}yz + \mathbf{k}zx$,
   (iii) $\mathbf{A} = \mathbf{i}yz + \mathbf{j}zx + \mathbf{k}xy$.
b) In those foregoing equations that can represent electrostatic fields, set $\mathbf{A} = \mathbf{E}$ and determine the volume density distribution required to produce such fields. Such fields usually have limited extent beyond which the equations do not hold. Ignore charges on the boundaries of such regions.

3–31 Assume that $\mathbf{E} = \mathbf{A}$ from (iii) of Problem 3–30 only for the region defined by a cube, with edges all of length l whose corners are at $(0, 0, 0)$ and (l, l, l). Find the surface charge σ on each of the two surfaces parallel to the xy-plane.

3–32 The Yukawa potential attributed to some spherical distribution of charge is $V = -kqe^{-ar/r}$. Find charge density ρ as a function of r.

3–33 Determine the charge distribution for the potential function $V = B\ln r$:
a) If r represents the r of conventional spherical coordinates.
b) If r represents the r of conventional cylindrical coordinates.

3–34 A spherical shell, with inner and outer radii of a and b, respectively, contains a volume distribution of charge of $\rho = \rho_0/r$. Inside and outside this shell $\rho = 0$.
a) Find $\mathbf{E}$ from $r = 0$ to $r = a$, from $r = a$ to $r = b$, and from $r = b$ to $r \to \infty$.
b) Find V at $r = b$, $r = a$, and $r = 0$.

3–35 A thin, nonconducting spherical shell of radius a has the potential $V_i = Mr\cos\theta$ inside and the potential $V_o = Ma^3\cos\theta/r^2$ outside. Find the charge distribution σ on the surface of this shell.

3–36 In the diagram of Fig. 3-4, any straight line from q can pass only twice through the surface of the general volume. Draw a diagram showing a different-shaped volume in which some straight lines from q will pass four times through its surface. Demonstrate that Gauss' law is also valid for this different-shaped volume.

3–37 A vector A has the value $\mathbf{i}xy + \mathbf{j}yz + \mathbf{k}xz$ throughout space. Calculate $\int \mathbf{A} \cdot d\mathbf{s}$ over the six sides of a rectangular box, with edges parallel to the x-, y- and z-axes and with corners at $(0, 0, 0)$ and $(1, 3, 2)$.

3–38 A circle of radius a lies in the xy-plane tangent to both the x-and y-axes. Calculate $\int \mathbf{A} \cdot d\mathbf{l}$ around this circle in the direction which directs $d\mathbf{l}$ in the plus x-direction at the point of tangency. Use (a) $\mathbf{A} = \mathbf{i}y + \mathbf{j}z + \mathbf{k}x$, and then (b) $\mathbf{A} = \mathbf{i}yz + \mathbf{j}xz + \mathbf{k}xy$.

3–39
a) Calculate the work done in traversing the closed path made by straight lines from $(0, 0, 0)$ to $(0, 20, 0)$ to $(0, 0, 15)$ and back to $(0, 0, 0)$, where the force is given by $\mathbf{F} = \mathbf{i}2y^2 + \mathbf{j}z^2 + \mathbf{k}2yz$.
b) Can this force be expressed as the gradient of a potential function? Give your reasons.

3–40
a) Use Stokes' theorem to evaluate $\int \mathbf{A} \cdot d\mathbf{l}$ around a path of straight lines from $(0, 2, 1)$ to $(0, 6, 1)$ to $(0, 6, 7)$ and back to $(0, 2, 1)$. Here $\mathbf{A} = \mathbf{i}3z + \mathbf{j}3x + \mathbf{k}3y$.
b) Verify (a) by direct integration along the path.

3–41 The magnetic field due to a certain distribution of currents is given by $\mathbf{H} = \mathbf{a}_\phi A e^{-r}/r$ in cylindrical coordinates. Test whether or not this field is conservative.

3–42 A new mathematical operation can often be best understood by applying it to simple problems for which the answer is known. The velocity and acceleration of a point moving at constant speed in a circle are well known, but their calculation can be used to illustrate the differentiation of unit vectors. Express the position of a point in cylindrical coordinates by $\mathbf{R} = \mathbf{a}_r b$, where $b =$ constant and $\partial\phi/\partial t = \omega$, a constant. Calculate $\mathbf{v} = \partial\mathbf{R}/\partial t$ and $\mathbf{a} = \partial^2\mathbf{R}/\partial t^2$ and express the latter in terms of $\mathbf{R}$.

Use rectangular coordinates in Problems 3–43 through 3–49:

3–43 Prove Eq. (3–20).
3–44 Prove Eq. (3–21).
3–45 Prove Eq. (3–22).
3–46 Prove Eq. (3–23).
3–47 Prove Eq. (3–24).
3–48 Prove Eq. (3–25).
3–49 Prove Eq. (3–26).

Chapter 4 SOLUTION OF ELECTROSTATIC PROBLEMS

In solving electrostatic problems, any one or several of the following quantities may be known or desired: charges, fields, potentials, forces, torques, and capacitances. All of these quantities can be calculated when the distribution of potentials is known. In particular, when expressions for V in terms of the coordinates have been obtained, such quantities are easily calculated.

Poisson's equation $\nabla^2 V = -\rho/\epsilon_0$, which follows directly from Gauss' law, provides the differential equation needed for obtaining such expressions for V. In the great majority of problems it is sufficient to obtain expressions for V only in those regions where $\rho = 0$. This is fortunate, since there are many more methods of solution available for Laplace's form of the equation, $\nabla^2 V = 0$, than for Poisson's more general form. It will be shown that a problem in electrostatics is solved when a solution of Laplace's equation has been found to satisfy the particular boundary conditions of the problem.

Because the number of solutions of Laplace's equation is unlimited, the task of finding the one solution satisfying the conditions of a particular problem may be arduous. For this reason, many types of solutions have been cataloged and special procedures have been developed for obtaining solutions for particular kinds of problems. Some of these special methods are so neat that it is easy to forget that they do, in fact, follow a more general procedure.

Some knowledge of the characteristics of materials is needed for the solution of any electrostatic problem. Consequently a fairly extensive discussion of the electric properties of insulators and conductors is given in the next chapters. For the problems treated in this chapter it is necessary to know only a few properties of metallic conductors. Like all materials in the solid state, metals consist of individual atomic nuclei, arranged in geometric patterns called lattices and surrounded by rapidly moving electrons. In the neutral or uncharged state there are as many electrons as there are protons in all of the atoms. Since electrons and protons have charges that are equal but of opposite sign, these materials normally produce no external electric fields. In metals a few electrons are free to drift quite easily through the lattice. In fact, electrons continue to drift toward surfaces of a metal until they have set up fields that exactly cancel the fields due to external charges. Thus there are never any electric fields in the interior of metals after static conditions have been established. Since there are no internal fields, there can be no potential differences within a metal and its potential must be constant.

4–1 Electrostatic solutions with Laplace's equation

The general method for solving problems by means of Laplace's equation can be understood best by considering some specific examples. A concentrated charge placed near a flat surface of a massive piece of metal is attracted toward that surface by the induced charge of opposite sign that collects on it. Situations that are more or less accurately described by this statement occur so frequently that it is desirable to be able to determine the magnitude of this attractive force. For example, when the metallic filament or cathode in a vacuum tube is heated, some electrons receive sufficient kinetic energy to escape from the surface. For a theoretical analysis of this phenomenon, the energy required for this escape must be known. The primary force opposing the escape is the attraction due to induced charges. Because the metal is made up of individual atoms arranged in a geometrical pattern called a lattice, the calculation of the attractive force near the surface is very complex. With an electron at greater distances, perhaps beyond ten lattice spacings, the surface may be treated as a plane. For relatively large distances the curvature of the surface may become important, but at such distances the force is so small that it can be neglected.

As originally stated, the problem of calculating the force on a concentrated charge near a plane surface of massive metal could be calculated by the integration methods of Section 2–7 if the distribution of the induced charge on the surface were known, but this distribution is not stated directly in the problem. However, when the potential distribution in a region, including both the concentrated charge and the metallic surface, is known, the density of the induced charge can be calculated. In fact, the force on the concentrated charge and its potential energy can then be calculated directly. Laplace's equation $\nabla^2 V = 0$ is ideally suited for determining the distribution of V in such situations. The general procedure for solving problems with Laplace's equation follows:

1. The region in which the potential must be known should be determined. This region then should be made part of a closed space, completely devoid of charges and completely surrounded by boundaries on which either V or $\partial V/\partial n$ can be specified.

2. The values of V or $\partial V/\partial n$ should be specified on every portion of the boundaries to provide the necessary boundary conditions.

3. A solution of $\nabla^2 V = 0$ should be found that is finite and continuous throughout the bounded space and that also satisfies all of the boundary conditions. Any solution that meets these requirements is the correct and the only solution, as is demonstrated in Section 4–3.

As stated, the real problem involves a concentrated charge near a very large surface. However, only negligible errors are introduced by assuming the plane to be infinite and the charge to be concentrated at a point, to simplify the analysis. The surface may be assumed horizontal, coinciding with the $z = 0$ plane and having zero potential. The zero potential on this surface may be achieved either with a grounded sheet of metal or with a massive metal that fills all space where z is negative. To complete the picture, a point charge q is located on the z-axis at a distance h above this plane.

Following the three steps of the general procedure, the force on q and its potential energy could be uniquely determined if the potential V could be completely specified in its neighborhood. However, to completely determine V, a closed space is needed on whose boundaries V is known. Now $V = 0$ over all the plane and over an imaginary hemispherical surface of infinite radius drawn about the origin. Since the potentials are known on these surfaces, they can constitute the outer boundaries of our closed space. This space can be made free of charge by surrounding q with an imaginary sphere of infinitesimal radius δ, on which the potential must approach $q/4\pi\epsilon_0\delta$ as $\delta \to 0$. This tiny sphere constitutes the inner boundary of the closed space, inside of which $\nabla^2 V$ is everywhere zero.

4–2 Method of images

The next step in the general procedure is to find a solution of $\nabla^2 V = 0$ that will satisfy the foregoing boundary conditions. If this were a problem that had never been solved, this step would require some ingenuity. A study of the solutions that have been obtained is essential, however, for developing the necessary ingenuity.

The plane attracts q downward. If the plane were replaced by a point charge of opposite sign located directly below q, the force would be in the same direction. Furthermore, if this charge had the magnitude $-q$ and were located on the z-axis a distance h below the $z = 0$ plane as shown in Fig. 4–1, the potential of this plane due to these two charges would be zero. Since the two charges are finite, the potential at infinity would also be zero. In addition, the $-q$ charge would be too far from the $+q$ charge to affect the potential on the inner boundary when δ is made sufficiently small. The

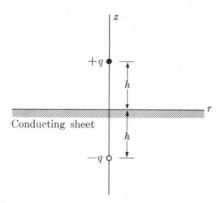

Fig 4–1 The potential above the conducting sheet due to $+q$ and the conducting sheet is identical to that produced by $+q$ and the image charge $-q$ in the absence of the sheet.

potential distribution within the bounded space due to these two charges, as calculated from Eq. (2–15), is

$$V = \frac{q}{4\pi\epsilon_0[(z-h)^2 + r^2]^{1/2}} + \frac{-q}{4\pi\epsilon_0[(z+h)^2 + r^2]^{1/2}}. \qquad (4\text{–}1)$$

This expression does satisfy all of the boundary conditions, namely, $V = 0$ when $z = 0$, $z = \infty$, or $r = \infty$, and $V = q/4\pi\epsilon_0\delta$ when δ is very small. As can be readily checked by direct differentiation, Eq. (4–1) also satisfies $\nabla^2 V = 0$ throughout the bounded region and is therefore the correct expression for the potential. The first term gives the potential due to q, and the second term, which will be designated by V', gives the potential due to the induced charge distribution on the metallic surface. This procedure for determining potentials is called the method of images, because $-q$ is placed at the location where the optical image of q would be in the metallic surface acting as a mirror. The procedure of satisfying boundary conditions by placing imaginary point charges outside the bounded space is called the method of images, whether or not these charges represent optical images. Image charges must not be placed inside the bounded regions, because V then becomes infinite at such points, so that $\nabla^2 V = 0$ cannot then be satisfied throughout the bounded region.

The force on q is given by $\mathbf{F} = -q\nabla V'$, where $\nabla V'$ must be evaluated at the point $r = 0$ and $z = h$, where q is located. This force can also be calculated directly from the Coulomb's law attraction of $-q$ for q, which is

$$F = \frac{-q^2}{4\pi\epsilon_0(2h)^2}. \qquad (4\text{–}2)$$

For an electron ten lattice spacings (3.6×10^{-10} m) from a copper surface,

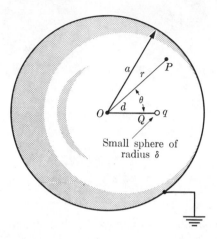

Fig. 4–2 Diagram illustrating the location of a point charge inside a hollow conducting sphere.

this so-called image force is

$$F = \frac{-e^2}{4\pi\epsilon_0(2h)^2} = 4.4 \times 10^{-10} \text{ N},$$

which would result in the relatively large acceleration of 4.9×10^{20} m s^{-2}. Since this image force is an important factor in confining electrons within conductors, it is discussed in more detail in Chapter 6.

The charge distribution on the metallic surface can also be calculated directly from Eq. (4–1). The density of surface charge, derived from Gauss' law, is just $\sigma = \epsilon_0 E = -\epsilon_0\, \partial V/\partial n$ evaluated near the surface of the conductor at $z = 0$. The surface-charge density on the plane is then

$$\sigma = \frac{-qh}{2\pi(h^2 + r^2)^{3/2}},$$

which has its maximum directly below the point charge where $r = 0$.

Many other problems can be solved by the image method. There is one that is especially useful for demonstrating the general procedure and that also provides a good example for the proof of the uniqueness theorem given in the next section. This problem calls for the calculation of the force acting on a charged particle somewhere inside a hollow conducting sphere that is connected to ground. A sphere of inner radius a and containing a point charge q located at a distance d from the center O is shown in Fig. 4–2.

Following the general procedure, we need an expression for the potentials in the region containing q. For the bounded space in which $\nabla^2 V = 0$, that region between an imaginary sphere of very small radius δ about q and the inner surface of the conducting sphere should be selected. No information is available and none is desired concerning the space outside,

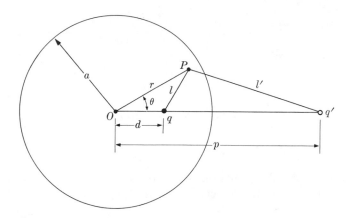

Fig. 4–3 A point charge q and its image q' produce the same potential in the region bounded by $r = a$ as would be produced by q and the hollow conducting sphere of Fig. 4–2.

since the conducting sphere acts as a perfect shield. Using the spherical coordinates shown in the diagram, the outer boundary condition is that $V = 0$ where $r = a$, and the inner boundary condition is that $V = q/4\pi\epsilon_0\delta$ as δ approaches zero.

   An expression for V that satisfies $\nabla^2 V = 0$ throughout the bounded space and that also fits these boundary conditions is needed. Expressions for the potentials of point charges always satisfy $\nabla^2 V = 0$ except at the charged points. This statement should be obvious from the fact that the Laplace equation was derived from Coulomb's law, but it can be demonstrated by direct differentiation of Eq. (2–15).

   In the actual problem, charges induced on the inner surface of the sphere attract q to the right. Symmetry considerations suggest that this effect could be produced by replacing the sphere with one or more image charges on the extension of the radial line through q. Since the simplest of the promising solutions should be tried first, just one image charge q' at the distance $p > a$ from O will be tried, as shown in Fig. 4–3. If q' is to replace the electron distribution on the sphere, it must combine with q to produce zero potential on the imaginary sphere of radius a shown in Fig. 4–3. According to Eq. (2–15), the potential inside the sphere is

$$V = \frac{1}{4\pi\epsilon_0}\left(\frac{q}{l} + \frac{q'}{l'}\right),$$

which will be zero on the inner surface if

$$\frac{q}{l} + \frac{q'}{l'} = 0 \tag{4–3}$$

for all values of θ when $r = a$.

Now when $r = a$,

$$l^2 = a^2 + d^2 - 2ad \cos \theta$$

and

$$l'^2 = p^2 + a^2 - 2pa \cos \theta,$$

and these can be written

$$l = a\sqrt{1 + (d/a)^2 - 2(d/a) \cos \theta}$$

and

$$l' = p\sqrt{1 + (a/p)^2 - 2(a/p) \cos \theta}.$$

By substituting the latter relations into Eq. (4–3) and setting $d/a = a/p$, the radicals containing θ can be factored out to leave

$$\frac{q'}{p} = -\frac{q}{a}.$$

The two charges shown in Fig. 4–3, q at the distance d from O and $q' = -qp/a$ at the distance $p = a^2/d$ from O, do combine to produce zero potential at all points on the imaginary sphere of radius a. The inner boundary condition is automatically satisfied, since q is the only charge in or near this tiny sphere. The potential due to q and q' of Fig. 4–3 is

$$V = \frac{q}{4\pi\epsilon_0}\left[\frac{1}{\sqrt{r^2 + d^2 - 2rd \cos \theta}} - \frac{p/a}{\sqrt{p^2 + r^2 - 2pr \cos \theta}}\right],$$

which gives the fields throughout space for this imaginary situation, as shown in Fig. 4–4. In the real problem the potentials and fields are known only inside the sphere, as shown in Fig. 4–5 but these fields are identical to those inside the sphere of Fig. 4–4. Again the force on q is easily calculated from the Coulomb attraction of q', which is

$$F = \frac{qq'}{4\pi\epsilon_0(p - d)^2} = \frac{-q^2ad}{4\pi\epsilon_0(a^2 - d^2)^2}.$$

In this problem the sphere was grounded, and this allowed opposite charges to flow in from the earth. If it were not grounded, then the charge distributing itself over the surface of the sphere would have to come from the sphere itself, not from the ground. The potential would still be constant over the surface of the sphere, but not zero. This problem can be solved by superposing a uniform charge over the outside surface.

4–3 Uniqueness theorem

In the two problems that were solved by the method of images, it was assumed that any single-valued and continuous function V of the coordinates that is a solution of $\nabla^2 V = 0$ throughout a bounded space and that also gives the correct potentials on the boundaries of this space gives the correct

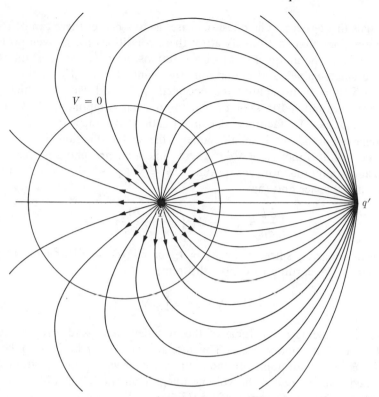

Fig. 4–4 Graph showing the lines of flux between the two point charges q and q' of Fig. 4–3. Near q' and q flux lines are radially symmetrical. However, in this diagram, only those flux lines arriving at q' from q are shown. The flux lines not shown near q' would originate from distant charges. The solid angle subtended at q' by lines of flux from q is simply given by $4\pi q/q'$. For the case illustrated, it is somewhat less than 2π.

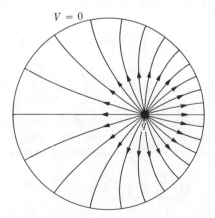

Fig. 4–5 Graph showing the lines of flux caused by the point charge q and the charge it induces on the hollow conducting sphere of Fig. 4–2.

potentials throughout. The question arises as to whether there can be two or
more functions that each satisfy all of these conditions for a given problem.
To investigate the uniqueness of such solutions, let it be assumed that there
are two single-valued and continuous functions, V_1 and V_2, each of which
satisfies $\nabla^2 V = 0$ throughout the bounded space and each of which gives
correct values either for V or $\partial V/\partial n$, the normal component of ∇V on all
the boundaries. To facilitate the investigation, two new functions of the
coordinates should be defined; namely, the scalar function $\phi = V_1 - V_2$
and the vector function $\phi \nabla \phi$. These functions are chosen because the
original assumptions now ensure that $\nabla^2 \phi = \nabla^2 V_1 - \nabla^2 V_2 = 0$ throughout
the bounded space and that

$$\phi \frac{\partial \phi}{\partial n} = (V_1 - V_2)\left(\frac{\partial V_1}{\partial n} - \frac{\partial V_2}{\partial n}\right) = 0$$

on all boundaries, since either $V_1 = V_2$ or $\partial V_1/\partial n = \partial V_2/\partial n$ on all these
boundaries. According to Gauss' theorem,

$$\int_\tau \nabla \cdot \phi \nabla \phi \, d\tau = \int_s \phi \nabla \phi \cdot d\mathbf{s}, \tag{4–4}$$

where τ represents the volume of the bounded space and s represents the
total area of its boundaries. The integral on the right is zero because
$\phi \nabla \phi \cdot d\mathbf{s} = 0$ at all points on the boundary over which this integral is to be
evaluated, and therefore the volume integral on the left is also zero. Ac-
cording to the vector identity of Eq. (3–25),

$$\nabla \cdot (\phi \nabla \phi) = \phi \nabla^2 \phi + (\nabla \phi)^2$$

and

$$\int_\tau \nabla \cdot (\phi \nabla \phi) \, d\tau = \int_\tau \phi \nabla^2 \phi \, d\tau + \int_\tau (\nabla \phi)^2 \, d\tau = 0.$$

Since $\nabla^2 \phi = 0$ throughout the bounded space,

$$\int_\tau (\nabla \phi)^2 \, d\tau = 0.$$

The integrand

$$(\nabla \phi)^2 = \nabla \phi \cdot \nabla \phi = \left(\frac{\partial \phi}{\partial x}\right)^2 + \left(\frac{\partial \phi}{\partial y}\right)^2 + \left(\frac{\partial \phi}{\partial z}\right)^2$$

cannot be negative at any point, since each of the three rectangular terms,
being squared, must either be positive or zero. For the integral to be zero,
the integrand then must be zero throughout τ; that is,

$$\left(\frac{\partial \phi}{\partial x}\right)^2 + \left(\frac{\partial \phi}{\partial y}\right)^2 + \left(\frac{\partial \phi}{\partial z}\right)^2 = 0$$

and

$$\frac{\partial \phi}{\partial x} = \frac{\partial \phi}{\partial y} = \frac{\partial \phi}{\partial z} = 0$$

throughout the bounded space. Since this requires ϕ to be constant, the two solutions can differ only by a constant. Whenever V is specified on any boundary, $\phi = 0$ on the boundary and ϕ must be zero throughout the bounded space. Within the volume τ, then, $V_1 = V_2$ everywhere and the assumption that two different solutions could satisfy the required conditions has led to an absurdity. Any single-valued and continuous function that satisfies $\nabla^2 V = 0$ throughout the bounded space and also gives the correct V or $\partial V / \partial n$ on the boundaries of that space constitutes the solution of that problem.

Frequently, problems are encountered in which part of the boundary of the closed space is a conducting surface s_0 on which neither V nor $\partial V / \partial n$ can be uniquely specified. If this surface is that of a conductor wholly inside the region of interest, it is sufficient to specify the total charge on this conductor and that the potential on this surface is constant. For example, in the image solutions of the previous section the charge q might reside on a small but finite conducting sphere. Since both the potential and the fields near its surface depend upon the surrounding charges as well as upon q, neither V nor $\partial V / \partial n$ can be uniquely specified until the problem is solved. However, by Gauss' law,

$$-\int_{s_0} \nabla V \cdot d\mathbf{s} = \frac{q}{\epsilon_0},$$

and on a conductor $V = \text{constant}$. Hence

$$\int_{s_0} \phi \, \nabla \phi \cdot d\mathbf{s} = (V_1 - V_2) \left[\int_{s_0} \nabla V_2 \cdot d\mathbf{s} - \int_{s_0} \nabla V_1 \cdot d\mathbf{s} \right]$$

$$= \frac{(V_1 - V_2)(q_2 - q_1)}{\epsilon_0} = 0,$$

and the integral on the right in Eq. (4–4) is still zero provided q is specified so that $q_1 = q_2$. Therefore a solution of Laplace's equation is unique if it satisfies at least one of the following conditions on each part of the surface bounding its closed region: (a) the surface potential V can be specified, (b) the normal component of the field given by $\partial V / \partial n$ can be specified, (c) the surface is an equipotential surface that completely surrounds a specified charge. Since it is desirable to have many solutions from which to choose, Laplace's equation has been the subject of much intensive study. Solutions are available for almost every conceivable geometry, and general methods have been developed for building solutions that will fit particular situations. One of the most useful theorems for this purpose is the one stating that *the sum of any number of individual solutions is itself a solution*, for if

$$V = V_1 + V_2 + \cdots + V_n$$

is the sum of n solutions, then

$$\nabla^2 V = \nabla^2 V_1 + \nabla^2 V_2 + \cdots + \nabla^2 V_n = 0.$$

The mathematical developments involving Laplace's equation are so exten-sive that a comprehensive treatment of any portion would be out of place here. Before outlining some general procedures in rectangular coordinates, it is desirable to derive a few more useful solutions and to demonstrate the methods for applying these to important physical problems.

4–4 Some solutions with spherical coordinates

When a problem has boundaries that are readily expressed in spherical coordinates, Laplace's equation may be written as

$$\nabla^2 V = \frac{1}{r^2} \frac{\partial}{\partial r}\left(r^2 \frac{\partial V}{\partial r}\right) + \frac{1}{r^2 \sin \theta} \frac{\partial}{\partial \theta}\left(\sin \theta \frac{\partial V}{\partial \theta}\right) + \frac{1}{r^2 \sin^2 \theta} \frac{\partial^2 V}{\partial \phi^2} = 0.$$

Many solutions of this equation are available.* The discussion here will be confined to those problems of this type that have symmetry about an axis, such that V is independent of ϕ. The equation then becomes

$$\frac{\partial}{\partial r}\left(r^2 \frac{\partial V}{\partial r}\right) + \frac{1}{\sin \theta} \frac{\partial}{\partial \theta}\left(\sin \theta \frac{\partial V}{\partial \theta}\right) = 0. \tag{4–5}$$

Solutions of this equation can be obtained by assuming that the variables can be separated and the potential can be written as $V = R(r)P(\theta)$, where R and P, respectively, are functions of r only and θ only. Substituting this value of V into Eq. (4–5), we obtain

$$P \frac{d}{dr}\left(r^2 \frac{dR}{dr}\right) + \frac{R}{\sin \theta} \frac{d}{d\theta}\left(\sin \theta \frac{dP}{d\theta}\right) = 0,$$

and division by RP gives

$$\frac{1}{R} \frac{d}{dr}\left(r^2 \frac{dR}{dr}\right) + \frac{1}{P \sin \theta} \frac{d}{d\theta}\left(\sin \theta \frac{dP}{d\theta}\right) = 0, \tag{4–6}$$

in which the first term is a function of r only and the second term is a function of θ only. Since variation of r can affect only the first term and variation of θ can affect only the second, each term must be constant, that is,

$$\frac{1}{R} \frac{d}{dr}\left(r^2 \frac{dR}{dr}\right) = K \tag{4–7}$$

and

$$\frac{1}{P \sin \theta} \frac{d}{d\theta}\left(\sin \theta \frac{dP}{d\theta}\right) = -K. \tag{4–8}$$

---

* T. M. MacRobert, *Spherical Harmonics*, E. P. Dutton, New York, 1927; E. Jahnke and F. Emde, *Tables of Functions with Formulae and Curves*, Dover Publications, New York, 1943.

Each equation, being a function of only one independent variable, is written as an ordinary differential equation. Under the assumption that V can be expressed as a power series in r, we should try $R = r^n$. Substitution in Eq. (4–7) then yields

$$n(n + 1) = K.$$

Substitution of $R = r^{-(n+1)}$ yields the same value for K and provides a second solution. Actually, this value* for K must be chosen to keep P finite when $\theta = 0$. The solution of a second order differential equation is then

$$R_n = A_n r^n + B_n r^{-(n+1)}, \tag{4–9}$$

where R_n is the general solution for Eq. (4–7) for a particular value of n, and A_n and B_n are arbitrary constants. Since the original Eq. (4–6) is valid for any positive integer from zero to infinity, a general solution consists of the sum of the $R_n P_n$ terms, and

$$V = \sum_{n=0}^{n=\infty} R_n P_n,$$

where P_n is the solution of Eq. (4–8) for a particular value of n. This equation may be written as

$$\frac{1}{P_n \sin \theta} \frac{d}{d\theta} \left(\sin \theta \frac{dP_n}{d\theta} \right) + n(n + 1) = 0,$$

and by setting $\mu = \cos \theta$ and $d\mu = -\sin \theta \, d\theta$, it becomes

$$\frac{d}{d\mu} \left[(1 - \mu^2) \frac{dP_n}{d\mu} \right] + n(n + 1)P_n = 0, \tag{4–10}$$

which is called Legendre's equation. Of the two solutions of this equation for the individual values of n, only one is finite for $\theta = 0$ and $\theta = \pi$. These finite solutions are called Legendre polynomials.

It will be sufficient here to show how the first three polynomials are obtained. When $n = 0$, then $P_0 = $ a constant obviously satisfies the equation. When $n = 1$, the equation becomes

$$\frac{d}{d\mu} \left[(1 - \mu^2) \frac{dP_1}{d\mu} \right] + 2P_1 = 0,$$

which is satisfied by $P_1 = B\mu$. When $n = 2$, the equation becomes

$$\frac{d}{d\mu} \left[(1 - \mu^2) \frac{dP_2}{d\mu} \right] + 6P_2 = 0,$$

---

* Solutions of Eq. (4–8) with nonintegral values of K become either infinite or undefined near $\theta = 0$ or π; H. Margenau and G. M. Murphy, *Mathematics of Physics and Chemistry*, Vol. 1, 2nd ed., D. Van Nostrand, Princeton, N.J., 1956, pp. 61–68.

Table 4–1

A SHORT TABLE OF LEGENDRE POLYNOMIALS

$$P_0 = 1$$
$$P_1 = \mu$$
$$P_2 = \tfrac{1}{2}(3\mu^2 - 1)$$
$$P_3 = \tfrac{1}{2}(5\mu^3 - 3\mu)$$
$$P_4 = \tfrac{1}{8}(35\mu^4 - 30\mu^2 + 3)$$
$$P_5 = \tfrac{1}{8}(63\mu^5 - 70\mu^3 + 15\mu)$$
$$P_6 = \tfrac{1}{16}(231\mu^6 - 315\mu^4 + 105\mu^2 - 5)$$
$$P_7 = \tfrac{1}{16}(429\mu^7 - 693\mu^5 + 315\mu^3 - 35\mu)$$

which is satisfied by $P_2 = C(3\mu^2 - 1)$. It is conventional to normalize the
P's by making them unity whenever $\mu = 1$, which determines the constants
A, B, C, etc. The first eight Legendre polynomials are listed in Table 4–1.

The solution for Eq. (4–6) can now be written in the following form as
an infinite series in which each individual term is also a solution:

$$V = \sum_{n=0}^{n=\infty} \left(A_n r^n + \frac{B_n}{r^{n+1}} \right) P_n. \tag{4–11}$$

Conducting sphere in a uniform field. Between two plates of a parallel-plate
capacitor the electric field is generally quite uniform for all points that are
not too close to the edges of the plates. Figure 4–6(a) shows lines of electric
field between two conducting planes on each of which z is constant. A small
conducting sphere placed between these plates will distort these field lines
as shown in Fig. 4–6(b). A calculation of the distortion in the field produced
by such a sphere can be useful in many problems. If the sphere is not large
enough to significantly affect the distribution of charges on the conducting
planes, this problem can be solved easily by means of Eq. (4–11). It is
convenient to assume that the field is produced by one infinite conducting
plane at $z = -Z$ with the potential $E_0 Z$, and another at $z = Z$ with the
potential $-E_0 Z$, where $Z \gg a$. Before the sphere was introduced, the
potential distribution between these planes was given by $V = -E_0 z$ or, in
the spherical coordinates of Fig. 4–6(b),

$$V = -E_0 r \cos \theta.$$

The placing of the sphere with its center at the origin, as shown in Fig. 4–6(b),
has created a space in which $\nabla^2 V = 0$ between the surface of the sphere at
$r = a$ and the infinite plates. By symmetry the conducting sphere must have
zero potential, and the boundary conditions become

$$V = 0 \qquad \text{at} \qquad r = a$$

and

$$V = -E_0 r \cos \theta \qquad \text{wherever} \qquad r \gg a.$$

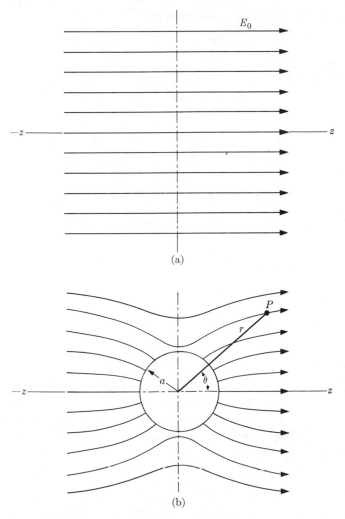

Fig. 4–6 (a) Uniform field E_0 pointing in the z-direction. (b) Distortion of field lines due to the presence of a conducting sphere. Point P is an arbitrary location in space which may be located with the spherical coordinates r and θ, since the field is independent of ϕ.

The second relation sets up the condition that the field shall not be distorted at great distances from the sphere. If a solution of Laplace's equation can be found that fits both of these boundary conditions, it is the only solution.

In examining the Legendre polynomials of Table 4–1, it appears that only the solutions for $n = 0$ and $n = 1$ can be expected to fit the boundary conditions, since all the other solutions contain higher powers of $\mu = \cos\theta$, which do not appear in the boundary conditions. Combining the first two Legendre polynomials with the corresponding terms from Eq. (4–9), we find that the expression for the potential becomes

$$V = A_0 + \frac{B_0}{r} + A_1 r \cos \theta + \frac{B_1}{r^2} \cos \theta.$$

Substitution into this equation of the boundary conditions as $r \to \infty$ yields $V = -E_0 r \cos \theta = A_0 + A_1 r \cos \theta$ and, since this must be true for all values of θ, $A_0 = 0$ and $A_1 = -E_0$. At the surface of the sphere where $r = a$, $V = 0$, and therefore

$$0 = \frac{B_0}{a} - E_0 a \cos \theta + \frac{B_1}{a^2} \cos \theta,$$

which can be true for all values of θ only if $B_0 = 0$ and $B_1 = E_0 a^3$. Hence the solution to this problem is

$$V = -E_0 r \cos \theta + \frac{E_0 a^3}{r^2} \cos \theta.$$

The field may be calculated directly from this potential:

$$\mathbf{E} = -\nabla V = \left(E_0 \cos \theta + 2 \frac{E_0 a^3}{r^3} \cos \theta \right) \mathbf{a}_r + \left(-E_0 \sin \theta + \frac{E_0 a^3}{r^3} \sin \theta \right) \mathbf{a}_\theta,$$

which can be expressed as

$$\mathbf{E} = E_0 \mathbf{k} + \frac{E_0 a^3}{r^3} (\mathbf{a}_r 2 \cos \theta + \mathbf{a}_\theta \sin \theta). \qquad (4\text{–}12)$$

It is interesting to compare this field expression with that due to a dipole, in Eq. (2–4), since this can be divided into the sum of the original uniform field plus the field of a dipole located at the origin, with a dipole moment of $4\pi\epsilon_0 a^3 E_0$. The dipole moment is proportional to the external field E_0 and to the volume of the sphere.

4–5 Some solutions with cylindrical coordinates

When the electrostatic problem exhibits cylindrical symmetry, Laplace's equation may be written in cylindrical coordinates:

$$\nabla^2 V = \frac{1}{r} \frac{\partial}{\partial r} \left(r \frac{\partial V}{\partial r} \right) + \frac{1}{r^2} \frac{\partial^2 V}{\partial \phi^2} + \frac{\partial^2 V}{\partial z^2} = 0.$$

Only those problems where V is independent of z will be considered here, so the equation becomes

$$\nabla^2 V = r \frac{\partial}{\partial r} \left(r \frac{\partial V}{\partial r} \right) + \frac{\partial^2 V}{\partial \phi^2} = 0.$$

If the variables are separable, the function $V = R(r)C(\phi)$ is a solution. Its substitution into the differential equation produces the relation

$$\frac{r}{R} \frac{d}{dr} \left(r \frac{dR}{dr} \right) + \frac{1}{C} \frac{d^2 C}{d\phi^2} = 0 \qquad (4\text{–}13)$$

Table 4–2

CYLINDRICAL SOLUTIONS TO LAPLACE'S EQUATION

$n = 0$	$V = \phi \ln r, \ln r, \phi$	
$n = +1$	$V = r \cos \phi,$	$V = r \sin \phi$
$n = -1$	$V = \dfrac{1}{r} \cos \phi,$	$V = \dfrac{1}{r} \sin \phi$
$n = +2$	$V = r^2 \cos 2\phi,$	$V = r^2 \sin 2\phi$
$n = -2$	$V = \dfrac{1}{r^2} \cos 2\phi,$	$V = \dfrac{1}{r^2} \sin 2\phi$

after division by RC. Since the first and second terms are respectively functions of r only and ϕ only, each term must be constant. As in the solution of Eq. (4–6), these constants can be determined by assuming $R = r^n$, the substitution of which gives the two ordinary differential equations

$$\frac{r}{R_n} \frac{d}{dr}\left(r \frac{dR_n}{dr} \right) = n^2 \tag{4–14}$$

and

$$\frac{1}{C_n} \frac{d^2 C_n}{d\phi^2} = -n^2. \tag{4–15}$$

The first is satisfied by either r^n or r^{-n} for R_n and the second is satisfied by either $\sin n\phi$ or $\cos n\phi$ for C_n, where n must be a positive integer if C_n is to be single-valued. A general solution for Eq. (4–13) must include

$$V = \sum_{n=1}^{n=\infty} (A_n r^n + B_n r^{-n}) \cos n\phi \tag{4–16}$$

and

$$V = \sum_{n=1}^{n=\infty} (A_n r^n + B_n r^{-n}) \sin n\phi.$$

The summations do not include $n = 0$ because when $n = 0$ Eqs. (4–14) and (4–15) can be integrated directly to give

$$R_0 = \ln r \quad \text{and} \quad C_0 = \phi.$$

Table 4–2 lists individual solutions* of Eq. (4–13). These same solutions can also be obtained by means of the complex variable procedure of Section 4–9.

---

* A more detailed discussion of these solutions may be found by looking up Bessel functions in mathematics texts such as W. Kaplan, *Advanced Calculus*, Addison-Wesley, Reading, Mass., 1952, and J. D. Murnaghan, *Introduction to Applied Mathematics*, John Wiley & Sons, New York, 1948.

Conducting cylinder in a uniform field. A conducting cylinder with its axis perpendicular to a uniform electric field causes distortions in that field similar to those caused by the conducting sphere discussed in the previous section. For example, a very long, fine copper wire stretched midway between two large plates of a parallel-plate capacitor will cause distortions in the field near the wire, like those shown in Fig. 4–7(b). The solutions of $\nabla^2 V = 0$ obtained in this section can be used to calculate the field around such a cylinder. Sufficient accuracy can be achieved by assuming that the

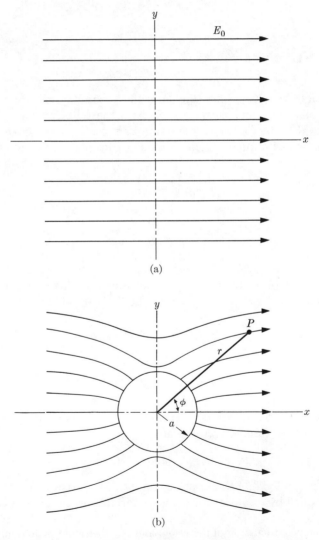

Fig. 4–7 (a) Uniform field E_0 pointing in the x-direction. (b) Distortion of field lines due to the presence of a conducting cylinder. The problem is independent of z, so that an arbitrary point in space may be located with the coordinates r and ϕ.

capacitor plates are infinite and widely separated. The uniform field shown in Fig. 4–7(a) before the cylinder is introduced can be produced by one infinite conducting plane at $x = -X$ with the potential E_0X and another at $x = X$ with the potential $-E_0X$, where $X \gg a$. The potential of the uniform field is then

$$V = -E_0x = -E_0r \cos \phi$$

in the cylindrical coordinates of Fig. 4–7(b). With the cylinder in place, $\nabla^2V = 0$ in the space between the planes and the cylinder.

   If the cylindrical axis coincides with the z-axis, the boundary conditions are

$$V = 0 \quad \text{at} \quad r = a$$

and

$$V = -E_0r \cos \phi \quad \text{wherever} \quad r \gg a.$$

Since $\cos \phi$ is the only function of ϕ in these boundary conditions, the solution may be assumed to be

$$V = Ar \cos \phi + \frac{B \cos \phi}{r}.$$

Substitution of the boundary conditions at $r \gg a$ gives

$$V = Ar \cos \phi = -E_0r \cos \phi, \quad \text{or} \quad A = -E_0,$$

and substitution of the boundary conditions at $r = a$ gives

$$V = -E_0a \cos \phi + \frac{B \cos \phi}{a} = 0, \quad \text{or} \quad B = E_0a^2.$$

Hence the solution is

$$V = -E_0r \cos \phi + E_0a^2 \frac{\cos \phi}{r}. \tag{4–17}$$

The field, determined directly from Eq. (4–17), is

$$\mathbf{E} = -\nabla V = \left(E_0 \cos \phi + E_0 \frac{a^2}{r^2} \cos \phi \right) \mathbf{a}_r$$

$$+ \left(-E_0 \sin \phi + E_0 \frac{a^2}{r^2} \sin \phi \right) \mathbf{a}_\phi,$$

which may be written in the form

$$\mathbf{E} = E_0\mathbf{i} + E_0 \frac{a^2}{r^2} (\mathbf{a}_r \cos \phi + \mathbf{a}_\phi \sin \phi). \tag{4–18}$$

The first term is just the uniform field that existed in the absence of the cylinder, while the second term is the field due to the cylinder. Figure 4–7(b) shows the field lines obtained from Eq. (4–18). It is interesting to note that on the surface of the cylinder where $r = a$, the field is $2E_0$ at $\phi = 0$ or π, and zero at $\phi = \pi/2$ or $3\pi/2$. The latter result should have been expected, since the tangential component of $\mathbf{E}$ must be zero.

The distribution of potential expressed by Eq. (4–17) is an exact solution only for an infinitely long conducting cylinder perpendicular to a uniform field extending throughout all space. Since these conditions cannot be achieved physically, Eq. (4–17) is not an exact expression for any real problem, although it may often express the potentials to accuracies as great or greater than can be determined by the best measuring techniques.

In all the problems thus far solved by means of Laplace's equation, geometrical and physical reasoning has been used as a guide for guessing the correct solutions, with the knowledge that none but the correct solutions would fit the boundary conditions. More mathematical methods are available. The next section gives a very brief outline of a general method for fitting solutions of Laplace's equation in rectangular coordinates to reasonable boundary conditions. However, the reader faced with difficult problems of this type should consult more advanced texts.

4–6 Laplace's equation in rectangular coordinates

In rectangular coordinates Laplace's equation is

$$\nabla^2 V = \frac{\partial^2 V}{\partial x^2} + \frac{\partial^2 V}{\partial y^2} + \frac{\partial^2 V}{\partial z^2} = 0. \tag{4–19}$$

This partial differential equation can be reduced to three ordinary differential equations by assuming a solution of the form $V = XYZ$, where X, Y, and Z are respectively functions of x only, of y only, and of z only. Substitution of this general solution gives

$$YZ \frac{d^2 X}{dx^2} + XZ \frac{d^2 Y}{dy^2} + XY \frac{d^2 Z}{dz^2} = 0,$$

which can be divided by XYZ to give

$$\frac{1}{X} \frac{d^2 X}{dx^2} + \frac{1}{Y} \frac{d^2 Y}{dy^2} + \frac{1}{Z} \frac{d^2 Z}{dz^2} = 0.$$

Because term one is a function of x, term two is a function of y, and term three is a function of z only, each of the three terms must be constant, and each can then be written in the form of an ordinary differential equation:

$$\frac{1}{X} \frac{d^2 X}{dx^2} = \alpha^2, \qquad \frac{1}{Y} \frac{d^2 Y}{dy^2} = \beta^2, \qquad \frac{1}{Z} \frac{d^2 Z}{dz^2} = \gamma^2, \tag{4–20}$$

where $\alpha^2 + \beta^2 + \gamma^2 = 0$.

Two-dimensional illustration. The procedure is most easily illustrated in two dimensions, although the extension to three-dimensional problems follows essentially the same pattern. If all functions are independent of z, the last constant is zero and $\beta^2 = -\alpha^2$. The general solutions for the first two equations are

$$X = A \cos \beta x + B \sin \beta x$$

and

$$Y = Ce^{\beta y} + De^{-\beta y}.$$

While the solution for Y has a form that is different from that for X, the roles of x and y obviously can be interchanged to fit specific boundary conditions. When the individual terms in a summation are all solutions of Laplace's equation, the summation is also a solution. Hence a general solution can be written as

$$V = \sum_n (A_n \cos \beta_n x + B_n \sin \beta_n x)(C_n e^{\beta_n y} + D_n e^{-\beta_n y}), \qquad (4\text{--}21)$$

with constants to be determined from the boundary conditions. The procedures for evaluating these constants are best illustrated by a specific example.

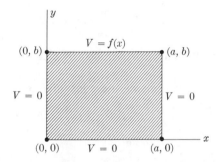

Fig. 4–8 A rectangular tube, infinite in the z-direction, with potentials specified on each boundary.

Evaluation of constants. The infinitely long rectangular tube shown in Fig. 4–8 is well suited to illustrate the general method for obtaining solutions of Laplace's equation. In the figure, three sides have the potential zero, while the fourth side has the potential $V = f(x)$. The solution of this specific problem actually gives the solution for the more general problem, where V is specified as some function not equal to zero on each of the four boundaries. The solution for such a general problem is just the sum of the solutions for four specific problems like the one considered here.

The general solution of Eq. (4–21) will give $V = 0$ on the right and left boundaries in Fig. 4–8 if each $A_n = 0$ and $\beta_n = n\pi/a$, where n is any positive integer from 0 to ∞. The condition that $V = 0$ along the x-axis boundary requires that $D_n = -C_n$, and the y-dependent part of the general solution then becomes $Y_n = 2C_n \sinh n\pi y/a$. Since now the B_n's provide the necessary constants, the constants C_n can be chosen to make each $Y_n = 1$ along the upper boundary where $y = b$; that is, a convenient choice is

$$Y_n = \frac{\sinh (n\pi/a)y}{\sinh (n\pi/a)b},$$

which does equal unity where $y = b$. Along the upper boundary, then,

$$V = \sum_n B_n \sin \frac{n\pi}{a} x = f(x). \tag{4–22}$$

The theory of Fourier's series shows that the B_n's can be found for any given $f(x)$. To find B_n, multiply Eq. (4–22) by $\sin (m\pi/a)x$, with m an integer, and integrate the resulting relations from $x = 0$ to $x = a$:

$$\sum B_n \int_0^a \sin \frac{n\pi}{a} x \sin \frac{m\pi}{a} x \, dx = \int_0^a \sin \frac{m\pi}{a} xf(x) \, dx.$$

The integral on the left is zero when $m \neq n$, but equals $a/2$ when $m = n$. Therefore

$$B_n = \frac{2}{a} \int_0^a \sin \frac{n\pi}{a} xf(x) \, dx,$$

and in principle each B_n can be determined by evaluating this integral, once the form of $f(x)$ is known.

To illustrate the evaluation of the constants B_n, a constant $f(x) = f_0$ can be chosen. Then

$$B_n = \frac{2f_0}{a} \int_0^a \sin \frac{n\pi x}{a} \, dx$$

or

$$B_n = \frac{2f_0}{a} \left[-\frac{a}{n\pi} \cos \frac{n\pi x}{a} \right]_0^a,$$

which gives $B_n = 0$ for even values of n but gives $B_n = 4f_0/n\pi$ for all odd values of n. The potential inside the rectangle of Fig. 4–8 is then

$$V = \sum_{n=1,3,5,\ldots}^{\infty} \frac{4f_0}{n\pi} \sin \frac{n\pi x}{a} \frac{\sinh (n\pi y/a)}{\sinh (n\pi b/a)}.$$

The techniques illustrated in the foregoing development provide general methods for solving Laplace's equation in rectangular coordinates. Similar methods are available for solving this equation in other coordinate systems.

4–7 Infinite series of images

The problems solved in Section 4–2 to illustrate the method of images were chosen from problems that require only a few images. Some useful problems require an infinite series of images, which produce potentials containing an infinite series of terms. Such potential expressions can be especially useful when the series converges rapidly. A capacitor consisting of a conducting sphere near a large conducting plane is an example of a useful geometry that can be solved with an infinite series of images.

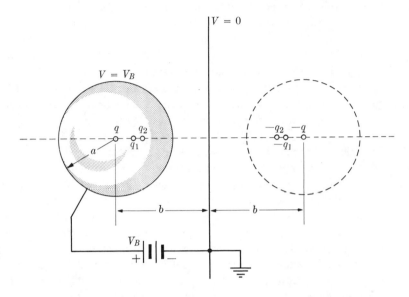

Fig. 4–9 Capacitor consisting of a conducting sphere near an infinite plane.

In developing this series it is instructive, first, to consider the plane in Fig. 4–9 to be far from the sphere. The potential near the sphere is then given by an image solution with a single charge q at its center. The value of q can be obtained from the expression for the potential at the surface of the sphere, namely,

$$V_B = \frac{q}{4\pi\epsilon_0 a}.$$

The capacitance of this isolated sphere is then $C = q/V_B = 4\pi\epsilon_0 a$.

Sphere and plane. As the conducting plane is brought nearer to the sphere, a charge of opposite sign appears on this plane and tends to reduce the potential of the sphere. With a constant voltage battery connected between the sphere and the plane, the potential of the sphere is kept constant by the additional charge it receives from the battery. If the final charge on the sphere is $Q > q$, the capacitance of the system is

$$C = \frac{Q}{V_B} = \frac{Q}{q} \, 4\pi\epsilon_0 a.$$

This depends upon the ratio Q/q, which can be obtained from the solution of the problem that involves an infinite series of image charges. From the uniqueness theorem, any system of point charges (inside the sphere or to the right of the plane) that produces $V = 0$ on the plane and $V = V_B$ on the surface of the sphere constitutes the correct solution. The six charges shown in Fig. 4–9 can be used to demonstrate how these boundary conditions are satisfied by an infinite series of charges.

Charge q alone produces $V = V_B$ on the surface of the sphere *but does not produce $V = 0$ on the plane*. Adding $-q$ at the distance b behind the plane produces $V = 0$ on the plane but destroys the constant potential on the surface of the sphere. Using the principles developed in Section 4–2 and Fig. 4–3, the potential produced on the surface of the sphere by $-q$ can be cancelled by placing an image charge $q_1 = (a/2b)q$ at the distance $d_1 = a^2/2b$ to the right of the center. Now, however, q_1 disturbs the $V = 0$ on the plane but, since q_1 is smaller than q, the discrepancy is not nearly as large as it was with q alone. A charge $-q_1$ can be introduced to the right of the plane to restore its potential to zero. Then the introduction of a charge $q_2 = q_1 a/(2b - d_1)$ at the distance $d_2 = a^2/(2b - d_1)$ to the right of the center of the sphere restores the potential of the surface of the sphere to V_B. Continuing this process creates an infinite series of charges ($q > q_1 > q_2 > q_3$, etc.) that can be shown to produce potentials expressed by an infinite series that converges rapidly with reasonable value for a and b. The total charge on the sphere is the sum of the image charges in it; that is,

$$Q = q + q_1 + q_2 + q_3 + \cdots$$

and, if $a/2b = \beta$, this can be written as

$$Q = q(1 + \beta + \beta^2 + \beta^3 + 2\beta^4 + 3\beta^5 + \cdots).$$

The capacitance of the system is then

$$C = 4\pi\epsilon_0 a(1 + \beta + \beta^2 + \beta^3 + 2\beta^4 + 3\beta^5 + \cdots). \tag{4–23}$$

Two spheres. This series of images can be used to calculate the capacitance of two metallic spheres each of radius a and with centers separated the distance $2b$. Since the sphere shown dotted on the right in Fig. 4–9 would have the potential $-V_B$, the capacitance would be given by

$$C = \frac{Q}{2V_B} = 2\pi\epsilon_0 a(1 + \beta + \beta^2 + \cdots).$$

Cylinder and plane. Situations occur frequently that can be closely approximated by a capacitor consisting of an infinitely long cylinder with its axis parallel to an infinite conducting plane. This problem is similar to that of the conducting sphere near an infinite plane that we have just treated. The image solution of this problem does not involve an infinite series, because the required potentials can be exactly reproduced with just two infinitely long but uniformly charged filaments. This image solution is left as an exercise. In Section 4–11 a solution involving complex variables is developed that includes this problem as one of its interesting applications.

4–8 Complex quantities

Complex numbers and complex variables are useful in solving many different kinds of physical problems. Here they are useful for finding solutions of Laplace's equation, especially for those two-dimensional problems in which the potentials can be considered to be independent of z. Complex quantities are especially useful for analyzing the properties of vectors that are confined to one plane. Historically, the vector analysis presented in Chapters 2 and 3 was developed in the 1880's by J. W. Gibbs from quaternions, which had been developed by W. R. Hamilton and others in their attempt to extend the vector properties of complex quantities to three-dimensional situations.

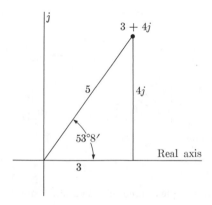

Fig. 4–10 A plot of the complex number $3 + 4j$ in the complex plane. Its absolute value is 5.

Complex numbers are algebraic sums of imaginary numbers and the real numbers that are the ordinary algebraic numbers. Imaginary numbers are real numbers multiplied by the symbol j, where j is defined by $j^2 = -1$. The number $3 + 4j$ is a complex number. The vector characteristics of these numbers stem from the properties they exhibit when they are plotted in what is called a complex plane. A complex plane, shown in Fig. 4–10, is defined by two rectangular coordinate axes in which the real parts of the complex numbers are plotted along the horizontal axis (usually the x-axis), while the imaginary parts are plotted along the vertical axis (usually the y-axis). Points located in this plane represent complex numbers whose absolute magnitudes are defined as the real distances of these points from the origin. The absolute magnitude of $3 + 4j$ is 5, and it is represented by a vector from the origin which is 5 units long and makes an angle of $\tan^{-1}\frac{4}{3}$ with the real axis.

In the complex plane, the number $+1$ is represented by a vector of unit length making an angle zero with the real axis. Multiplication of this number by j gives the complex number $+j$, which amounts to a 90° rotation about the origin of the original vector. Multiplying again by j produces -1, which amounts to another 90° rotation of this unit vector. In fact, the multiplication of any complex number by j produces a 90° rotation without changing the absolute magnitude of the original number. This general property of numbers plotted in complex planes will be found to be particularly useful in the ac circuit analysis of Chapter 11.

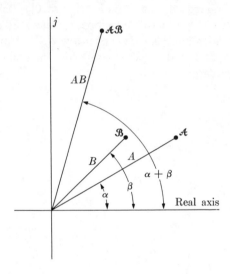

Fig. 4–11 Plots in the complex plane of the complex numbers

$$\mathcal{A} = A(\cos\alpha + j\sin\alpha) \qquad \text{and} \qquad \mathcal{B} = B(\cos\beta + j\sin\beta),$$

together with their product $\mathcal{A}\mathcal{B} = AB[\cos(\beta + \alpha) + j\sin(\beta + \alpha)]$.

The product of any two complex numbers is another complex number, whose vector representation makes an angle with the real axis that is the sum of the angles made by the original two vectors. Any complex number can be written as $\mathcal{A} = A(\cos\alpha + j\sin\alpha)$, where A is its absolute magnitude and α is the angle it makes with the real axis. When two complex numbers are written as $\mathcal{A} = A(\cos\alpha + j\sin\alpha)$ and $\mathcal{B} = B(\cos\beta + j\sin\beta)$, their product, as shown in Fig. 4–11, is

$$\mathcal{A}\mathcal{B} = AB(\cos\alpha + j\sin\alpha)(\cos\beta + j\sin\beta)$$
$$= AB[(\cos\alpha\cos\beta - \sin\alpha\sin\beta) + j(\sin\alpha\cos\beta + \cos\alpha\sin\beta)]$$
$$= AB[\cos(\alpha + \beta) + j\sin(\alpha + \beta)],$$

which proves the statement. This proof is even more obvious when exponential expressions are used.

The exponential form for the complex number $\mathcal{A}$ is

$$\mathcal{A} = Ae^{j\alpha}.$$

To demonstrate this, $e^{j\alpha}$ can be expanded into a series in powers of α:

$$e^{j\alpha} = 1 + j\alpha - \frac{\alpha^2}{2!} - j\frac{\alpha^3}{3!} + \frac{\alpha^4}{4!} + j\frac{\alpha^5}{5!} - \cdots,$$

and these terms may be rearranged as follows:

$$e^{j\alpha} = \left[1 - \frac{\alpha^2}{2!} + \frac{\alpha^4}{4!} - \cdots\right] + j\left[\alpha - \frac{\alpha^3}{3!} + \frac{\alpha^5}{5!} - \cdots\right].$$

The first bracket contains the power series expansion for $\cos\alpha$, and the second that for $\sin\alpha$. Hence

$$e^{j\alpha} = \cos\alpha + j\sin\alpha$$

and because this represents a dimensionless vector of unit magnitude making an angle α with the positive real axis it is called a unit vector. Now the product of the complex vectors $\mathcal{A}$ and $\mathcal{B}$ can be written as

$$\mathcal{A}\mathcal{B} = (Ae^{j\alpha})(Be^{j\beta}) = ABe^{j(\alpha+\beta)},$$

which is identical to the result previously obtained by use of trigonometric identities. In general, multiplication of a complex vector $\mathcal{C} = Ce^{j\gamma}$ by a unit vector $e^{j\theta}$ results in a rotation of the first vector through an angle θ without changing its length:

$$e^{j\theta}\mathcal{C} = e^{j\theta}Ce^{j\gamma} = Ce^{j(\theta+\gamma)}.$$

Complex conjugate. In the algebraic manipulation of expressions containing complex quantities it is frequently desirable to introduce a new complex quantity, which is called the complex conjugate of some quantity in these expressions. The *complex conjugate* of the quantity $\mathcal{A} = a + jb$ is designated by $\mathcal{A}^*$ and is defined by

$$\mathcal{A}^* = a - jb.$$

The complex conjugate of any complex quantity is obtained by changing the sign of every j that appears in the expression for the original quantity. Complex conjugates have many properties, but two of the most important are that the product of the original quantity with its conjugate is real, and the sum of the original and its conjugate is real; for example,

$$\mathcal{A}\mathcal{A}^* = a^2 + b^2$$

and

$$\mathcal{A} + \mathcal{A}^* = 2a.$$

4–9 Functions of a complex variable for two-dimensional problems

All real problems in electrostatics have geometries that are three-dimensional. However, a very large number of these problems can be approximated with great accuracy by using two-dimensional or even one-dimensional models. In this use of the word "dimensional" it is customary to think in terms of rectangular coordinates, in which a function independent or almost independent of one of the three coordinates may be treated as two-dimensional and a function almost independent of two coordinates may be treated as one-dimensional. A problem to be solved in two dimensions can be drawn on flat paper or on a blackboard, but every mark on the paper represents a physical item extending to infinity on each side of the paper: circles represent cylinders, straight lines represent flat sheets, and dots represent filaments, all of infinite length.

For two-dimensional problems there is a very general method for finding solutions of Laplace's equation which involves the use of a complex variable. Two different potential functions, $G(x, y)$ and $H(x, y)$, which are related to some analytic function ϕ of $Z = x + jy$ by the equation

$$G + jH = \phi(Z), \tag{4–24}$$

will be considered. Whenever two potential functions are related to each other in this manner, each one is necessarily a solution of Laplace's equation, and the two have certain reciprocal properties that will be demonstrated.

To show that both G and H satisfy Laplace's equation, it will first be demonstrated that $\phi(Z)$ is always a solution. The relations $\partial Z/\partial x = 1$ and $\partial Z/\partial y = j$ follow directly from the definition of Z. With these relations it is easily shown that

$$\frac{\partial \phi}{\partial x} = \frac{\partial \phi}{\partial Z} \frac{\partial Z}{\partial x} = \phi' \quad \text{and} \quad \frac{\partial^2 \phi}{\partial x^2} = \frac{\partial \phi'}{\partial Z} \frac{\partial Z}{\partial x} = \phi'',$$

as well as

$$\frac{\partial \phi}{\partial y} = \frac{\partial \phi}{\partial Z} \frac{\partial Z}{\partial y} = j\phi' \quad \text{and} \quad \frac{\partial^2 \phi}{\partial y^2} = j \frac{\partial \phi'}{\partial Z} \frac{\partial Z}{\partial y} = -\phi'',$$

where ϕ' and ϕ'' are defined by

$$\phi' = \frac{\partial \phi}{\partial Z} \quad \text{and} \quad \phi'' = \frac{\partial^2 \phi}{\partial Z^2} = \frac{\partial \phi'}{\partial Z}.$$

Adding

$$\frac{\partial^2 \phi}{\partial x^2} = \phi'' \quad \text{to} \quad \frac{\partial^2 \phi}{\partial y^2} = -\phi''$$

results in

$$\frac{\partial^2 \phi}{\partial x^2} + \frac{\partial^2 \phi}{\partial y^2} = 0,$$

which is just Laplace's equation for ϕ in two dimensions. Applying this to Eq. (4–24) gives

$$\nabla^2\phi = \nabla^2 G + j\nabla^2 H = 0,$$

which can be true only if $\nabla^2 G$ and $\nabla^2 H$ each equal zero, that is, only if G and H are each solutions of Laplace's equation. This proof that G and H are solutions of Laplace's equation is valid everywhere in space where the first and second derivatives of G and H exist, but not where either function approaches infinity, as is illustrated in the next section.

Solutions for the two-dimensional Laplace equation can be obtained by selecting any analytic functions of the variable Z, each of which will produce two real solutions G and H.

The simplest function of Z is just $bZ = bx + jby$, from which two simple functions, $G = bx$ and $H = by$, are obtained. Obviously, G represents the potential for a uniform field of magnitude $-b$ in the x-direction, whereas H represents a uniform field of magnitude $-b$ in the y-direction. Another simple function is $\phi(Z) = bZ^2 = b(x^2 - y^2) + j(2bxy)$, which gives $G = b(x^2 - y^2)$ and $H = 2bxy$. If G is chosen for the potential function, the potential distribution can be illustrated by giving G various constant values and plotting the equipotential surfaces obtained from these relations. The result is a family of hyperbolas. When the H function is treated in the same manner, a second family of hyperbolas results, and the lines of $H = $ constant cut each of the $G = $ constant curves in such a manner that all intersections are at right angles, as shown in Fig. 4–12. When one set of curves cuts another set of curves in this manner, the two sets are said to be *orthogonal*. It should be noted that the G and H functions for $\phi(Z) = bZ$ are also orthogonal. This behavior will be shown to be general for all G and H functions that can be obtained from $\phi(Z)$.

To show this, ϕ can be differentiated with respect to x and y separately, to yield

$$\frac{\partial\phi}{\partial x} = \frac{\partial G}{\partial x} + j\frac{\partial H}{\partial x} = \phi' \quad \text{and} \quad \frac{\partial\phi}{\partial y} = \frac{\partial G}{\partial y} + j\frac{\partial H}{\partial y} = j\phi',$$

from which it can be seen that

$$j\frac{\partial G}{\partial x} - \frac{\partial H}{\partial x} = \frac{\partial G}{\partial y} + j\frac{\partial H}{\partial y}.$$

Equating the real and imaginary parts, we find

$$-\frac{\partial H}{\partial x} = \frac{\partial G}{\partial y} \tag{4–25}$$

and

$$\frac{\partial G}{\partial x} = \frac{\partial H}{\partial y}, \tag{4–26}$$

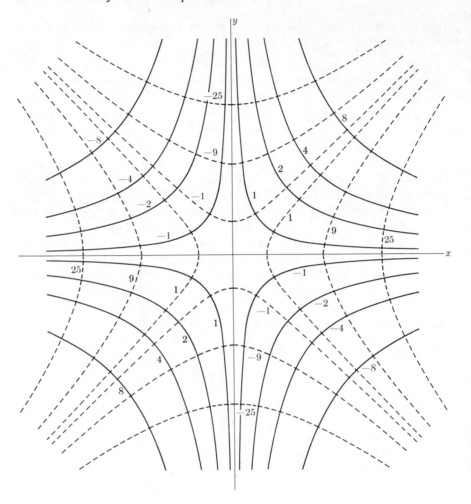

Fig. 4–12 Graph of $G/b = (x^2 - y^2) =$ constant and $H/2b = xy =$ constant. G/b is represented by dashed lines and $H/2b$ is represented by solid lines. The constant values selected for G/b and $H/2b$ are indicated on the curves.

which are called the Cauchy-Riemann relations. The foregoing analysis assumed that $\phi(Z)$ was analytic. Actually, these relations provide a test as to whether or not a given $\phi(Z)$ is analytic. In regions where the Cauchy-Riemann relations are satisfied and where G, H, and their first derivatives are continuous, $\phi(Z)$ is analytic. Multiplying Eqs. (4–25) and (4–26) together and transposing, we obtain

$$\frac{\partial G}{\partial x}\frac{\partial H}{\partial x} + \frac{\partial G}{\partial y}\frac{\partial H}{\partial y} = 0, \tag{4–27}$$

which is just $\nabla G \cdot \nabla H = 0$. Whenever the dot product of two nonzero vectors is zero, the two vectors are at right angles; therefore the gradients of G and H are perpendicular to each other at all points. Since the gradients are always perpendicular to their equipotential surfaces, the equipotential surfaces of G and H must intersect each other at right angles.

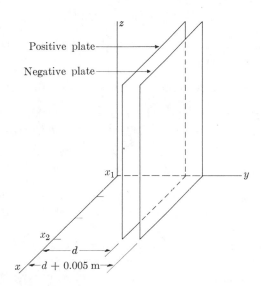

Fig. 4–13 Two parallel conducting plates charged to a potential difference of 100 volts are located perpendicular to the y-axis, as shown.

These two functions exhibit another useful relationship at all points; namely, $|\nabla G| = |\nabla H|$, which follows immediately from the Cauchy-Riemann relations. Upon squaring and adding Eqs. (4–25) and (4–26), we obtain

$$\left(\frac{\partial G}{\partial x}\right)^2 + \left(\frac{\partial G}{\partial y}\right)^2 = \left(\frac{\partial H}{\partial x}\right)^2 + \left(\frac{\partial H}{\partial y}\right)^2.$$

This may be rewritten as

$$|\nabla G|^2 = |\nabla H|^2, \qquad (4\text{–}28)$$

which is true only if $|\nabla G| = |\nabla H|$. Equations (4–27) and (4–28) are both very useful in solving two-dimensional problems involving Laplace's equation.

As a simple illustration, consider two large, parallel conducting plates with a difference in potential of 100 V. The plates may be perpendicular to the y-axis and cut this axis at $y = d$ and $y = d + 0.005$ m, so that they are 0.005 m apart, with the positive plate at $y = d$, a shown in Fig. 4–13. In the space between these plates $\nabla^2 V = 0$ and the potential can be represented by the solution of Laplace's equation, $H = by$. The boundary condition at the positive plate gives

$$H_+ = 100 \text{ V} + V_0 = bd$$

and that at the negative plate gives

$$H_- = V_0 = b(d + 0.005 \text{ m}),$$

where V_0 is introduced because the absolute potential of the plates was not specified. Subtraction yields

$$100 \text{ V} = -(0.005 \text{ m})b, \quad \text{or} \quad b = -20{,}000 \text{ V m}^{-1}.$$

The potential between the plates is $V = H = -(20{,}000 \text{ V m}^{-1})y$, the electric field in the space is $\mathbf{E} = -\nabla V = \mathbf{j}20{,}000 \text{ V m}^{-1}$, and the charge density on the positive plate is

$$\sigma = \epsilon_0 E = \epsilon_0 \, 20{,}000 \text{ V m}^{-1} = 0.18 \ \mu\text{C m}^{-2}.$$

In practical cases the plates of the capacitor are never infinite, and it is often desirable to obtain the approximate value of the total charge on one of the plates. For example, if the plates extend from $x_1 = 0$ to $x_2 = 0.3$ m in the x-direction, and from $z_1 = 0$ to $z_2 = 1.0$ m in the z-direction, the total charge on the positive plate is approximately $0.18 \ \mu\text{C m}^{-2} \times 0.3 \text{ m}^2 = 0.054 \ \mu\text{C}$. The actual charge is less than this, because the charge density is somewhat lower near the edges. The same value for the charge could have been obtained from the difference between the G's at the two edges of the positive plate, that is, $\epsilon_0(G_2 - G_1) = (bx_2 - bx_1)\epsilon_0 = 0.3b\epsilon_0 = 0.054 \ \mu\text{C}$. This procedure for calculating charges* follows directly from the relations between G and H. Since $\sigma = \epsilon_0 E_n$ and $|\mathbf{E}| = |\nabla H| = |\nabla G|$ is constant between x_1 and x_2 in this problem, the areal density of charge is

$$\sigma = \epsilon_0 |\nabla G| = \epsilon_0 \left| \frac{G_2 - G_1}{x_2 - x_1} \right|,$$

where G_1 and G_2 are the values of G at x_1 and x_2, respectively. Therefore the charge per meter of distance in the z-direction is

$$\lambda = \sigma(x_2 - x_1) = \epsilon_0(G_2 - G_1).$$

This method is too complex for so simple a problem, but with a more complex $\phi(Z)$, this relation, which has general validity, is very useful.

---

* An ambiguity in the signs of λ, σ, and q is inherent in calculations of their magnitudes on surfaces of conductors that coincide with equipotential surfaces determined mathematically. When charges are positive with the bulk of the conductor on one side of the equipotential, they are negative with the bulk of the conductor on the other side. The expression $\sigma = \epsilon_0 E_n = -\epsilon_0 \, \partial V / \partial n$ from Eq. (3–6) is not ambiguous.

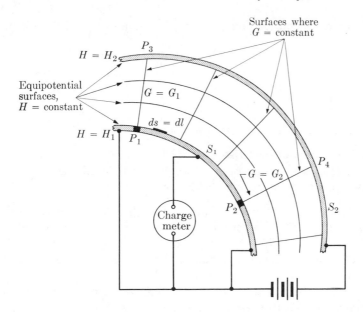

Fig. 4–14 Lines of constant G and H between two curved cylindrical sheets S_1 and S_2.

For this demonstration a general problem like that shown in Fig. 4–14 should be considered, where the curved cylindrical sheets S_1 and S_2 extend to plus and minus infinity in the z-direction. These sheets are charged to potentials of H_1 and H_2, respectively, and it is desired to find the charge per unit length residing on the region of S_1 that extends from P_1 to P_2. The charge per unit area is $\sigma = \epsilon_0 E_1$, where $\mathbf{E}_1 = -\nabla H_1$ is the field evaluated along the surface S_1. The charge per unit length between P_1 and P_2 is given by

$$\lambda = \int_{P_1}^{P_2} \sigma \, dl = \int_{P_1}^{P_2} \epsilon_0 |\nabla H_1| \, dl,$$

since the element of area ds, being one meter long, is equal to the element of length dl. Using the relation $|\nabla H| = |\nabla G|$, the equation for λ becomes

$$\lambda = \int_{P_1}^{P_2} \epsilon_0 |\nabla G| \, dl,$$

and if ∇G points in the direction $d\mathbf{l}$, the quantity $|\nabla G| \, dl = \nabla G \cdot d\mathbf{l} = dG$. The integral then becomes

$$\lambda = \int_{G_1}^{G_2} \epsilon_0 \, dG = \epsilon_0 (G_2 - G_1),$$

where G_1 and G_2 are the values of G at P_1 and P_2, respectively.

Since the potential difference between the sheets is given by $H_2 - H_1$ and since the charge per unit length on S_1 between P_1 and P_2 is given by

$\epsilon_0(G_2 - G_1)$, a general relation for the capacitance per unit length of this capacitor is

$$C_l = \frac{\epsilon_0(G_2 - G_1)}{H_2 - H_1}.$$ (4–29)

4–10 Some characteristics of electrostatic fields

There are some characteristics of all electrostatic fields that are most easily visualized in two dimensions. For example, the characteristics of all equilibrium points in electrostatic fields are clearly shown in Fig. 4–12.

Earnshaw's theorem. If the solid lines in Fig. 4–12 represent equipotential surfaces, the electric field is zero at the origin. A test charge placed at this point would be in equilibrium, but not in stable equilibrium. While the zero potential of this point is a minimum with respect to quadrants 1 and 3, it is a maximum with respect to quadrants 2 and 4. Such points have been given the descriptive name "saddlepoints." A theorem by Earnshaw states that *all equilibrium points in electrostatic fields are saddlepoints.* This is easily visualized by noting that if the point is surrounded by a small closed surface, the net flux out of it must be zero by Gauss' law; that is, field lines that enter through one portion of the surface must leave through some other portion. While there is no field and therefore no force on a test charge placed exactly at the equilibrium point, there is a force on it when it is slightly displaced. This force tends to carry the test charge away from that region rather than to restore it to the equilibrium position. *Charges cannot be in stable equilibrium in electrostatic fields.* The characteristic pattern for flux lines near equilibrium points is shown in Fig. 4–15, where the field is produced by two conducting sheets made to coincide with the two hyperbolic curves, marked 4 in Fig. 4–12, and maintained at the same positive potential. While this two-dimensional equilibrium point is the infinite line representing the z-axis, in real three-dimensional situations equilibrium points are always mathematical points.

Guard rings. A simple capacitor consists of two conducting surfaces insulated from each other, so that they may have different potentials in a static field. If the two surfaces are made to coincide with any two of the equipotentials that have different numbers in Fig. 4–12, the capacitance can be calculated by means of Eq. (4–29). The mathematical expressions for the shapes of the two conducting surfaces must be known if the capacitance is to be calculated exactly. The mathematical expressions defining the surfaces plotted in Fig. 4–12 extend to infinity, as do many such expressions. It is impossible to build capacitors with conducting sheets extending to infinity as required by these expressions. Furthermore, the calculated capacitance is usually infinite for two such complete sheets. Finite sections of these

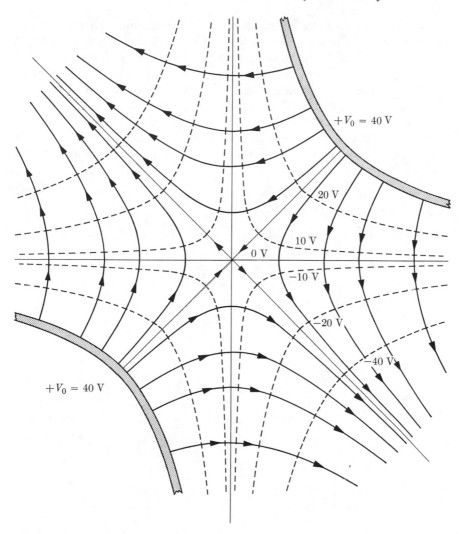

$+V_0 = 40$ V

20 V

10 V

0 V

-10 V

-20 V

-40 V

$+V_0 = 40$ V

Fig. 4–15 Typical electrostatic field lines and equipotential surfaces near an equilibrium point. Potentials always have "saddlepoint" characteristics in these regions.

surfaces can be made to have finite capacitances, but the calculations cannot be accurate because of edge effects.

A technique for eliminating most of the inaccuracies caused by edge effects, called the "guard-ring technique," has been developed. As shown in Fig. 4–14, that section of surface S_1 between P_1 and P_2 and extending one meter in the z-direction has been chosen for one conductor of a capacitor. A fine saw cut separates this section and keeps it insulated from the rest of

S_1, whose shape conforms to the mathematical expression for some distance beyond the cut. Connections are shown in Fig. 4–14 for producing a single potential on all of S_1, while measuring only that charge that flows to the isolated section. It is not necessary to physically separate the section of S_2 between P_3 and P_4, since its charge is automatically equal and opposite to that on S_1 between P_1 and P_2. The portion of S_1 outside the saw cut is called a "guard ring," because in the earliest applications of this technique, this portion, whose only function is to control the potential distribution, was in the shape of a ring. A guard ring of this design is shown in Fig. 3–13 for Problem 3–13.

4–11 Eccentric cylinder problems

Although there are thousands of solutions of the Laplace equation, only a few can be presented here because of space limitation. Anyone faced with problems that cannot be solved with the solutions given here should consult reference books that cover solutions for many other geometrical shapes.

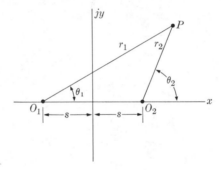

Fig. 4–16 Plot in complex plane of $r_1 e^{j\theta_1} = Z + s$ and $r_2 e^{j\theta_2} = Z - s$. This plot is useful in interpreting the function $\phi(Z) = A \ln [(Z + s)/(Z - s)]$.

Two right circular cylinders whose axes are parallel but not coincident occur in many practical applications. Such problems can be solved with the following complex function, which incidentally serves to illustrate the usefulness of complex variables for solving such problems. This function is

$$F(Z) = A \ln \frac{Z + s}{Z - s}. \tag{4–30}$$

To obtain functions useful for solving problems this expression should be separated into its real and imaginary parts and plotted in the complex plane, as shown in Fig. 4–16. The numerator is given by the line $O_1 P$, since

$$Z + s = (x + s) + jy = r_1 e^{j\theta_1},$$

and the denominator is given by the line O_2P, since

$$Z - s = (x - s) + jy = r_2 e^{j\theta_2}.$$

The function then can be written as

$$F(Z) = A \ln \frac{r_1 e^{j\theta_1}}{r_2 e^{j\theta_2}} = A \ln (r_1/r_2) - jA(\theta_2 - \theta_1).$$

This function of Z has thus provided the two functions

$$G = A \ln (r_1/r_2) \quad \text{and} \quad H = -A(\theta_2 - \theta_1),$$

where either G or H can be used as potential functions. While these bipolar coordinates are convenient for specifying the G and H functions, actual solutions are best obtained with rectangular coordinates.

When G is taken to be the potential function, families of equipotentials are obtained by setting G equal to a series of different constant values. It is convenient to determine these constant values for G by defining $r_1/r_2 = m$ and $G = A \ln m$, where each value of m determines a particular potential surface. This method of choosing the constant G's is used to facilitate the remaining analysis. Expressions for the equipotential surfaces can be obtained from

$$m^2 = \frac{r_1^2}{r_2^2} = \frac{(x + s)^2 + y^2}{(x - s)^2 + y^2}. \tag{4-31}$$

Multiplying by the denominator, expanding, collecting terms, and dividing by $(m^2 - 1)$, we obtain

$$x^2 - 2\left(\frac{m^2 + 1}{m^2 - 1}\right) xs + s^2 + y^2 = 0,$$

which can be written

$$\left(x - \frac{m^2 + 1}{m^2 - 1} s\right)^2 + y^2 = \left(\frac{2ms}{m^2 - 1}\right)^2, \tag{4-32}$$

as can be verified by expanding both sides. A series of equipotential surfaces can be obtained by plotting Eq. (4–30) with a series of values for m, and each of these equipotentials will be a circle of radius $2ms/(m^2 - 1)$ with its center at $x = (m^2 + 1)s/(m^2 - 1)$. The potential of each circle is then obtained from $G = A \ln m$.

In Fig. 4–17 circles have been plotted for $m = 0, \frac{1}{4}, \frac{1}{3}, \frac{1}{2}, 1, 2, 3, 4,$ and ∞. The circles for $m = 0$ and $m = \infty$ have zero radii with centers at $(-s, 0)$ and $(s, 0)$, respectively. These two points, marked O_1 and O_2 in Fig. 4–16, are called singular points or poles, because their potentials are $-\infty$ and ∞. For $m = 1$ the circle has infinite radius and is just the median plane through $x = 0$, as can be seen from the fact that, in Fig. 4–16, $r_1 = r_2$ for all points on the median plane where $m = r_1/r_2 = 1$. Mathematically,

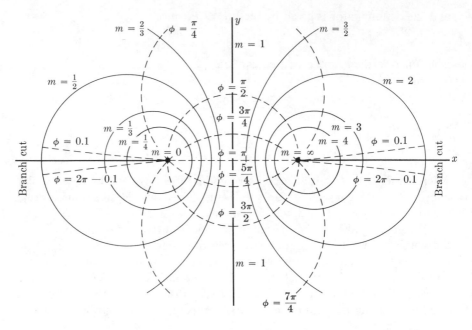

Fig. 4–17 The function $F(Z) = A \ln [(Z + s)/(Z - s)]$ may be separated into its real and imaginary parts, $G = A \ln m$ and $H = -A\phi$. Dashed lines indicate curves for constant H, while solid lines are curves for constant G.

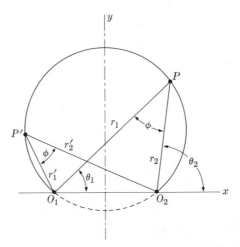

Fig. 4–18 Arcs of circles passing through O_1 and O_2 represent curves of constant H. This fact is illustrated by the figure, since the angle ϕ enclosed by O_1PO_2 is independent of the location of P on the solid portion of the circle, and $H = -A\phi$.

all traces in the xy-plane represent surfaces extending to infinity in the positive and negative z-directions, and each circle represents an infinitely long cylinder.

The surfaces for different constant values of H are sections of cylinders with circular arc traces in the xy-plane, with centers on the y-axis. Each circular arc extends from O_1 to O_2, as shown by the dashed curves in Fig. 4–17. Proof that these arcs of circles do represent constant values of H in the xy-plane is readily obtained from Fig. 4–18. Now $H = -A(\theta_2 - \theta_1) = -A\phi$, where $\phi = \theta_2 - \theta_1$ is the angle between PO_1 and PO_2. From a well-known theorem in geometry, the angle between $P'O_1$ and $P'O_2$ is also ϕ, provided P' is a point on the circle passing through P, O_1, and O_2. This curve of constant $H = -A\phi$ extends only from O_1 through P' and P to O_2. The remainder of each circle represents another curve of constant H for which $H = -A\phi'$ and $\phi' = \pi - \phi$.

To illustrate the use of these solutions, the capacitance per meter, C_l, can be calculated for an infinitely long copper cylinder of 3-cm radius whose axis is parallel to an infinite conducting plane and 5 cm from it. The expression for the plane is given by $m = 1$ and, since $G = A \ln m$, the potential of the plane is zero. If the electrical connections in a real situation do not produce zero potential on the plane, *the constant potential of the plane can be added to all of the potentials* calculated here. Adding a constant to all of the potentials in any system does not affect capacitances or any of the fields.

To satisfy the boundary conditions of this problem, it is necessary to find values for m and s required to provide a circle of radius a having its center a distance d from the median plane of Fig. 4–17. The two equations needed for calculating this m and this s can be obtained from Eq. (4–32). They are

$$\frac{m^2 + 1}{m^2 - 1} s = d \qquad \text{and} \qquad \frac{2ms}{m^2 - 1} = a.$$

Dividing the first by the second and simplifying, we obtain

$$m^2 - 2md/a + 1 = 0. \qquad (4\text{–}33)$$

The two solutions for m are

$$m_1 = d/a - \sqrt{(d/a)^2 - 1}$$

and

$$m_2 = d/a + \sqrt{(d/a)^2 - 1}.$$

The product of these two m's is $m_1 m_2 = 1$. They represent circles of the same radius spaced symmetrically with respect to the y-axis, as can be seen on Fig. 4–17. For the conducting plane that coincides with the y-axis, $m = 1$.

The potential difference between the cylinder and the plane is then

$$V = G_2 - G_1 = A(\ln m_2 - \ln 1) = A \ln m_2,$$

where A is a constant to be determined by the actual potential difference that is maintained between these two conductors. The total charge per unit length in the z-direction is obtained by integrating σ, the surface density, around the cylinder. Starting from the x-axis on the right in Fig. 4–17 and progressing counterclockwise around the circle to the starting point gives

$$\lambda = \oint \sigma \, dl = \oint \epsilon_0 E \, dl = \oint \epsilon_0 |\nabla G| \, dl.$$

Furthermore, from Eq. (4–28)

$$\oint \epsilon_0 |\nabla G| \, dl = \epsilon_0 \oint |\nabla H| \, dl = \epsilon_0 \int_H^{H'} dH = \epsilon_0(H' - H) = \epsilon_0 2\pi A,$$

since $H = -A\phi$, and ϕ ranges from 0 to 2π. The charge per unit length on the cylinder is therefore $\epsilon_0 2\pi A$. Since the function H has the same range of values along the plane (the y-axis), the charge on it is equal but opposite to that on the cylinder. The capacitance per unit length is

$$C_l = \frac{\epsilon_0(H_2 - H_1)}{G_2 - G_1} = \frac{2\pi\epsilon_0}{\ln m_2}. \tag{4–34}$$

The capacitance per unit length for a geometry in which two conducting cylinders each of radius a are parallel and spaced the distance $2d$ apart is easily calculated:

$$C_l = \frac{\epsilon_0(H_2 - H_1)}{G_2 - G_1} = \frac{2\pi\epsilon_0}{\ln m_2/m_1} = \frac{\pi\epsilon_0}{\ln m_2}. \tag{4–35}$$

These two relations, Eqs. (4–34) and (4–35), are used in Chapter 12 for the capacitances of transmission lines. The general relations developed here are also used in Chapter 8 for obtaining expressions for the inductance of transmission lines needed in Chapter 11. For calculating inductances in these cases, the solid lines in Fig. 4–17 represent the magnetic flux lines, while for calculating capacitances we have used the dashed lines to represent electric flux lines.

Solutions with specified cylinders. Any problem like the foregoing that involves a cylinder near a plane is easily solved, because $m = 1$ for a plane and it is necessary only to find m for the cylinder. Other problems that can be solved with this complex function may be more difficult, because both of the m's may need to be determined from the geometry. While this is simple in principle, common techniques for determining these m's often result in fourth-order equations that must be solved. Fortunately, the m's can be determined without solving fourth-order equations.

As a general problem of this type, consider two cylinders, of radii a and b, with distance d between their centers. Designating the m's for the cylinders by m_a and m_b respectively, the unknown quantities are m_a, m_b, and the s in Eq. (4–32). This equation shows that the given geometrical quantities can be expressed by the following three equations:

$$a = \frac{2m_a s}{m_a^2 - 1},$$

$$b = \frac{2m_b s}{m_b^2 - 1},$$

and

$$d = d_a - d_b = \frac{m_b^2 + 1}{m_b^2 - 1} s - \frac{m_a^2 + 1}{m_a^2 - 1} s.$$

Since the first two equations are easily solved for m_a and m_b in terms of s, it is desirable to first solve the last equation for s. It is easy to show that $d_a^2 = s^2 + a^2$ and $d_b^2 = s^2 + b^2$. The last equation can then be written as

$$d = \sqrt{s^2 + b^2} - \sqrt{s^2 + a^2}.$$

By defining $d/a = D$, $b/a = B$, and $s/a = S$ this can be simplified to

$$D = \sqrt{S^2 + B^2} - \sqrt{S^2 + 1},$$

which can be written as

$$2\sqrt{(S^2 + 1)(S^2 + B^2)} = B^2 + 2S^2 + 1 - D^2.$$

Finally, this can be squared and solved for S^2 to give

$$S^2 = \left(\frac{B^2 - D^2 - 1}{2D}\right)^2 - 1. \tag{4–36}$$

When a problem is stated in terms of a, b, and d, s can be obtained from Eq. (4–36) and m_a and m_b can then be obtained from the first two of the original three simultaneous equations.

For the common problem in which $b > a$ and $(b + a) > d$, the solution can be obtained either with both m's greater than one or with both less than one. The expressions for the m's are

$$m_a = \sqrt{S^2 + 1} \pm S$$

and

$$m_b = \sqrt{(S/B)^2 + 1} \pm S/B.$$

Singular points, or poles. As pointed out, the function

$$F(Z) = A \ln \frac{Z + s}{Z - s} = A \ln \frac{r_2}{r_1} - jA(\theta_2 - \theta_1)$$

has two singular points (poles), at $r_1 = 0$ and at $r_2 = 0$, where the potential $G = \ln(r_2/r_1)$ is infinite. At such points neither Laplace's equation nor the Cauchy-Riemann relations are valid. In particular, when there are singular points where G becomes infinite, the integral $\oint \nabla H \cdot dl$ is zero only when the path of integration does not make a loop around any of these poles. This can be seen from Stokes' theorem:

$$\oint \nabla H \cdot dl = \int_s \nabla \times \nabla H \cdot ds,$$

where the integral on the right is identically zero only where H is differentiable, not where it is discontinuous, as at the poles. In complex variable analysis it is shown that $\oint \nabla H \cdot dl = \pm 2\pi A$ whenever the path of integration does make a loop around a pole. A region can be defined in which $\oint \nabla H \cdot dl = 0$ everywhere by making a cut along any line joining the poles and prohibiting the paths of integration from crossing this cut. In Fig. 4–17 this effect is achieved by making cuts to infinity along the x-axis from each pole. The function H is therefore multiple-valued. This last relation has been used, in fact, in obtaining Eq. (4–33), the charge per unit of length on a cylinder. This mathematical property of functions containing singular points will be found to be very valuable when magnetic fields due to electric currents are studied in Chapter 8.

REFERENCES

Jeans, J. H., *The Mathematical Theory of Electricity and Magnetism*, Cambridge University Press, Cambridge, 1925.

Maxwell, J. C., *Electricity and Magnetism*, *Vol. I*, Academic Reprints, Stanford, California, 1953.

Panofsky and Phillips, *Classical Electricity and Magnetism*, Addison-Wesley, Reading, Mass , 1955.

Pipes, L. A., *Applied Mathematics for Engineers and Physicists*, McGraw-Hill, New York, 1946.

Smythe, W. R., *Static and Dynamic Electricity*, McGraw-Hill, New York, 1939.

PROBLEMS

4–1 The charge q near the infinite conducting plane in Fig. 4–1 induces on that plane a surface density of charge which depends upon the radial distance from the z-axis.

a) Derive an expression for the total induced charge within a circle of radius r.

b) Let θ be an angle, with its apex at q, between the negative z-axis and any point on the circle of radius r. For the circle that encloses the total charge $-q/2$, calculate $\tan \theta$.

c) Determine the value of θ for the circle that encloses $-0.9q$.

4-2 Show by direct differentiation that $\nabla^2 V = 0$ for the V given in Eq. (4–1). If there are any points in space where this relation does not hold, explain the apparent discrepancy.

4-3
a) Calculate the field $\mathbf{E} = -\nabla V$ directly from the value of V in Eq. (4–1).
b) Calculate the value of $\mathbf{E}$ just above the conducting plane, from the expression obtained in (a).
c) Calculate the value of $\mathbf{E}$ just above the conducting plane, by adding vectorially the fields due to q and its image. Compare this with the field obtained in (b).

4-4 Remembering that the image charge in Fig. 4–1 does not exist and that the actual fields are due to q and the induced charge, obtain an expression for the field below the surface of the plane.

4-5 Obtain an expression for σ, the density of the charge induced on the inner surface of the hollow sphere of Fig. 4–2, in terms of q, a, d, and θ.

4-6 Find the density of the charge induced on the surface of the sphere in Fig. 4–6(b) in terms of E_0, a, and θ.

4-7 Equation (2–15) expresses the distribution of potential due to any distribution of point charges.
a) Show by direct differentiation that $\nabla^2 V = 0$ for this distribution.
b) Are there any regions where $\nabla^2 V = 0$ is not satisfied? Identity any such regions and explain the physical meaning of the existence of these regions.

4-8 A concentrated charge $+q$ is placed a distance $p = 2a$ from the center of an insulated conducting sphere of radius a. This sphere was originally uncharged and is not grounded. Calculate the charge densities σ_n and σ_f at the two points on the spherical surface that are, respectively, nearest to and farthest from the charge $+q$.

4-9 A thin conducting plane passes through the origin of coordinates perpendicular to the z-axis and extends to infinity in the xy-plane. Charges q_1 and q_2 are placed at $(0, 0, a)$ and at $(0, 0, -a)$, respectively.
a) Find the net surface charge on the sheet as a function of the distance r from the z-axis, using cylindrical coordinates.
b) What is the force on q_1?

4-10 Two semi-infinite conducting planes, one on which $y = x$, and another on which $y = -x$, pass through the z-axis. A charge q is placed at $(a, 0, 0)$. Find V in the wedge-shaped region where $x > 0$, as a function of x, y, z.

4-11 A nonconducting sphere of radius a contains a uniform volume density of real charges; i.e., the total charge is $q = 4\pi a^3 \rho/3$. The center of the sphere is on the z axis at $(0, 0, 2a)$ and an infinite conducting plane passes through the origin in the xy-plane. Find the electric field at the following two points on the z-axis: (a) at $(0, 0, \delta)$ where δ is negligible compared to a; and (b) at $(0, 0, 2a - r)$, where $r < a$.

4-12 According to one modern theory, the center of a thunderstorm usually includes an electric dipole with a concentration of negative charges $-q$ at a height h above the earth's surface (treated as an infinite conducting plane) and an equal concentration of positive charges $+q$ directly above the $-q$ but at the height $h(1 + \delta)$. Make the earth's surface an xy-plane and pass the z-axis through the charges. Near the origin on the earth's surface the induced charges are positive, but at some distance r from this origin the induced charges change sign. Find r where the surface-charge density is zero, assuming that the dipole is (a) an ideal one h meters above the earth's surface, or that it is (b) the real dipole described above. [*Note*. The value of r should be approximated on the assumption that $\delta \ll 1$. If a binomial series is used for this, discard all terms multiplied by δ^3 and higher powers of δ.]

4–13 A nonconducting sphere of radius a has a charge q uniformly distributed over its surface. It is centered at $(0, 0, b)$, where $b > a$. The xy-plane through the origin contains a grounded metallic sheet of infinite extent. Find the potential in cylindrical coordinates (a) inside the sphere, and (b) outside the sphere wherever z is positive.

4–14 A thin spherical shell of metal has a radius b and is grounded. This contains a thin spherical shell of nonconducting material with a charge q uniformly distributed over its surface. The inner shell has a radius a and its center is displaced a distance d from the center of the metallic sphere, where $d < a < b$. Write the potential (a) between the spheres, and (b) inside the smaller sphere.

4–15 A hollow metallic sphere has inner and outer radii a and b respectively. It has no net charge and is not grounded. Two point charges each of magnitude $+q$ are placed inside the sphere on the z-axis, where the origin is at the center of the sphere. These charges are at $z = d$ and $z = -d$. Write the potential in spherical coordinates (a) outside the sphere, and (b) inside the sphere.

4–16 A slab of insulating material extends from $y = -b$ to $y = b$ and is infinite in the x- and z-directions. It contains a volume distribution of total charge $\rho = Ae^{y/b}$.

a) Find $\mathbf{E}$ as a function of y, given that $\mathbf{E} = \mathbf{j}Ab/\epsilon_0$ at $y = 0$.

b) Find V as a function of y, given that $V = 0$, at $y = 0$.

4–17 An infinitely long conducting cylinder of radius a has an infinitely long filament with the charge density λ in $C\,m^{-1}$ stretched parallel to its axis. This filament is d meters from the cylindrical axis.

a) Show that the fields and potentials can be obtained from an image solution using just one infinite filament with the charge density $-\lambda$ for the image.

b) Find the distance p of the image filament from the cylindrical axis.

4–18 Consider two concentric and infinitely long cylinders of copper. In the space between them $\epsilon = \epsilon_0$, and the inner and outer radii are a and b, respectively. The outer cylinder is grounded and the inner one is charged to 100 V. Show that either $V_1 = A/r + B$ or $V_2 = C \ln r + D$ satisfy the boundary conditions. Which potential function is correct for this problem? Prove your statement and show that the other function is incorrect.

4–19

a) Obtain Eq. (4–10) in the text by carrying out the indicated change of variables in the previous equation.

b) Show by direct substitution that P_4 in Table 4–1 is a solution of the corresponding Legendre equation.

4–20 The inner surface of a nonconducting spherical shell of radius a has a potential distribution given by MP_2 (Table 4–1).

a) Find the potential $V(r, \theta)$ inside this shell.

b) Could this potential be given by $MP_2 r^4/a^4$?

c) By $MP_2 a^3/r^3$?

Give your reasons for your answers to both (b) and (c).

4-21 When a finite quantity of charges $\sum q$ (all of the same sign) are grouped together in a finite region, the potential at large distances due to these charges can be shown to approach $V = \sum q/4\pi\epsilon_0 r$, where the large distance r is measured from the interior of the region. Since the magnitude of this potential is decreased by adding charges of opposite sign to the region, the potential due to any finite group of mixed charges must decrease with distance at least as rapidly as $1/r$. Show that for finite charges in a finite region $\int_s V\nabla V \cdot d\mathbf{s}$ approaches zero as r approaches infinity.

4-22 There are many problems in electrostatics in which the charge density is not zero throughout the region of interest. In those special cases in which the charge distribution $\rho = \rho(x, y, z)$ is known throughout the bounded region:

a) Show that the uniqueness theorem holds for solutions of $\nabla^2 V = -\rho/\epsilon_0$.

b) Show also that the function $V = V_p + \sum_l V_l$ is a solution of Poisson's equation if V_p is a solution of this equation and the V_l's are each solutions of Laplace's equation.

4-23 The integral $V = \int_r \rho \, d\tau/4\pi\epsilon_0 r$ is a general solution of $\nabla^2 V = -\rho/\epsilon_0$ that can be combined with solutions of $\nabla^2 V = 0$ to satisfy specific boundary conditions whenever the function $\rho = \rho(x, y, z)$ is known throughout the bounded region. Consider a nonconducting sphere of radius a to be placed inside a hollow conducting sphere of inner radius b, where the distance between the centers of the spheres is h. With the conducting sphere connected to ground:

a) Find the potential distributions inside the two spheres when the inner sphere has a uniform surface charge density σ.

b) Find the potential distribution inside the two spheres when the inner sphere has a uniform volume charge density ρ.

4-24 A solid metallic sphere of radius a is connected to ground by a long fine wire that maintains its potential at zero, and a point charge q is placed a distance $p > a$ from its center.

a) Find the expression for V in terms of spherical coordinates having their origin at the center of the sphere.

b) Find an expression for σ, the density of charge on the surface of the sphere.

c) Find the force on q due to the charged sphere.

d) Would any of the foregoing results be different if the sphere were hollow rather than solid?

4-25 In Problem 4–24 assume that the conducting sphere was not connected to ground but was isolated so that it was neutral before the charge q was placed at the distance p from its center. For this new situation obtain the results called for in parts (a), (b), (c), and (d) of Problem 4–24. In addition, (e) find the potential of the sphere, and (f) show that $l/p = (1 - \lambda^2)^{1/3}$, where l is the distance from q to the $\sigma = 0$ points on the sphere and $\lambda = a/p$.

4-26 As in the problem discussed in Section 4–2, a hollow spherical shell of metal has external and internal radii b and a, respectively, and has a point charge q at a distance $d < a$ from its center. How would the solution given there be modified if, instead of being connected to ground, the sphere had been isolated and neutral before q was introduced?

a) Find the modified expressions for V both inside and outside the sphere.

b) Find new expressions for σ on both the inside and outside surfaces.

c) Find the new force on q due to these surface charges.

4-27 Two concentric spherical shells have inner and outer radii a and b, respectively. At $r = a$, $V = C \cos \theta$ and at $r = b$, $V = D$. Find the potentials between these shells in terms of the known constants C, D, a, b, and the spherical coordinates.

4-28 A capacitor consists of concentric cylinders of infinite length with inner and outer radii a and b, respectively. The potentials of these two cylinders are V_a and zero, respectively. The potential in the space between these cylinders can be represented by either $V_1 = V_a a(b - r)/r(b - a)$ or $V_2 = (V_a \ln r/b)/\ln a/b$, provided there may be a distribution of charges in the space. Calculate the space charge distribution (a) for V_1 and (b) for V_2.

4-29 A ring of radius a lies in the xy-plane with its center at the origin. It carries a total charge q uniformly distributed. The potential along the z-axis is given by $kq/(z^2 + a^2)^{1/2}$. Find the potential everywhere in space due to this charged ring. To do this write the general solution in spherical harmonics and then determine the constants needed to give the correct potential along the z-axis. Separate solutions are needed for $r < a$ and for $r > a$. Neither of these series will converge for $r = a$.

4-30 Show that the function $F(z) = (x + jy)^n = (re^{j\theta})^n$ leads to the cylindrical harmonics of Section 4–5.

4-31 Find the capacitance, C_l, per meter of length between an infinite conducting plane and an infinitely long cylinder of radius a, whose center is d meters from the plane, where $d/a = \frac{5}{4}$.

4-32 Two semi-infinite conducting planes are perpendicular to each other and are joined along the z-axis from $z = -\infty$ to $z = \infty$. These planes extend from $x = 0$ to $x = \infty$ in the xz-plane, from $y = 0$ to $y = \infty$ in the yz-plane, and are connected to ground. A point charge q is located at the (x, y, z) coordinates $(a, b, 0)$. Find expressions for (a) the charge distribution on the yz-plane, and (b) the force acting on the point charge q.

4-33 One excellent method for checking the correctness of complicated solutions is to determine whether they provide correct results in limiting situations that have been calculated by other methods. Show that the image solution of Fig. 4–3 approaches the image solution of Fig. 4–1 when q approaches very close to the surface of the sphere, without coming into contact with it.

4-34 In Section 4–4 the potentials near a conducting sphere in a uniform field are calculated. A uniform field at the origin of coordinates could be created by placing very large point charges far from the origin, so that the relations obtained in Problem 4–24 could be used. Suppose a uniform field to be created by placing one point charge $+q$ at $z = -p$ and another point charge $-q$ at $z = +p$, where the magnitude of q depends upon p in such a manner as to make $q/4\pi\epsilon_0 p^2 = E_0/2$ a constant. Show that the potentials near a conducting sphere centered at the origin in this uniform field can be obtained from those calculated in Problem 4–24(a).

4-35 A horizontal metallic surface in the xy-plane is grounded and is so large that it can be considered infinite. An ideal dipole of moment $\mathbf{p}$ is located on the z-axis at $z = h$ and points in the xz-plane at the angle β with the z-axis. Find expressions in cylindrical coordinates for the potential V above the plane and for the surface charge density σ on the plane, (a) when $\beta = 0°$, and (b) when $\beta = 90°$.

4–36 A positive point charge q placed outside an insulated metallic sphere is attracted toward the sphere. If a is the radius of the sphere and d is the distance of q from its center, find the net charge Q on the sphere that is required to cancel the attraction and produce zero force on q.

4–37 Suppose the point charge q, inside the hollow conducting sphere of Fig. 4–2, is replaced by an ideal dipole of moment **p**. Find expressions for V inside the sphere and for σ on the inside surface of the sphere, (a) when **p** is directed upward perpendicular to OQ, and (b) when **p** is directed to the right parallel to OQ.

4–38 A plane metallic surface is horizontal, is connected to ground, and is so large as to be considered infinite. An infinitely long filament that is a distance h from the plane and parallel to it carries a uniform linear charge density λ. Employ an image solution to find (a) the charge density on the plane, and (b) the force per unit length on the charged filament.

4–39 A hollow conducting cylinder of internal radius a is infinitely long and is connected to earth. A very fine wire, charged to λ in C m^{-1}, is stretched parallel to the cylindrical axis but is a distance h, in meters, from this axis. Find the force per meter of length on the wire.

4–40 A nonconducting spherical shell of radius a has a potential distribution on its surface that is given by $V = A \cos \theta$ in spherical coordinates.
a) Find the potential inside this sphere.
b) Find the potential outside this sphere.
c) Find the charge density σ on the surface of this sphere.

4–41 A metallic sphere of radius a is suspended in a perfect vacuum. Draw a straight line from its center at O to a point M which is d meters from O. Place a dipole of moment **p** at M perpendicular to OM. For the plane defined by **p** and OM:
a) Calculate a general expression for V outside the sphere, in terms of p, a, d, and the coordinates r and θ, measured from O and the line OM respectively.
b) Obtain a general expression for the R-components of field E_R, and
c) Show that the density of charge induced on the surface of the sphere in this plane is given by

$$\sigma = \frac{-p \sin \theta}{\tau_s}\left[\frac{\gamma^3(1 - \gamma^2)}{(1 + \gamma^2 - 2\gamma \cos \theta)^{5/2}}\right],$$

where $\gamma = a/d$ and $\tau_s = 4\pi a^3/3$, the volume of the sphere.
[*Note*. The expressions for the potential and the field are much simpler when written in terms of symbols like r, which is the distance from M to the field point, since r is easily defined in terms of a, d, R, and θ.]

4–42 The distance between the centers of two metallic spheres, each of radius a, is d. If the total charges on these spheres are Q and $-Q$ respectively:
a) Calculate the electric field at the point midway between them.
b) Calculate the difference of potential between the two spheres.

4–43 A two-dimensional capacitor is constructed as shown in Fig. 4–19. The conducting sheet that is connected to the positive terminal of the battery coincides with the mathematical surface given by $x^3 - 3xy^2 = A^3$, where A^3 is a positive constant, and the conducting sheet connected to the negative terminal coincides with $x^3 - 3xy^2 = 0$. The portion of this latter sheet that is connected through G is electrically isolated from the remainder by small breaks at (X, Y) and $(X, -Y)$.

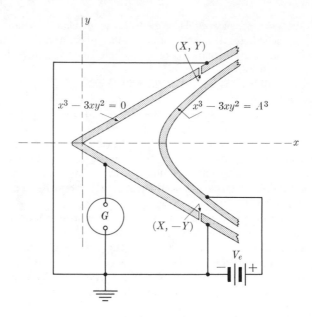

Figure 4–19

a) For the capacitor that is determined by the charge that would flow through G on charging, find the capacitance per unit length in terms of A, X, and Y.

b) Find the charge density on the positive sheet where it cuts the x-axis at $x = A$.

c) Find the charge density on the isolated V-shaped section as a function of x and find the total charge on this section.

4–44 In some problems it has been assumed that a fine wire, used to assure that two conductors have the same potential in a static field, would not otherwise disturb the potentials of the field. Such a wire would disturb the potentials if and only if it has charges distributed on it. The magnitude of the disturbance caused by a fine wire can be estimated by solving a concentric cylinder problem with a constant potential difference between the cylinders. Keeping the radius of the outer cylinder constant at b, determine how the charge density λ on the inner cylinder varies as its radius r approaches zero.

4–45 A capacitor consists of two infinitely long conducting cylinders, one inside the other, having different axes. Its geometry and dimensions are such that the complex function of Section 4–9 can be used for the analysis. The values of m for the outer and inner cylinders are 3 and 4, respectively. Calculate (a) the capacitance C_l per meter of length, and (b) the dimensions and eccentricity of the cylinders. (c) Can the inside and outside radii of each cylinder be determined in meters, from the information given here? (d) Can you draw any general conclusions concerning the dependence of C_l upon the absolute dimensions of two-dimensional capacitors?

4–46 To draw a family of eccentric circles like those shown in Fig. 4–17, some method for determining the radius a and the distance d of the center of each circle from the median plane is needed.

a) Prove that $s^2 + a^2 = d^2$, where s is the distance of either pole from the median plane.

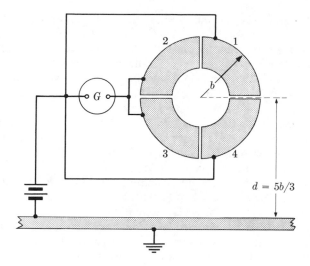

$d = 5b/3$

Figure 4–20

b) A construction circle of radius s can be drawn about the origin, through which all the eccentric circles must pass. A good distribution of these eccentric circles is obtained by marking equidistant points on one quadrant of the construction circle and then drawing one eccentric circle through each of these points. To draw the circle passing through one of these points, designated by P, connect P with the origin O by a straight line making the angle β with the x-axis. Show that the radius a of this eccentric circle is the length of the straight line drawn from P to the x-axis perpendicular to OP. Show also that this determines the center of this eccentric circle.

c) Show that the value of m for this eccentric circle is given by $m = \csc \beta \pm \text{ctn } \beta$.

d) Show that the two values for m (m_1 and m_2) given by the equation in (c), correspond to equal cylinders symmetrically spaced on opposite sides of the median plane. Show also that $m_1 m_2 = 1$.

4–47 Figure 4–20 shows a capacitor consisting of an infinite conducting plane and an infinitely long cylinder. The cylinder has been cut into four quadrants, 1, 2, 3, and 4, which, although insulated from one another, can be connected electrically in any desired manner. The diagram shows quadrants 1 and 4 connected as "guard rings," so that only those charges flowing to quadrants 2 and 3 are measured by G. Neglecting the widths of the saw cuts, calculate the capacitance (a) when quadrants 1 and 4 are the guard rings (as shown), and (b) when quadrants 1 and 2 are the guard rings.

Lines of electric field flux extend from each quadrant to that portion of the plane on which an equal and opposite charge resides. (c) Determine the region on the plane where the flux lines from quadrant 3 terminate.

4–48 A transmission line consists of two copper cylinders of radius a, with their centers separated a distance D. Find the capacitance per meter, C_l.

4–49 Capacitors like the concentric-cylinder capacitor of Fig. 3–13 have been constructed as standards. How much error would be produced by an error of $d = a\,\delta$ in centering the cylinders, where a is the radius of the inner cylinder and

$\delta \ll 1$ is a dimensionless fraction. Assume the inner radius of the outer cylinder to be b.

4–50 What is the capacitance per meter of length between two infinite cylinders which are both 16 mm in diameter and parallel with 34 mm between centers?

4–51 Suppose that in Problem 2–15 one copper sphere of radius 0.5 cm is located at P_1 and an identical sphere is located at P_2. Use simple approximations that in this case should be accurate to within one percent.

a) If neither sphere has a net charge, what are their respective potentials?

b) If the two spheres are now connected by a fine copper wire, what is the resultant potential of the sphere at P_1?

c) What is the resultant charge on the sphere at P_1?

[*Note.* An exact solution requires the combining of solutions from both Section 4–2 and Section 4–7. For Section 4–7, $\beta = 0.5/400 \ll 1$, so that even its first power and all higher powers should be neglected here.]

4–52 An exact solution for Problem 4–51 is interesting though more complex. Treat the charges on the y-axis as point charges of q at $(0, 0, 0)$ and $-q$ at $(0, a, 0)$. Place P_1 at $(0, 0, b)$ and P_2 at $(0, a, b)$ with spheres of radius r and let $a^2 + b^2 = c^2$. Calculate parts (a), (b), and (c) of Problem 4–51.

d) Substitute numbers from Problem 4–51 in the result from (c). Calculate this only to eight significant figures.

4–53 In Fig. 4–17 any circle and the y-axis can represent a capacitor consisting of an infinite cylinder and an infinite plane, with the capacitance C_l per meter of length. If the plane is separated along a z-axis through the origin, the capacitance of the upper half becomes $C_l/2$. This can be divided into two capacitors, each of which has the capacitance $C_l/4$, by producing another separation of the plane along another z-axis. Determine where this separation must be made.

4–54 A nonmetallic cylinder of radius a is centered on the z-axis and extends from $z = -\infty$ to $z = +\infty$. The potential at its surface is $G \sin 2\phi$, due to its surface charges. Find the potential inside and outside the cylinder.

4–55 Plotted in the complex plane, 1, j, -1, and $-j$ represent unit vectors that divide the plane into quadrants. Since the fourth power of each is $+1$, these constitute four solutions to the equation $z^4 = 1$.

a) Write three solutions for $z^3 = 1$ in component form and show they are solutions.

b) Prove that there are n solutions to the equation $z^n = 1$ and that they divide the complex plane into n equal sectors.

4–56 The electrostatic potential within a two-dimensional rectangular space is given by $V = x - y \sin \pi x$, being independent of z. This potential is known only within the rectangle whose corners are at $(0, 0)$, $(3, 0)$, $(3, 1)$ and $(0, 1)$.

a) Which, if any, of the sides of this rectangle could coincide with the surface of a conductor? Determine the potential these conductors must have.

b) Calculate the values of σ that must appear on these conducting surfaces.

c) Calculate the volume density of charge ρ throughout this rectangle.

Chapter 5   DIELECTRIC MATERIALS

A perfect dielectric is a material whose internal electric charges are so tightly bound that no electric current can be conducted through it. It is therefore a perfect insulator. While perfect dielectrics do not exist in nature, the properties of real dielectrics are best understood in terms of the properties of perfect dielectrics. Also, many insulating materials have conductivities at room temperature which are less than 10^{-20} of the conductivity of copper, and such conductivities may be treated as zero for most purposes.

While the electric charges are not free to move through these materials, the positive and negative charges may be displaced relative to each other in the presence of an electric field. If a knowledge of the exact displacement of each charged particle were necessary for the solution of electrostatic problems involving dielectrics, such solutions would be impossible. Fortunately, the fields in and around a dielectric may be treated as though they originate from a continuum of charge rather than from individual charges, provided the dielectric is sufficiently extensive to contain a large number of charges. This size limitation has little practical significance, since a solid particle as small as 10^{-3} mm on a side contains more than 10^{12} charged particles.

While a knowledge of the electronic structure of matter is not necessary for the solution of most practical problems involving dielectrics, some knowledge is indispensable for the solution of more sophisticated problems in solid-state physics. Since this also provides a better understanding of the macroscopic properties of dielectrics, a brief introduction to the structure of solids is presented in this chapter.

5–1   The atomic structure of matter

The planetary model of an atom in which negatively charged electrons circle about a very heavy, positively charged nucleus was proposed by Rutherford in 1911 to explain the scattering of alpha particles passing through matter. The predominant small-angle scattering was attributed to numerous collisions with the light electrons, while the occasional large-angle scattering was attributed to very heavy, positively charged nuclei of small diameter. The large-angle nuclear scattering was well described by the Coulomb inverse square law down to distances as small as 10^{-15} m, and so it was assumed that the nuclei were no larger than this.

Bohr-Sommerfeld model of the atom. By 1916 the Bohr-Sommerfeld model of the atom had been developed, in which the planetary electrons are confined to move in well-defined elliptical orbits whose ellipticities and distances from the nucleus are specified by certain quantum conditions. The total energy of an orbital electron is the sum of its potential energy, due to the coulomb attraction to the nucleus, and its kinetic energy, associated with the angular velocity required to keep the electron from spiraling into the nucleus. A major assumption of this model is that electrons circling the nucleus in the specified orbits, or "shells," lose no energy and hence are able to remain in the orbit indefinitely. This is in direct contradiction to classical electromagnetic theory, in which any accelerating charged particle emits energy in the form of electromagnetic radiation, as will be shown in Chapter 12.

It is assumed that energy may be emitted by orbital electrons only in discrete amounts in the process of dropping from one atomic shell to a lower-energy shell. Quantum rules specify the number of electrons which may occupy a given shell, so that, in general, the lower-energy orbits are filled and electrons from higher orbits are unable to make the transition. If the orbital electrons are thermally excited, or otherwise caused to jump from lower shells to higher shells, they may then spontaneously drop back to the lower-energy shell and emit electromagnetic radiation with a characteristic frequency. This frequency is determined by the quantum condition

$$W = hf,$$

where W is the energy difference between the shells, $h = 6.625 \times 10^{-34}$ J $\cdot$ s is Planck's constant, and f is the frequency of the radiation. Selecting the electron orbits to obey the condition that the angular momentum has values of $nh/2\pi$, where n may have values of $1, 2, 3, \ldots$, ensures that the theoretically predicted frequencies for radiation resulting from the transition of electrons from one shell to a lower shell will be in close agreement with the characteristic atomic spectra. Furthermore, the successive filling of the atomic shells, in accordance with the specified quantum conditions, results in the same periodicity as is exhibited by the periodic table.

Since its conception, the Bohr-Sommerfeld model of the atom has provided a quantitative or semiquantitative explanation for much experimental data. However, the detailed explanation of many experimental results has had to wait for the development of wave-mechanical calculations.

Wave mechanics. According to the wave-mechanical calculations first proposed by Schroedinger in 1926, an electron cannot be located at an exact point in space, but must be represented by a probability distribution. The quantity actually calculated, known as a wave function, is a function of the coordinates x, y, and z, and is frequently represented by the Greek letter ψ. The probability of an electron being at any point in space is given by the value of $|\psi|^2$ at that point, where $|\psi|$ is the absolute value of the complex wave function. The most probable location for the electron corresponds to the point or curve in space where $|\psi|^2$ is maximum. Since the probability of a given electron being somewhere in space is necessarily unity, the wave function must also have the property that

$$\iiint |\psi|^2 \, dx \, dy \, dz = 1$$

when the integration is performed over all space.

Wave-mechanical solutions are undertaken because they yield results not obtainable by classical models or by the semiclassical quantum model of Bohr and Sommerfeld. For those problems where the classical or semi-classical quantum models are adequate, however, the wave-mechanical calculations must yield essentially the same results. It should not be surprising, therefore, that many features of the Bohr-Sommerfeld model of the atom can be derived directly from wave mechanics. A calculation of the electron wave functions for various atoms, for example, reveals that the locus of the maximum value for $|\psi|^2$ is essentially the same as in the electron orbits proposed by Bohr and Sommerfeld.

For problems in which the simple atomic model gives basically the same results as wave mechanics, it is convenient to talk in terms of the simple model because it is easier to visualize than are the abstract wave functions. A general understanding of the properties of dielectrics can be obtained from the simple atomic model even though the details of the interactions between the atoms and electrons can be explained only in terms of wave-mechanical calculations. Even in semiclassical discussions, however, the orbital electrons are often referred to as the electron cloud or charged cloud, in recognition of the fact that the spatial locations of electrons are known only in terms of a probability distribution.

Solid materials. In any solid material the atoms are regularly spaced with respect to each other in a geometric pattern which is characteristic of the material and which is referred to as its crystal structure or crystal lattice. The electronic orbits of the atoms in a solid are very little changed from those of free atoms, with the exception of the outermost orbits, where large perturbations are produced by the proximity of neighboring atoms. In some materials, notably metals, the perturbations are of such form that the electrons in outer orbits become free to wander through the lattice. Such materials are referred to as conductors. In most solids, however, all electrons are relatively tightly bound to their atomic nuclei located on the lattice sites and are thus unable to conduct electric current. These solids are known as insulators or dielectrics.

The binding in these materials may be divided somewhat crudely into two types, *covalent* and *ionic*. In the first case the coulomb force between a positive nucleus and its orbital electrons is sufficiently strong that the number of electrons orbiting each nucleus in the solid is the same as for the free atom. Each atom is therefore electrically neutral, and the binding energy between atoms is of a quantum-mechanical origin. This type of bonding frequently occurs between identical atoms in a solid or between different elements found near the center of the periodic table. Some typical examples are diamond, sulfur, germanium, indium antimonide, and silicon carbide.

In an ionic bond, the quantum forces, which tend to give lower energies for completely filled electron shells, fill the outer shell of one atom by removing electrons from the slightly filled outer shell of a neighboring atom. The atom which gains the extra electron or electrons becomes a negative ion, while the neighboring atom becomes a positive ion. The binding energy between the ions is then supplied primarily by the coulomb attraction. Ionic bonding frequently occurs between two atoms when one has a nearly filled outer shell while the other has a nearly empty outer shell. Some examples of this ionic bonding are sodium chloride, lithium fluoride, silver chloride, magnesium oxide, and calcium oxide.

5–2 Mechanisms of polarization

Regardless of which type of bonding dominates in a given crystal, the gross charge distribution within the lattice will be the same. Positive nuclei will be regularly spaced throughout the lattice, and the electrons will also be regularly distributed, with a much greater concentration near each nucleus than in the intervening regions. The application of an electric field will displace the electrons surrounding each nucleus a very small distance opposite to the field, so that each nucleus and its electron cloud form a small electric dipole. The displacement produced by a given field depends upon the binding energy between the electron cloud and the nucleus and is a characteristic of the particular dielectric.

In ionic crystals a second form of polarization takes place simultaneously with electron displacement, in which each positively charged ion, made up of the nucleus and its surrounding electron cloud, is displaced in the direction of the field, while the negatively charged ions are displaced in the opposite direction. The magnitude of this effect and its importance relative to electron displacement varies from one crystal to another.

A third mechanism of polarization can occur in a few materials that contain permanent electric dipoles. In the absence of any externally applied electric field, these dipoles are randomly oriented throughout the crystal by thermal vibrations of the lattice, so that there is no net polarization. An externally applied electric field causes polarization by creating a greater alignment parallel to the field. The origin of these internal dipoles is in the nonsymmetrical quantum-mechanical bonding forces which occasionally occur between two or more atoms. A very familiar example of this is the water molecule, whose molecular structure is illustrated in Fig. 5–1. The nonsymmetrical location of the positively charged hydrogen ions produces a dipole moment pointing to the right in this figure. The resulting molecule is said to be polar, since it exhibits the properties of a small electric dipole.

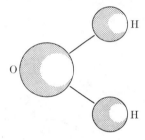

Fig. 5–1 The nonsymmetrical structure of the dipolar water molecule gives rise to a concentration of positive charge to the side of the oxygen atom where the two hydrogen atoms are bonded.

Polarization and polarizability. A substance is said to be polarized when the vector sum of all the dipole moments within it is greater than zero. It would be meaningless to define the polarization at mathematical points within a material. However, *polarization at a physical point is defined as the vector sum of the dipole moments per unit volume within the space defined by this physical point.* This polarization is designated by **P**, and it is a vector function of the coordinates. Whenever **P** is directly proportional to the average electric field **E**, within the physical point, the ratio $P/\epsilon_0 E = \chi$ is called the polarizability. The field's tendency to create dipoles is opposed by quantum-mechanical forces, while its tendency to line up permanent dipole moments is opposed by heat vibrations. The latter type of polarization is, therefore, strongly dependent upon temperature, while the former is almost independent of temperature.

Determination of the mechanism of polarization. The temperature dependence of the polarizability originating from permanent dipoles is used to distinguish it from that of electronic or ionic origin. This dependency is well described by the Langevin-Debye equation for the polarizability,

$$\chi = \frac{C}{T},$$

where T is the absolute temperature and C is a constant dependent upon the material.

Further evidence as to the origin of electric polarization in a given substance can be obtained from its response to variations in the frequency of an alternating electric field applied to it. The frequency dependence of a dielectric which has dipolar, ionic, and electric charge displacement for electric field frequencies below the microwave region is represented in Fig. 5–2. Above the microwave region, the permanent electric dipoles do not have time to orient themselves relative to the field, and their contribution disappears.

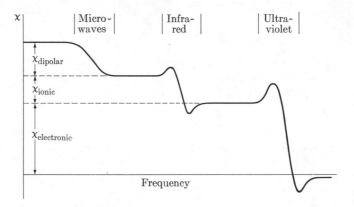

Fig. 5–2 Schematic frequency dependence of the several contributions to the polarizability. (After Charles Kittel, *Introduction to Solid State Physics*, 2nd ed., John Wiley & Sons, Inc., 1956.)

Above the infrared region, the ionic contribution drops out; and finally, for frequencies higher than those of ultraviolet light, the orbital contribution vanishes. Just before vanishing, the orbital and ionic contributions each pass through a resonance which results in the rapid fluctuations shown in the figure.

The usual method of measuring the polarizability of a dielectric is to compare the capacitance of a capacitor containing the substance with that when it is evacuated. The ratio of these two capacitances determines the polarizability. The capacitance is usually obtained by placing the capacitor in a resonant circuit in a manner which is described in more detail in the chapter on alternating currents. At microwave frequencies and above, measurements utilize a relationship between the index of refraction for electromagnetic radiations and the polarizability of the material. As is shown in Chapter 12, the index of refraction is proportional to the square root of the polarizability at high frequencies.

Static polarization. The exact mechanism of the polarization is important in some applications because it governs the temperature dependence and frequency response of the dielectric; and it is, of course, important in any detailed study of the structure of the material. However, the gross effect of the polarization is independent of the mechanism for static fields and for alternating fields up to frequencies near a million cycles per second.

To understand this effect, some more detailed definitions are needed. A polarization model will be utilized in which the negatively charged cloud surrounding each nucleus is shifted relative to it. Consider the two-dimensional representation of atoms in a solid bar, in Fig. 5–3. In the absence of an electric field the negatively charged cloud and the positively charged nucleus have the same center for each atom, as shown in Fig. 5–3(a). With the application of a field **E**, Fig. 5–3(b), the negatively charged cloud shifts opposite to the field by an amount which will be designated by $-\mathbf{R}$. It is actually more convenient to think of the positive nucleus as being moved a distance $\mathbf{R}$ with the field.

Assume for the moment that the charge displacements are the same for each atom or, in other words, that the polarization is uniform throughout the bar. The interior of the bar retains its electric neutrality, since the negative charge that moved into any volume is just offset by the negative charge that moved out of this volume, provided the volume is large compared with an atomic volume. This is illustrated by the dashed rectangle in Fig. 5–3(b), which contains an equal number of positive and negative charges. Electric neutrality is not maintained, however, on the ends of the bar, where there are no charges of opposite sign available to cancel the displaced charges. If σ_p is the areal charge density induced on the right-hand surface, the total charge there is $\sigma_p A$, where A is the cross-sectional area. An equal but opposite charge $-\sigma_p A$ resides on the left.

The charge on the ends is independent of the length of the bar and the total dipole moment of the bar is given by the product of one charge times the distance of separation of the charges, $\sigma_p A l$. If ρ_1 represents the charge density per unit volume of the displaceable outer electrons and R is the average of their diplacements, then $\sigma_p = \rho_1 R$, and the total dipole moment is $\rho_1 R A l$, which is proportional to the volume. The polarization at every point in the bar can be written in vector notation as

$$\mathbf{P} = \rho_1 \mathbf{R}, \qquad (5\text{–}1)$$

which is a general relation even for nonuniform polarizations. When the polarization is perpendicular to a surface of the dielectric, as it is at the ends of the bar in Fig. 5–3, the surface density of polarization charges σ_p is equal

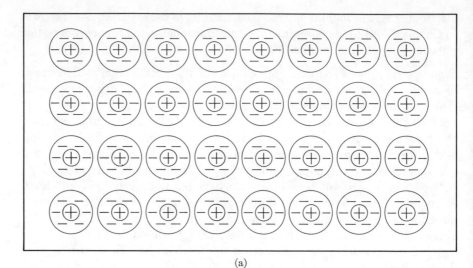

(a)

(b)

Fig. 5–3 (a) A bar of nonpolar dielectric in the absence of electric fields, in which a negative electron cloud neutralizes the positive charge of each nucleus. (b) The same dielectric uniformly polarized toward the right. The electron clouds have shifted equal distances to the left, leaving excess plus charges on the right end and excess negative charges on the left end, with the interior still neutral.

to **P**. In any case, in fact, $\sigma_p = P_n$, where P_n is the component of **P** normal to the surface of the dielectric. For those materials in which **R** is proportional to the average electric field within each physical point, the constant of

Table 5-1

POLARIZABILITIES OF SOME SUBSTANCES

Helium	0.0000684
Oxygen	0.000523
Water	79.0
Quartz	3.3
Sulfur	3.0

proportionality is $\chi\epsilon_0/\rho_1$, and the polarization can be expressed in its usual form as

$$\mathbf{P} = \chi\epsilon_0\mathbf{E}. \tag{5-2}$$

Values of χ, which is called the polarizability or the electric susceptibility of the material, that have been determined electrostatically at room temperatures are given in Table 5–1 for a few substances.

The magnitude of the average displacement $\mathbf{R}$ can be obtained from the special case of pure dry sulfur, which is an excellent insulator. It has 16 electrons per atom, of which 4 are much less tightly bound to the nucleus than are the remaining 12. Since the number of sulfur atoms per cubic meter is 3.8×10^{28}, the density of the 4 displaceable negative charges is 15×10^{28} electrons·m^{-3} or -24×10^9 C m^{-3}. The displacement of this density of negative charges against the field is equivalent to a displacement of $\rho_1 = 24 \times 10^9$ C m^{-3} with the field. If the bar in Fig. 5–3 were composed of sulfur and the average internal fields throughout the bar could be maintained at 1000 V/m, the polarization given by Eq. (5–2) would be

$$P = 3.0 \times 8.85 \times 10^{-12} \times 1000 = 26.6 \times 10^{-9} \text{ C m}^{-2}.$$

Using Eq. (5–1), the relative displacement of the electronic cloud from the nucleus then would be 1.1×10^{-18} m, which is less than a thousandth of the nuclear diameter. Nevertheless, this small displacement of the relatively large density of charges would produce a surface-charge density of magnitude 26.6×10^{-9} C m^{-2} or 16.5×10^{10} electrons·m^{-2} on each end of the bar.

These calculations and conclusions are based upon a bar of homogeneous dielectric material that is assumed to be uniformly polarized. In practice it is difficult to achieve this uniformity even when the material is homogeneous. The presence of charges on the end surfaces tends to depolarize regions near these surfaces. Often the material is not homogeneous, and its polarizability varies from point to point. Such inhomogeneities usually result in polarization charges being distributed throughout the volume of the material. The volume density of polarization charge will be designated by ρ_p. Note that ρ_p will always be much smaller than ρ_1, since ρ_p is obtained from the algebraic sum of two nearly equal and opposite densities of magnitude ρ_1. The polarization charge density ρ_p can be observed experimentally but ρ_1 cannot. The quantity ρ_1 has been introduced only as an aid to analysis, and will be abandoned as soon as general relations have been established.

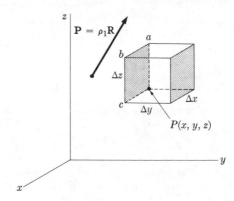

Fig. 5–4 On polarization of a dielectric, the charge moving into and out of an elementary volume leaves a net charge density of $\rho_p = -\nabla \cdot \mathbf{P}$.

5–3 Maxwell's first relation

To investigate the magnitude of ρ_p, it is necessary to calculate the difference between the charge moving into and out of a small elementary volume like any one of those shown in Figs. 3–6, 3–7, or 3–8. The volume in rectangular coordinates shown in Fig. 5–4 will be used with the point P on one corner as in Fig. 3–6. The charge moving in through the shaded surface on the left is $\rho_1 R_y \, \Delta x \, \Delta z$, and that moving out through the shaded surface on the right is

$$+\rho_1 R_y \, \Delta x \, \Delta z + \left(\frac{\partial}{\partial y}\right)(\rho_1 R_y \, \Delta x \, \Delta z) \, \Delta y.$$

The net charge remaining in the element due to this process is the charge entering one side minus the charge leaving the other, or

$$-\left(\frac{\partial}{\partial y}\right)(\rho_1 R_y \, \Delta x \, \Delta z) \, \Delta y.$$

The elementary volume has three such pairs of surfaces, and the net charge remaining inside, due to the charges moving through all six surfaces, is just the sum of three expressions similar to this one. The net charge remaining inside is then

$$\left[-\frac{\partial}{\partial y}(\rho_1 R_y) - \frac{\partial}{\partial z}(\rho_1 R_z) - \frac{\partial}{\partial x}(\rho_1 R_x)\right] \Delta x \, \Delta y \, \Delta z.$$

In this expression, $(\partial/\partial y)(\rho_1 R_y \, \Delta x \, \Delta z) \, \Delta y$ has been written as

$$\left(\frac{\partial}{\partial y}\right)(\rho_1 R_y) \, \Delta x \, \Delta y \, \Delta z \qquad \text{because} \qquad \left(\frac{\partial}{\partial y}\right)(\Delta x \, \Delta z) = 0.$$

A similar situation obtains for the other two partial derivatives.

The polarization charge density at the physical point defined by this elemental volume is obtained by dividing this expression by $\Delta x \, \Delta y \, \Delta z$. The resulting expression may be written in vector notation as

$$\rho_p = -\boldsymbol{\nabla} \cdot (\rho_1 \mathbf{R}) = -\boldsymbol{\nabla} \cdot \mathbf{P}, \tag{5–3}$$

which is the same equation that would have been obtained by using the general volume element of Fig. 3–8 instead of that of Fig. 5–4.

In Section 3–2 a general expression, Eq. (3–5), was developed in which ρ expressed the charge density obtained by dividing the algebraic sum of all the charges in a volume element by its volume. However, the polarization charges are harder to measure than are the net charges on charged dielectrics or charges that move through conductors. Therefore the total charge density, which will now be designated by ρ_t, is usually divided into two parts, $\rho_t = \rho_p + \rho$, where ρ_p is the polarization charge density and ρ includes all charges that are not specifically polarization charges. Of course, the charges included in ρ_p are the same as those included in ρ but those of ρ_p are tightly bound into atoms. To distinguish between them, the charges that are relatively free to move through the material are frequently given the misleading name "real" charges, since they are the ones that cause electric conductivity. With this modification the differential form of Gauss' law found in Eq. (3–7) can be written as

$$\boldsymbol{\nabla} \cdot \epsilon_0 \mathbf{E} = \rho_t = \rho_p + \rho.$$

Substituting the expression for ρ_p from Eq. (5–3) and transposing, we obtain

$$\boldsymbol{\nabla} \cdot (\epsilon_0 \mathbf{E} + \mathbf{P}) = \rho$$

or

$$\boldsymbol{\nabla} \cdot \mathbf{D} = \rho, \tag{5–4}$$

where $\mathbf{D}$ is defined by

$$\mathbf{D} = \epsilon_0 \mathbf{E} + \mathbf{P}. \tag{5–5}$$

This defines a new vector function of the coordinates that is called the *displacement vector* $\mathbf{D}$, which was introduced by Maxwell. Equation (5–4) is called *Maxwell's first field relation* in differential form. This can be integrated over any general volume τ surrounded by a closed surface s, to give

$$\int_\tau \rho \, d\tau = \int_\tau \boldsymbol{\nabla} \cdot \mathbf{D} \, d\tau = \int_s \mathbf{D} \cdot d\mathbf{s},$$

using the divergence theorem. *The integral form of Maxwell's first relation
is usually written as*

$$\int_s \mathbf{D} \cdot ds = \int_\tau \rho \, d\tau. \tag{5–6}$$

Physically, both forms of this relation state that the net flux of $\mathbf{D}$ out of any
volume is equal to the algebraic sum of the "real" charges within that volume,
no matter how large or small the volume.

The vectors $\mathbf{D}$ and $\mathbf{P}$ are defined only at physical points and are very
useful for dealing with real materials. It would be meaningless to define them
at mathematical points. On the other hand, an electric field can, in principle,
be defined at a mathematical point as the limit of the force per unit test charge,
as the test charge at this point approaches zero. However, to be consistent
the vector $\mathbf{E}$ in Eq. (5–5) should be defined at a physical point. *Unless otherwise specified, the* $\mathbf{E}$ *used in this text will be defined as the volume average
throughout a physical point of the electric fields defined at mathematical
points.*

5–4 Dielectric constants

In general, all solids consist of crystals that are somewhat more easily
polarized in one direction than in another; that is, they are not isotropic.*
Usually, in such materials $\mathbf{P}$ is not parallel to $\mathbf{E}$, and therefore $\mathbf{D}$ is not
parallel to either $\mathbf{P}$ or $\mathbf{E}$. Such effects are most important in single crystals
of those substances that are strongly anisotropic. In many substances the
differences in polarizabilities from one direction to another are small and,
furthermore, most solids are made up of a multitude of randomly oriented
small crystals. Therefore the majority of ordinary dielectrics can be treated
as isotropic materials in which $\mathbf{P}$ is proportional to $\mathbf{E}$ and also parallel to it.
This condition was tacitly assumed when Eq. (5–2) was written. When Eq.
(5–2) and Eq. (5–5) are combined, we have

$$\mathbf{D} = \epsilon_0 \mathbf{E} + \epsilon_0 \chi \mathbf{E} = \epsilon_0 (1 + \chi) \mathbf{E},$$

which can be written as

$$\mathbf{D} = K\epsilon_0 \mathbf{E} = \epsilon \mathbf{E}, \tag{5–7}$$

by defining $K = 1 + \chi$ and $\epsilon = K\epsilon_0$. The constant K is usually called the
"dielectric constant" of the isotropic material and ϵ is the permittivity.

Anisotropic crystals. Dielectric constants cannot be expressed as simply as in
Eq. (5–7) for anisotropic single crystals or for groups of such crystals that

---

* In addition to anisotropy, some substances show hysteresis effects. That is, when
they have been polarized by an electric field, they may retain some polarization
even after the field is removed.

are preferentially oriented. In these cases, the general expression* for **D** is given by

$$D_x = \epsilon_{11}E_x + \epsilon_{12}E_y + \epsilon_{13}E_z,$$

$$D_y = \epsilon_{21}E_x + \epsilon_{22}E_y + \epsilon_{23}E_z,$$

$$D_z = \epsilon_{31}E_x + \epsilon_{32}E_y + \epsilon_{33}E_z,$$

where the nine dielectric constants can be reduced to six by means of the relations $\epsilon_{12} = \epsilon_{21}$, $\epsilon_{13} = \epsilon_{31}$, and $\epsilon_{23} = \epsilon_{32}$, which can be proved. These general expressions for the rectangular components of **D** also hold for isotropic substances, where $\epsilon_{11} = \epsilon_{22} = \epsilon_{33} = \epsilon$ and the remaining constants are zero. Substances may be nearly isotropic in static or low-frequency electric fields and may become strongly anisotropic at high frequencies. For example, the anisotropic nature of quartz crystals at optical frequencies† is responsible for the double refraction and the polarization phenomena these crystals exhibit in visible light.

In general, liquids are isotropic at all frequencies, but some liquids can be made anisotropic to high-frequency fields by application of either a steady electric field or a steady magnetic field. The anisotropy to visible light produced by steady electric fields is called the Kerr effect, and that produced by steady magnetic fields is called the Faraday effect. Such effects are also exhibited by some solids.

Whenever the dielectric constants for an anisotropic material are true constants, independent of the magnitude or direction of **E**, the analysis can be somewhat simplified in special cases. For example, when a capacitor containing such a dielectric is being charged, the field at any point in the dielectric changes its magnitude but not its direction, so that $\mathbf{E} = \gamma\mathbf{E}_f$, where $\mathbf{E}_f$ is the final field at this point and γ varies from zero to unity during the charging. At the same time, the value of **D** at any point is given by $\mathbf{D} = \gamma\mathbf{D}_f$, where $\mathbf{D}_f$ is the final value of **D** at this point. Thus the angle between **D** and **E** remains constant throughout the charging process. This simplifying relation is used for the discussion of energy relations in Section 5–10.

5–5 Boundary conditions with dielectrics

In order to solve electrostatic problems involving dielectric materials, the boundary conditions between two different dielectrics or between a dielectric and a metal must be obtained. A very small cylindrical pillbox can

---

* Georg Joos, *Theoretical Physics*, 1st ed., pp. 330–362, G. E. Stechert and Company, New York, 1934.

† Jenkins and White, *Fundamentals of Optics*, 3rd ed., pp. 535–553. McGraw-Hill, New York, 1957.

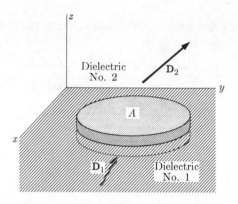

Fig. 5–5 Gaussian pillbox with its lower half in Dielectric No. 1 and its upper half in Dielectric No. 2.

be constructed as shown in Fig. 5–5, with one of the flat circular surfaces in dielectric No. 1 and the other in dielectric No. 2, both being very close to the bounding surface represented by the xy-plane.

The displacement vector $\mathbf{D}$ may be represented by $\mathbf{D}_1$ and $\mathbf{D}_2$ in the two media, and their respective components normal to the bounding surface may be represented by D_{1n} and D_{2n}. The boundary conditions can then be obtained by applying Eq. (5–6),

$$\int_s \mathbf{D} \cdot d\mathbf{s} = \int_\tau \rho \, d\tau.$$

to this pillbox, where the integral on the right is zero because there are no real charges inside the pillbox. By making the area of the cylindrical surface very small compared with the two flat circular surfaces, the contribution to the integral on the left from the cylindrical surface can be made as negligibly small as desired. The only contributions then will come from the two flat surfaces, each of area A. These areas too may be made so small that $\mathbf{D}$ is constant over them, so that the left integral of Eq. (5–6) gives

$$-D_{1n}A + D_{2n}A = 0,$$

and the boundary condition becomes

$$D_{1n} = D_{2n}. \tag{5–8}$$

Another boundary condition may be obtained from the fact that $\mathbf{E}$ is conservative, so that $\oint \mathbf{E} \cdot d\mathbf{l} = 0$ for any closed curve that can be drawn in an electrostatic field. A closed rectangular loop can be chosen as shown in Fig. 5–6, where $b \gg a$, although b may be made so small that the field in each medium may be considered uniform over the parts of the path that lie in it. With these assumptions the normal and tangential components of the electric field $\mathbf{E}_1$ in dielectric No. 1 can be designated E_{1n} and E_{1t}, while the

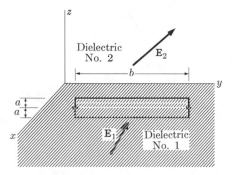

Fig. 5–6 Rectangular path for the line integration $\oint \mathbf{E} \cdot d\mathbf{1} = 0$, with half of the path in each of the adjoining dielectrics.

corresponding components of $\mathbf{E}_2$ can be designated by E_{2n} and E_{2t}. Starting from the upper right corner and progressing counterclockwise, the line integral is

$$-E_{2t}b - E_{2n}a - E_{1n}a + E_{1t}b + E_{1n}a + E_{2n}a = 0,$$

which gives

$$E_{2t} = E_{1t}. \tag{5-9}$$

Equations (5–8) and (5–9) provide the boundary conditions needed for solving problems involving interfaces between two dielectrics.

   In the special case where the boundary is between a conductor and a dielectric, $\mathbf{E}$ is zero everywhere in the conductor. Therefore, since $E_t = 0$ in the metal, it is necessarily zero in the dielectric also. While $E_n = 0$ in the metal, it is not zero in the dielectric, because a real charge resides at the bounding surface of the metal. From the Maxwell relation it can be seen that if D_n is the component of $\mathbf{D}$ in the dielectric normal to the conducting surface, then

$$D_n = \sigma, \tag{5-10}$$

where σ is the surface density of real charge on the conductor. Typical fields near boundaries between different kinds of media are shown in Fig. 5–7. On crossing the boundary separating the two isotropic dielectrics of Fig. 5–7(a) the field changes direction toward the normal, because dielectric No. 1 has been assumed to have greater polarizability than dielectric No. 2. The fields $\mathbf{D}$ and $\mathbf{E}$ remain parallel to each other in each isotropic medium, while $D_{1n} = D_{2n}$ and $E_{1t} = E_{2t}$. At the boundary between a conductor and the isotropic dielectric of Fig. 5–7(b), $\mathbf{D}$ and $\mathbf{E}$ must both be normal to the surface, because they must be parallel and $E_t = 0$. In this case the conductor has a surface density of charge $\sigma = D$. At the boundary between the isotropic and the anisotropic dielectric shown in Fig. 5–7(c), $\mathbf{D}$ and $\mathbf{E}$ are parallel in the first dielectric but not in the second. Their directions and magnitudes

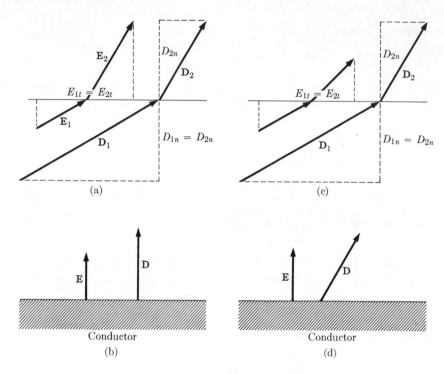

Fig. 5–7 Electric vectors at a surface separating two media: **D** and **E** are parallel in linear isotropic media but not in anisotropic media. The **E** vector is always perpendicular to a conducting surface. (a) Isotropic to isotropic dielectric media. (b) Conductor to isotropic dielectric medium. (c) Isotropic to anisotropic dielectric media. (d) Conductor to anisotropic dielectric medium.

must be such as to satisfy both boundary conditions; namely, $D_{1n} = D_{2n}$ and $E_{1t} = E_{2t}$. Finally, at the boundary between a conductor and the non-isotropic dielectric, illustrated in Fig. 5–7(d), **E** must be normal because $E_t = 0$, but **D** may be at some angle to the surface, since it is only necessary that $D_n = \sigma$. Obviously, if the surface separating two dielectrics has the real charge density σ, then $D_{1n} - D_{2n} = \sigma$.

These conditions also may be used to show that the surface density of polarization charge is given by $\sigma_p = P_n$, where P_n is the component of **P** parallel to the outward-drawn normal to the surface. It is simpler to integrate Eq. (5–3) directly over the volume of the pillbox in Fig. 5–5, to obtain

$$\int_\tau \rho_p \, d\tau = -\int_\tau \mathbf{\nabla} \cdot \mathbf{P} \, d\tau = -\int_s \mathbf{P} \cdot d\mathbf{s}.$$

Again making the cylindrical surfaces small compared with A and making A small enough so that the fields are uniform over the flat surfaces, this relation becomes $\sigma_p A = (P_{1n} - P_{2n})A$, or

$$\sigma_p = P_{1n} - P_{2n},$$

for the net polarization-charge density on the surface of separation. If dielectric No. 2 is replaced with a vacuum, the relation becomes

$$\sigma_p = P_n.$$

5–6 Uniqueness theorem with dielectric materials

Problems involving stationary groups of charges and charged conductors in empty space have been discussed in Chapter 4, where the uniqueness of the solutions for these problems has also been treated. In most of the real problems in electrostatics, the spaces are not empty but contain one or more dielectric materials. The introduction of dielectric materials into these problems necessarily modifies the problem-solving procedures, which in turn requires a reappraisal of the uniqueness of the resulting solutions. A perfectly general discussion would need to consider anisotropic and non-homogeneous dielectric materials, which may also exhibit hysteresis effects. However, the little that can be gained from maintaining such perfect generality does not justify the additional complications that would be introduced. For practical purposes it is sufficient to consider two homogeneous and isotropic materials with different dielectric constants occupying a closed space which is surrounded by some surface with known boundary conditions. For this discussion an air space or an empty space may be considered to be one of the dielectrics.

Figure 5–8 shows two dielectric materials, with dielectric constants K_1 and K_2, occupying a closed space on the surface of which the boundary

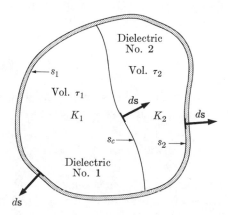

Fig. 5–8 Two linear dielectrics in a volume of general shape which is enclosed by a surface on which the electrostatic boundary conditions are known.

conditions are known. The outer surface in contact with dielectric No. 1 is designated by s_1, that in contact with dielectric No. 2 is designated by s_2, and the common surface separating the two is designated by s_c. The two dielectrics occupy volumes τ_1 and τ_2, respectively. If the boundaries have been chosen so that $\rho = 0$ in both τ_1 and τ_2,

$$\mathbf{V} \cdot \mathbf{D} = \mathbf{V} \cdot (-\epsilon \mathbf{V} V) = 0$$

throughout both regions, including the common boundary s_c. However, since ϵ changes abruptly at this boundary, $\nabla^2 V \neq 0$ on s_c, although $\nabla^2 V = 0$ in both τ_1 and τ_2. It is usually necessary to obtain one mathematical expression V_1 for the potentials in τ_1 and another expression V_2 for the potentials in τ_2. The boundary conditions on s_1 and s_2 must be satisfied by V_1 and V_2 respectively, and the boundary conditions developed in the last section must be satisfied along s_c by the two functions together. Following a procedure similar to that of Section 4–3, it will be assumed that two other functions of the coordinates, V_1' and V_2', meet these requirements and that two new functions $\phi_1 = V_1 - V_1'$ and $\phi_2 = V_2 - V_2'$ can be defined.

Throughout dielectric No. 1, $\nabla^2 V_1 = \nabla^2 V_1' = 0$ and hence $\nabla^2 \phi_1 = 0$. Likewise, throughout dielectric No. 2, $\nabla^2 V_2 = \nabla^2 V_2' = 0$, and hence $\nabla^2 \phi_2 = 0$. On s_1 either $\phi_1 = 0$ or $\mathbf{V} \phi_1 \cdot d\mathbf{s} = (\mathbf{V} V_1 - \mathbf{V} V_1') \cdot d\mathbf{s} = 0$, while on s_2 either $\phi_2 = 0$ or $\mathbf{V} \phi_2 \cdot d\mathbf{s} = (\mathbf{V} V_2 - \mathbf{V} V_2') \cdot d\mathbf{s} = 0$. On the common boundary, $V_1 = V_2$, $V_1' = V_2'$, and the normal components of $\mathbf{D}$ are continuous.

Considering dielectric No. 1 first, Gauss' theorem can be written as

$$\int_{s_1} \phi_1 \mathbf{V} \phi_1 \cdot d\mathbf{s} + \int_{s_c} \phi_1 \mathbf{V} \phi_1 \cdot d\mathbf{s} = \int_{\tau_1} \mathbf{V} \cdot (\phi_1 \mathbf{V} \phi_1) \, d\tau,$$

and because of the boundary conditions on s_1 the first integral is zero. Furthermore, the volume integral on the right can be written as

$$\int_{\tau_1} \mathbf{V} \cdot (\phi_1 \mathbf{V} \phi_1) \, d\tau = \int_{\tau_1} \phi_1 \nabla^2 \phi_1 \, d\tau + \int_{\tau_1} (\nabla \phi_1)^2 \, d\tau,$$

where the first integral on the right is zero because $\nabla^2 \phi_1 = 0$ throughout τ_1. The Gauss' theorem expression then reduces to

$$\int_{s_c} \phi_1 \mathbf{V} \phi_1 \cdot d\mathbf{s} = \int_{\tau_1} (\nabla \phi_1)^2 \, d\tau. \tag{5–11}$$

The analysis for the space inside τ_2 is similar to that just given for τ_1, but the $d\mathbf{s}$ in Gauss' theorem is conventionally chosen as positive along the outward-drawn normal of any closed space. If $d\mathbf{s}$ is chosen as pointing to the right along the surface s_c, as shown in Fig. 5–8, to keep the same element of area for both integrations, it will point into rather than out of τ_2, so that the surface integral for ϕ_2 must be given a negative sign:

$$-\int_{s_c} \phi_2 \nabla \phi_2 \cdot d\mathbf{s} = \int_{\tau_2} (\nabla \phi_2)^2 \, d\tau. \tag{5–12}$$

Multiplying Eq. (5–11) by K_1, Eq. (5–12) by K_2, and adding the two, we obtain

$$\int_{s_c} (\phi_1 K_1 \nabla \phi_1 - \phi_2 K_2 \nabla \phi_2) \cdot d\mathbf{s} = \int_{\tau_1} K_1 (\nabla \phi_1)^2 \, d\tau + \int_{\tau_2} K_2 (\nabla \phi_2)^2 \, d\tau. \tag{5–13}$$

Now since $V_1 = V_2$ and $V_1' = V_2'$ along the common boundary s_c, then $\phi_1 = \phi_2$ along s_c, so that ϕ_1 and ϕ_2 can be factored out of the integrand of the surface integral. Furthermore, the continuity of D_n along this boundary guarantees that

$$K_1 \nabla \phi_1 \cdot d\mathbf{s} - K_2 \nabla \phi_2 \cdot d\mathbf{s} = 0$$

along s_c, so that the surface integral of Eq. (5–13) is zero. Now $(\nabla \phi_1)^2$ is either positive or zero throughout τ_1 and $(\nabla \phi_2)^2$ is either positive or zero throughout τ_2, so that the two remaining integrals of Eq. (5–13) are separately zero. Following the arguments given in Section 4–3, these integrals can be zero only if ϕ_1 is constant in τ_1 and ϕ_2 is constant in τ_2, but since $\phi_1 = 0$ on s_1 and $\phi_2 = 0$ on s_2, ϕ_1 must be zero throughout τ_1 and ϕ_2 must be zero throughout τ_2. Thus the assumption that more than one set of solutions can be found to satisfy the conditions of the problem has led to absurdity, and any set of solutions that does satisfy all of the conditions is unique.

5–7 Dielectric sphere in a uniform field

A sphere made from a homogeneous and isotropic dielectric becomes polarized in an electric field. The effects produced by placing such a sphere in a uniform field are readily calculated with the assumption that $\mathbf{D} = \epsilon \mathbf{E}$ and $\epsilon = K\epsilon_0$ for the material of the sphere.

The origin of spherical coordinates can be placed at the center of this sphere of radius a, with the positive z-axis pointing in the direction of the field. Since there will be fields inside as well as outside the dielectric sphere, the potential inside will be designated by V_i and that outside by V_o. For values of r between zero and a, $\nabla^2 V_i = 0$, and for r between a and infinity, $\nabla^2 V_o = 0$. The boundary conditions at $r = a$, given by Eqs. (5–8) and (5–9), respectively, are

$$-\epsilon \frac{\partial V_i}{\partial r} = -\epsilon_0 \frac{\partial V_o}{\partial r} \tag{5–14}$$

and

$$-\frac{\partial V_i}{\partial \theta} = -\frac{\partial V_o}{\partial \theta}. \tag{5–15}$$

An outer boundary for the region can be chosen where $r \gg a$, sufficiently far from the sphere to make its influence on the uniform field negligible. On this outer boundary, then,

$$V_o = -E_0 z = -E_0 r \cos \theta. \tag{5–16}$$

As in the conducting sphere problem of Section 4–4, symmetry suggests that the following solutions of Laplace's equation should fit the boundary conditions:

$$V_o = A_0 + \frac{B_0}{r} + A_1 r \cos \theta + \frac{B_1}{r^2} \cos \theta$$

and

$$V_i = C + F r \cos \theta.$$

No terms with r in the denominator are included in V_i, because these would be infinite at $r = 0$. The boundary condition, Eq. (5–16), at $r \gg a$ gives

$$A_0 + A_1 r \cos \theta = -E_0 r \cos \theta,$$

and since this must hold for all values of θ, $A_1 = -E_0$ and $A_0 = 0$. The other two boundary conditions are more easily satisfied if these constants are introduced immediately, to give

$$V_o = \frac{B_0}{r} - E_0 r \cos \theta + \frac{B_1}{r^2} \cos \theta.$$

The conditions at the sphere's surface are satisfied by performing the differentiations of Eqs. (5–14) and (5–15) and then substituting $r = a$ into the equations that are obtained. Equation (5–14), after dividing through by ϵ_0, becomes

$$\frac{B_0}{a^2} + E_0 \cos \theta + \frac{2B_1}{a^3} \cos \theta = -KF \cos \theta$$

and Eq. (5–15) becomes

$$-E_0 a \sin \theta + \frac{B_1}{a^2} \sin \theta = Fa \sin \theta.$$

The fact that these equations must be valid for all values of θ requires that

$$B_0 = 0,$$

$$E_0 + \frac{2B_1}{a^3} = -KF \quad \text{and} \quad -E_0 a + \frac{B_1}{a^2} = Fa.$$

The two remaining unknowns, B_1 and F, can now be obtained from the last two equations, from which

$$B_1 = E_0 a^3 \frac{K-1}{K+2} \quad \text{and} \quad F = -E_0 \frac{3}{K+2}.$$

The two potentials can now be written as

$$V_o = -E_0 r \cos \theta + E_0 a^3 \frac{K-1}{K+2} \frac{\cos \theta}{r^2} \tag{5-17a}$$

and

$$V_i = -E_0 \frac{3}{K+2} r \cos \theta. \tag{5-17b}$$

Since these satisfy Laplace's equation and the boundary conditions, they are the correct solutions. The potential V_o shows that the field outside the sphere is the sum of the original uniform field plus the field of an ideal dipole at the origin. From Eq. (2–24), the equivalent dipole moment is

$$p = \epsilon_0 E_0 4\pi a^3 \frac{K-1}{K+2}. \tag{5-18}$$

It is interesting that the equivalent dipole moment is proportional to the applied field and to the volume of the sphere, as was the case for the conducting sphere in Section 4–4.

The external and internal fields can be obtained from the gradients of the potentials of Eqs. (5–17a) and (5–17b). The external field is

$$\mathbf{E} = -\nabla V_o = E_0 \mathbf{k} + E_0 \frac{K-1}{K+2} \frac{a^3}{r^3} [\mathbf{a}_r \, 2 \cos \theta + \mathbf{a}_\theta \sin \theta]$$

or

$$\mathbf{E} = E_0 \mathbf{k} + E_0 \frac{K-1}{K+2} \frac{a^3}{r^3} \mathbf{d}_f,$$

and the internal field is

$$\mathbf{E}_i = -\nabla V_i = E_0 \frac{3}{K+2} \mathbf{k}. \tag{5-19}$$

The field inside the sphere is a uniform field smaller than E_0, since $K > 1$. Expressions for the displacement vector outside and inside the sphere may be obtained by multiplying $\mathbf{E}$ by ϵ_0 and $\mathbf{E}_i$ by $K\epsilon_0$. This gives

$$\mathbf{D} = D_0 \mathbf{k} + D_0 \frac{K-1}{K+2} \frac{a^3}{r^3} [\mathbf{a}_r \, 2 \cos \theta + \mathbf{a}_\theta \sin \theta]$$

and

$$\mathbf{D}_i = E_0 \epsilon_0 \frac{3K}{K+2} \mathbf{k}.$$

Since K is greater than unity in a dielectric medium, D_i is greater than the applied displacement D_0, while E_i is less than the applied field E_0, as shown in Fig. 5–9. Physically, the internal reduction in $\mathbf{E}$ is caused by the polarization surface charge induced on the sphere, which opposes the applied field. The increase in $\mathbf{D}$ results from the requirement that lines of $\mathbf{D}$ be con-

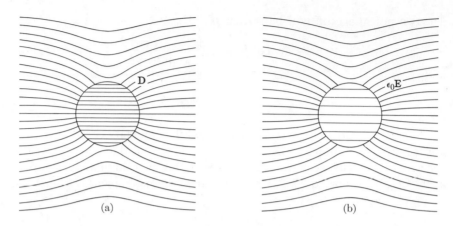

Fig. 5–9 Dielectric sphere with $K = 6$ in a uniform electric field. (a) The **D** lines are continuous and have greatly increased density inside the sphere. (b) Outside the sphere $\epsilon_0\mathbf{E} = \mathbf{D}$, but inside the sphere $\epsilon_0\mathbf{E}$ is considerably reduced by the polarization charges on the surface.

tinuous throughout space when no real charge is present. With the external **D** lines converging on the sphere, the **D** lines within the sphere must be more dense.

A check on the foregoing results can be obtained by substituting $K = \infty$, which will be found to give the relations obtained in Section 4–4 for a conducting sphere in a uniform field. In purely static fields conductors behave like dielectrics with an infinite dielectric constant. A further check is obtained by substituting, for a vacuum, the value of $K = 1$, which makes the effect due to the sphere disappear. Checks of this sort are invaluable for eliminating mistakes in calculations.

5–8 Dielectric cylinder in a uniform field

An infinitely long right-circular cylinder, placed with its axis perpendicular to a uniform electric field, modifies that field in much the same way as does the sphere of the last section. If the cylinder has radius a and is made of the same homogeneous and isotropic dielectric as was the sphere, the modification of the field will be qualitatively the same though numerically different. The z-axis of the coordinate system may be set to coincide with the cylindrical axis, while the x-axis is set to point in the direction of the uniform field whose strength is E_0. With these choices, the usual cylindrical coordinates of Fig. 2–4 can be used, as they were for the conducting cylinder problem of Section 4–5.

The solution for this problem follows the form used in the previous section. The boundary condition equations at $r = a$ are similar to Eqs.

(5–14) and (5–15):

$$-\epsilon \frac{\partial V_i}{\partial r} = -\epsilon_0 \frac{\partial V_o}{\partial r}$$

and

$$-\frac{\partial V_i}{\partial \phi} = -\frac{\partial V_o}{\partial \phi}.$$

The boundary condition equation at $r \gg a$ is similar to Eq. (5–16):

$$V_o = -E_0 x = -E_0 r \cos \phi.$$

From the cylindrical harmonics of Table 4–2, the following potential representations are chosen as being most likely to fit the boundary conditions:

$$V_o = Ar \cos \phi + \frac{B \cos \phi}{r}$$

and

$$V_i = Cr \cos \phi.$$

No terms with r raised to negative exponents are used for V_i because they become infinite at $r = 0$. Terms of the form $\sin n\phi$ or $\cos n\phi$ (where $n > 1$), as well as the $\ln r$ term, have been omitted because the boundary condition equation at $r \gg a$ does not suggest the need for these terms. If such terms were included in the potential expressions now, later work would prove their coefficients to be zero.

Application of the boundary condition equations to the assumed solutions gives

$$A = -E_0, \qquad B = \frac{K-1}{K+1} a^2 E_0, \qquad \text{and} \qquad C = -\frac{2E_0}{K+1}.$$

Thus the boundary conditions are satisfied by the assumed solutions that also satisfy Laplace's equation, and are, therefore, the correct solutions. The potentials then are

$$V_o = -E_0 r \cos \phi + \frac{(K-1)a^2 E_0}{(K+1)r} \cos \phi$$

and

$$V_i = -\frac{2}{K+1} E_0 r \cos \phi = -\frac{2E_0 x}{K+1}.$$

The external and internal fields can be obtained from the gradients of these potentials. The external field is

$$\mathbf{E} = E_0 \mathbf{i} + E_0 \frac{K-1}{K+1} \frac{a^2}{r^2} (\mathbf{a}_r \cos \phi + \mathbf{a}_\phi \sin \phi)$$

and the internal field is

$$\mathbf{E}_i = E_0 \frac{2}{K + 1} \mathbf{i}. \tag{5–20}$$

It is left as an exercise to obtain the displacement vector and to compare these results with those for the conducting cylinder in Section 4–5.

5–9 Microscopic electric fields

All discussions thus far have utilized the concept of average fields rather than considering the complex subject of fields near individual atoms. A detailed treatment of this subject is necessarily beyond the scope of this text; however, the nature of such fields can be determined for crystal lattices with cubic symmetry.

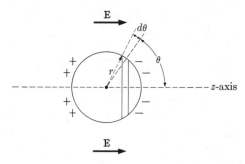

Fig. 5–10 Fictitious hollowed-out sphere of radius r is shown centered about an atom in the dielectric.

Consider a dielectric material completely filling all space, as is shown in Fig. 5–10. The macroscopic field in the dielectric is $\mathbf{E}$ but it is desired to find the microscopic field $\mathbf{E}_a$ that acts on an individual atom. The exact form of $\mathbf{E}_a$ depends upon the electric fields of the surrounding polarized atoms, being the vector sum of $\mathbf{E}$ and the fields due to all the atoms within the dielectric. In calculating the magnitude of this field it is desirable to draw an imaginary sphere around one atom to separate the distant atoms from the near neighbors as shown in Fig. 5–10. It is necessary that r be large compared with the distance between atoms. The atoms outside the sphere can be treated as a continuum, while the atoms inside the sphere must be treated individually.

The field acting on the central atom can then be expressed by

$$\mathbf{E}_a = \mathbf{E}_1 + \mathbf{E}_2,$$

where $\mathbf{E}_1$ is the field in the fictitious hollow sphere of radius r due to the external field $\mathbf{E}$ and the surrounding polarized dielectric, and $\mathbf{E}_2$ is the sum of the dipole fields of all the atoms within the sphere. If a real spherical cavity were introduced into the dielectric, the polarization of the atoms remaining near the cavity would be distorted. Since no real cavity exists, the calculation of the field inside the fictitious cavity must be based upon a uniform polarization for all atoms. With the field and the polarization directed to the right in Fig. 5–10, the inner surface of the spherical cavity has the polarization-charge density $\sigma_p = P_n = -P \cos \theta$. The field at the center of this cavity due to this charge density can be obtained by writing the field due to the ring element of charge, shown in the diagram, and integrating over the surface of the sphere from $\theta = 0$ to $\theta = \pi$. Since the field due to the ring element is along the z-axis, it is given by

$$dE_p = \frac{\sigma_p 2\pi r^2 \cos \theta \sin \theta \, d\theta}{4\pi\epsilon_0 r^2} .$$

Substituting $\sigma_p = -P \cos \theta$, simplifying, and integrating, we obtain

$$E_p = -\frac{P}{2\epsilon_0} \int_0^\pi \cos^2 \theta \sin \theta \, d\theta = \frac{P}{3\epsilon_0} .$$

The field $\mathbf{E}_1$ is the sum of the applied field and the field due to the distant atomic dipoles; that is,

$$\mathbf{E}_1 = \mathbf{E} + \frac{\mathbf{P}}{3\epsilon_0} .$$

To this must be added the field $\mathbf{E}_2$, due to the atomic dipoles within the fictitious sphere.

The electric potential due to these atoms may be written from Eq. (2–24) as

$$V = \sum_i \frac{p_i \cos \theta_i}{4\pi\epsilon_0 r_i^2}$$

$$= \frac{p}{4\pi\epsilon_0} \sum_i \frac{z_i}{(x_i^2 + y_i^2 + z_i^2)^{3/2}} ,$$

where the sum is taken over all atoms within the sphere and each atom is assumed to be polarized by the same amount, p. The z-component of $\mathbf{E}_2$ is given by

$$E_{2z} = -\frac{\partial V}{\partial z}$$

$$= \frac{p}{4\pi\epsilon_0} \sum_i \frac{3z_i^2 r_i - r_i^3}{r_i^6} .$$

This is zero in a cubic-crystal lattice because, by symmetry,

$$\sum_i \frac{x_i^2}{r_i^5} = \sum_i \frac{y_i^2}{r_i^5} = \sum_i \frac{z_i^2}{r_i^5}$$

and therefore

$$\sum_i \frac{r_i^3}{r_i^6} = \sum_i \frac{3z_i^2 r_i}{r_i^6}.$$

The x- and y-components are

$$E_{2x} = -3 \sum_i x_i z_i r_i^{-5} \quad \text{and} \quad E_{2y} = -3 \sum_i y_i z_i r_i^{-5},$$

which are also zero because in a cubic lattice there is a negative value corresponding to each positive value for x, y, or z, so that the sum over all values of the products $x_i z_i$ and $y_i z_i$ must be zero.

The microscopic field on an individual atom in a crystal lattice with cubic symmetry has thus been shown to be

$$\mathbf{E}_a = \mathbf{E}_1 = \mathbf{E} + \frac{\mathbf{P}}{3\epsilon_0},$$

where $\mathbf{E}$ is the uniform macroscopic field through the dielectric. This simple calculation shows that the polarizability of individual atoms is somewhat less than that of the bulk material, since $\mathbf{E}_a$ is larger than $\mathbf{E}$. This result is important in the study of the electronic properties of solids. In particular, it can be shown that lack of cubic symmetry in many dielectrics results in a nonzero value for $\mathbf{E}_2$, and a corresponding nonlinear or even nonisotropic enhancement of the polarization of the material. Some properties of nonlinear and nonisotropic dielectrics are discussed in Sections 5–5 and 5–12.

5–10 Energy relations in static electric fields

As in all other branches of physics, energy relations in electrostatics constitute a powerful tool for solving many problems. It is, therefore, desirable to obtain an energy expression to describe any general configuration of charges and dielectrics in space. For most purposes, an expression for the energy required to produce a given change in the charge distribution or in the geometry is sufficient. A general expression for the energy required to change electric and magnetic fields as a function of time and coordinates will be discussed in Chapter 10; this section treats only the energies required to make slow changes in electrostatic configurations. However, these results will be useful for establishing the more general case.

A basic concept which is to be demonstrated is that the work required to make changes in electrostatic situations can be expressed in terms of the changing electric fields, thus strongly suggesting that all electrostatic energy resides in the electric field. This concept is a great aid to visualization of energy problems, especially in electromagnetic waves, and it does not lead to inconsistencies. Before treating the most general case, the charging of a simple capacitor will be discussed.

Work done in charging capacitors. The simplest capacitor is one with two parallel plates of metal of the same area A, placed very close together compared with their linear dimensions. The region between the plates may either be evacuated or filled with a homogeneous medium, not necessarily isotropic. Ignoring the region near the edges of the plates, the electric field $\mathbf{E}$ between the plates is uniform and is perpendicular to their surfaces. If d is the separation and V is the potential difference between the plates, $V = Ed$. The facing surfaces of the plates have uniform charge densities σ on the one with higher potential, and $-\sigma$ on the other, and from Eq. (5–10) the normal component of $\mathbf{D}$ is $D_n = \sigma$. If a very small charge δq is transported from the plate with lower potential to the one with higher potential, the work done by the battery or other device affecting the transfer is $\delta W = V \, \delta q$. Since $q = A\sigma$, so that $\delta q = A \, \delta\sigma = A \, \delta D_n$, the work done can be stated as

$$\delta W = A \, d \, E \, \delta D_n.$$

The volume of the space occupied by these uniform fields is $\tau = A \, d$ and hence

$$\delta W = \tau E \, \delta D_n = \tau \mathbf{E} \cdot \delta \mathbf{D}.$$

The work done in changing the charge on this simple capacitor has been expressed in terms of the fields and the total volume occupied by them. For this special case, the work done per unit volume to increase the charge by δq is $\mathbf{E} \cdot \delta \mathbf{D}$.

This expression is true also in the more complicated case of charging a spherical capacitor. Such a capacitor may consist of a conducting sphere of radius a, concentrically surrounded by a conducting spherical shell of inner radius b. The space between the spheres may be filled either with a homogeneous dielectric or with one whose properties vary only with the radius. Rather than calculating the work done in the same manner as for the parallel-plate capacitor, the work can be calculated from the expression

$$\delta W = \int \mathbf{E} \cdot \delta \mathbf{D} \, d\tau$$

to see if it leads to inconsistencies.

The diagram of Fig. 5–11 shows a thin spherical shell of radius r and thickness dr which can be used as a general volume element $d\tau = 4\pi r^2 \, dr$, since the fields and the dielectric properties depend only upon r. Applying Gauss' law to the space surrounded by the volume element gives $q = 4\pi r^2 \, D_n$, which leads directly to $\delta D_n = \delta q / 4\pi r^2$, and since $\mathbf{E}$ is necessarily normal to this equipotential surface, the expression for the work done becomes simply

$$\delta W = \int_{r=a}^{r=b} E \, \frac{\delta q}{4\pi r^2} \, 4\pi r^2 \, dr,$$

which reduces to

$$\delta W = \delta q \int_{r=a}^{r=b} E \, dr = \delta q \, V,$$

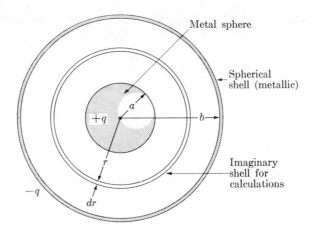

Fig. 5–11 Concentric-sphere capacitor containing a dielectric medium whose properties vary only with the radius.

where V is the potential difference between the metallic spheres. Since this is just the expression for the work done in transporting a charge δq across a potential difference of V, no inconsistency has resulted from using the assumed expression for work done in terms of the fields.

Note that this development is valid when the outside sphere is made very large; that is, it is valid when $b \to \infty$ and the sphere of radius a becomes an isolated and charged sphere. In this case, the total work required to transfer δq from infinity to the charged sphere is obtained by integrating $\int \mathbf{E} \cdot \delta \mathbf{D} \, d\tau$ from $r = a$ to $r \to \infty$. The development is also valid when the inner radius a is reduced to very small though finite dimensions. However, when the radius is reduced to the dimensions of an atomic nucleus, $\mathbf{P}$ becomes meaningless, and $\mathbf{D}$ becomes $\epsilon_0 \mathbf{E}$.

These examples suggest that the work done in creating certain electrostatic configurations can be calculated on the assumption that all electrostatic energy resides in the field and that the work required to increase the fields at any point is given by $\mathbf{E} \cdot \delta \mathbf{D}$. To prove that this is true in all cases requires a more general treatment.

Work required to produce arbitrary field configurations. The determination of the energy required to produce any arbitrary electrostatic field configuration involves conductors, nonhomogeneous and nonisotropic media with an arbitrary distribution of charges.

For the sake of definiteness, consider problems in which all of the charge is contained within a finite region of space, which can be represented by a real charge density ρ, a function of the coordinates. There may be polarization charges associated with dielectrics. Let us calculate the energy of the system by calculating the work required to put together the specified charge distribution.

First, calculate the work required to produce a small change $\delta\rho$ in the charge density at each point. In general, $\delta\rho$ will be different for each point in the space. For a volume $d\tau$, the change in charge in the volume element will be $\delta\rho\, d\tau$. If the potential in that element is V, then the work required to bring this charge from infinity is $V \delta\rho\, d\tau$, assuming $V = 0$ at infinity. The work required to add $\delta\rho$ to the charge density throughout the volume is

$$\delta W = \int_\tau V\, \delta\rho\, d\tau.$$

With Maxwell's first relation, Eq. (5–4), this can be written

$$\delta W = \int_\tau V\, \delta(\mathbf{\nabla} \cdot \mathbf{D})\, d\tau = \int_\tau V\mathbf{\nabla} \cdot (\delta\mathbf{D})\, d\tau,$$

where $\delta\mathbf{D}$ is the change in displacement at each point. From the vector identity found in Eq. (3–20),

$$\mathbf{\nabla} \cdot (V\, \delta\mathbf{D}) = \mathbf{\nabla}V \cdot \delta\mathbf{D} + V\mathbf{\nabla} \cdot (\delta\mathbf{D}).$$

Using this and Gauss' theorem, we obtain

$$\delta W = \int_\tau \mathbf{\nabla} \cdot (V\, \delta\mathbf{D})\, d\tau - \int_\tau \mathbf{\nabla}V \cdot (\delta\mathbf{D})\, d\tau$$

or

$$\delta W = \int_s V\, \delta\mathbf{D} \cdot d\mathbf{s} - \int_\tau \mathbf{\nabla}V \cdot (\delta\mathbf{D})\, d\tau.$$

The surface integral in this equation is negligible. It should be calculated over a closed surface s completely surrounding the region of interest. For our earth it can be taken over a spherical surface with a radius r much larger than that of the earth. According to the results of Problem 5-41, any net charge in this earth region will produce $\mathbf{D}$'s and V's that fall off at least as rapidly as r^{-2} and r^{-1} respectively. Since $s = 4\pi r^2$, the surface integral is proportional to r^{-1}, which approaches zero as r becomes very large. This is obviously valid for other regions in space. Thus

$$\delta W = -\int_\tau \mathbf{\nabla}V \cdot (\delta\mathbf{D})\, d\tau = \int_\tau \mathbf{E} \cdot \delta\mathbf{D}\, d\tau.$$

This can be interpreted as saying that it requires an amount of work

$$\delta w = \mathbf{E} \cdot \delta\mathbf{D} \tag{5–21}$$

per unit volume to change the displacement at any point by $\delta\mathbf{D}$. The expression $\mathbf{E} \cdot \delta\mathbf{D}$ represents the change in energy density in the volume.

5–11 Total energy to create static electric fields

The total volume density of energy required to create a field at a point can be obtained directly by integrating Eq. (5–21):

$$w = \int_0^D \mathbf{E} \cdot d\mathbf{D} \tag{5–22}$$

where it is assumed that $\mathbf{D} = 0$ initially. The actual integration can be performed, however, only if the functional relationship between $\mathbf{D}$ and $\mathbf{E}$ is known.

Linear isotropic dielectrics. In the special case where the dielectric is linear and isotropic at a given point, so that $\mathbf{D} = K\epsilon_0\mathbf{E}$, with K being a true constant for the material, the volume density of energy required to produce the field may be expressed by

$$w = \int_0^E K\epsilon_0 E \, dE = \tfrac{1}{2}K\epsilon_0 E^2, \tag{5–23}$$

where $K > 1$ in dielectric materials and $K = 1$ in a vacuum. In a non-homogeneous material, K will have different values at different points, but if K is independent of $\mathbf{E}$, as is assumed here, all the energy of Eq. (5–23) is recovered when $\mathbf{E}$ goes to zero. The constancy of K implies that the forces opposing charge displacement in the dielectric are of the "elastic" type, in which displacements are directly proportional to the field.

Of special interest is the fact that energy exists in the field even in a vacuum, where $K = 1$. More energy is required to create a field $\mathbf{E}$ of given intensity in a dielectric than in a vacuum. The *additional* energy required is called the *intrinsic energy*. In a linear isotropic dielectric this is given by

$$w_{\text{int}} = \tfrac{1}{2}(K - 1)\epsilon_0 E^2 = \tfrac{1}{2}\chi\epsilon_0 E^2.$$

Nonlinear isotropic dielectrics. A nonlinear dielectric is one in which the polarizabilities change with changing electric field. Frequently, a dielectric will exhibit rather large polarizations in weak fields, although larger fields do not produce proportionately larger polarizations. The polarization then is said to saturate at these higher fields. A typical curve of D versus E for such a material is shown in Fig. 5–12(a). The volume density of work done in increasing D from 0 to D_1 is given by Eq. (5–22) as

$$w = \int_0^{D_1} E \, dD,$$

where the vector notation is omitted because $\mathbf{D}$, $\mathbf{E}$, and $d\mathbf{D}$ are always parallel in an isotropic medium between parallel plates. The integral of $E \, dD$ from 0 to D_1 for this material is the shaded area to the left of the curve. The curve for a linear dielectric having the same values of E_1 and D_1 is shown in Fig.

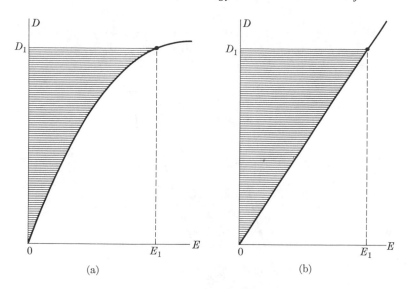

Fig. 5–12 Curves of D versus E for isotropic dielectric materials. The shaded areas represent the work per volume required to produce the displacement D_1 (a) in nonlinear materials and (b) in linear materials.

5–12(b), where the work required to produce the field is represented by the shaded area to the left of the curve. In this example, the volume density of energy required to produce the fields D_1 and E_1 is considerably larger in the linear dielectric than for the nonlinear example of Fig. 5–12(a), which points up the fact that Eq. (5–23) is applicable only when the polarizability is independent of the field. The difference between these two energy densities is due to the action of nonelastic forces within the material.

Hysteresis effects. Many nonlinear dielectric materials exhibit hysteresis effects, as shown in Fig. 5–13; that is, after being initially polarized to a value D_1 in a field E_1, a residual polarization represented by point b remains even after E has been decreased to zero. Increasing the electric field in the reverse direction finally reduces D to zero at point c, and when the field becomes $-E_1$ the displacement becomes $-D_1$, as indicated by the point d which is equivalent to point a in the first quadrant. As the field is slowly* reversed again, this time from $-E_1$ to E_1, the displacement takes on values from points d to e to f and returns to D_1 at point a. As the field E is repeatedly reversed from $+E_1$ to $-E_1$ and back again, the value of D traces out the hysteresis loop of Fig. 5–13.

---

* For these statements to be true the field changes must be slow enough for any heat energy released within the material to be dissipated. If the temperature increases, the properties of the dielectric usually change.

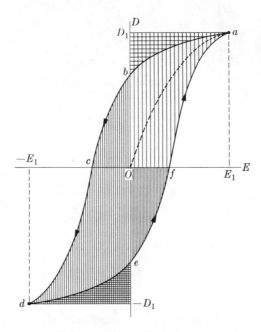

Fig. 5–13 Hysteresis loop for isotropic ferroelectric materials. Areas representing the volume density of work done on the materials are shaded vertically, while horizontal shading indicates the volume energy regained.

The hysteresis effect is caused by irreversible polarization effects within the material which dissipate energy in the form of heat. This loss of energy can be calculated by evaluating Eq. (5–21) over one complete cycle. The integral of $E\,dD$ evaluated from f to a gives the area in the first quadrant enclosed by curves connecting the points f, a, D_1, and O, which is represented by vertical shading. In going from a to b, E is positive but dD is negative, so that the resulting area represented by horizontal shading lines in the first quadrant must be subtracted from the first area obtained. The net area after this subtraction is just the area enclosed by the hysteresis loop in the first quadrant. In the second quadrant both E and dD are negative, so that the resulting area represented by vertical shading must be added to the area already obtained. After the integrations from c to d, d to e, and e to f have been performed and added to the previous results, the total area will be just the area enclosed by the complete hysteresis loop. The total volume density of energy in J m^{-3} dissipated within the material is therefore given by the area enclosed by the hysteresis loop in the D versus E coordinate axes, with D in C m^{-2} and E in V/m.

Dielectric materials exhibiting hysteresis effects are known as *ferroelectric materials*, because their polarization versus electric field curves are so much like the magnetization versus magnetic field curves of ferromagnetic materials like iron. These materials usually have a rather complex chemical formula and crystalline structure, as for example Rochelle salt,

$$NaK(C_4H_4O_6)\cdot4H_2O,$$

and barium titanate, $BaTiO_3$. Interest in these materials has developed only in recent years, while interest in ferromagnetism has existed for centuries. Because a great deal more is known about ferromagnetic than ferroelectric materials, a detailed study of hysteresis effects will be postponed until the study of magnetic materials in Chapter 9.

Linear anisotropic dielectrics. Materials that polarize more easily in some directions than in others are anisotropic. Such materials were discussed briefly in Section 5–4, where the nine dielectric constants were defined. While these many constants are important in high-frequency fields, electrostatic problems can usually be solved by more simple methods. If the space between the plates of a parallel-plate capacitor is filled with a homogeneous but anisotropic dielectric material, the boundary conditions require that the **E** vector be normal to the plates, although the **P** and **D** vectors are usually directed at angles to the normal. At any point the final fields can be designated by $\mathbf{E}_f$, $\mathbf{P}_f$, and $\mathbf{D}_f$, and since **E** does not change direction, it can be written as $\mathbf{E} = \gamma\mathbf{E}_f$, where γ increases from zero to unity during the charging process. With linear dielectrics the other two fields also can be written as $\mathbf{P} = \gamma\mathbf{P}_f$ and $\mathbf{D} = \gamma\mathbf{D}_f$. When γ increases by an infinitesimal $d\gamma$, **D** increases by $d\mathbf{D} = \mathbf{D}_f\, d\gamma$ and the work per unit volume required to produce this increase is given by

$$dw = \mathbf{E} \cdot d\mathbf{D} = \mathbf{E}_f \cdot \mathbf{D}_f\gamma\, d\gamma.$$

Integration gives

$$w = \int_0^1 \mathbf{E}_f \cdot \mathbf{D}_f\gamma\, d\gamma = \frac{\mathbf{E}_f \cdot \mathbf{D}_f}{2}, \qquad (5\text{–}24)$$

which is a general expression for the work required to set up the fields $\mathbf{E}_f$ and $\mathbf{D}_f$ in any linear dielectric, since it includes Eq. (5–23) as a special case. The energy represented by Eq. (5–24) is all recovered when a capacitor with a dielectric material that polarizes linearly is slowly discharged. Further study shows that the energy given by Eq. (5–24) can be identified as the "free energy" for thermodynamic studies.

5–12 Electric dipoles

Ferroelectric materials having D versus E curves like the one shown in Fig. 5–12 remain polarized after the external field is removed, in somewhat the same way that a permanent magnet remains magnetized. The condition is usually so unstable that this polarization does not last long.

Electrets. More or less permanent polarization can be produced in certain waxes by first melting and then freezing them in a strong electric field. The permanent dipoles become oriented in the molten state and this orientation becomes frozen in on solidification. These permanently polarized objects are the electric equivalent of the permanent magnet, and are called *electrets*.

The electric field produced by an electret is the vector sum of the fields produced by all of the individual dipoles, which is just the field of a dipole of larger dimensions. For example, suppose a cylindrical rod of dielectric material is uniformly polarized parallel to its axis. The effects of all the internal charges cancel and the field is just that produced by the polarization charges appearing on the two end surfaces. The electric field and potential at a considerable distance from this rod are very closely approximated by Eqs. (2–4) and (2–24), but near such large dipoles the potential must be calculated from

$$V = \int_\tau \frac{\rho_p \, d\tau}{4\pi\epsilon_0 r}$$

integrated over the dipole, where $\rho_p = -\boldsymbol{\nabla} \cdot \mathbf{P}$ is the actual distribution of polarization charges and r is the distance of $d\tau$ from the general field point.

Dipoles in electric fields experience forces and torques that can be calculated from $\mathbf{F} = \int_\tau \mathbf{E}\rho_p \, d\tau$, where $\mathbf{E}$ is the field at $d\tau$ that is obtained by ignoring the field of the dipole itself. These forces and torques can also be calculated from the energy of the dipole in the field, $U = \int_\tau V\rho_p \, d\tau$, where V is the potential at $d\tau$, and the potential of the dipole is ignored. Here the dipole is considered rigid and the energy required to create it is not included. In the special case where large "point" charges, q and $-q$, are separated a very small distance $\mathbf{l}$, so that the dipole moment $\mathbf{p} = q\mathbf{l}$ is finite, the potential at $-q$ may be V while that at $+q$ is $V + \delta V$. The energy of the dipole is then

$$U = q(V + \delta V) - qV = q\,\delta V.$$

When $\mathbf{l}$ is very small

$$\delta V = \boldsymbol{\nabla} V \cdot \mathbf{l},$$

and then

$$U = \boldsymbol{\nabla} V \cdot q\mathbf{l} = -\mathbf{E} \cdot \mathbf{p} = -Ep \cos \theta, \tag{5–25}$$

where θ is the angle between $\mathbf{E}$ and $\mathbf{p}$. If the torque on the dipole is L, the work required to produce an infinitesimal rotation $d\theta$ is

$$-L \, d\theta = \left(\frac{\partial U}{\partial \theta}\right) d\theta,$$

or

$$L = -\frac{\partial U}{\partial \theta} = -Ep \sin \theta. \tag{5–26}$$

This same expression is obtained from a consideration of the forces on the

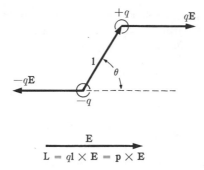

Fig. 5–14 A small electric dipole in a uniform field, showing that the torque is $L = p \times E$.

individual charges shown in Fig. 5–14, where the force Eq, parallel to the field, is separated a perpendicular distance $l \sin \theta$ from an equal but opposite force. The torque due to this couple is $(-Eq)(l \sin \theta) = -Ep \sin \theta$, which may also be expressed by $L = p \times E$. In a uniform field the two forces are exactly equal and opposite, leaving no net force on the dipole. In a non-uniform field, however, these forces are not equal and opposite and there is a net force. This is best calculated from the energy of the dipole. If F is the force exerted by the field on the dipole, the work required to move the dipole an infinitesimal distance $d\mathbf{l}$ is

$$-F \cdot d\mathbf{l} = dU = \nabla U \cdot d\mathbf{l},$$

or from Eq. (5–25)

$$F = -\nabla U = \nabla(E \cdot p). \tag{5–27}$$

Since p is constant, this force is proportional to the gradient of the component of E that is parallel to the positive direction of p. The force points in the direction in which this component of E increases most rapidly.

PROBLEMS

5–1 Two oppositely charged conducting plates, having numerically equal quantities of charge per unit area, are separated by a dielectric 5 mm thick, of dielectric constant 3. The resultant electric intensity in the dielectric is 10^6 V/m. Compute (a) the free charge per unit area on the conducting plates, (b) the induced charge per unit area on the surfaces of the dielectric.

5–2 The paper dielectric in a paper and foil capacitor is 0.005 cm thick. Its dielectric constant is 2.5 and its dielectric strength is 50×10^6 V/m.

a) What area of paper, and of tinfoil, is required for a 0.1-μF capacitor?

b) If the electric intensity in the paper is not to exceed one-half the dielectric strength, what is the maximum potential difference that can be applied across the capacitor?

5–3 The capacitance of a variable radio capacitor can be changed from 50 pF to 950 pF by turning the dial from $0°$ to $180°$. With the dial set at $180°$ the capacitor is connected to a 400-V battery. After charging the capacitor is disconnected from the battery and the dial is turned to $0°$.
a) What is the charge on the capacitor?
b) What is the potential difference across the capacitor when the dial reads $0°$?
c) What is the energy of the capacitor in this position?
d) How much work is required to turn the dial, if friction is neglected?

5–4 An air capacitor, consisting of two closely spaced parallel plates, has a capacitance of 1000 pF. The charge on each plate is 1 μC.
a) What is the potential difference between the plates?
b) If the charge is kept constant, what will be the potential difference between the plates if the separation is doubled?
c) How much work is required to double the separation?

5–5 A 20-μF capacitor is charged to a potential difference of 1000 V. The terminals of the charged capacitor are then connected to those of an uncharged 5-μF capacitor. Compute (a) the original charge of the system, (b) the final potential difference across each capacitor, (c) the final energy of the system, and (d) the decrease in energy when the capacitors are connected.

5–6 A large plane parallel capacitor is half filled with a uniform and homogeneous dielectric having the dielectric constant K. The conducting surfaces $x = -a$ and $x = a$ have potential V and $-V$, respectively, and $\epsilon = \epsilon_0$ where $-a < x < 0$, and $\epsilon = K\epsilon_0$ where $0 < x < a$.
a) Find E and D where $-a < x < 0$.
b) Find E and D where $0 < x < a$.
c) Locate all charges and specify whether they are real or polarization charges.

5–7 Calculate the radius of the first Bohr orbit of the hydrogen atom, by equating the Coulomb force between the electron and the nucleus to the centripetal force. Using the quantum condition that the angular momentum is $h/2\pi$, show that this radius is approximately one-half angstrom.

5–8 A very simple calculation for the binding energy per molecule of an NaCl crystal assumes that the total energy arises from the Coulomb attraction between the Na^+ ion and the nearest Cl^- ion. Using the known ionic separation of 2.81×10^{-10} m, show that this simple calculation gives a result close to $\frac{2}{3}$ of the measured value of 12.8×10^{-19} J/molecule.

5–9 Two homogeneous and isotropic dielectrics are separated by a plane interface. This interface is given a uniform density of real charges $\sigma(C\ m^{-2})$.
a) Are the normal components of **D** equal on the two sides of the surface?
b) If not, what are the boundary conditions on **D** at this interface? Presume the fields to proceed from the first to the second dielectric.

5–10 A conducting sphere of radius a is surrounded by a dielectric of inner radius a and outer radius b whose dielectric constant is K. Consider four different field points: (1) just inside the dielectric at r slightly greater than a, (2) inside the dielectric at r slightly less than b, (3) just outside the dielectric at r slightly greater than b, and (4) at some general point a distance r from the center of the sphere where $r > b$.
a) At each of these four points, calculate the displacement **D**, the electric field **E**, the polarization **P**, and the potential V, when q is the charge on the sphere.
b) Plot V and E versus r.

5-11

a) Calculate σ_p on the inner and outer surfaces of the dielectric shell in Problem 5–10.

b) Show that the field **E** inside this dielectric shell can be calculated from two points of view, namely, from $\mathbf{E} = \mathbf{D}/K\epsilon_0$ or from the knowledge of the magnitude and location of the polarization charges, otherwise ignoring the presence of the dielectric.

5-12 Assume that the dielectric surrounding the charged sphere in Problem 5–11 has a dielectric constant given by $K = 4e^{-[(r/a)-1]^2}$.

a) Calculate **D**, **P**, **E**, and ρ_p as functions of r inside the dielectric.

b) Calculate the total polarization volume charge if $b = 2a$, and compare this with the algebraic sum of the polarization surface charges on the inner and outer surfaces.

c) Plot ρ_p versus r.

5-13 Assume that the dielectric constant in Problem 5–10 has the radial variation given by $K = f(r)$ and calculate **D**, **P**, **E**, and ρ_p as functions of r inside the dielectric.

5-14

a) Calculate the capacitance of a capacitor consisting of concentric metallic spheres, with the space between filled with a dielectric of dielectric constant K. Assume that the space has inner and outer radii a and b, respectively.

b) Let $b - a = d$ remain constant while b and a increase so that $b \gg d$. Calculate the approximate capacitance in terms of the area of the sphere, $A = 4\pi b^2$, and compare your expression with that for a parallel-plate capacitor.

5-15

a) Calculate the capacitance of a capacitor made up of concentric metallic cylinders insulated from each other by a material whose dielectric constant is K. The outer radius of the inner cylinder is a, the inner radius of the outer cylinder is b, and the length of these cylinders is l, where $l \gg (b - a)$.

b) Find the capacitance of this same capacitor when the insulator fills only part of the space between the two concentric cylinders extending from $r = a$ to $r = \sqrt{ab}$.

5-16 One of the boundary conditions at the interface between two dielectric materials is $D_{1n} - D_{2n} = 0$, as is proved in Section 5–5. Use this same method to find the value of $E_{1n} - E_{2n}$ in terms of the components of the **P**'s.

5-17 One of the boundary conditions at the interface between two dielectric materials is $E_{1t} - E_{2t} = 0$, as is proved in Section 5–5. Use this same method to find the value of $D_{1t} - D_{2t}$ in terms of the components of the **P**'s.

5-18 Assume that both of the dielectrics in Fig. 5–6 are isotropic and homogeneous, with dielectric constants K_1 and K_2, respectively. For points very near the interface between the two, calculate the angle θ_2 between the field $\mathbf{E}_2$ and the z-axis, when the field $\mathbf{P}_1$ is directed at the angle θ_1 to the z-axis.

5-19 Calculate σ_p on the surfaces of (a) the sphere of Section 5–7, and (b) the cylinder of Section 5–8. Compare these results when $K \to \infty$ with the values for σ on the surfaces of conductors of the same shape.

5-20 Write expressions for **E** and **P** inside and for **E** outside the dielectric sphere of Section 5–7 and the dielectric cylinder of Section 5–8. Compare these expressions as $K \to \infty$ with the corresponding expressions for conductors of the same shape.

5-21 As has been pointed out, a charged particle is attracted toward a nearby conducting surface by the charge it induces upon that surface. Likewise, such a particle is attracted toward a nearby dielectric surface. To estimate this latter attraction, assume a point charge $+q$ on the z-axis at $z = h$ in space that is otherwise empty wherever z is positive but is filled with a homogeneous and isotropic dielectic of constant K wherever z is negative.

a) Show that the potential V_o for the region where z is positive is given by placing an image charge q' at $z = -h$, while the potential V_i for the region where z is negative is obtained by replacing the $+q$ charge with an image q'' at $z = h$.

b) Calculate the image charges q' and q''.

c) Calculate the force on the $+q$ charge.

d) Calculate σ_p on the surface of the dielectric in the xy-plane as a function of $r = \sqrt{x^2 + y^2}$.

e) Calculate the total polarization charge on the xy-plane.

5-22 Replace the point charge in Problem 5–21 with an ideal dipole of moment **p**, directed at an angle β to the positive z-axis.

a) Determine the locations and magnitudes of the images required to calculate V_o wherever z is positive and V_i wherever z is negative.

b) Calculate the torque acting on the original dipole.

c) Calculate the force acting on the original dipole.

5-23 Replace the point charge in Problem 5–21 with an infinitely long filament that has a uniform charge per unit length λ, where this filament is parallel to the x-axis and passes through $z = h$. Find (a) the force per unit length on the filament, and (b) the value of σ_p on the xy-plane.

5-24 Suppose that a certain wax for making electrets can retain a uniform polarization **P**, parallel to the axis, when this wax is formed into a right-circular cylinder of radius a and length $l \gg a$. Calculate the direction and magnitude of the electric fields due to such a cylinder (a) at the center of the cylinder, and (b) at two points on the axis near the positively charged end, one just inside and the other just outside the wax cylinder.

5-25 By warming and then freezing certain kinds of wax spheres in uniform electric fields, these spheres can be given the uniform polarization $P\mathbf{k}$, which results in internal and external fields described by $\mathbf{D}_i = C\mathbf{k}$ and $\mathbf{D}_o = F\mathbf{d}_f/a^3$. Here a is the radius of the sphere, C and F are constants. The vectors $\mathbf{k} = \mathbf{a}_r \cos \theta - \mathbf{a}_\theta \sin \theta$ and $\mathbf{d}_f = \mathbf{a}_r 2 \cos \theta + \mathbf{a}_\theta \sin \theta$ in spherical coordinates. Use the boundary conditions to express C and F in terms of P and a.

5-26 An isotropic dielectric of infinite extent lies in a uniform electric field $\mathbf{E} = E_o\mathbf{i}$. It contains a cylindrical cavity of radius a with its axis parallel to the z-axis; i.e., perpendicular to the field. Find the electric field in the cavity.

5-27 A spherical cavity of radius a is made in a uniform and isotropic dielectric in which a uniform electric field $\mathbf{E} = kE_o$ had been established. Find the electric field in the cavity.

5-28 As an aid to the understanding of electric and magnetic fields inside materials, Kelvin invented two imaginary cavities: a thin needle-shaped cavity cut parallel to the field, and a thin disklike cavity cut perpendicular to the field. For homogeneous and isotropic materials:

a) Prove that in the first cavity the field $E_c = E$, while in the second $E_c = D/\epsilon_0$.

b) Find the field E_c in a spherical cavity of radius a, and in a cylindrical cavity of radius a whose axis is perpendicular to the field.

5–29 A spherical shell of inner radius a and outer radius $2a$ has a volume density of total charge given by $\rho_t = c(r - 3a/4)$ with no other charges near it.
a) Find E as a function of r from $r = a$ to $r = \infty$.
b) Find V as a function of r from $r = 2a$ to $r = \infty$.

5–30 The space between two concentric spherical shells of metal is filled with a dielectric in which $\epsilon = \epsilon_0 r/a$. The inner and outer surfaces of this dielectric have radii of a and b, respectively. There is a charge q on the inner metallic shell.
a) Find D in the dielectric.
b) Find σ at $r = a$ and at $r = b$.
c) Find σ_p at $r = a$ and at $r = b$.
d) Find ρ_p in the dielectric.
e) From your answers to (c) and (d), show that the sum of all the polarization charges is zero.

5–31 Solve Problem 5–30 assuming that $\epsilon = \epsilon_0 b/r$ for the dielectric material.

5–32 A capacitor consists of concentric spherical shells of metal. The inner and outer radii of the space between them are a and b, respectively. An excellent insulating oil is poured to exactly fill the lower half of this space. For this oil $K = \epsilon/\epsilon_0$ is a constant. With the potentials V on the inner sphere and zero on the outer sphere:
a) Calculate E in the air ($\epsilon = \epsilon_0$) and the oil.
b) Calculate σ on the inner sphere where it contacts the oil and where it contacts the air.
c) Calculate the total charge q on the inner sphere.
d) Calculate the capacitance of the capacitor.

5–33 Assume a space containing two isotropic dielectrics which in spherical coordinates have the following constants and potentials: Where $0 < r < a$, $\epsilon = \epsilon_1$, and $V_1 = Br\theta$ and where $a < r < b$, $\epsilon = \epsilon_2$ and $V_2 = Ba^2\theta/r$.
a) Find ρ_1 where $r < a$.
b) Find ρ_2 where $a < r < b$.
c) Use a Gaussian pill-box to determine the surface density of real charges σ at $r = a$.

5–34 A parallel plate capacitor consists of two large circular disks of metal spaced a distance $3d$ apart, where $3d \ll a$ and a is the radius of the disks. The space between these disks contains a disk of dielectric material of radius a, thickness $2d$ and dielectric constant K. In the remaining space $\epsilon = \epsilon_0$. Neglecting edge effects:
a) Find the capacitance.
b) Find the energy stored in the dielectric with charges of $+Q$ and $-Q$ on the two disks.
c) Find the total energy of the capacitor with these charges.

5–35 A parallel plate capacitor consists of two sheets of aluminum foil each 2 m² in area and separated by insulating paper 0.04 mm thick. Charges of 354 μC on one and -354 μC on the other of the two foils produce a difference of potential of 100 V between them. Since the paper is not a perfect insulator, 10 pC leaks through it in 400 s. How much energy does this charge deliver to the paper?

5-36 A plane parallel capacitor consists of circular disks of radius a, separated by a dielectric disk of radius a, thickness $d \ll a$ and dielectric constant K. It is originally connected to a battery which provides the potential difference V. Call its capacitance KC.

a) With the battery disconnected calculate the work required to remove the dielectric.

b) Calculate the work required to remove the dielectric disk, if the battery is connected to the capacitor.

5-37 An ideal dipole $\mathbf{p} = \mathbf{j}3 + \mathbf{k}4$ in C m is located at the origin. Without the dipole the electric field is $\mathbf{E} = \mathbf{i}(12 + y) + \mathbf{j}(4 + z)$.

a) Find the torque on $\mathbf{p}$.   b) Find the force on $\mathbf{p}$.

5-38 A point charge q is located at the origin and an ideal dipole $\mathbf{p} = p(\mathbf{j} \sin \alpha + \mathbf{k} \cos \alpha)$ is located at $(0, 0, h)$ in rectangular coordinates.

a) Calculate the torque on the dipole.

b) Calculate the force on the dipole.

5-39 One ideal dipole $\mathbf{p} = p\mathbf{k}$ is at the origin of coordinates and another $\mathbf{p} = p\mathbf{a}_r$ is located at the general point in spherical coordinates (r, θ, ϕ).

a) Calculate the torque on the second dipole.

b) Calculate the force on the second dipole.

5-40 Problems 2–43 and 2–44 call for the calculation of equipotential surfaces near two point charges of $+4q$ and $-q$ that are l m apart. The potential surfaces are altered very little by assuming that these point charges reside on conducting spheres both of radius $a \simeq l/20$. Calculate within 1% the work required to add δq to each conducting sphere by bringing these infinitesimal charges from an infinite distance.

5-41 An arbitrary distribution of charges is confined within a finite space τ. Prove that a coordinate system can be located with its origin within this finite space so that $Vr = k \int \rho \, d\tau = kq$ for values of r that are constant and large compared to the linear dimensions of this finite space. Also show that $D_r 4\pi r^2 = q$ for the same constant and large values of r. [*Hint:* Set up the general expression for V in rectangular coordinates with an arbitrary origin. Use the approximation that $(1 + \delta)^n \simeq 1 + n\delta$ when $\delta \ll 1$.]

5-42 The plates of a parallel-plate capacitor in vacuum have charges $+Q$ and $-Q$ and the distance between the plates is x. The plates are disconnected from the charging voltage and pulled apart a short distance dx.

a) What is the change dC in the capacitance of the capacitor?

b) What is the change dW in its energy?

c) Equate the work $F \, dx$ to the increase in energy dW and find the force of attraction F between the plates.

d) Explain why F is not equal to QE, where E is the electric intensity between the plates.

5-43 A certain parallel-plate capacitor consists of two metal disks, each of radius a, which are separated by the distance d. Assume that $d \ll a$, so that the electric field can be considered to be confined to the evacuated space between the plates.

a) With a charge of $+q$ on one plate and $-q$ on the other, calculate the electrostatic energy of the charged capacitor.

b) From the work required to increase the plate separation from d to $d + \delta d$, where $\delta d \ll d$ and q remains constant, calculate the electrostatic force on the positively charged plate.

c) Instead of keeping q constant, assume that the separation of the plates is increased from d to $d + \delta d$ while the plates are kept at constant potentials by being connected to the terminals of a storage battery. How does the work done in this case compare with that calculated in (b)? If the results are different, account for this difference.

5–44 A region containing many small conducting spheres in an electric field acts like a dielectric, since the spheres become induced dipoles.

a) Calculate the effective dielectric constant for the following geometrical arrangement, assuming that the field around each sphere is not affected significantly by the presence of the other spheres. Consider a simple cubic lattice; i.e., a space completely filled with imaginary cubes having edges in common, each of which is $4a$ in length. Further, assume that metallic spheres, each of radius a, are located with their centers at each corner.

b) Correct your calculation by taking account of the fields of the surrounding dipoles, including those of the distant spheres.

5–45 According to relativity theory, the total energy of a particle at rest is mc^2, where m is the rest mass. Hence for an isolated charged particle the electrostatic energy, which depends upon its radius, must be less than mc^2. This sets a minimum for the radii of the elementary charged particles. Calculate the minimum radii for the following particles: (a) the proton, (b) the nucleus of an atom having Z protons and Z neutrons, and (c) the electron. Compare these radii with the recently measured radii of the proton (0.7×10^{-15} m) and the atomic nuclei ($1.2A^{1/3} \times 10^{-15}$ m), for which the number of nucleons is $A = 2Z$.

Chapter 6 STEADY CURRENTS

Electric currents may exist in any medium in which electric charges are free to migrate. Ionized gases and ionic liquids, as well as many solid materials, are good conductors. The fundamental principles governing conductivity apply equally well to all conductors; however, in this chapter special attention is given to solid conductors because of their greater practical importance.

The crystal structure and distribution of electrons within a solid conductor are similar to those of nonconductors as described in Chapter 5. The basic difference is that all electrons in nonconductors are tightly bound to their respective nuclei, while in conducting materials a few electrons from the outer atomic shells are free to wander through the crystal. These electrons are referred to as "free electrons" or, more properly, as conduction electrons because they transport electric current through the solid.

6–1 Metallic conductors

The number of conduction electrons per atom in a metal depends on the electronic structure of the atoms and on the degree of perturbation of the outer electron shells when the atoms are placed in the crystal lattice. An examination of the periodic table reveals that the elements Li, Na, K, Cu, Rb, Ag, Cs, and Au have completely filled inner shells, with only one additional electron in the outer shell. In these simple metals one would expect intuitively that the one outer electron would become a conduction electron in the metallic solid. That the number of conduction electrons per atom is approximately one in these metals has been verified experimentally by Hall measurements of the type described in Chapter 7. Intuitive estimates of the number of conduction electrons in other metals are less likely to be correct, since most metals have (on the average) a fractional number of conduction electrons per atom. The transition elements Ni, Co, Pd, Rh, and

176

most of their alloys, for example, each have approximately 0.6 conduction electrons per atom. The number of current carriers per atom in metals is always of the order of magnitude of one, while in semiconductors it is very much lower, being of the order of 10^{-9} in germanium at room temperature. The very small number of carriers in semiconductors results in interesting electrical properties, some of which are discussed in later chapters.

Thermal vibrations within a metal keep the conduction electrons constantly moving through the lattice, with velocities as high as 10^6 m s^{-1} even at room temperature. Since this motion is random, it does not give rise to a net electric current. Application of an electric field results in a slower drift velocity superimposed upon these random thermal velocities in a direction opposite to the electric field. The resulting electric current, by convention, is taken to be in the direction of the field.

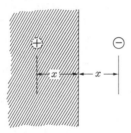

Fig. 6–1 The image force is largely responsible for the attraction of an electron back toward the metallic surface from which it has escaped.

With conduction electrons moving so rapidly through the metal at all times, it is logical to inquire what prevents them from continuing their motion out into space and thus escaping from the metal. A simple calculation of the electrostatic forces acting on an escaping electron quickly reveals why this does not take place. Assume that an electron has escaped from the plane surface of a metal by a distance x, as illustrated in Fig. 6–1. Then, using an image solution of the type discussed in Section 4–2, the electron is attracted back to the surface of the metal by the force

$$F = \frac{e^2}{4\pi\epsilon_0(2x)^2},$$

that is, its electric potential energy relative to the outside world is given by

$$\mho = -\frac{e^2}{16\pi\epsilon_0 x}. \tag{6–1}$$

According to this equation, the work required to remove an electron would be infinite, because $U \rightarrow \infty$ as $x \rightarrow 0$. Actually, Eq. (6–1) is accurate only for distances large compared with the lattice spacings or for distances large compared with 10^{-9} m. For regions closer to the metal than this, the surface may no longer be treated as a continuum of charge but, instead, individual electronic interaction must be considered. An exact solution would require quantum mechanics; however, a better approximation can be obtained by adding an experimentally determined constant to the denominator of Eq. (6–1). The potential energy function for the electrons then becomes

$$\mathcal{U} = \frac{e^2}{16\pi\epsilon_0 x + (e^2/W)}, \tag{6–2}$$

where W is known as the work function and is obviously the work required to remove an electron from $x = 0$ to $x \to \infty$.

When the temperature of the metal is increased, some electrons may receive enough energy to penetrate the potential barrier into space. This is called thermionic emission and accounts for the electrons emitted from the heated cathode of an electronic vacuum tube. A fairly accurate theoretical equation for thermionic emission has been calculated by Richardson and Dushman, using Fermi-Dirac statistics. Their equation is

$$J = AT^2 \exp\left(-W/kT\right),$$

where J is the emitted current density, T is the temperature of the metal in degrees Kelvin, and k is the Boltzmann constant, 1.38×10^{-23} J/deg. Some experimentally determined values for A and W are $A = 32 \times 10^4$ ampere/m^2 deg^2 and $W = 8.5 \times 10^{-19}$ J in platinum, and $A = 75 \times 10^4$ ampere/m^2 deg^2 and $W = 7.1 \times 10^{-19}$ J in tungsten. The energy required for an electron to be emitted from the surface is decreased when the metal is charged to a negative potential; this type of emission is called the Schottky effect.

The electronic forces responsible for preventing the escape of electrons from the surface of a metal are also responsible for making the ensemble of conduction electrons behave as a very slightly compressible fluid. A bunching of electrons within a metal creates strong electric forces which tend to smooth out the distribution. Some appreciation of the magnitude of the potential that is generated by small changes in electron concentrations can be obtained from the fact that a copper sphere 0.1 m in diameter becomes charged to 1000 V when only 10^{-15} of its conduction electrons are removed.

This fundamental property of conduction electrons in metals is responsible for the hydraulic analogy which likens steady electric currents in wires to the flow of water through gravel-filled pipes. When properly used, this analogy is an excellent one, although recently it has fallen into disfavor because it has been so frequently misused. All analogies necessarily have limited applicability and the use of any analogy without pointing out its limitations is likely to lead to misconceptions. An excellent feature of the analogy is the similarity between potentials and pressures, since, with steady flow, potential differences and pressure differences are both scalar functions of the coordinates. Furthermore, pressure differences can be defined by the work per unit volume done on the water flowing between two pressures, just as potential differences are defined by the work per unit charge done on the charges flowing between two points. Serious limitations do show up, however, when the flow is intermittent or alternating, since there are no good counterparts in the hydraulic analogy for the phenomena of the charging of capacitors or of the magnetic fields associated with currents.

6–2 Ohm's law

When the potential difference across a conductor is V and the current through the conductor is I, the ratio

$$R = \frac{V}{I} \tag{6-3}$$

defines the resistance R of the conductor. Ohm's law states that this ratio is independent of the magnitudes and directions of V and I and depends only upon the physical properties and the geometry of the conductor. Conductors for which this law is accurate are called ohmic conductors and all others are called nonohmic. This law is quite accurate for most metals under standard conditions, and it is valid for most conductors and semiconductors within certain limitations. One rather obvious limitation is that the conduction electrons cannot move with a velocity greater than the velocity of light. This is no practical limitation in copper, which has one conduction electron per atom or about 10^{29} per cubic meter, but in germanium, with as few as 10^{19} conduction electrons per cubic meter, this limitation can be important. The current density in any material can be expressed by $\mathbf{J} = -nNe\mathbf{v}$, where N is the number of atoms per unit volume, n is the number of conduction electrons per atom, e is the charge on an electron, and $\mathbf{v}$ is the average drift velocity. Using $nN = 10^{19}$, $e = -1.6 \times 10^{-19}$, and $v = 3 \times 10^8$ (all in mks units), a maximum current density of 5×10^8 A m^{-2} is obtained for germanium. Thus, regardless of the potential applied to a germanium wire with 0.1 mm^2 cross section, no more than a 50-A current can be generated. In actual fact, the limiting current is considerably lower than this; at high voltages breakdown in the crystal occurs, freeing more electrons to conduct current. These considerations are mentioned here to point up the fact that the ohmic behavior of materials should not be taken for granted. All materials fail to be ohmic under some conditions and some materials are ohmic only under very limited conditions. Some special examples of nonohmic materials are discussed in a later chapter.

In metallic conductors, where Ohm's law is particularly accurate, the law can be obtained by elementary considerations of the forces on free electrons. Newton's law alone would predict an unlimited velocity increase for conduction electrons under the influence of an electric field. A damping term is needed and is physically reasonable in view of the interaction between the electrons and the crystal lattice. Assuming this damping to be proportional to the velocity (a friction-type damping), the equation of motion becomes

$$e\mathbf{E} = m\frac{d\mathbf{v}}{dt} + \frac{m}{t_r}\mathbf{v},\qquad(6\text{–}4)$$

where m/t_r is the damping constant, e is the charge on an electron, and $\mathbf{v}$ is a drift velocity superimposed on the random thermal velocities. The solution of this differential equation is

$$\mathbf{v} = \frac{et_r}{m}\mathbf{E}[1 - \exp(-t/t_r)]\qquad(6\text{–}5)$$

and the reason for using the symbol t_r in the damping parameter becomes evident, since it represents a relaxation time. For times long compared with t_r, the drift velocity is simply $\mathbf{v} = (et_r/m)\mathbf{E}$, and the current density is

$$\mathbf{J} = nNe\mathbf{v} = \frac{nNe^2t_r}{m}\mathbf{E}.\qquad(6\text{–}6)$$

Making the substitution $\rho^* = m/nNe^2t_r$, the above equation reduces to

$$\mathbf{E} = \rho\mathbf{J},\qquad(6\text{–}7)$$

which is the general form of Ohm's law for conductors not moving in magnetic fields. For the simple case of a linear conductor of length l and cross-sectional area A, Eq. (6–7) becomes $V = IR$, where $R = \rho l/A$, $I = JA$, and $V = El$.

Because of the relationship between t_r and ρ, it is possible to determine the value for t_r in Eq. (6–5). Using values for copper of $m = 0.9 \times 10^{-30}$ kg, $e = 1.6 \times 10^{-19}$ C, $\rho = 1.7 \times 10^{-8}$ Ω·m, $n = 1$ electron/atom, and $N = (6.025 \times 10^{23}$ atoms/mole) $\times$ (1/63.5 mole/g) $\times$ (8.89 $\times$ 10⁶ g/m³) = 8.5×10^{28} atoms/m³, a relaxation time of $t_r \cong 2 \times 10^{-14}$ s is obtained. This indicates that the momentum of the conduction electrons should not cause them to lag the applied voltage except at very high frequencies approaching that of visible light, and it is experimentally well substantiated that energy losses in waveguides may be calculated accurately using the assumption that $t_r \ll 1/f$ for frequencies as high as $f = 10^{10}$ cycles·s⁻¹ or more.

---

* The symbol ρ is used universally both for resistivity and for volume density of charge, since this duplication seldom causes confusion. Where ρ's must be used together, ρ_c designates charge density.

Temperature dependence of Ohm's law. The damping parameter introduced in Eq. (6–4) has been justified on the classical basis of interactions between the conduction electrons and the positive nuclei of the metallic crystal. Even in an absolutely perfect single crystal with no imperfections, no impurities, and no thermal vibrations, the drift velocity of conduction electrons would be retarded by continual collisions with the nuclei of the lattice. In this respect, wave mechanics is in direct contradiction with classical mechanics.

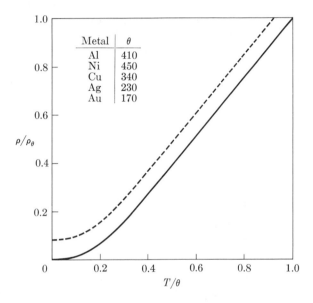

Fig. 6–2 The solid line is Gruneisen's calculation for the variation of resistivity with temperature, in which θ is the Debye temperature and ρ_0 is the resistivity of the metal at Debye temperature. The dashed curve, which is a plot of ρ for Cu + 0.2% Ni divided by ρ_0 for pure Cu, illustrates Matthiessen's rule. Representative values of θ, in degrees Kelvin, obtained from specific heat measurements are listed in the figure.

A wave-mechanical calculation treats the crystal lattice as a periodic potential function with the periodicity of the lattice spacings. No damping of the motion of the conduction electrons is obtained so long as this potential remains strictly periodic. However, at temperatures above absolute zero, the thermal vibrations of the crystal destroy the periodicity and a finite resistance results. A theoretical calculation of the temperature variation for a pure metal has been made by Gruneisen and its result is plotted in Fig. 6–2. The constant θ is the characteristic Debye temperature of the material and may be determined by specific heat measurements, while ρ_θ is the resistivity of the

Table 6–1

RESISTIVITIES AND TEMPERATURE COEFFICIENTS OF SOME COMMON MATERIALS
AT ROOM TEMPERATURES

Element	Resistivity, $\Omega \cdot m$	Temperature coefficient per C°	Alloy or compound	Resistivity, $\Omega \cdot m$	Temperature coefficient per C°
Silver	1.6×10^{-8}	3.75×10^{-3}	Constantan 45%Ni–55%Cu	49×10^{-8}	0.5×10^{-5}
Gold	2.2×10^{-8}	$3.4 \ \times 10^{-3}$	Nichrome, Ni-Cr-Fe	112×10^{-8}	40.0×10^{-5}
Copper	1.7×10^{-8}	$4.1 \ \times 10^{-3}$			
Aluminum	2.7×10^{-8}	$4.0 \ \times 10^{-3}$	Bakelite	10^{10}	-0.06
Tungsten	5.3×10^{-8}	$4.7 \ \times 10^{-3}$	Glass	10^{12}	-0.07
Nickel	6.9×10^{-8}	$3.4 \ \times 10^{-3}$			
Carbon	3.5×10^{-3}	-0.03	Mica	10^{15}	-0.07
Sulfur	10^{15}				

metal at its Debye temperature. At high temperatures the resistivity is proportional to T, and at low temperatures it is proportional to T^5. Some experimentally determined values of θ for various metals are listed on the curve. This curve is in excellent agreement with experimental observations on many metals.

For metals containing a small amount of impurity, a modification to Gruneisen's curve must be made. This modification, which is known as Matthiessen's rule, states that the scattering of conduction electrons by impurity atoms is temperature-independent, so that the total resistivity of the material may be expressed by

$$\rho \ (\text{total}) = \rho \ (\text{Gruneisen}) + \rho \ (\text{impurity}),$$

where ρ (impurity) is a temperature-independent resistivity due to impurity scattering. The change in resistivity for copper with 0.2% nickel is illustrated by the dashed curve in Fig. 6–2. In commercially pure metals, the small number of impurities present has little effect on the resistivity at room temperature, amounting to 0.1% to 1.0%. The effect does become important, however, at low temperatures where the resistivity of pure metals approaches zero.

Because of the almost linear change of resistivity with temperature near room temperature, it is possible to express this variation by

$$\rho(t) = \rho(20°C)[1 + \alpha(t - 20)]. \tag{6–8}$$

Room-temperature resistivities of some common materials are listed in

Table 6–1, along with their temperature coefficients. Impurity scattering in Cu can be increased by adding more Ni until ρ (impurity) becomes the dominant term, but Matthiessen's rule becomes less and less accurate. Certain alloys of Cu and Ni have been used in making resistors for measuring instruments because of their relatively high resistivities and low temperature coefficients.

The temperature coefficients for nonmetals are typically larger than those for metals and are of opposite sign, as indicated by the nonmetals in Table 6–1. To understand this, consider the expression for resistivity as obtained for metallic conduction,

$$\rho = \frac{m}{nNe^2t_r}.$$

This expression is just as applicable for semiconductors as it is for metals and, as in metals, t_r decreases with increasing temperatures. However, in semiconductors the conductivity is limited more by the small number of current carriers than by the size of t_r. Some electrons which are bound to lattice sites at low temperatures are freed at higher temperatures by thermal vibrations of the lattice. Thus the decreasing resistivity with increasing temperature in nonmetals reflects an increase in n, the number of current carriers, rather than a change in the relaxation time.

Superconductivity. A dramatic deviation from the resistivity versus temperature curve of Fig. 6–2 occurs in some materials which exhibit a phenomenon called superconductivity, discovered by Kamerlingh Onnes in 1911. Below a critical temperature the resistivity is identically zero for these materials; at slightly higher temperatures, the normal Gruneisen-Matthiessen relation is followed. This phenomenon, for which there is no classical analogy, results from a cooperative interaction between pairs of conduction electrons. It is also accompanied by perfect diamagnetism, that is, $B = 0$ or $\mu = 0$ in a superconductor; and in fact, as Meissner and Ochsenfeld discovered in 1933, a material cooled below its critical temperature while in the presence of a magnetic field will expel that field as it passes through the critical temperature. A related phenomenon is the ability of an externally applied magnetic field, if strong enough, to penetrate a superconductor and destroy its superconductivity.

In Table 6–2 are listed critical temperatures T_c for some common superconductors as well as the critical field H_c, required to destroy the superconductivity at absolute zero. The Nb$_3$Al-Nb$_3$Ge alloy was reported by B. T. Matthias in 1967 and has the highest T_c so far observed.

In 1956 D. A. Buck[*] proposed that the magnetic field generated by current pulsed through one superconducting wire could be used to destroy

---

[*] D. A. Buck, "The Cryotron—A Superconductive Computer Component," *Proc. IRE*, **44**, 482–493 (April 1956).

Table 6–2

CRITICAL TEMPERATURES AND MAGNETIC FIELDS FOR
SOME COMMON SUPERCONDUCTORS

Material	T_c, °K	H_c
Al	1.2	99
$H_g(\alpha)$	4.2	412
In	3.4	293
Nb	9.2	1,940
Pb	7.2	803
Sn	3.7	305
V	5.3	1,020
Nb_3Al	18.0	—
Nb_3Sn	18.1	>200,000
Nb_3Al-Nb_3Ge	20.1	—

the superconductivity of an adjacent wire. Thus, a current-controlled switch could be constructed. Numerous switches and complicated circuitry have been modeled since then; however, the millions of development dollars invested to date have failed to produce an economically attractive use for this concept. This lack of commercial success to date is as much the result of the fantastic progress in semiconductor devices as it is the result of technological problems in fabricating superconducting devices.

A practical use for superconductivity in recent years has been in generating intense magnetic fields for research projects. Electric current in coils of superconducting wire produces magnetic fields according to the relations developed in Chapter 8. Niobium-tin compounds will support over 10^9 A m^{-2} in fields of 150,000 gauss (15Wb·m^{-2}), and by use of such compounds fields of that size have been produced over a large enough volume to perform useful experiments.

While a number of problems in superconducting circuits can be solved by setting $\rho = 0$ and $B = 0$ in the superconductor, there are enough subtleties in the field that a person should consult detailed treatments of the subject before attempting to do so.†

6–3 Energy and power conversions

Motions of charges in electric fields usually result in conversions of energy from one form to another. A charged particle in free space loses or

---

† See for example V. L. Newhouse, "*Applied Superconductivity*," John Wiley & Sons, New York, 1964, "*Superfluids, Macroscopic Theory of Superconductivity*," Vol. 1, Fritz London, Dover, 1961, or "Resource Letter Scy-1 on Superconductivity *Am. J. Phys.* **32**, 2 (Feb 1964).

gains kinetic energy. Conduction electrons within a conductor gain energy which is transmitted to the atoms on lattice sites as thermal vibrations. The work done by an electric field on a charge moving through it is given by

$$W = \int \mathbf{F} \cdot d\mathbf{l} = \int q\mathbf{E} \cdot d\mathbf{l},$$

integrated over the path of the moving charge, q. For a charge moving with the field the integrand is positive, showing that the electric field energy is being converted to some other form. However, for a charge moving in the opposite direction the integrand is negative, and the energy conversion is reversed.

The power being delivered to a charge moving with the velocity $\mathbf{v}$ is given by

$$p = q\mathbf{E} \cdot \mathbf{v},$$

and the total power being delivered to all the charges in an element of volume $d\tau$ is

$$dP = \Sigma\, p_i = \Sigma\, q_i \mathbf{E}_i \cdot \mathbf{v}_i,$$

where the sum is performed over all the charges in the volume element. When this expression is applied to conduction electrons in a volume small enough so that they all have the same drift velocity and are all acted upon by the same field $\mathbf{E}$, the above equation becomes

$$dP = Nne\mathbf{E} \cdot \mathbf{v}\, d\tau,$$

where N is the number of atoms per unit volume and n is the average number of conduction electrons per atom. Using the expression $\mathbf{J} = Nne\mathbf{v}$ for current density, the power delivered to an element of volume may be expressed as

$$dP = \mathbf{E} \cdot \mathbf{J}\, d\tau.$$

A positive sign again indicates that electric power is being converted to some other form. The total electric power being converted within any volume τ is

$$P = \int_\tau \mathbf{E} \cdot \mathbf{J}\, d\tau. \tag{6–9}$$

Joulean heat. The current density in an ohmic conductor is $\mathbf{J} = \mathbf{E}/\rho$, provided the conductor is isotropic. According to Eq. (6–9) the power converted into heat within a volume τ of such a conductor is

$$P = \int_\tau \rho\mathbf{J} \cdot \mathbf{J}\, d\tau = \int_\tau J^2\rho\, d\tau, \tag{6–10}$$

and the heat produced within this volume is called joulean heat. In any single conductor carrying a current I, imaginary surfaces can be drawn perpendicular to $\mathbf{J}$ so that $I = \int_s J\, ds$. Then, if $d\tau = ds\, dl$, with dl measured parallel to **J**, Eq. (6–9) becomes

$$P = \int_s J \, ds \int_l E \, dl = IV = I^2 R,$$

when IR from Ohm's law is substituted for V. This is the well-known relation for joule heating in an ohmic resistor.

6–4 Sources of electric energy and power

Electromagnetic generators with rotating armatures, storage batteries and other chemical cells, thermocouples, and the so-called static generators that include the Van de Graaff generator are all capable of converting some other form of energy into electric energy. Each of these devices can be made to produce steady currents in networks of conductors. To do so, they must continually convert some form of energy to electric energy to balance the electric energy which is continually being converted back to some other form of energy, for example to heat within the conductors or to mechanical energy by the motors. The integral of Eq. (6–9) is therefore negative when performed over just the volume occupied by one of these devices. However, there may be regions within these devices where the integrand is positive, i.e., where the energy conversions proceed in the opposite direction, for example the joulean heat produced in the windings of an electromagnetic generator or the mechanical energy delivered to that portion of the belt of a Van de Graaff generator that carries a small residual charge out of the charge-collecting sphere.

Van de Graaff generator. The Van de Graaff generator is ideal for illustrating these power conversions. As shown in Fig. 6–3, this device consists of a conveyor belt of insulating material, which can be operated mechanically to transport charges to a conducting sphere against the forces of a strong electric field. Since very large potential differences can be obtained with this device, it is commonly used for accelerating elementary charged particles for scientific experiments. In the figure it is assumed that a steady stream of positively charged particles is projected down the evacuated tube from C to O, arriving at O with relatively large kinetic energies.

The fine needlepoints at A are connected to ground through a low-voltage source. Depending on the polarity of this source, electrons will either be sprayed onto the belt from the points or be withdrawn from the belt by the points. This latter case is electrically equivalent to spraying positive charges onto the belt. With mechanical power turning the lower pulley clockwise, this positive charge is carried up to the insulated metallic sphere. The appearance of positive charges on the sphere means that electrons have been withdrawn from the sphere through the needlepoints at B, and sprayed onto the belt. These partially neutralize the belt's positive charge, but complete neutralization does not occur because electrons move across the gap only while the field between the belt and the points is sufficiently strong. Although a background of positive charges remains on the belt, the number of electrons removed per second from the belt at A, under steady-state conditions, equals the number returned per second to the belt at B. For this condition to be maintained, there must be an equal current from the sphere to ground, consisting of positively charged particles proceeding down the tube from C to O, plus some leakage currents through imperfect insulation.

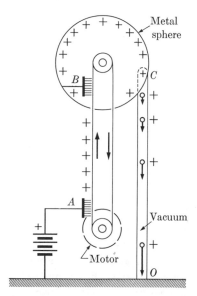

Fig. 6–3 A Van de Graaff generator projecting positively charged particles down an evacuated tube from C to O.

The background charge remaining on the belt absorbs some energy from the field on the downward motion of the belt but returns this energy as it travels upward. Only the additional charges added at A and removed at B absorb net energy from the belt. Neglecting the losses due to leakage currents, this energy reappears as kinetic energy of the accelerated particles. The power being delivered to the belt can be calculated from Eq. (6–9). On the upward-moving portion of the belt $\mathbf{E} \cdot \mathbf{J}\, d\tau = -EI_t\, dl$, which is negative because $\mathbf{E}$ and $\mathbf{J}$ are oppositely directed, while on the downward-moving portion $\mathbf{E} \cdot \mathbf{J}\, d\tau = EI_r\, dl$, which is positive because $\mathbf{E}$ and $\mathbf{J}$ are in the same direction. The currents are defined by $I_t = I + I_r$, where I_t is the total current directed upward on the left and I_r is the residual current carried by the background charges remaining on the belt after it passes B. The net power delivered to the belt is

$$P = -\int_A^B EI_t \, dl + \int_B^A EI_r \, dl$$

$$= -I\int_A^B E \, dl - I_r\int_A^B E \, dl + I_r\int_B^A E \, dl,$$

which may be written in vector notation as

$$P = I\int_A^B \mathbf{E} \cdot d\mathbf{l} + I_r \oint \mathbf{E} \cdot d\mathbf{l} = -I\mathcal{E},$$

where in this case

$$-\int_A^B \mathbf{E} \cdot d\mathbf{l} = \mathcal{E}$$

is the emf of the device and

$$\oint \mathbf{E} \cdot d\mathbf{l} = 0$$

when integrated around the path of the belt in this electrostatic field.

Storage batteries and other chemical cells. When a battery is supplying a steady current for a network of conductors, the current in the battery is directed from the terminal of low potential to that of high potential. Chemical forces are responsible for moving the charges against the internal electrostatic field, and in doing this work, chemical energy is converted to electric energy. It is possible to reverse the direction of electric current in any chemical cell by sufficiently increasing the electric potential by means of an external emf. The increased electrostatic potential is then strong enough to cause a net flow of charge against the chemical forces, and electric energy is converted to chemical energy. Chemical energy created in a dry cell in this manner is not in a form that can be recovered by reversing the current, but in a storage battery the process is almost completely reversible for moderate currents. When electric charges are forced to flow against the chemical forces, electric energy is converted to chemical energy and the battery is said to be charging. The battery is said to be discharging when this chemical energy is converted to electric energy by forcing an electric current through an external circuit.

Thermocouples. The simplest thermocouple consists of a wire of one metal with its ends joined to those of a wire of some other metal to form a closed loop. In general, if the two junctions are maintained at different temperatures there will be a current in the loop that depends upon the nature of the two metals and the temperatures of the two junctions. Such thermocouples can be used to maintain steady currents in a network of resistors by cutting one of the wires of the simple loop into two parts and joining the resulting loose ends to the network.

If there are temperature differences throughout the network there may be many thermocouples to consider. However, if all but the original two junctions are maintained at a single temperature, these original junctions will provide the only emf and it will depend upon the nature of the original two metals and the temperatures of the two original junctions. Heat energy is then continually absorbed at the hotter of these two junctions and converted into electric energy. However, electric energy is also continually converted back into heat at the colder of these two junctions, and this heat must be continuously removed from the cold junction if its low temperature is to be maintained.

Although the magnitudes of the emf's available from thermoelectric junctions are very low, these devices are now used in places where weight and space are at a premium, e.g., in space ships. They have been most widely used for measuring temperatures at otherwise inaccessible locations.

Electromagnetic generators. Electric dynamos and other direct-current electromagnetic generators will be discussed in detail in Chapter 7, after the fundamental principles have been presented.

6–5 Definition and origin of emf

The emf of any device is defined as the power which the device converts to the electromagnetic form, divided by the current through the device, when this device is supplying electromagnetic power to external circuits. In most cases this is equivalent to the potential difference across the terminals of the device when the current equals zero; the emf is then easily measured with a very-high-resistance voltmeter or, better, with an electronic voltmeter that draws no current. It is interesting that the term emf came from the words "electromotive force" even though it is not really a force.

In a single closed loop consisting of resistors in series which are also in series with a source of steady emf,

$$\oint \mathbf{E} \cdot d\mathbf{l} = 0,$$

provided there are no magnetic fields changing with time.* This can be written as

$$\sum_j IR_j - \mathcal{E} = 0,$$

where $\mathcal{E}$ is the emf of the source and the summation is taken over all the resistive elements including the internal resistance of the source of emf.

The paradox concerning sources of emf and the $\mathbf{E}'$ *field.* The fact that, in sources of emf, charges must move against existing electric fields has created

---

* E. M. Pugh, "Conservative Fields in d-c Circuits," *Am. J. Phys.* **29**, 484 (1961).

an apparent paradox. It did not appear that such motion was consistent with Maxwell's four equations and the Lorentz force equations discussed in Section 7–1. To resolve this apparent paradox a number of authors have postulated the existence of a new field capable of forcing charges in the direction they travel in these emf sources. This postulated field is often designated by $\mathbf{E}'$. However, this postulation assumes that the $\mathbf{E} \cdot \mathbf{J}$ term of Eqs. (6–9) and (7–16) must never be negative, which would eliminate all sources of electromagnetic power. As discussed in Chapter 10, it would leave many sinks but no sources for the very useful Poynting vector.

The difficulty comes from the mechanical principles that have been tacitly assumed, rather than from Maxwell's equations. Although Newton's laws of mechanics are remarkably accurate approximations in most ordinary situations, they are never strictly correct. They become useless for electrons moving about atomic nuclei. Here quantum mechanical principles must be used. In fact, these quantum principles are essential for the understanding of the behavior of many sources of emf. In each of these sources quantum mechanical principles show why charges do move against the existing electric fields. Of the four sources of emf previously described, the thermocouple probably best illustrates these quantum mechanical considerations.

Mechanics of thermocouples. As is stated in Section 6–4, the simplest form of the thermocouple consists of two wires of different materials joined together at their ends to form a circuit. In general, when the two junctions are at different temperatures, electric charges flow around this circuit. The emf's causing this current are due to two different phenomena, each of which is capable of causing electronic charges to move against forces due to electric fields.

The first phenomenon, called the Thomson effect, results from thermal agitation of conduction electrons, causing an emf per unit length which is dependent on the thermal gradient. Under the influence of these temperature gradients, conduction electrons diffuse toward one end of each wire and thus create electric fields in each of them. If the wires were separated and the thermal gradients maintained, electrons would continue to flow against these fields until the potential differences between the ends were equal to the individual emf's. Because this Thomson effect depends on the material, the emf's in the two wires are in general not equal and a net emf exists in the circuit. These emfs are each dependent on the line integrals of their individual temperature gradients from junction to junction so that the net emf due to this effect is dependent on the temperatures of the two junctions.

The second phenomenon results from differences in the maximum energy of conduction electrons in different materials at the same temperature. When two such materials are intimately joined, the electrons having the higher energies may spill over into the material with lower-energy electrons. This quantum mechanical process lowers the total energy of the system. However, the loss of electrons from one previously neutral body leaves it positively charged while the other becomes negatively charged. A potential difference, called the contact potential, and a resultant electric field are established. Electrons have moved against this field. This phenomenon depends only on the materials and temperature of the junction. When another junction between these two materials is established at a different temperature, the net emf in the circuit due to this phenomenon will equal the difference between the two contact potentials.* This second phenomenon also depends only on the two materials and the temperatures at the junctions. Thus the contact potentials at the junctions combine with the Thomson effect in the wires to produce the emfs observed with thermocouples.

When electric charges flow in simple thermocouples, the energy of the hotter junction decreases and that of the colder junction increases. For steady operation a quantity of heat must be subtracted from the colder junction and a larger amount of heat must be added to the hotter junction to keep the junction temperatures constant. The net result is that some of the thermal energy is converted to electrical energy by the thermocouple. Thermocouples are also used now to refrigerate regions that are inaccessible to conventional types of refrigeration, by means of an external source of emf that drives current against the internal emf.

The absorption and emission of heat by such junctions was discovered by Peltier after Seebeck had discovered thermocouples to be sources of emf. Thus the thermocouple phenomenon as a whole is called the Seebeck effect while the absorption or emission of heat at a junction is called the Peltier effect. The effectiveness of both the Thomson effect and the contact potential is due to the great differences in the physical properties of different materials. In fact, the effectiveness of all sources of emf depends strongly on the enormous differences in the physical and chemical properties found among different elements and materials. The understanding of these differences had to wait for the introduction of quantum mechanics, because no significant progress could be made with Newtonian mechanics. Since the introduction of quantum mechanics, great strides toward the understanding of all physical and chemical properties have been achieved. However, the job is far from finished. Each property has involved some new field for investigation and many articles and books have been published on these subjects. The *American Journal of Physics* has issued pamphlets listing source material for many different subjects. These pamphlets, which can be found in practically all technical libraries, provide excellent bibliographies.

On the mechanics of other sources of emf. The explanation of the emf from chemical cells is more complex than that for thermocouples although it does

---

* Figure 11–12(c) shows the contact potential as the area under the curve, for two materials that differ only by their impurities.

involve similar principles. Instead of converting heat energy the chemical cell converts chemical energy to the electromagnetic form. Charges do flow against the existing electric fields.

The Van de Graaf and dynamo generators convert mechanical energy to the electromagnetic form. In the first type, the mechanical energy is carried to the charges by the belt; in the second, by wires moving through magnetic fields. The belt must be a good insulator and the wires should be good conductors. In either case quantum mechanical principles are needed to understand the existence of these stable solids. Both generators move charges against electric fields.

6–6 Conservation of charge

One of the best-established laws of nature is that electric charge is always conserved. The net charge flowing per second into any arbitrary volume τ through its bounding surface s equals the rate of increase of charge in that volume. Mathematically, this means that

$$-\int_s \mathbf{J} \cdot d\mathbf{s} = \int_\tau \frac{\partial \rho_c}{\partial t} \, d\tau, \tag{6–11}$$

where $\mathbf{J}$ is the current density in $C\ m^{-2}\ s^{-1}$ and ρ_c is the charge density $C\ m^{-3}$. Gauss' theorem can be used to transform the surface integral into a volume integral:

$$-\int_\tau \mathbf{\nabla} \cdot \mathbf{J} \, d\tau = \int_\tau \frac{\partial \rho_c}{\partial t} \, d\tau.$$

Since τ is an arbitrary volume, the integrands of these two-volume integrals can be equated to give

$$\mathbf{\nabla} \cdot \mathbf{J} + \frac{\partial \rho_c}{\partial t} = 0, \tag{6–12}$$

which is usually referred to as the conservation of charge equation. Because of its importance it will be referred to many times throughout this text.

6–7 Steady current-density solutions

In the simple cases where there are steady current densities within homogeneous and isotropic media, relatively simple solutions are available. Under these conditions the conservation of charge, Eq. (6–12), gives $\mathbf{\nabla} \cdot \mathbf{J} = 0$ because $\partial \rho_c / \partial t = 0$. Combining this with Ohm's law, Eq. (6–7), gives $\nabla^2 V = 0$. Therefore problems involving steady currents in such media can be solved with Laplace's equation, by the methods discussed in Chapters 4 and 5. For example, the techniques allow us to calculate the flow pattern, the electric fields, and the total resistance of the carbon washer in Fig. 6–4.

Considering all of the devices shown in Fig. 6–4, however, the steady currents produced by the chemical cell also pass through the copper cylinder, through the carbon washer and back through the nickel cylinder to the cell. At both of the metal-carbon junctions ρ changes very rapidly but not discontinuously from the ρ of one material to that of the other. Real discontinuities do not exist in nature.

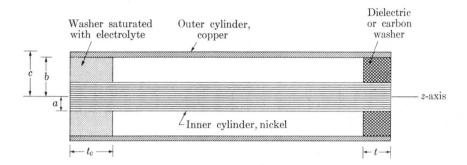

Fig. 6–4 Concentric cylinders of nickel and copper form an electrochemical cell with the electrolyte at the left of this diagram. These cylinders also constitute a transmission line for either the capacitor or the resistor found at the right of the diagram.

Steady currents in dc circuits may pass through many different materials with widely varying physical properties. Here we treat only those materials that are ohmic with isotropic dielectric properties. For such materials the resistivity ρ and the dielectric constant $\epsilon = K\epsilon_0$ have no directional properties but may vary with position in space. Values for K are not well known for conductors but they can hardly vary beyond the range between one and ten and are probably close to one. On the other hand even at room temperatures ρ may vary between 10^{-8} and 10^{15} Ω m, as shown in Table 6–1. Thus in discussing properties based upon the product $\rho\epsilon$, the effect of K might well be ignored. It will be retained here for completeness.

We continue to use the symbol ρ_c to designate real charge density and to distinguish it from the resistivity ρ.

Maxwell's first relation in Eq. (5–4) can be written

$$\rho_c = \mathbf{\nabla} \cdot \mathbf{D} = \mathbf{\nabla} \cdot (-\epsilon \mathbf{\nabla} V),$$

since $\mathbf{D} = \epsilon \mathbf{E} = -\epsilon \mathbf{\nabla} V$. However,

$$-\epsilon \mathbf{\nabla} V = \epsilon \rho \mathbf{J} = \mathbf{D},$$

so that $\rho_c = \mathbf{\nabla} \cdot (\epsilon \rho \mathbf{J})$. This can be expanded by using Eq. (3–20) to give

$$\rho_c = \epsilon \rho \, \mathbf{\nabla} \cdot \mathbf{J} + \mathbf{J} \cdot \mathbf{\nabla}(\epsilon \rho). \tag{6–13}$$

When ϵ and ρ are true constants, as we assume them to be within the carbon washer, the last term in this equation is zero. The first term on the right is zero under steady-state conditions so that $\rho_c = 0$ throughout the carbon washer.

However, at the junction between carbon and copper $\nabla(\epsilon\rho)$ is very large and practically parallel to **J**. Real charges do collect at such junctions. A general solution for such junctions is obtained by using a vector **n** normal to the interface in the general direction of **J**. At such an interface between two materials with very different resistivities $\nabla(\epsilon\rho)$ is normal to the interface and can be written **n** $d(\epsilon\rho)/dn$. Then

$$\rho_c = \mathbf{J} \cdot \nabla(\epsilon\rho) = J_n \frac{d(\epsilon\rho)}{dn}.$$

Now J_n has the same value on both sides of the interface, since $\nabla \cdot \mathbf{J} = 0$. Therefore the surface density of charge is

$$\sigma = \int \rho_c \, dn = J_n \int_{(\epsilon\rho)_1}^{(\epsilon\rho)_2} \frac{d(\epsilon\rho)}{dn} \, dn$$

or

$$\sigma = J_n(\epsilon_2\rho_2 - \epsilon_1\rho_1), \tag{6–14}$$

where the subscripts 1 and 2 designate the properties of the two materials that meet at this interface.

In Fig. 6–4 $\epsilon_2\rho_2$ is so very much larger than $\epsilon_1\rho_1$ that the latter should be ignored. In the problem of Fig. 6–4 J is the current density in the carbon at $r = b$ and hence $\rho_2 J = E_b$. Thus

$$\sigma_b = J\epsilon_2\rho_2 = \epsilon_2 E_b,$$

which is the value of σ_b that would be calculated for a perfect dielectric for which $\epsilon = \epsilon_2$. Obviously in this case the charge in the junction is most easily calculated from the dielectric point of view. However, Eq. (6–14) is very useful for calculating either the space distribution of real charges in a non-homogeneous conductor or the surface distribution between materials having only mildly different values of $(\epsilon\rho)$.

We will now solve the problems suggested by Fig. 6–4. It will be simplest to solve the problem involving the dielectric washer first. Strictly speaking, the potential distributions with the dielectric must differ from those with the semiconductor. In the carbon washer the current densities are almost exactly radial inward, being confined to the carbon. On the other hand, lines of **D** with the dielectric are not confined to the washer but fringe out into space on both sides unless the K of the dielectric is very large. We will assume that $K > 30$ so that this fringing effect can be ignored.

The space inside the dielectric washer can be considered two-dimensional because the fringing flux of **D** existing in the space to the right of the dielectric is small enough to be ignored with this large dielectric constant. With the nickel cylinder at one potential and the copper cylinder at another, lines of electric flux pass between the two cylinders. By the symmetry of the problem, these must be radial lines drawn perpendicular to both cylinders, and the density of these lines must decrease as $1/r$ between the cylinders. A potential function is to be found whose gradient has the form of $1/r$. Such a potential is represented by

$$V = G \ln r,$$

which is readily shown to satisfy Laplace's equation. To be the correct solution, it must also satisfy the boundary conditions $V = V_{\text{Ni}}$ at $r = a$ and $V = V_{\text{Cu}}$ at $r = b$. These conditions yield two equations:

$$V_{\text{Ni}} = G \ln a$$

and

$$V_{\text{Cu}} = G \ln b,$$

from which $V_B = V_{\text{Cu}} - V_{\text{Ni}} = G \ln b/a$ and

$$V = \frac{V_B \ln r}{\ln b/a}.$$

The field in the dielectric is then

$$\mathbf{E} = -\nabla V = -\frac{V_B \mathbf{a}_r}{r \ln b/a},$$

and the charge densities on those surfaces of the nickel and copper cylinders that are in contact with the dielectric are

$$\sigma_{\text{Ni}} = \epsilon E_{\text{Ni}} = \frac{\epsilon V_B}{a \ln b/a}$$

and

$$\sigma_{\text{Cu}} = -\epsilon E_{\text{Cu}} = -\frac{\epsilon V_B}{b \ln b/a}$$

respectively. Since the magnitude of the charge where either cylinder is in contact with the dielectric is $2\pi t \epsilon V_B / \ln (b/a)$, the capacitance of this section is given by

$$C = \frac{q}{V_B} = \frac{2\pi \epsilon t}{\ln b/a}. \qquad (6\text{--}15)$$

If the dielectric is now replaced with a carbon washer of identical shape, the region where $\nabla^2 V = 0$ and its boundary conditions remain the same. The potential distribution in the washer is still

$$V = \frac{V_B \ln r}{\ln b/a}$$

and the field is still

$$\mathbf{E} = -\frac{V_B \mathbf{a}_r}{r \ln b/a}.$$

However, there is now a current density $\mathbf{J} = \mathbf{E}/\rho$, and the total current through the washer can be obtained by integrating over either metallic surface in contact with the washer to obtain

$$I = \int \mathbf{J} \cdot d\mathbf{s} = \frac{V_B 2\pi t}{\rho \ln b/a}.$$

The resistance of the washer is then

$$R = \frac{V_B}{I} = \frac{\rho \ln b/a}{2\pi t}.$$

For this geometry the product $RC = \rho\epsilon$, although the accuracy of this expression depends upon the accuracies of the individual expressions for R and C. The expression for R cannot be mathematically exact unless ρ for the air space is infinitely larger than that for carbon. Likewise, the calculation for C could be mathematically exact only if K for the dielectric were infinitely larger than that for the air space. Since the ratio of resistivities will be near 10^{18}, the expression for R is sufficiently accurate for all purposes. However, the ratio of the dielectric constants will seldom be as large as 30, so that the expression for C is much less accurate. The error in the product stems from the fact that the potential distributions are not quite the same in the dielectric washer as in the carbon washer.

The expression $RC = \rho\epsilon$ can be shown to have general validity only when the potential distribution in an isotropic and linear dielectric is identical with that in the same shape of semiconductor, which is both isotropic and ohmic. The chief advantage of the similarities in the expressions for these two types of problems lies in the fact that a current problem can sometimes be solved by electrostatic methods, using assumptions that would be unrealistic for the electrostatic problem but not for the current problem. For example, to correlate the Hall effects of ferromagnetic materials with their magnetic properties,* it was desirable to make simultaneous measurements of the Hall emf, the magnetic fields, and the magnetization within a rectangular bar of iron. For the magnetic measurements the fields should be parallel to a long bar, but high current densities perpendicular to these fields were needed. Figure 6–5 shows how a high current density was achieved, perpendicular to the length of the bar and constant along the line HH'. A relatively large current I entered the bar through the copper wedge in contact with its upper surface, spread out through the bar, and left it through the copper wedge contacting the bottom surface. In the neighborhood of the line HH', this produced a relatively large and fairly uniform current density, whose magnitude had to be calculated accurately. Since $\mathbf{E} = -\boldsymbol{\nabla}V$ in an

---

* Emerson M. Pugh, "Hall Effect and the Magnetic Properties of Some Ferromagnetic Materials," *Phys. Rev.* **36**, 1505 (1930), Fig. 1.

electrostatic field and $\mathbf{J} = -\nabla V/\rho$ in this situation, the problem could be solved if $\mathbf{E}$ could be determined for some fictitious electrostatic problem with the same boundary conditions on V and ∇V.

The actual current density pattern is two-dimensional in the bar, with the wedge contacts acting as source and sink. Electrostatically, these contacts can be simulated by charged filaments of infinite length. If the bar also is assumed to be infinitely long, the boundary conditions require that ∇V be parallel to the y-axis along both the upper and the lower surfaces of the bar, where $\mathbf{J}$ must parallel the surface. These boundary conditions can be satisfied easily for an electrostatic problem with an infinite series of equally charged filaments each infinitely long; that is, with filaments parallel to the x-axis, through $z = 0, \pm d, \pm 2d, \pm \cdots \pm nd \pm \cdots$, where the filaments must be positively charged when n is odd and negatively charged when n is even.

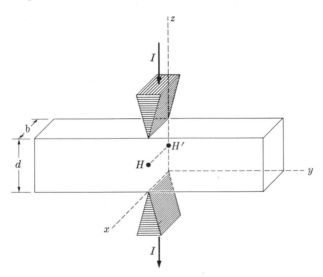

Fig. 6–5 A rectangular bar of ferromagnetic metal designed for measuring its Hall effect and its magnetic properties simultaneously. The sample current I passes through the bar from the upper to the lower copper wedge.

This infinite array of image filaments is symmetrically spaced above and below the xy-plane, in which one negatively charged filament passes through the origin. Symmetry requires that everywhere in this xy-plane the field $\mathbf{E}$ is directed parallel to the y-axis toward this negative filament. Since it is infinite, the array is also symmetrical about a parallel plane through the positive filament at $z = d$. Fields in this plane are parallel to it but directed away from the positive filament. These images then guarantee that currents based upon them will be properly directed. To find the magnitudes of the current densities, the effective charge per unit length, q_l, of the filaments must be determined. It is not difficult to show that $q_l = \rho \epsilon I_l$, where $I_l = I/b$ is the current per unit length through one of the wedge contacts.

From this set of image charges the field **E** at HH' can be written as an infinite series that fortunately converges quite rapidly, and from the limit of its convergence it can be shown that the current density at HH' is just $J = I/bd$. This solves the problem, since J can be determined from the easily measured quantities I, b, and d.

For a bar of finite length the problem can be solved by extending the pattern of image filaments throughout space, as shown in Fig. 6–6, where the pattern extends to infinity in both the y- and z-directions. Because they are two-dimensional, these problems can also be solved by means of complex variables.

6–8 Kirchhoff's laws

Kirchhoff stated two laws, both of which are indispensable for solving problems concerning dc currents in networks of conductors. When the extremely low compressibility of the "fluid" of conduction electrons in metals is considered, the first law seems so obvious for steady currents that it hardly needs to be stated. These laws are:

1. The algebraic sum of the n charges flowing out of any junction is zero; that is, if $I_1, I_2, I_3, \ldots , I_n$ are the currents out of a junction, then

$$\sum_{i=1}^{i=n} I_i = 0. \tag{6–16}$$

2. For any closed loop that can be drawn in a network, the algebraic sum of current through each resistor times its resistance for all of the resistors in a closed loop is equal to the algebraic sum of all the emf's encountered in traversing this loop. This may be expressed mathematically as

$$\sum_j I_j R_j = \sum_j \mathcal{E}_j. \tag{6–17}$$

The first law can be proved mathematically from the conservation of charge, Eq. (6–12):

$$\int_s \mathbf{J} \cdot d\mathbf{s} = -\int_\tau \frac{\partial \rho}{\partial t} \, d\tau.$$

The junction should be surrounded by some imaginary surface s enclosing a volume τ, as shown in Fig. 6–7. Under steady conditions the volume integral is zero and the surface integral gives Eq. (6–16).

The second law can be demonstrated by following the procedure of Section 6–4, where the integral

$$\oint \mathbf{E} \cdot d\mathbf{l} = 0$$

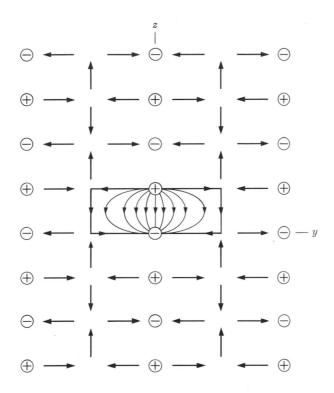

Fig. 6–6 System of line filament images, electrically charged, that are required to produce the electric fields·found in the iron bar of Fig. 6–5. The current densities in this bar are parallel and proportional to these fields. The filaments are infinitely long and the array extends to infinity in the y- and z-directions.

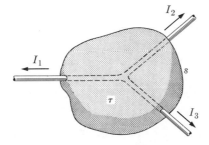

Fig. 6–7 Currents I_1, I_2, and I_3 leaving a junction enclosed within an arbitrary volume. Kirchhoff's first law that $I_1 + I_2 + I_3 = 0$ can be proved by applying Gauss' law $\int_s \mathbf{J} \cdot d\mathbf{s}$ to the bounding surface s.

around the enclosed loop of a network was divided into two integrals: (1) through all the resistors but excluding all sources of emf to obtain

$$\int \rho \mathbf{J} \cdot d\mathbf{l} = \sum_j I_j R_j,$$

and (2) through all of the sources of emf in the loop to obtain

$$\int \mathbf{E} \cdot d\mathbf{l} = \sum_j (-\mathcal{E}_j + I_j R_{ij}),$$

where R_{ij} represents the internal resistance of the source whose emf is $\mathcal{E}_j$. Therefore

$$\sum_j I_j (R_j + R_{ij}) = \sum_j \mathcal{E}_j,$$

which can be written simply as

$$\sum_j I_j R_j = \sum_j \mathcal{E}_j,$$

provided the internal resistance of each source is added to the resistance of the series segment of the network in which it resides.

6–9 Solution of dc network problems

The solution of dc problems involving networks of resistors connected together usually can be simplified by first calculating the equivalent resistance of some of the combinations of resistors. Whenever a group of resistors within a larger network is connected to that network by just two wires, this group can be replaced by a single resistor without affecting the rest of the network. The resistance of the single resistor required to replace the group is called its equivalent resistance. For example, in Fig. 6–8 the group of resistors enclosed by the smaller of the dashed-line rectangles can be replaced by a single resistor without affecting the currents in the rest of the network. Furthermore, the group enclosed by the larger dashed rectangle also can be replaced without changing the current through the battery.

When resistors are connected in series in a dc circuit, there is the same current I in each and the difference of potential across the group is the sum of the differences of potential across the separate wires. If $R_1, R_2, \ldots, R_n$ are the resistances of resistors connected in series and R_s is their equivalent resistance, then, from Eq. (6–17),

$$I R_s = I R_1 + I R_2 + \cdots + I R_n$$

and

$$R_s = R_1 + R_2 + \cdots + R_n.$$

When a group of resistors are joined so that one end of each is connected to one common junction while the other end of each is connected to another, the resistors are in parallel and the potential difference V is the same across all resistors. However, there are different currents, I_1, I_2, I_3, etc., through each resistor, so that the total current, according to Eq. (6–16), is $I_p = I_1 + I_2 + \cdots + I_n$. Defining R_p as the equivalent resistance of these resistors in parallel, this becomes

$$\frac{V}{R_p} = \frac{V}{R_1} + \frac{V}{R_2} + \cdots + \frac{V}{R_n}$$

and

$$\frac{1}{R_p} = \frac{1}{R_1} + \frac{1}{R_2} + \cdots + \frac{1}{R_n}. \tag{6–18}$$

The network problem of Fig. 6–8 is easily solved by the method of equivalent resistances. The equivalent resistance of the three resistors connected in parallel between b and c, by Eq. (6–18), is just 1 Ω. Since this is in series with a 5-Ω resistor, the equivalent resistance of the group enclosed by the small rectangle is 6 Ω. The effective resistance within the large rectangle between a and c is just 2 Ω, since the 6-Ω resistance in the small rectangle and the 3-Ω resistor are in parallel. The network then becomes a simple series circuit containing an emf of 24 V and a total resistance of $2 + 1 + 1 = 4\,\Omega$, which gives a current through the battery of 6 A. The current through any separate series segment can now be determined by simple applications of these two laws. For example, since the voltage drop from a to c is 6 A $\times$ 2 $\Omega = 12$ V, the current through the middle 3-Ω resistor is 12 V/3 $\Omega = 4$ A, and that through the 5-Ω resistor is 6 A $- 4$ A $= 2$ A.

Complex networks. The simple network of Fig. 6–8 has been reduced by the method of equivalent resistances to a problem in a single unknown current.

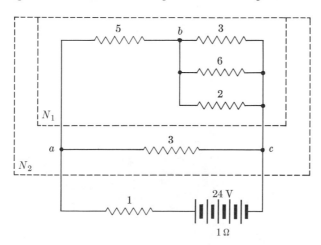

Fig. 6–8 A direct-current circuit which is completely solvable by the effective resistance method.

With many networks, there will be more than one unknown current after all possible equivalent resistances have been calculated. In general, there will be one unknown current for every remaining series branch. If N is the number of unknowns, it will be necessary to obtain N independent equations from Kirchhoff's two laws for the solution. Since Kirchhoff's junction equation (6–16) is very simple, the number of unknowns can be decreased conveniently by writing results from each junction equation directly on the network diagram. For example, at a junction like that in Fig. 6–7, the unknown I_3 can be eliminated by writing instead $-(I_1 + I_2)$. At each junction one unknown current can be eliminated, although the equation for the last junction considered is never independent of the rest. If J is the number of junctions, the number of unknowns can usually be reduced to $N - J + 1$. This number of equations must be obtained by applying Kirchhoff's closed loop equation (6–17) to $(N - J + 1)$ closed loops. Each branch of the network must be included in at least one loop equation. These general methods for solving network problems are demonstrated in the next section by solution of the potentiometer and the Wheatstone bridge problems.

6–10 Direct-current instruments

Many of the instruments for measuring properties of dc networks employ some form of the d'Arsonval galvanometer, whose characteristics are treated more extensively in the next chapter. A schematic drawing of one type of these galvanometers, which is made portable by supporting the coil assembly with its pointer in jeweled bearings, is shown in Fig. 6–9. Current through the coil, in the nearly constant magnetic field, causes it to

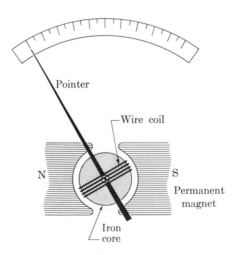

Fig. 6–9 Schematic drawing of a galvanometer pivoted in jeweled bearings.

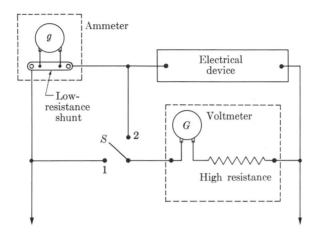

Fig. 6–10 A diagram to show how resistors are combined with a sensitive pivot-type galvanometer to produce either an ammeter or a voltmeter. It also shows two ways for connecting these instruments to measure the current through and the potential difference across an electrical device.

turn against the restoring torque of a coiled spring, not shown in the diagram. If the instrument is well designed the deflection of the pointer from zero to full scale is proportional to the current in the coil. As it stands, such an instrument will measure currents up to its capacity, which may be 10^{-3} A, although these instruments have been built to give full-scale deflections with only 5×10^{-6} A. It is neither feasible nor desirable to construct galvanometer coils to carry very much larger currents.

The laws of equivalent resistances presented in the previous section make it possible to combine resistors with galvanometers to make instruments for the direct reading of wide ranges of currents, and of potential differences. By adding a dry cell, resistance can also be read directly. For precision measurements, resistors are needed with accurately known resistances that remain constant over reasonable ranges of temperature and for a considerable time. Fortunately, resistors are available that satisfy these requirements reasonably well.

Ammeters and voltmeters. An ammeter connected in series to measure the current through some electrical device is shown in Fig. 6–10. It consists of a galvanometer whose coil resistance is R_g connected in parallel with a resistor or shunt often of very low resistance R_p, because this ammeter must carry a large fraction of the current being measured. The equivalent resistance of the ammeter is

$$R_A = \frac{R_p R_g}{R_g + R_p}$$

and, if I_g is the current in the coil, the total current through the ammeter is

$$I = \left(\frac{R_p + R_g}{R_p}\right)I_g.$$

For a galvanometer scales calibrated in terms of I_g, the ratio inside the parentheses provides the factor for determining I. It is desirable for R_A to be small compared with the resistances in series with it, for otherwise the ammeter causes too much disturbance to the circuit into which it is connected.

The voltmeter consists of one of these same sensitive galvanometers of resistance R_G, connected in series with a resistor having a very high resistance R_S. The potential difference measured by this voltmeter is the difference of potential across its terminals:

$$V = (R_S + R_G)I_G.$$

Because V is proportional to I_G, it is also proportional to the deflection. The total resistance of the voltmeter,

$$R_V = R_S + R_G,$$

should be very large, to avoid disturbing the circuits into which it is connected.

When these instruments are connected into networks to determine currents or potential differences, they become part of the network and alter its characteristics. To obtain precision results the networks must be analyzed with their measuring instruments in place. For example, two ways are shown in Fig. 6–10 for connecting the ammeter and voltmeter to obtain the current through the device and the potential difference across it. If I_1 and V_1 are the readings of the instruments when S is switched to contact 1 and I_2 and V_2 are these readings with S at contact 2, the following equations can be shown to give I, the current through the device, and V, the potential difference across it:

With S at 1,

$$I = I_1 \qquad \text{and} \qquad V = V_1\left(\frac{R}{R_A + R}\right),$$

and with S at 2,

$$I = I_2\left(\frac{R_V}{R + R_V}\right) \qquad \text{and} \qquad V = V_2.$$

Which connection is best depends upon the relative values of the resistances involved. A good rule is that 1 is best when $R/(R_A + R) \simeq 1$ and 2 is best when $R_V/(R + R_V) \simeq 1$.

Multiple-purpose meters. Meters for measuring both current and voltage may be purchased under various trade names. These instruments contain one galvanometer and several resistors that can be switched into series or into parallel with the galvanometer. Those of high resistance provide for measurements of several ranges of voltage, while those of low resistance provide for measurements of several ranges of current. Frequently these instruments also contain a battery which may be switched into series with the galvanometer. Its readings will then be inversely proportional to the resistance included in the circuit. With the use of proper calibrating resistors within the instrument, it can be adjusted to read the value of any resistance connected between its terminals.

The major advantage of the instruments discussed in this section is that they give rapid, direct readings of the desired quantity. In some applications, however, their precision is not good enough. The current drawn from the circuit to produce the galvanometer deflection is the major source of error in some applications; but even when this is not a serious problem, there is still the limitation in the accuracy with which a scale may be read. For applications in which considerable accuracy is required a null reading instrument is superior to a deflection instrument. Many circuits have been devised for specific applications,* but those most commonly used are the potentiometer and the Wheatstone bridge.

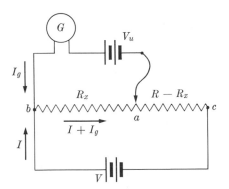

Fig. 6–11 Slidewire potentiometer with current designations suitable for easy solution of the circuit.

Potentiometer. The basic circuit for the potentiometer is shown in Fig. 6–11. The zigzag line from b to c represents a resistance wire of uniform cross section and total resistance R. The arrow touching the line at point a represents a sliding electric contact that divides R into two portions, R_x and $R - R_x$. The unknown difference of potential, V_u, that is to be measured is connected in series with a galvanometer as shown. By sliding the contact a along the resistance R, the galvanometer deflection may be reduced to zero and the

---

* For a discussion of several such circuits, see W. C. Michels, *Advanced Electrical Measurements*, D. Van Nostrand, Princeton, N.J., 1957.

instrument is then said to be balanced. This determines the value of V_u when no current is passing through it, as can be shown by solving the circuit as a network problem.

This problem fortunately provides an excellent illustration of the methods outlined in the previous section. The number of series branches is $N = 3$, but two junctions reduce the required number of unknown currents to $N - J + 1 = 2$. The directions for the unknown currents may be chosen arbitrarily, since, when the assigned direction is wrong, the solution will give the correct magnitude but with a minus sign. The current I_g through the galvanometer is of primary interest; its direction is designated by the arrow in Fig. 6–11. The other unknown current will be designated by I and chosen to pass through the potentiometer battery labeled V. The current equation for either junction now shows that the current through R_x is $I + I_g$. The two independent equations that are required can be obtained from the closed loop containing V_u, G, and R_x, and the closed loop containing V and R. If we let R_g be the internal resistance of the galvanometer, the resulting equations are

$$I_g R_g + (I + I_g)R_x = V_u$$

and

$$(I + I_g)R_x + I(R - R_x) = V.$$

These may be rewritten as

$$I_g(R_g + R_x) + I R_x = V_u$$

and

$$I_g R_x + I R = V,$$

and an immediate solution is found by the determinant method:

$$I_g = \frac{\begin{vmatrix} V_u & R_x \\ V & R \end{vmatrix}}{\begin{vmatrix} R_g + R_x & R_x \\ R_x & R \end{vmatrix}} = \frac{RV_u - VR_x}{RR_g + R_x(R - R_x)}. \tag{6–19}$$

To obtain the unknown potential difference V_u, the bridge is balanced by adjusting R_x until $I_g = 0$. If I_g is set equal to zero in Eq. (6–19) the following simple relation is obtained:

$$V_u = V \frac{R_x}{R}. \tag{6–20}$$

This suggests that the accuracy of the unknown voltage V_u is determined by the accuracy of V, R_x, and R, and that this accuracy is independent of R_g. This is not quite true. The accuracy with which the potentiometer can be balanced depends upon how much current I_g is required to visibly deflect the galvanometer. For maximum accuracy a high-sensitivity galvanometer is desirable; it is also desirable to have R_g small, since by Eq. (6–19) this greatly affects the value of I_g for a given unbalance. Most potentiometers are provided with an extra resistance to be connected in series with R_g to reduce sensitivity until a rough balance is obtained. The extra resistance is then removed for greater accuracy.

In the foregoing discussion it has been assumed that V and R are constant and known. In practice, a standard cell, whose emf V_s is accurately known from a Bureau of Standards certificate, is used for comparison. This cell is substituted into the circuit of Fig. 6–11 in place of the unknown V_u. Now if the galvanometer reads zero when R_x has been changed to R_s, the expression for V_s is $V_s = V(R_s/R)$. Division into Eq. (6–20) then yields

$$V_u = V_s \frac{R_x}{R_s},$$

which gives the unknown voltage in terms of the voltage of the standard cell and the ratio between two resistances.

Wheatstone bridge. Electrical measurements quite generally involve the comparison of the quantity to be measured with a standard. In principle, all electrical measurements can be made directly from the mechanical units of length, mass, and time, but such measurements are made by the Bureau of Standards in setting up their master standards. The standards used in other laboratories have been standardized by comparison with a master standard.

The Wheatstone bridge shown in Fig. 6–12, or some modification of this bridge,* is commonly used for the accurate comparison of unknown resistances with some standard designated by R_s. Like the potentiometer, the Wheatstone bridge employs the galvanometer as a null-reading instrument only.

The function of the galvanometer is to indicate when the circuits are balanced, so that there is no current in it. A great advantage of null methods is that they do not require a calibrated galvanometer. From Fig. 6–12 the requirement that there is no current in the galvanometer imposes two conditions on the circuit: first, the difference of potential between b and c is zero; and second, there is a single current I_s through ab and bd, and another single current I_R through ac and cd. From this analysis, elementary texts show that $R_x = R_s(1 - n)/n$. While this is the correct relation for a balanced bridge, it provides no information concerning the accuracy with which this balance can be obtained.

An expression for the current I_g through the galvanometer is needed to determine the accuracies obtainable with a Wheatstone bridge. Such an

---

* W. C. Michels, *Advanced Electrical Measurements*, D. Van Nostrand, Princeton, N.J., 1957.

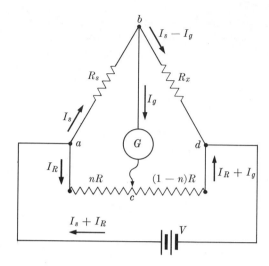

Fig. 6–12 Slidewire Wheatstone bridge with current designations suitable for easy solution of the circuit.

expression can be obtained by the methods outlined in the previous section. Since the bridge network has six series branches and four junctions, the number of unknown currents can be reduced to $N - J + 1 = 6 - 4 + 1 = 3$. In Fig. 6–12, currents I_g, I_s, and I_R have been assumed through G, R_s, and nR respectively, while junction equations have been used to determine the values of the currents in the remaining branches. The three unknown currents I_s, I_R, and I_g can be found from three independent equations for three closed loops such as *abca*, *bdcb*, and *acdVa*. The three equations obtained from these loops can be solved for I_g by means of determinants.

Although the result is complicated,* it can be considerably simplified and made more meaningful by assuming that only a small unbalance is produced by increasing the resistance in the R_x branch from $R_s(1 - n)/n$ to $R_s(1 + \delta)(1 - n)/n$, where $\delta \ll 1$. The solution resulting from this approximation is

$$I_g = \frac{n\,V\delta}{R_s + nR + R_g/(1 - n)}.$$

Study of this relation shows that best results are obtained when all the five branches have nearly equal resistances.

6–11 Routine solution of networks

The solution of dc networks can be reduced to a routine procedure that includes some automatic checks on the correctness of the equations. The

---

* The general solution may be found in L. Page and N. I. Adams, *Principles of Electricity*, 2nd ed., pp. 169–172, D. Van Nostrand, 1949.

procedure can be illustrated with either the circuit of Fig. 6–11 or that of 6–12, but the symmetry of the resultant equations can be seen more easily with the circuit parameters shown in Fig. 6–13, which is essentially the same network as that in Fig. 6–12. Since there must be three unknown currents, the three currents I_1, I_2, and I_3 can be assumed to circulate in the three loops of the network as shown. Loop equations are then written for each of the three loops, in which the resistances are collected with each current that they multiply. This gives

$$I_1(R_1 + R_5 + R_3) - I_2 R_5 - I_3 R_3 = 0,$$
$$-I_1 R_5 + I_2(R_2 + R_4 + R_5) - I_3 R_4 = 0,$$
$$-I_1 R_3 - I_2 R_4 + I_3(R + R_3 + R_4) = V,$$

which show a characteristic symmetry that helps to avoid blunders. The solution of these equations for any of the three unknown currents can be written as the ratio obtained by dividing one third-order determinant by another. The determinant for the denominator of this ratio, which is the same for each of these solutions, is given by

$$\begin{vmatrix} (R_1 + R_5 + R_3) & -R_5 & -R_3 \\ -R_5 & (R_2 + R_4 + R_5) & -R_4 \\ -R_3 & -R_4 & (R + R_3 + R_4) \end{vmatrix},$$

which is symmetric about the diagonal. The numerator for calculating I_1, I_2, or I_3 is obtained from this determinant by replacing column 1, column 2, or column 3, respectively, in this determinant with the terms on the right of the original three equations. With experience it is not necessary to write the separate equations, since the determinant for the denominator can be written directly from the diagram and the determinants for the numerators can be constructed directly from this denominator.

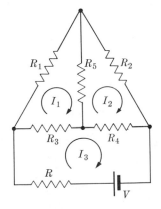

Fig. 6–13 Typical network, showing how the routine method of solution can be applied.

This system of solution may have one disadvantage if only the galva-nometer current of Fig. 6–12 is desired. With the routine solution, $I_g = I_1 - I_2$ can be obtained from the two separate solutions for I_1 and I_2, while with the system of solution used in the previous section I_g was obtained directly from only one solution. However, the advantages in using a routine system with automatic checks on the correctness of the solutions will usually outweigh this apparent advantage in having to solve for only one of the un-knowns.

Actually, an additional check can be introduced, if the network is such that it can be drawn on a flat surface with no intersecting branches. In the determinant of the denominator, all of the off-diagonal terms can be made negative by choosing loop currents that are either all clockwise, as in Fig. 6–13, or all counterclockwise.

PROBLEMS

6–1 A 660-W electric iron is plugged into a 110-V outlet. Calculate the number of electrons per second passing through the iron.

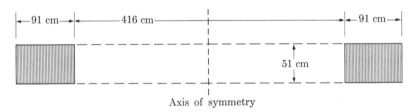

Figure 6–14

6–2 One of the coils which energizes the magnet for the *C-MU* cyclotron at Saxon-burg, Pa. has the dimensions given in Fig. 6–14, where the dimensions refer to the outer edges of the copper. The coil lies in a plane which is perpendicular to the cylindrical axis of symmetry, as shown in the diagram. The coil is made of 24 turns of copper conductor having a cross section 51 cm in height (the full height of the coil) and 3.4 cm in width (the direction of the coil radius). A space of 0.4 cm between each of the turns is occupied by some insulation and the channels for the cooling oil which flows upward between the turns. In operation, the coil carries a direct current of 20,000 A, and the temperatures of the cooling oil are 25°C as it enters and 30°C as it leaves the coil.
a) What is the resistance of this coil in operation and what voltage is required to operate it?
b) If the oil has a specific heat of 0.5 cal/gm C°, how much must flow through the coil? State the approximations you use, and indicate the approximate accuracy of your answer.

6–3 A 100-V dc generator with 1 Ω internal resistance is used to charge a 66-V storage battery with 2 Ω internal resistance. A 5-Ω rheostat is connected in series with these two to control the charging current.

a) Find the power delivered by the generator.
b) Find the heat loss in each part of the circuit.
c) Find the chemical energy stored in the battery during 3 hr of charging.

6-4 Two of the currents and all of the resistances of a certain network are shown in Fig. 6–15. Batteries A and B have negligible internal resistance.

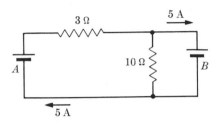

Figure 6–15

a) Find the current through the 10-Ω resistor.
b) Find the power delivered to the 3-Ω resistor. In what form does this energy appear?
c) Find the emf's $\mathcal{E}_A$ and $\mathcal{E}_B$ of the two batteries.
d) Find the power delivered to battery B. In what form does this energy appear?

6-5 Batteries A and B are connected to send current in opposite directions through a 14-Ω resistor. The current through A is 5 A and its internal resistance is 5 Ω. The current through B is 3 A and its internal resistance together with that of its connectors is 3 Ω. Find the current through the 14-Ω resistor and the emfs of the two batteries.

6-6 A dc generator using permanent magnets has an emf of 72 V and an internal resistance of 2 Ω. It is connected through a resistance of 3 Ω to charge a storage battery with 48 V emf and 1 Ω internal resistance.
a) Calculate the charging current and the chemical energy stored in the battery in 4 min and 10 s.
b) Neglecting friction, calculate the mechanical power needed to operate the generator.

6-7 A storage battery with an emf of 48 V, a resistor with the resistance 8 MΩ, and a leaky capacitor with the capacitance 20 μF and the resistance 12 MΩ are connected in series through a knife switch. One MΩ = 10^6 Ω.
a) Find the current I through the battery the instant after the switch is closed.
b) Find I after the capacitor ceases to charge.
c) Calculate the resistivity of the capacitor insulation if $\epsilon = 6\epsilon_0$.

6-8 The emf of a dry cell measured 1.56 V on a potentiometer and 1.24 V on a voltmeter of resistance 300 Ω. What is the internal resistance of the cell?

6-9 A suspended-coil wall galvanometer G is connected in the circuit shown in Fig. 6–16 with a dry cell, a fixed resistor $R = 10^5$ Ω, and two dial-decade resistors R_s and R_p. When $R_s = 0$ and $R_p = 25$ Ω, G deflects 15 cm. When $R_s = 70$ Ω and $R_p = 50$ Ω, G deflects 20 cm. Find the galvanometer resistance R_G and its sensitivity in A/mm deflection.

6-10 A certain simple potentiometer like that shown in Fig. 6–11 has movable contacts at both a and b. Contact a has 15 positions, each 10 Ω apart. Contact b

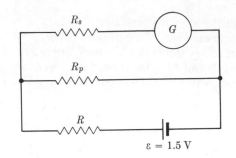

Figure 6–16

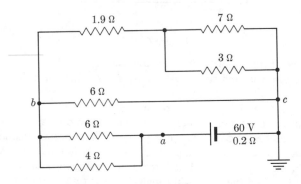

Figure 6–17

moves on a slidewire having 11 Ω total resistance. The maximum resistance of R_x is thus 161 Ω. The slidewire is divided into 220 divisions but the settings can be estimated to $\frac{1}{5}$ of a division. When a dry cell with $V = 1.5$ V is used, the total resistance is adjusted to $R = 1500$ Ω.

a) To what accuracy can an unknown voltage of approximately 0.0012 V be measured with this instrument?

b) If the only available galvanometer has a sensitivity of 5×10^{-6} A/cm and a resistance of 200 Ω, with what accuracy can the unknown voltage of (a) be determined? Deflections of 0.1 mm are just observable.

c) Suppose an unknown emf ($\sim$0.0012 V) with an internal resistance of 500 Ω is being measured, using the galvanometer described in (b). With what accuracy can this voltage be read?

6–11 In Fig. 6–17 the numbers give the resistances in ohms of the resistors under them. Calculate (a) the current through the battery whose emf is 60 V, (b) the current through the 6-Ω resistor between b and c, and (c) the actual potentials at a and b.

6–12 A flexible cable for conducting electricity consists of N strands of copper wire, each of which has the cross-sectional area a in m² and the length l in meters. The strands are insulated from each other except at the two ends, where they are all connected.

a) Calculate the effective resistance of this cable in terms of N, a, l, and ρ_{Cu}.

b) The actual length of this flexible cable is only $L = 0.95l$, because the individual strands are wound spirally. Consider a solid copper wire designed to have the same resistance with *the same length L* as this flexible cable. Calculate the ratio τ_s/τ_0, where τ_s = volume of copper in the strands and τ_0 = volume of copper in the solid wire.

6–13 A simple series circuit consists of an 18-V battery with an internal resistance of 0.1 Ω in series with two resistors whose resistance 0.4 Ω and 6.0 Ω are independent of temperature or current. A semiconductor, whose resistance in ohms is given by $R = \sqrt{18/I}$, where I is the current through it in amperes, is connected across the 6-Ω resistor. Calculate I.

6–14 Because the crystal structures of Cu and Ni are so nearly identical, one would expect the mobilities of the free electrons in the two materials to be similar. Mobility is defined by $\mu = \sigma/Nne$, where σ is the conductivity, N is the number of atom·m^{-3}, n is the number of conduction electrons per atom, and e is the electronic charge. Calculate μ_{Cu}/μ_{Ni} at room temperature and give a qualitative explanation of its value. Problem 6–12 shows that this ratio must be considered to be an average value.

6–15 There is some evidence that the ferromagnetic nature of nickel affects the scattering of conduction electrons in such a manner that half of these have far more mobility than the other half. Calculate the ratio of the mobilities of the conduction electrons of copper to the mobility of the more mobile half of the conduction electrons in Ni on the assumption that these more mobile conduction electrons in nickel have 10 times the mobility of the other half. [*Hint:* The conduction by the two groups of conduction electrons can be treated as through they constitute parallel circuits.]

6–16 Early strings of Christmas tree lights had eight bulbs in series connected across 110 V. With filaments at high temperatures these bulbs burned out frequently. Some individuals found that when two strings were connected in series the bulbs were bright enough and lasted almost indefinitely. On the assumption that the resistances of the bulbs remain the same, compare the watts output for each bulb on the string of eight with that for each bulb on the string of sixteen and also compare the total power used by these two strings. Actually, the power used by the string of sixteen is much greater and the bulbs are much brighter than this simple calculation indicates. Explain why these statements are true.

6–17 The power source for a research magnet on the third floor of a laboratory is a dc generator located in the basement 300 ft away (this distance is the length of the conduit necessary to carry the wires). The generator terminal potential difference may be regulated between 200 and 300 V, and the magnet, which has a total resistance of 0.50 Ω, may need up to 250 V across its terminals. Make a recommendation as to the material and size of wire to be used to connect the magnet to the generator.

6–18 For pumping water, 2500 W of power must be transmitted to a motor from a generator 1000 m away. The transmission line, consisting of two copper wires of equal diameter, should not dissipate more than 6% of the power delivered to the motor.

a) Calculate the minimum diameter to the copper wires if the motor requires 110 V across its terminals.

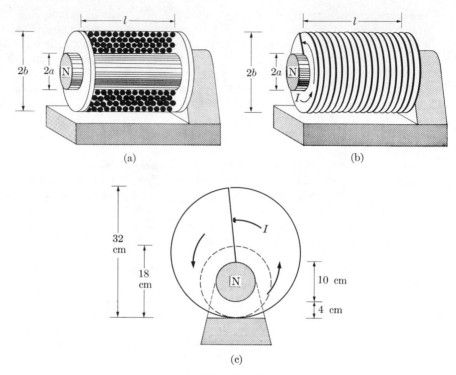

(a)   (b)

(c)

Figure 6–18

b) Calculate the minimum diameter of the copper wires if the motor requires 240 V across its terminals.

c) What voltage must be generated in each case?

6–19 Figure 6–18(a) shows a cross section of a coil of wire on a spool, whose shape and size are limited by the magnet for which it is designed. In Chapter 9 it is shown that the effectiveness of such a coil depends upon the number of its ampere-turns, which is limited by the rate heat can be lost from its surfaces. Would many turns of fine wire be superior to fewer turns of larger wire?

a) Calculate the number of ampere-turns as a function of the spool dimensions (a, b, and l), the wire diameter d, the space factor k, and the power P dissipated. The space factor k is the ratio of the volume of copper in the coil to the volume of the space occupied by the coil.

b) Does the equation obtained in (a) suggest the diameter of the wire that should be used?

6–20 A perfect capacitor (capacitance C) with infinite resistance is in parallel with a high resistance R_c. This combination is connected in series with a storage battery with the emf ε, a resistor with the large resistance R, and a switch. Find the expression for the current through R as a function of the time t after the switch is closed.

This circuit is equivalent to that of a leaky capacitor whose leak resistance and capacitance are R_c and C respectively. With large resistances the effects of self-inductance covered in Chapter 8 can be ignored. Such effects must be included with small resistances.

6–21 A large space, which can be assumed to be infinite with the homogeneous and isotropic resistivity ρ, has a uniform current density J_0 in the z-direction. A small sphere of radius a, embedded in the center of this space, has the resistivity 3ρ. Find the current density in the center of this sphere.

6–22 Calculate the real surface charge density σ on the surface of the sphere in the preceding problem, assuming the internal and external dielectric constants are given by $7\epsilon_0$ and $7\epsilon_0/3$, respectively.

6–23 A small conducting sphere of radius a is suspended h m above an infinite conducting plane, where $h \gg a$. A current I is supplied to the sphere by means of a battery whose negative terminal is connected to the plane. A dielectric for which $\epsilon = 3\epsilon_0$ and $\rho = 10^{15}\ \Omega\,m$ fills the space above the plane. Suppose the z-axis of cylindrical coordinates is perpendicular to the plane and passes through the center of the sphere. Place the origin an infinitesimal distance above the plane.
a) Find **J** at $r = r$ and $z = 0$.   b) Find the charge on the sphere.

6–24 Referring to Fig. 4–17, two long cylindrical conductors a and b eccentric to each other can be determined by $m_a = 4$ and $m_b = 3$.
a) In terms of s find the radii of these two cylinders.
b) In terms of s find the distance between their axes.
c) If the space between them is filled with semiconducting material of resistivity ρ, find the conductance per meter of length between them.

6–25 Find the real charge at the junction between the carbon washer and the copper cylinder of Fig. 6–4, assuming the potential difference between the copper and nickel is V and $\epsilon = \epsilon_0$. What would this charge be if $\epsilon = 10\epsilon_0$?

6–26 A large number N of identical storage cells is available, each with emf $\mathcal{E}$ and internal resistance r, to furnish the current for an electromagnet having a resistance R.
a) Calculate the current in the winding when the N cells are connected into p parallel strings, each having s cells in series.
b) Determine the relationship between R, p, N, and r for maximum current.
c) If $R = 0.5\ \Omega$, $\mathcal{E} = 2.2\ V$, $r = 0.1\ \Omega$, and $N = 36$, what value of p would you choose to obtain maximum current? Designate this value as p_m.
d) Calculate the current, using $p = p_m$, $p = p_m + 1$, and $p = p_m - 1$.
e) Calculate the current per cell for each case in (d) and state whether this information should influence your choice of p.

6–27 For finding the current density distributions in the very long bar of ferromagnetic material shown in Fig. 6–5, a method is outlined in the text for using an electrostatic solution involving an infinite series of imaginary line charges.
a) Calculate the charge per unit length λ required for these line charges in terms of I, ρ, and b, which are respectively the current through the bar, its resistivity, and its width.
b) Write the expression as an infinite series for the current density J along line HH' through the middle of a very long bar.
c) Show that J approaches I/ab.

6–28 Two identical stainless steel spheres of radius a buried deep in the ground are separated by the distance $l \gg a$. By means of insulated wires connected to each sphere a difference of potential V is maintained between them. Assuming the earth to have a uniform resistivity ρ, calculate the resistance between the two spheres. The expression for R should contain l, to show how large l must be before its magnitude can be neglected.

6–29 A stainless steel sphere of radius a is buried in the ground having a uniform resistivity ρ, near a very large sheet of stainless steel. If the normal distance between the center of the sphere and the plane is $l/2$, the resistance can be expressed in terms of an infinite series containing increasing powers of a/l. Calculate this resistance, retaining terms in the series up to $(a/l)^4$.

6–30 Calculate the resistance between the two spheres of Problem 6–28, retaining terms in the series up to $(a/l)^4$.

6–31 Suppose that the very large sheet of stainless steel in Problem 6–29 is bent at right angles, so that one part coincides with the xy-plane while the other part coincides with the xz-plane. Find the resistance between this bent sheet and a stainless steel sphere of radius a which is buried at the distance $l/2$ from each of the planes, assuming that $a \ll l$.

6–32 A certain dc transmission line consists of two large copper wires of radius a, nickel-plated to reduce corrosion. These are buried in the ground a distance $2s$ apart.

a) Assuming that the ground has a uniform resistivity $\rho \gg \rho_{Cu}$, calculate the conductance per meter of length between the wires.

b) Assuming that the potential difference between the wires is a constant V, calculate the power lost in heating the ground when the length of the line is l.

6–33 The spiral copper winding shown in Fig. 6–18(b) has advantages over the more conventional wire winding shown at (a) in that figure. When a fixed number of ampere-turns NI must be crowded into the space shown, the design in (b) provides less heat loss than the design in (a) and this heat loss is more easily dissipated.

a) Calculate the resistance per turn of the spiral if the copper sheet has a thickness t.

b) Calculate the total resistance of a spiral of N turns that will fill the space shown if the insulation between sheets makes the space factor k the same as that in Problem 6–19. This resistance should be expressed in terms of N, a, b, l, k, and ρ_{Cu}.

c) With a fixed power input P, calculate the ratio of the ampere-turns produced by the designs of Fig. 6–18(a) and (b). Calculate the numerical ratio for $b = 1.8a$.

[*Note:* The design of Fig. 6–18(b) has the disadvantages of being much harder to fabricate and of requiring a power source with a high current capacity and a low emf.]

6–34 Magnetic fields of electromagnets like those shown in Fig. 6–18 depend strongly upon the number of ampere-turns that can be passed through the space between the pole piece and the base. In the (c) part of the figure is a design of spiral copper winding that has less resistance per turn than does the spiral design in (b). With this design more ampere-turns are produced with the same power and the joulean heat is more easily dissipated than with either of the coils shown at (a) or (b).

a) Using the design and dimensions shown in (c) of Fig. 6–18, calculate the resistance per turn for copper sheet of thickness t.

b) Suppose the three coil designs shown in Fig. 6–18 are being considered for use in the iron frame of the magnet shown there. With constant power calculate the ratios of the ampere-turns for the designs of (b) and (c), each versus that for the design in (a).

6–35 One method for measuring the resistance of a galvanometer is known as Kelvin's method. The galvanometer whose resistance R_g is to be measured replaces R_x in the Wheatstone bridge shown in Fig. 6–12, while G is replaced by a circuit-closing key. How should such a bridge be balanced? Calculate R_g if the bridge balances with $R_s = 200 \ \Omega$ and with $n = 0.65$.

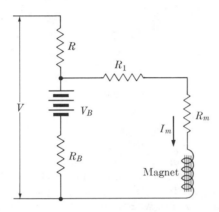

Figure 6–19

6–36 Many modern experiments require large electromagnets with constant magnetizing currents. Direct connection to dc power lines causes too much fluctuation. Storage batteries are better, but they must be monitored constantly because their voltages drift lower as they become discharged. The so-called floating-battery circuit, shown in Fig. 6–19, provides a remarkably constant current with a relatively small storage battery. Find an expression for the fractional variation in magnet current, $\delta I_m / I_m$, in terms of the circuit constants, and the fractional variation in the line voltage, $\delta V / V$. Discuss the case where the time average of the current through the battery is zero, so that V_B remains constant. (Actually storage batteries require a very small charging current I_c to keep their emf's constant. Best regulation is obtained when the average battery current is I_c.) State the circuit constants needed for best regulation and decide whether these constants are practical. Regulation will be improved by the inductance of the magnet that is ignored here.

6–37 An electromagnet has two coils, each with 1-Ω resistance, that can be connected either in series or in parallel. It is desired that each coil carry an 85-A current that is as steady as possible. Direct current lines with either 220 V or 110 V are available but each has a 2% voltage fluctuation. Two storage batteries are available, each with 88 V emf and 0.044 Ω internal resistance.

a) Design two different floating-battery circuits like the one shown in Fig. 6–19, one for use when coils are in series and the other for the parallel connection. Determine the values of the resistances needed in R and R_1 for each design.

b) Calculate the fractional fluctuations in I_m when the coils and the batteries are used in series.

c) Calculate the fractional fluctuations in I_m when the coils and the batteries are used in parallel.

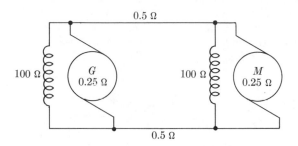

0.5 Ω

100 Ω   G 0.25 Ω   100 Ω   M 0.25 Ω

0.5 Ω

Figure 6–20

6–38 A machine called a dynamo, which is designed to operate as a dc motor, can usually be used as a dc generator. Suppose that two identical dynamos, G and M, are used as a generator and a motor respectively, to transport power from a gasoline engine to a distant water pump. As shown in Fig. 6–20, the direct current generated by G flows to and from M along two wires which constitute the transmission line. The dynamos have wire-wound armatures that rotate between poles of electromagnets. Assume that in each dynamo an emf is generated that is proportional to the rate of rotation of the armature, i.e., 0.1 V per rev/min. Since this can be true only if the field is constant, it will be assumed that the electromagnet is saturated as long as the current in its windings exceeds 0.1 A. (Actually it is impossible to maintain a constant field with this simple arrangement. The current through the armature affects the field and must be compensated to achieve a constant field.) Each field winding has 100 Ω resistance and is connected directly across the dynamo terminals, as shown in Fig. 6–20. Suppose that the armature of G rotates at 1200 rev/min and that of M at 1000 rev/min. Calculate (a) the currents through the field windings and through the armatures of G and M, (b) the power delivered by the gasoline motor to the generator and the power delivered to the water pump, if all mechanical losses can be neglected, and (c) the over-all efficiency in transmitting power from gasoline motor to water pump, neglecting all mechanical losses.

6–39 A uniform current density J is directed parallel to the y-axis of rectangular coordinates through a slab of nonuniform semiconductor. This slab is d meters thick, between two large copper plates both parallel to the xz-plane. The resistivity of the slab is given by $\rho = \rho_1 + [(\rho_2 - \rho_1)/d]y$. Assuming that the dielectric constant is given by $\epsilon = \epsilon_0$:

a) Find the charge density ρ_c in the slab.

b) Find the potential difference between the plates.

Chapter 7 ELECTROMAGNETIC INDUCTION

Both magnetic and electrostatic phenomena were observed by ancient man, as is evidenced by references to these phenomena in some of the earliest writings. Discussions of the properties of natural magnets are found in the writings of Thales of Miletus (600 B.C.), Socrates, and Plato. However, knowledge of magnetism was still only qualitative in 1600 A.D. when Gilbert summarized the known facts in his *De Magnete*.

Quantitative studies did not begin until the demonstration by Coulomb in 1785 that the forces between magnetic poles obeyed an inverse-square law. It was soon discovered that magnetic poles were unlike electric charges, in that poles could not be isolated. Furthermore, in 1820 Oersted established a relationship between moving charges and magnetism when he discovered that a compass needle could be deflected by an electric current. These facts stimulated Ampere to devise a theory which presumed that the magnetic properties of all materials were due to the circulation of submicroscopic electric currents. It is now well established that most, if not all, of these magnetic properties are due to such currents. These currents, now called "amperian currents," consist chiefly of electrons spinning about an axis or orbiting about the positively charged nuclei.

The observed magnetic forces can now be ascribed solely to forces between charges in motion. The magnetic fields produced by moving charges will be treated in Chapter 8; the forces on charges moving through magnetic fields are discussed here. Thus magnetic fields are defined here in terms of the forces on moving charges.

7–1 Lorentz force equation and field definitions

Electric charges moving through magnetic fields experience forces in addition to the electric field force expressed by $\mathbf{F} = q\mathbf{E}$. The total force acting on a charge in the presence of both electric and magnetic fields is given by the Lorentz force-on-a-charge equation

$$\mathbf{F} = q(\mathbf{E} + \mathbf{v} \times \mathbf{B}),\qquad(7\text{–}1)$$

where $\mathbf{B}$ is a vector field referred to as the magnetic induction and $\mathbf{v}$ is the velocity of the charge q moving through the field.

Equation (7–1) was originally a generalization of the empirical law governing the force on current-carrying wires in magnetic fields. This generalization has now been justified by many experiments involving charged particles and has been shown to be consistent with other theoretical considerations. It is therefore accepted as the fundamental law from which other relations can be derived and serves as a convenient definition for both $\mathbf{E}$ and $\mathbf{B}$ when written in the form

$$\mathbf{E} + \mathbf{v} \times \mathbf{B} = \lim_{q \to 0} \frac{\mathbf{F}}{q}.\qquad(7\text{–}2)$$

Here q is the magnitude of a small test charge in coulombs, $\mathbf{F}$ is the force on the charge in newtons, and $\mathbf{v}$ is the velocity of the charge in meters per second. By making $\mathbf{v} = 0$, Eq. (7–2) defines $\mathbf{E}$ in volts per meter, and with $\mathbf{E}$ determined, the charge can be given a convenient velocity to define $\mathbf{B}$. Just as electric flux lines can be drawn to represent electric fields, magnetic flux lines can be drawn to represent magnetic fields. Maxwell's second relation, which will be introduced in Section 7–4, shows that the intensity and the direction of the $\mathbf{B}$ field at every point in space can be represented by the density and direction, respectively, of continuous flux lines. *The mks unit for $\mathbf{B}$, therefore, is expressed in Wb m^{-2}, and the mks unit of magnetic flux is the weber.* It is shown in Section 7–3 that this definition leads to practical methods for measuring $\mathbf{B}$ in terms of emf's induced in search coils.

Considerable confusion may result from any attempt to apply Eq. (7–1) or (7–2) to coordinate systems other than those fixed to the earth. For example, a charge moving through a large uniform magnetic field in the absence of an electric field experiences a force $q\mathbf{v} \times \mathbf{B}$; however, if a new coordinate system is chosen in which the charge is at rest, then in the new system $\mathbf{v} = 0$, and no force would appear to be acting on q. Such dilemmas are discussed in Chapter 13, Relativity. Equation (7–2) and Maxwell's four field equations are strictly valid only when all of the quantities are measured in the observer's system of coordinates. Under such circumstances these five equations cover rigorously the whole field of electromagnetic phenomena.

Many devices and experiments make use of the forces described by Eq. (7–1). The cyclotron and Hall effect are chosen as two illustrations of the use of this equation; other illustrations may be found among the problems.

Cyclotron. A cyclotron consists of two copper dees, labeled D_1 and D_2 in Fig. 7–1, placed between two pole faces of a very large electromagnet. The two dees could be made by constructing a giant copper pillbox and then halving it with a plane cut that includes the central axis. The two dees are then enclosed in an evacuated container to allow free motion of the charged particles.

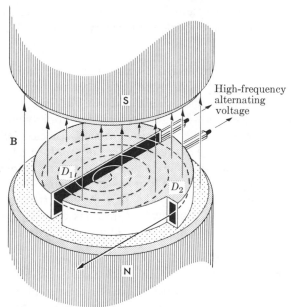

Fig. 7–1 Copper dees for a cyclotron. The dashed line indicates the path of a proton being accelerated to high energies. The uniform magnetic field **B** between the large poles of an electromagnet makes the proton travel arcs of circles within the dees.

A beam of protons or other charged particles is injected into one dee near the axis, with its velocity parallel to the faces of the dees. Within a dee $E = 0$, and each particle in the beam travels in a circular arc under the influence of the force $q\mathbf{v} \times \mathbf{B}$. Since this force is perpendicular to **v**, the magnitude of **v** does not change, and the kinetic energy of the particle remains constant as it traverses its path inside either dee. By equating the Lorentz force qvB to the centrifugal force mv^2/r, the radius of the path of the particle is found to be proportional to v, that is, $r = mv/qB$. The time required for a particle to complete one revolution is $t_r = 2\pi r/v = 2\pi m/qB$, which is independent of both r and v. A potential difference between the dees is caused to alternate at a frequency of $f_r = 1/t_r = qB/2\pi m$, so that those particles that happen to arrive at the separation between the dees when the potential difference is a maximum in the direction that will increase their velocities complete the next half revolution in time to emerge from this dee just when the potential difference is maximum in the opposite direction. Each time such a particle passes across the separation, its energy is increased by qV_m, which is evidenced by an increase in both velocity and path radius. The maximum velocity obtained, v_m, depends upon the maximum radius r_m available within the uniform field, and is given by $v_m = r_m qB/m$. A localized field, either electric or magnetic, is usually provided to cause the beam of charged particles to emerge along a predetermined path, shown near D_2 in the diagram.

Synchrocyclotron. This calculation is valid only for velocities that are small compared with the velocity of light, because of the relativistic increase in the mass of the particle as it approaches the velocity of light. Higher velocities and higher energies can, however, be achieved by properly controlling either the frequency or the field. The pole pieces at the Carnegie-Mellon University Research Center are about 12 ft in diameter and produce a field of approximately 2 Wb m^{-2}. Since protons in this cyclotron reach one-tenth of the velocity of light on paths only one foot in diameter, the larger diameters are utilized by accurately controlling the frequency of the potential applied across the dees. With this modification, protons can be given energies up to $4 \cdot 10^8$ eV, and such a cyclotron is called a synchrocyclotron.

Strong focusing. Since this cyclotron was constructed in 1956 there have been many improvements which have enormously increased the energies that can be given to charged particles. Chiefly these improvements increase the radius of the circular path. This is done by replacing the one very large magnet with many smaller magnets carefully placed upon the circumference of a large circle. The space between the pole faces of these magnets must have just enough taper to keep the charged particles moving in the median plane. This has been called "strong focusing." It is desirable to have the particles injected into the circular orbit at velocities close to the velocity of light. This can be done with early types of accelerators like the synchrocyclotron just described. When particles have velocities close to that of light, their kinetic energies can be increased greatly with infinitesimal increases in their velocities as is shown in Chapter 13.

Hall effect. When a conductor is placed in a magnetic field with a current that is perpendicular to the field, a voltage is induced across the conductor, perpendicular to both the current and the magnetic field. This voltage, known as the Hall voltage, is rather simply explained by means of Eq. (7–1).

A conductor of length l, width w, and thickness t oriented with respect to the *xyz*-axes as shown in Fig. 7–2 may carry an electric current in the *x*-direction through a magnetic field directed along the *z*-axis. The force on the conduction electrons in such a conductor is given by Eq. (7–1). Under steady-state conditions it is impossible for electrons to move in the *y*-direction, so that the net force in that direction must be zero. Equation (7–1) then reduces to

$$F_y = 0 = q(E_y - v_x B_z),$$

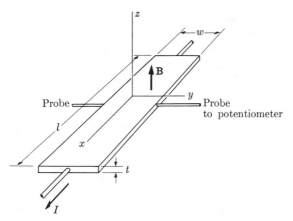

Fig. 7–2 Hall effect sample, a thin rectangular strip (length l, width w, thickness t) in the xy-plane with field **B** parallel to the z-axis. A potentiometer connected to the probes measures the Hall voltage.

where v_x is the average velocity of the conduction electrons flowing opposite to the current. With N being the number of atoms per m^3, m the number of conduction electrons per atom, e the charge of the electron in coulombs, the current is given by

$$I_x = Nmev_xwt,$$

where e and v_x are both negative for electronic conduction. The potential difference across the width of the conductor is $V_y = E_yw$; therefore the expression for the force in the y-direction reduces to

$$F_y = 0 = e\left(\frac{V_y}{w} - \frac{I_xB_z}{Nmewt}\right)$$

or

$$\frac{V_ywt}{wI_x} = \frac{B_z}{Nme} = RB_z.$$

The expression on the left is just F_y/J_x, which has been called the Hall resistivity, ρ_H. The common form of the expression is then

$$\rho_H = RB,$$

where $R = 1/Nme$ is the Hall coefficient, which depends on the nature of the material in the sample. This expression for the Hall resistivity ρ_H gives the number of conduction electrons per atom in terms of easily measurable quantities. In view of the grossly oversimplified assumptions used in deriving this relation, it is surprising that it quite accurately predicts the Hall voltages observed in many metallic elements and in many semiconductors. As should be expected from their position in the periodic table, the following metals exhibit Hall voltages which indicate that each has approximately one conduction electron for every atom in the solid: Na, K, Cu, Ag, and Au. Semi-conductors, on the other hand, have been shown to have a very small fraction of one conduction electron per atom.

One of the most important uses of Hall effect measurements in non-ferromagnetic materials has been to determine the nature of the conduction bands in these materials. Early measurements showed that both the magnitudes and the directions of the measured Hall emf's depended on the materials in which they were measured. Those materials that produced emf's in the direction predicted by the classical theory for conduction by electrons were called "normal," while those producing emf's in the opposite direction were called "abnormal." However, there appeared to be as many "abnormal"as "normal" materials. These designations now have been replaced by *n* for the negative or electronic type of conduction and *p* for the positive or non-electronic type of conduction. This latter type of conduction is now called "conduction by holes."

The quantum mechanical solution for this problem is quite complex. Fortunately there are available excellent approximations for conduction bands that are either nearly empty or nearly full. Both approximations give the same expression for the ordinary Hall coefficient, namely, the expression just obtained from the classical theory. A nearly empty conduction band exhibits electronic type conduction, and R is negative since e is negative. Kittel* gives an excellent discussion of the Hall effect in which he clearly explains why the vacancies (holes) in a nearly filled conduction band produce the type of conduction to be expected of positively charged electrons. These results are also confirmed by cyclotron resonance experiments.†

The Hall effect is also used in measuring magnetic fields, as described in Section 7–4. Since in ferromagnetic and even in some strongly paramagnetic materials, the Hall effect depends on their magnetic properties, the Hall effects in these materials are treated in Section 9–17.

7–2 Forces on current-carrying wires

The total force acting on a current-carrying wire can be thought of as the sum of the forces acting on all of the current carried within the wire. Consider a small element of wire of length dl carrying the current I, in which ρ is the density of conduction carriers (negative for electrons), v is the average velocity of the carriers in the direction $d\mathbf{l}$, and A is the cross-sectional area of the wire perpendicular to dl. The total conduction charge in the element is given by $\rho A\, dl$, so that the $q\mathbf{v}$ needed for Eq. (7–1) is $q\mathbf{v} = \rho A v\, \mathbf{dl} = I\, \mathbf{dl}$.

---

* Charles Kittel, *Introduction to Solid State Physics*, 2nd ed., pp. 296–301, John Wiley & Sons, New York 1956.
† *Ibid.*, pp. 371–379.

In the usual case in which the net charge on the wire is zero, there is no force on the wire due to **E**. Equation (7–1) then gives the expression for the force acting on an element of wire carrying a current:

$$d\mathbf{F} = I\,d\mathbf{l} \times \mathbf{B}. \tag{7–3}$$

The force d**F** on each element is perpendicular to both **B** and $d\mathbf{l}$. The total force on a long wire can be obtained by integrating Eq. (7–3) over the length of it. In dealing with wires, it is best to let $d\mathbf{l}$ indicate the direction of the current instead of I, since total current is not a vector, as is illustrated by the fact that the total current I through a surface s of any shape is given by

$$I = \int_s \mathbf{J} \cdot d\mathbf{s}.$$

Any volume element $d\tau$ in which there is a current density **J** and a magnetic field **B** is acted on by a force $d\mathbf{F}$, obtained directly from Eq. (7–1):

$$d\mathbf{F} = \rho\,d\tau\,\mathbf{v} \times \mathbf{B} = \mathbf{J} \times \mathbf{B}\,d\tau. \tag{7–4}$$

This equation readily reduces to Eq. (7–3) for the case of an element of wire of length $d\mathbf{l}$.

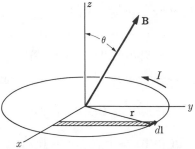

Fig. 7–3 Current-carrying loop of wire in the xy-plane experiences a torque about the x-axis in the uniform **B** field.

Torques on plane current-carrying loops. The windings for electric motors often consist of many individual coils of wire, each of which approximates a plane loop of wire carrying current in a closed path. Simple discussions of galvanometer characteristics usually treat the galvanometer coil as a plane loop. Even electrons moving in orbits about their nuclei can be treated as closed current loops for many purposes. When the current is the same at all points on the loop and the magnetic field is uniform, the loop experiences a torque but no net force.

In calculating the torque on the current loop, the shape of the loop will be assumed arbitrary, although confined to some plane in space and residing in a uniform field directed at any angle. The analysis is facilitated by drawing a rectangular coordinate system, as shown in Fig. 7–3, with its origin coinciding with the center of area of the loop, its xy-plane coinciding with the plane of the loop, and with the magnetic field $\mathbf{B}$ in the yz-plane. The force on the element $d\mathbf{l}$ is then

$$
\begin{aligned}
d\mathbf{F} &= I \, d\mathbf{l} \times \mathbf{B} \\
&= I(\mathbf{i} \, dx + \mathbf{j} \, dy) \times (\mathbf{j}B_y + \mathbf{k}B_z) \\
&= I(\mathbf{i}B_z \, dy - \mathbf{j}B_z \, dx + \mathbf{k}B_y \, dx).
\end{aligned}
$$

With $\mathbf{B}$ constant, the net force on the rigid loop is zero because $\oint dx = \oint dy = \oint dz = 0$. This can be obtained more directly from

$$
\mathbf{F} = \oint I \, d\mathbf{l} \times \mathbf{B} = -I\mathbf{B} \times \oint d\mathbf{l},
$$

where $\oint d\mathbf{l}$ is zero, since it is a vector sum representing the net displacement of a point traveling once around the closed path and returning to its starting point.

The total torque acting on this rigid loop is obtained by calculating moments of force about the origin for all of the elements in the loop and obtaining their vector sum. From Section 2–14, the moment of force acting on a typical element may be expressed by

$$
\begin{aligned}
d\mathbf{L} &= \mathbf{r} \times d\mathbf{F} = I\mathbf{r} \times (d\mathbf{l} \times \mathbf{B}) \\
&= I[d\mathbf{l}(\mathbf{r} \cdot \mathbf{B}) - \mathbf{B}(\mathbf{r} \cdot d\mathbf{l})].
\end{aligned}
$$

By introducing $\mathbf{r} = \mathbf{i}x + \mathbf{j}y$ and $B_x = 0$, the two terms in the brackets can be expanded into

$$
\mathbf{B}(\mathbf{r} \cdot d\mathbf{l}) = \mathbf{B}(x \, dx + y \, dy)
$$

and

$$
d\mathbf{l}(\mathbf{r} \cdot \mathbf{B}) = B_y(\mathbf{i}y \, dx + \mathbf{j}y \, dy),
$$

thus giving a total of four terms. Three of these terms make no contribution to the torque, since $\oint x \, dx = \oint y \, dy = 0$ around any closed loop. The remaining term is the area integral $\oint y \, dx$, which can be evaluated from its value in each of the four quadrants. In the first two quadrants y is positive and dx is negative, so that the integration of the element of area shown in Fig. 7–3 for that part of the loop lying to the right of the x-axis is just the negative of the area between the curve and the x-axis. In the third and fourth quadrants y is negative and dx is positive, so that completion of the integration yields the negative of the total area of the loop. Representing this area by A, we find

$$
\mathbf{L} = \oint \mathbf{i}IB_y y \, dx = \mathbf{i}IB_y \oint y \, dx = -\mathbf{i}IAB_y,
$$

a torque that can be written as

$$
\mathbf{L} = -\mathbf{i}p'_m B \sin \theta = \mathbf{p}'_m \times \mathbf{B}, \tag{7–5}
$$

where by definition* $\mathbf{p}'_m = \mathbf{k}IA$. Equation (7–5) is then identical in form to the torque that would act on a true magnetic dipole of moment $\mathbf{p}'_m$ in a magnetic field $\mathbf{B}$. This results in the very useful concept that *closed current loops act like magnetic dipoles in respect to the torques they experience in a magnetic field.*

7–3 Conductors moving in magnetic fields

When conductors move in magnetic fields, emf's are generated and electric currents are produced in closed circuits to which they are connected. The power required to produce these currents comes from the mechanical power needed to move the conductors through the magnetic fields. The law of conservation of energy requires that the current produced by this motion be in such a direction as to give rise to forces opposing the motion.

Lenz' law. A more general statement applicable to all magnetic induction processes was given by Lenz. It states that *an induced emf is directed so as to tend to oppose the cause producing it.* If the moving conductors are not connected to closed circuits there will be no current and no opposition; however, the emf's are always directed so that if a current exists there will be opposition.

Electric dynamos. Simple dc motors and generators have such similar constructions that they can be used interchangeably, and they are both frequently referred to as dynamos. Coils of insulated wire wound on an iron core make up the rotor or armature that rotates in a constant magnetic field between the poles of an electromagnet. An old style machine with coils wound around a hollow iron core is illustrated schematically in Fig. 7–4. The principles involved are essentially the same as for more modern machines, which are somewhat more difficult to illustrate clearly. Details of the wiring are shown only for the top half of the rotor, and the pole pieces and collector brushes are shown disassembled for clarity.

The armature coil consists of a single wire wrapped many times around the hollow armature, but each loop of the coil is connected electrically to armature segments, as shown in the figure. As the armature rotates, these segments make electrical contact only at the sides, where they slide by the collector brushes. In effect, these brushes produce two parallel circuits, one consisting of the coils on top that are connected in series and are moving to the right, and the other consisting of the coils on the bottom that also are connected in series but are moving to the left.

As the outer wires of the coils pass the north pole, the electrons within these wires experience a force $\mathbf{F} = q\mathbf{v} \times \mathbf{B}$, where $\mathbf{v}$ is the velocity of the

---

* Later in this text it will be found convenient to define a magnetic dipole moment by $\mathbf{p}_m = \mu_0 \mathbf{p}'_m$.

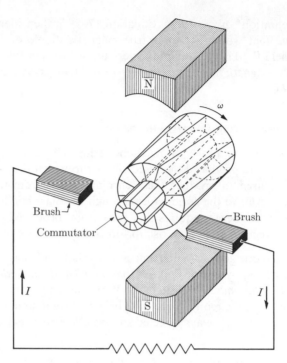

Fig. 7–4 Electric dynamo (dc generator or motor) with a hollow cylindrical iron core for a rotor. For clarity, the poles are separated from the rotor and the brushes are separated from the commutator.

wire.* Since it is perpendicular to **B** and to the motion of the wire, this force causes a drift of the conduction electrons toward the commutator segments, producing a positive current in the opposite direction, as indicated in Fig. 7–4. The inner wires of the coils are so effectively shielded from the magnetic field by the iron core that they do not contribute significantly to the induction process. The conduction electrons in those outer wires that are passing the south pole are directed away from the commutator segments and constitute a current that also is directed toward the brush on the right. The upper and lower circuits then contribute equally to the current in the external circuit. The direct current produced by such a dynamo would contain a small ac ripple caused by the changing conditions as the brushes slip from one commutator segment to another.

It is easily verified that the dynamo will act as a motor when current is forced through the windings by an external circuit. It is necessary only to consider the torques exerted on the closed current-carrying loops of wire in a magnetic field, as derived in Section 7–2.

---

* The drift velocity of the conduction electrons should be added, though this is usually negligible.

Simple linear generator. Rather than analyze the dynamo in more detail, it is more instructive to study the emf induced in a single conductor consisting of a straight metal rod moving through a uniform magnetic field with constant velocity **v**. Such a rod is shown moving along two frictionless metal rails in Fig. 7–5. Each of the charges in the rod experiences a force $q(\mathbf{v} \times \mathbf{B})$, but only the conduction electrons are free to drift through the rod. These electrons drift until sufficient charges have accumulated near the ends of the rod (or onto the rails in the figure) to set up an electric field **E** that counteracts the magnetic force. Positive charges appear on rail 1 and negative charges on rail 2. Because the electric field between the rails cancels the magnetic force on the charges in the rod, $\mathbf{F} = e(\mathbf{E} + \mathbf{v} \times \mathbf{B}) = 0$ when the conduction electrons stop drifting. Actually, the statement of Ohm's law for stationary ohmic resistors in Eq. (6–7) should be modified for moving resistors to

$$\mathbf{E} + \mathbf{v} \times \mathbf{B} = \rho \mathbf{J}. \tag{7–6}$$

Since the free charges in the rod cannot drift at arbitrary angles to the rod but must move along it, **J** must be parallel to the axis of the moving rod. Therefore, the left-hand side of Eq. (7–6) must also be parallel to this axis. Since in general $\mathbf{v} \times \mathbf{B}$ will not be parallel to this axis, **E** must be automatically adjusted so that the vector sum on the left side is parallel to the rod axis. This adjustment takes place as follows: When electrons cannot drift past the cylindrical surface of the rod, they collect there to cancel components normal to its axis.

If in Fig. 7–5 the rails are long and a constant velocity **v** has been achieved with $\mathbf{J} = 0$ and $\mathbf{E} = -\mathbf{v} \times \mathbf{B}$, then the voltmeter will read $V = \int \mathbf{E} \cdot d\mathbf{l} = \int \mathbf{B} \cdot \mathbf{v} \times d\mathbf{l}$, both integrated along the rod from rail 2 to rail 1. Now assume that the total resistance of the rails and rod is negligible

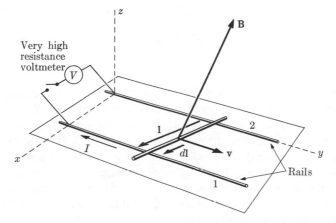

Fig. 7–5 Simple linear generator—a conducting rod sliding with velocity **v** along conducting rails in a uniform magnetic field **B**.

and that the voltmeter is replaced with a resistor having the resistance R. The current in the circuit then will be $I = V/R = JA$, where A is the cross-sectional area of the rod. This current in the rod produces a force $\mathbf{F} = I \int d\mathbf{l} \times \mathbf{B}$, which must be balanced mechanically to maintain the constant velocity. The mechanical power to be supplied is

$$-\mathbf{F} \cdot \mathbf{v} = -I \int \mathbf{B} \cdot \mathbf{v} \times d\mathbf{l}. \tag{7–7}$$

From the definition of emf $\mathcal{E}$ in Section 6–5,

$$\mathcal{E} = -\int \mathbf{B} \cdot \mathbf{v} \times d\mathbf{l} \tag{7–8}$$

integrated from rail 2 to rail 1, which is also the potential difference measured by the voltmeter in Fig. 7–5. Even if the rod also has a resistance R_i, this potential difference V gives the emf whenever the total current approaches zero.

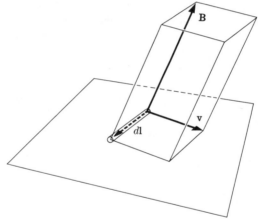

Fig. 7–6 Element of wire $d\mathbf{l}$ moving with velocity $\mathbf{v}$ and sweeping magnetic flux at the rate of $\mathbf{B} \cdot d\mathbf{l} \times \mathbf{v}$ per second.

Faraday's law. In general, $d\mathcal{E} = -\mathbf{B} \cdot \mathbf{v} \times d\mathbf{l}$ gives the emf induced in any small element of wire that moves through a field of intensity $\mathbf{B}$ with velocity $\mathbf{v}$. As can be seen from Fig. 7–6, the area swept per second by the element of wire is $\mathbf{v} \times d\mathbf{l}$ and the flux through this swept area is $\mathbf{B} \cdot \mathbf{v} \times d\mathbf{l}$. Consequently, the emf induced in any closed circuit in motion in a magnetic field is given by Eq. (7–8), which is just $-d\phi/dt$ when only wires are moving. Whenever the motion of a closed circuit in a fixed magnetic field consists only of the motion of some or all of the insulated wires that make up that circuit, the induced emf is given by

$$\mathcal{E} = -d\phi/dt. \tag{7–9}$$

Equation (7–9) is Faraday's law of induction. It is often used as the starting point for treating all magnetically induced emf's. While it is strictly true for all such cases *when it is properly interpreted*, it has often been misinterpreted.

Measurement of **B**. A very common method of measuring magnetic induction is the search-coil method, which correctly utilizes the principles of Faraday's law. A coil is wound with N turns of wire, each enclosing a well-defined area such that the vector sum of these areas is Ns, and this vector also defines the axis of the coil. The coil is placed in the field with its axis parallel to that component of **B** which is to be measured. The coil is connected in series with some device for measuring the total charge that flows through the circuit, such as the ballistic galvanometer or fluxmeter we shall discuss in Section 7–8. When the search coil is rapidly removed from the field to a region where **B** is zero, the resulting emf induced in the coil causes a transient current through the instrument which results in the passage of a total charge q that is proportional to the flux originally threading the circuit. Using Faraday's law we obtain

$$q = \int \frac{dq}{dt}\, dt = \int \frac{\mathcal{E}}{R}\, dt = -\int \frac{d\phi}{R\, dt}\, dt = -\frac{\phi}{R},$$

where R is the total resistance of the circuit. If the circuit is so constructed that no flux but that through the coil threads it, then

$$\phi = B_a N s,$$

where B_a is the average of the components of **B** parallel to s within the N turns. The charge flow through the circuit is then

$$q = B_a N s / R,$$

from which B_a may be determined. In principle **B** at a point can be approximated to any desired accuracy by reducing s. If neither the magnitude nor the direction of **B** is known, three independent observations in three mutually perpendicular directions will be needed.

The technique described gives the fundamental principle, although many modifications are possible. For example, the flux through the coil can be changed from ϕ to 0 by merely rotating the coil 90° whenever the direction of **B** is known. In another modification the coil is rotated at a constant angular velocity and measures the alternating voltage produced. It is easy to show that when the axis of rotation is perpendicular to the field $B = \mathcal{E}_m / \omega N s$, where $\mathcal{E}_m$ is the maximum emf in volts and ω is the angular velocity of the coil in radians per second.

Recently the Hall effect described in Section 7–1 is being used more and more for measuring **B**. A small Hall sample is placed normal to the field to be measured. The expression for the Hall voltage V_y given in that section can be written

$$V_y/I_x = (1/Net)B_z = RB_z,$$

where N is the number of free electrons or holes per unit volume of the sample. If V_y is measured with a potentiometer and the current through it is either in a series circuit with I_x or is otherwise proportional to it, the observed B_z will be independent of fluctuations in I_x. Usually semiconductors are used for the sample, because in them N is small and the resultant R is large. The chief handicap stems from the temperature dependence of R.

7–4 Maxwell's second relation

The method of measuring $\mathbf{B}$ by moving a search coil makes it easy to determine values for this vector near electric currents or inside and around magnetized materials. For example, the average direction and magnitude of $\mathbf{B}$ at the center of a permanent bar magnet can be determined by winding a search coil about the middle of the bar, connecting it to a fluxmeter, and then slipping it off the bar. The large deflection that results shows that the magnitude of $\mathbf{B}$ is quite large inside the magnet and that it points towards the positive end (usually designated as the north or north-seeking pole). The complete pattern of $\mathbf{B}$ values in and around bar magnets has been determined by techniques of these sorts. The lines, or tubes, of flux pass through the bar from points near the negative end and pass out into space at points near the positive end. In the surrounding space they follow curved paths from the positive to the negative end, and are continuous throughout space, with no beginning and no ending. Many experiments have shown that this pattern is typical of magnetic induction fields from all conceivable sources, and that the lines of magnetic induction are everywhere continuous, having no points of origin or termination.

The situation is correctly expressed by Maxwell's second relation in integral form:

$$\int_s \mathbf{B} \cdot d\mathbf{s} = 0, \tag{7–10}$$

where s is a completely closed surface surrounding a volume τ of perfectly general shape and location. The differential form of Maxwell's second relation is obtained by using the Gauss mathematical theorem to transform the surface integral of Eq. (7–10) into a volume integral:

$$\int_s \mathbf{B} \cdot d\mathbf{s} = \int_\tau \nabla \cdot \mathbf{B} \, d\tau = 0,$$

and since this is a general relation for all volumes,

$$\nabla \cdot \mathbf{B} = 0, \tag{7–11}$$

which is the differential form of Maxwell's second relation. This relation shows that the **B** fields can always be represented by continuous lines of magnetic flux. In Section 3–2 the integral form of Gauss' law,

$$\epsilon_0 \int_s \mathbf{E} \cdot d\mathbf{s} = \int_\tau \rho \, d\tau,$$

was applied to the typical tube of flux shown in Fig. 3–5 to show that electric flux originates on positive charges and ends on negative charges. When the arguments presented there are applied with Eq. (7–10) to the magnetic case, it is found that the magnetic flux must be continuous throughout space, since there are no real magnetic poles to provide either a source or a sink.

This experimental evidence for Maxwell's second relation can hardly be regarded as a formal proof but, strictly speaking, there is no formal proof for any of Maxwell's field relations. As with most of the fundamental laws of nature, the validity of these laws rests on the fact that their predictions agree with the results of carefully performed experiments. Since the results of most experiments have been checked with more elementary and possibly less fundamental relations, it is important to determine how closely the predictions of the elementary relations agree with those of Maxwell's relations. It is therefore demonstrated in Section 8–1 that Maxwell's second relation can be obtained directly from the Biot-Savart law provided it is assumed that all magnetic fields, including those due to magnetized materials, are generated by the flow of electric charges in closed paths.

7–5 Maxwell's third relation

One method described in the previous section for measuring the magnetic induction inside a permanent bar magnet consists of measuring the charge induced to flow through a coil when the coil is slipped off the magnet. If, instead, the coil is held fixed and the magnet is removed from the coil, the charge that flows through the meter is $q = \phi/R$, as before, where ϕ is the flux passing through the circuit before the magnet is removed. The correspondence between these cases was first observed by Faraday, and Eq. (7–9)

$$\mathcal{E} = -\frac{d\phi}{dt}$$

which expresses the induced emf at any instant of time for either case in terms of the rate of change of flux passing through the coil, is known as Faraday's law.

It should be noted, however, that there is a basic difference between the two experiments. With the observer in the laboratory coordinate system, the first case constitutes a coil moving in a time-independent field, while the second case constitutes a stationary coil in a field that changes with time.

While Eq. (7–9) can be used to describe either case, occasionally a problem arises which suggests the need for caution. The homopolar generator discussed in the following section represents such a problem. Equation (7–9) can hardly be misinterpreted when applied to stationary circuits in time-dependent magnetic fields.

However, Faraday's law can be used to obtain a more general expression, known as Maxwell's third relation. Note that the magnetic flux ϕ through a closed loop of wire is equal to the flux through *any* imaginary surface s whose periphery coincides with the wire. If **B** is the field at the element $d\mathbf{s}$ of this surface, then

$$\phi = \int_s \mathbf{B} \cdot d\mathbf{s}.$$

The emf induced in a stationary circuit is given by Faraday's law:

$$\mathcal{E} = -\frac{d\phi}{dt} = -\int \dot{\mathbf{B}} \cdot d\mathbf{s},$$

where $\dot{\mathbf{B}} = \partial \mathbf{B}/\partial t$. The current induced in the wire loop alters the value of **B**, but the equation is valid so long as **B** represents the resultant field due to all cases. The fact that charges do flow in the wire indicates that there must be an electric field. For conduction along fine wires of cross section a in series,

$$\mathcal{E} = \int i \, dR = \int Ja \, \frac{\rho \, dl}{a} = \int J\rho \, dl,$$

which is consistent with the more general expression

$$\mathcal{E} = \oint \rho \mathbf{J} \cdot dl = \oint \mathbf{E} \cdot dl$$

integrated around the circuit. The line integral for the induced electric field is seen to be independent of the value of ρ and is valid for any closed path whether or not wires are present. Combining the two expressions for $\mathcal{E}$, we find that

$$\oint \mathbf{E} \cdot dl = -\int_s \dot{\mathbf{B}} \cdot d\mathbf{s}, \qquad (7\text{–}12)$$

which has general validity and is called Maxwell's third relation in integral form. The differential form of Maxwell's third relation is obtained by using Stokes' theorem to transform the line integral into a surface integral; that is,

$$\oint \mathbf{E} \cdot dl = \int_s \boldsymbol{\nabla} \times \mathbf{E} \cdot d\mathbf{s} = -\int_s \dot{\mathbf{B}} \cdot d\mathbf{s}.$$

Since s represents any general surface, the integrands of the two surface integrals must be equal, and therefore

$$\boldsymbol{\nabla} \times \mathbf{E} = -\dot{\mathbf{B}}. \qquad (7\text{–}13)$$

is the differential form of Maxwell's third relation.

The special case of a stationary circuit in which a magnetic field is changing with time has been used here to justify Eq. (7–13). This equation, however, has general validity. In particular it is valid in a vacuum with no conductors present. As mentioned in the last paragraph of Section 7–4, the validity of each of Maxwell's relations rests on the agreement of its predictions with the results of many carefully performed experiments.

When the curl of a vector field is zero, the field is conservative. Thus where $\dot{\mathbf{B}} = 0$, as in dc networks, $\mathbf{E}$ is conservative. Generally $\mathbf{E}$ is not conservative and cannot be expressed as the gradient of a scalar potential. In Section 8–5, however, it is shown that $\mathbf{E}$ can be expressed in terms of both a vector and a scalar potential.

7–6 Induced emf's

The third field relation of Maxwell, given in integral form by Eq. (7–12) and in differential form by Eq. (7–13), applies specifically to changing fields in stationary circuits. For the special case in which a circuit encloses a plane area $\mathbf{A}$ in a uniform magnetic field $\mathbf{B}$, where it is assumed that both $\mathbf{A}$ and $\mathbf{B}$ may be time-dependent, Faraday's law gives the correct result and may be written as

$$\mathcal{E} = -\frac{d\phi}{dt} = -\frac{d}{dt}(\mathbf{B} \cdot \mathbf{A})$$

or

$$\mathcal{E} = -\mathbf{A} \cdot \frac{d\mathbf{B}}{dt} - \mathbf{B} \cdot \frac{d\mathbf{A}}{dt}.$$

The first and second terms on the right-hand side of this equation can be obtained from Eqs. (7–12) and (7–8) respectively, which proceed directly from Maxwell's third relation and the Lorentz force-on-a-charge equation, respectively.

The *very important general expression* for magnetically induced emf's is obtained directly from these two relations. It is

$$\mathcal{E} = -\int \dot{\mathbf{B}} \cdot d\mathbf{s} - \oint \mathbf{B} \cdot \mathbf{v} \times d\mathbf{l}. \tag{7–14}$$

Corson* has shown that when proper attention is paid to the fact that Faraday's law involves a total differential, it can be expanded to give Eq. (7–14). *The symbol* $\mathbf{v}$ *in this equation must be interpreted as the velocity relative to the observer's coordinates of some real material at the point where the element* $d\mathbf{l}$ *is located.* Since all real materials in motion contain many charges that move with them, this interpretation is consistent with the derivation of the second term in Eq. (7–14) from the Lorentz force equation (7–6).

---

* D. R. Corson, "Electromagnetic Induction in Moving Systems," *Am. J. Phys.*, **24**, 126 (1956).

Unfortunately an incorrect interpretation is commonly used with Faraday's law. This interpretation can be stated as follows: The emf magnetically induced in a given circuit is equal to the time rate of change of magnetic flux through that circuit. It is presumed, of course, that this rate of change can be brought about both by changes in field intensities and by movements of parts of the circuit to include more or less flux. Actually both interpretations do lead to the same predictions for circuits consisting only of small diameter wires. The difficulty arises primarily when, with extended conductors, ϕ is not well defined because the circuit is not well defined. The fallacy of this last interpretation is clearly revealed in the following study of the homopolar generator.

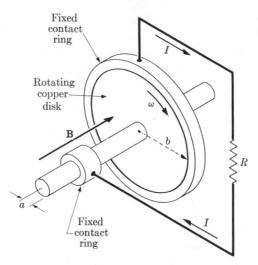

Fig. 7–7 Homopolar generator—a copper disk rotating in a uniform magnetic field. The heavy-copper contact ring assures that the induced current is radial.

Consider a flat disk of copper mounted on an axle so that it can be turned by mechanical power, and assume that some means can be devised for making good electric contact between the periphery of this rotating disk, shown in Fig. 7–7, and the fixed copper ring surrounding it. The ring is assumed to be so thick that its electrical resistance is negligible compared with that of the resistor R through which it is electrically connected to the axle of the disk. If the disk is now caused to rotate in a uniform magnetic field $\mathbf{B}$ which is perpendicular to its surface, an emf is induced which produces a current through the resistor R even though the total magnetic flux through the closed circuit remains strictly constant. The first interpretation predicts an induced emf of $\frac{1}{2}B\omega(b^2 - a^2)$, where $\mathbf{B}$ is the uniform magnetic field, ω is the angular velocity of the disk, b is its radius, and a is the radius of its axle. The second interpretation predicts no emf. Experiments show the first prediction to be correct.

Then* has performed several experiments to demonstrate the fallacy of the second interpretation of Faraday's law. We have applied Eq. (7–14) with the correct interpretation to all of Then's experiments. It correctly predicts all of his experimental results although the second interpretation fails in every case. In Section 13–2 on *relativity* this Eq. (7–14) also is used with the correct interpretation to solve "the puzzle."

7-7 Electric power conversion

Probably the most important single aspect of electromagnetic phenomena is the ease with which electromagnetic energy is converted into other forms of energy and vice versa. The fundamental relation needed for these conversions can be obtained from consideration of the power delivered to a charge q by the electric and magnetic fields in which it resides. If $\mathbf{F}$ is the force on q and $\mathbf{v}$ is its velocity, the power delivered is

$$P = \mathbf{F} \cdot \mathbf{v} = q(\mathbf{E} + \mathbf{v} \times \mathbf{B}) \cdot \mathbf{v}$$

or

$$P = q\mathbf{E} \cdot \mathbf{v}, \tag{7-15}$$

since $\mathbf{v} \times \mathbf{B} \cdot \mathbf{v} = 0$. In terms of electric currents, the volume density of power delivered to a charge density ρ moving with velocity $\mathbf{v}$ is

$$p = \mathbf{f} \cdot \mathbf{v} = \rho\mathbf{v} \cdot \mathbf{E} = \mathbf{J} \cdot \mathbf{E}, \tag{7-16}$$

where $\mathbf{J}$ is the current density. These two equations, which were derived and used in Chapter 6 for steady-state conditions, have here been shown to have general validity. Wherever these expressions differ from zero, conversions of power are taking place between electrical and other forms of energy. Where the sign of the product is positive, electric power is converted to other forms. Where this sign is negative, the conversion proceeds in the opposite direction. Two dynamos similar to the one pictured in Fig. 7–4 are often connected by a transmission line. The armature of one may be rotated by a steam or gasoline engine to generate power, which is carried by the electric current in the line to the other dynamo acting as a motor. Charges are made to flow opposite to the force of the electric field in the generator but not in the motor. In principle, two Faraday disks like the one in Fig. 7–7 can be used for this purpose, and because the detailed analysis is easier with the disks, we shall consider such a system.

In Fig. 7–8 the disk on the left generates power for the motor disk on the right. From Eq. (7–14), the emf of either disk rotating at ω radians per second is

$$\mathcal{E} = \int_a^b B\omega r \, dr = \tfrac{1}{2}B\omega(b^2 - a^2).$$

---

* J. W. Then, "Experimental Study of the Motional Electromotive Force," *Am. J. Phys.* **30**, 411 (1962).

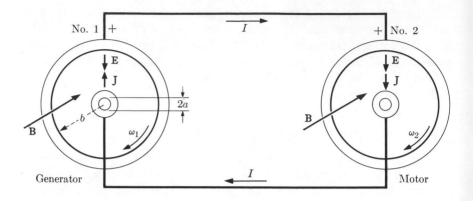

Fig. 7–8 Homopolar generator supplying power through a transmission line to a homopolar motor.

If the total resistance of the two wires of the transmission line is R_l and the internal resistance of each disk is R_i, the current I is given by

$$I(R_l + 2R_i) = \mathscr{E}_1 - \mathscr{E}_2.$$

Multiplying through by I to obtain the power relations, we obtain

$$I^2 R_l + I^2 R_i + I^2 R_i = \mathscr{E}_1 I - \mathscr{E}_2 I.$$

The five terms taken in order represent (1) the heat loss in the line, (2) the heat loss in disk No. 1, (3) the heat loss in disk No. 2, (4) the total power generated by disk No. 1, and (5) the electric power converted into mechanical power in disk No. 2 to overcome the torque due to friction and the mechanical load. A heavier load slows the rotation of the motor disk and reduces its back emf $\mathscr{E}_2$. If $\mathscr{E}_1$ is maintained constant, this reduction in $\mathscr{E}_2$ increases I and therefore the product $\mathscr{E}_2 I$ increases, since $I = (\mathscr{E}_1 - \mathscr{E}_2)/(R_l + 2R_i)$. The total electric power delivered by No. 1 to the rest of the circuit, $\mathscr{E}_1 I - I^2 R_i$, can be obtained by integrating $-\int \mathbf{E} \cdot \mathbf{J} \, d\tau$ over its disk, since $\mathbf{E}$ and $\mathbf{J}$ are there oppositely directed. Furthermore, the total power converted into mechanical energy and heat in No. 2, $\mathscr{E}_2 I + I^2 R_j$, is obtained by integrating $\int \mathbf{E} \cdot \mathbf{J} \, d\tau$ over its disk, where $\mathbf{E}$ and $\mathbf{J}$ are parallel.

7–8 Galvanometers

The moving-coil galvanometer, whose essential features are shown in Fig. 7–9, has been so improved in design and construction that it has practically replaced all other types of galvanometer. It has become an essential component in a multitude of instruments designed to measure or detect a great variety of physical quantities. For example, it is an important com-

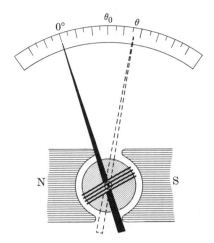

Fig. 7–9 Coil with angular position indicator for a typical moving-coil galva-nometer. The magnetic flux through this coil is given by $\phi = \phi_m(\theta - \theta_0)$, where $\phi_m = BA$, as discussed in the text, and is not a maximum flux.

ponent in instruments for measuring currents from 10^{-11} A to 10^6 A, potential differences from 10^{-8} V to 10^6 V, magnetic fluxes from 10^{-6} Wb m^{-2} to 10^4 Wb m^{-2}, etc.

Usually the coil of a galvanometer is held in its zero position by elastic forces. In a good portable instrument the coil may be mounted in jeweled bearings and held in its zero position by spiral springs that also serve to connect the coil to external circuits. A less portable instrument may have the coil suspended between two taut metal filaments that supply the elastic torque and also provide the conducting leads. The most sensitive galvanom-eters are not portable. Their coils are suspended by single metallic fila-ments that are made very long and thin to reduce the restoring torque. This single filament provides one conducting lead to the coil, but two are needed. The second lead is usually a light metal ribbon that contributes very little to the restoring torque on the coil.

For observing angular positions of the coils, metal pointers are rigidly attached to the coil in most portable instruments, while light mirrors are fastened to the coils of the more sensitive instruments. For these latter instruments many unique optical devices have been devised and are being used for determining angular deflections with great accuracy.

The circuit shown in Fig. 7–10 is adequate for representing most circuits that might be connected to a galvanometer. The resistance R_g includes that of the galvanometer and any control resistors connected in series with it, while R includes the internal and external resistances associated with the source of emf. Effects produced by eliminating the shunt can be determined by merely making $R_s = \infty$.

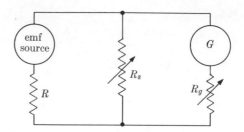

Fig. 7-10 Typical circuit connected to a galvanometer. Variations in R_s and R_g provide flexible controls of the sensitivity and damping characteristics.

In writing the differential equation for the galvanometer, use will be made of the relation derived in Section 7–2 for the torque on a closed loop of wire. The geometry of the permanent magnet is so designed that the flux lines are nearly radial for all angles within the allowed coil deflection, so that the magnitude of the torque applied to the coil is always $NiAB$, where N is the number of turns in the coil, i is the instantaneous current through the coil, A is its area, and B is the field of the permanent magnet. Referring to Fig. 7–10, the current through the coil can be expressed as the sum of the current through the coil induced by the source of emf plus the current through the coil induced by the motion of the galvanometer coil through the field B. The expression is

$$i = i_e + i_c = \frac{\mathcal{E}}{R_e} + \frac{\mathcal{E}_c}{R_c},$$

where i_e and i_c both represent currents through the galvanometer coil and the effective resistances R_e and R_c defined by this equation can be shown to be

$$R_e = R + R_g + \frac{R_g R}{R_s} \tag{7-17}$$

and

$$R_c = R_g + \frac{R_s R}{R_s + R} \tag{7-18}$$

by using simple circuit analysis like that presented in Section 6–9. By defining $\phi_m = BA$, the differential equation for the rotation of the galvanometer coil through an angle θ is found to be

$$G\frac{d^2\theta}{dt^2} + 2G\alpha'\frac{d\theta}{dt} + k\theta = N\phi_m\left(\frac{\mathcal{E}}{R_e} + \frac{\mathcal{E}_c}{R_c}\right), \tag{7-19}$$

where G is the axial moment of inertia of the coil, k is the spring constant, and $2G\alpha'$ is a convenient expression for the damping constant. If the coil is in equilibrium when $\theta = \theta_0$ and if the flux through it is zero in this position, then at the deflection angle θ this flux is given by $\phi = \phi_m(\theta - \theta_0)$. As the coil swings it produces the emf

$$\mathcal{E}_c = -N\frac{d\phi}{dt} = -N\phi_m\frac{d\theta}{dt},$$

and the differential equation of motion may then be rewritten as

$$\frac{d^2\theta}{dt^2} + 2\alpha\frac{d\theta}{dt} + \omega_0^2\theta = S\mathcal{E}, \qquad (7\text{-}20)$$

where $\alpha = \alpha' + N^2\phi_m^2/2GR_c$, $\omega_0^2 = k/G$, and $S = N\phi_m/GR_e$. The general solution for this second-order differential equation with constant coefficients is the particular integral plus the complementary function. The particular integral θ_p depends on the time-dependent nature of $\mathcal{E}$, while the complementary function θ_c does not. The solution for θ_c is defined as the general solution for the differential equation with the right side set equal to zero. This solution may be obtained by trying a solution of the form $\theta = e^{\mathcal{D}t}$, from which $\mathcal{D}^2 + 2\alpha\mathcal{D} + \omega_0^2 = 0$. This quadratic equation yields the two solutions

$$\mathcal{D}_1 = -\alpha + \sqrt{\alpha^2 - \omega_0^2} \quad \text{and} \quad \mathcal{D}_2 = -\alpha - \sqrt{\alpha^2 - \omega_0^2}. \qquad (7\text{-}21)$$

Since the solution for a second-order differential equation must have two arbitrary constants, the solution is

$$\theta_c = \mathcal{A}e^{\mathcal{D}_1 t} + \mathcal{B}e^{\mathcal{D}_2 t}, \qquad (7\text{-}22)$$

where each of the constants $\mathcal{A}$, $\mathcal{B}$, $\mathcal{D}_1$, and $\mathcal{D}_2$ may be complex. The nature of θ_c depends upon whether $\sqrt{\alpha^2 - \omega_0^2}$ is real or imaginary.

Underdamped galvanometer. When the radical in Eq. (7–21) is imaginary, it can be written as

$$j\sqrt{\omega_0^2 - \alpha^2} = j\beta.$$

Then Eq. (7–22) can be written as

$$\theta_c = Ae^{-\alpha t}\sin(\beta t + \delta), \qquad (7\text{-}23)$$

where β and the arbitrary constants A and δ can be chosen to be real. The resultant oscillatory motion is gradually damped out by the exponential term, as illustrated by curve U in Fig. 7–11. Assuming that α' may be neglected, Eq. (7–20) shows that $\alpha = N^2\phi_m^2/2GR_c$, and the galvanometer is underdamped when

$$R_c > \frac{N^2\phi_m^2}{2G\omega_0}.$$

Overdamped galvanometer. When the radical in Eq. (7–21) is real, it can be written as

$$\sqrt{\alpha^2 - \omega_0^2} = \gamma,$$

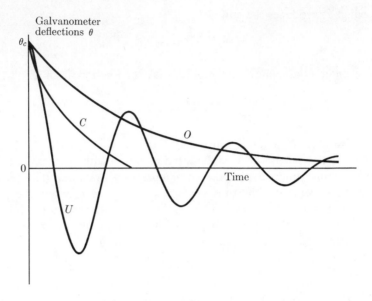

Fig. 7–11 Curves showing how the deflection θ of a galvanometer coil decays toward zero with different types of damping. The underdamped, overdamped, and critically damped cases are shown by the curves marked *U*, *O*, and *C*, respectively.

and then Eq. (7–22) can be expressed as

$$\theta_c = Be^{-\alpha t} \sinh{(\gamma t + \delta)}, \tag{7–24}$$

where γ and the arbitrary constants B and δ are real. When $\alpha \gg \omega_0$ damping is so complete that not only do no oscillations occur, but it requires a long time for the coil of the galvanometer to reach its equilibrium position. Again assuming α' to be negligible, the galvanometer is overdamped when

$$R_c < \frac{N^2 \phi_m^2}{2G\omega_0}.$$

Critically damped galvanometer. A galvanometer is said to be critically damped when $\alpha = \omega_0$. If α' may be neglected, critical damping can be achieved by making

$$R_c = \frac{N^2 \phi_m^2}{2G\omega_0}.$$

Since critical damping represents the transition between the underdamped and the overdamped conditions, the critically damped expression for θ_c can be obtained by letting either β or γ approach zero. In either case,

$$\theta_c = Ce^{-\alpha t}(t + D), \tag{7–25}$$

where the arbitrary constants C and D are real.

Each of equations (7–23), (7–24), and (7–25) decays exponentially with time, but Eq. (7–25) decays the most rapidly, as can be seen from the plots of these expressions in Fig. 7–11. "Critical-damping resistance," R_x, is defined as the value of R_c for critical damping less the actual resistance of the galvanometer.

Steady-current galvanometer. The general solution of Eq. (7–20) contains a term θ_p that does depend upon $\mathcal{E}$. When a steady potential difference $\mathcal{E} = V$ is supplied by the source in Fig. 7–10, this term is obtained by assuming $\theta_p = K$, which satisfies the differential equation and gives $\omega_0^2 K = SV$, or

$$K = \frac{SV}{\omega_0^2} = \frac{N\phi_m}{G\omega_0^2} \frac{V}{R_e}.$$

Hence

$$\theta_p = \frac{N\phi_m}{G\omega_0^2} \frac{V}{R_e} = \frac{N\phi_m}{G\omega_0^2} I_g,$$

where I_g is the final steady current through the galvanometer. The angle θ_p is the steady deflection remaining after θ_c has decayed to zero. The sensitivity of the galvanometer to current depends upon

$$\frac{\theta_p}{I_g} = \frac{N\phi_m}{G\omega_0^2} = \frac{N\phi_m}{k}. \tag{7–26}$$

Whenever possible, the galvanometer should be used in the critically damped condition, since final deflections are then achieved most rapidly.

Ballistic galvanometer. Galvanometers are often described as ballistic galvanometers when the first throw is being used to determine the total flow of charge, although this term was originally applied only to underdamped galvanometers used in this fashion. If the total flow is completed with sufficient rapidity, the first throw is proportional to the total charge regardless of the nature of the damping. This could be demonstrated by solving the differential equation in two stages. It is simpler, however, to recognize that a brief pulse of current through the coil gives it the angular momentum

$$G \frac{d\theta}{dt} = \int N\phi_m i \, dt = N\phi_m q.$$

Thereafter the coil deflection is controlled by Eq. (7–20) with $\mathcal{E} = 0$, which means that thereafter $\theta_p = 0$. Since θ_p is zero for the subsequent motion, the expression for θ obtained from Eq. (7–22) is

$$\theta = Ae^{D_1 t} + Be^{D_2 t},$$

regardless of the damping conditions. From the boundary conditions that at $t = 0$, $\theta = 0$ and $d\theta/dt = N\phi_m q/G$, we obtain the two equations

$$0 = A + B \quad \text{and} \quad \frac{N\phi_m q}{G} = AD_1 + BD_2.$$

With the values for $\mathcal{D}_1$ and $\mathcal{D}_2$ given in Eq. (7–21), these yield

$$\mathcal{A} = -\mathcal{B} = \frac{N\phi_m q}{2G\sqrt{\alpha^2 - \omega_0^2}},$$

from which we obtain

$$\theta = \frac{N\phi_m q}{2G\sqrt{\alpha^2 - \omega_0^2}}(e^{\mathcal{D}_1 t} - e^{\mathcal{D}_2 t}). \tag{7–27}$$

This general expression takes the form of Eq. (7–24) when $\alpha > \omega_0$, the form of Eq. (7–23) when $\alpha < \omega_0$, and the form of Eq. (7–25) when $\alpha = \omega_0$. In each case the function passes through at least one maximum. If the first maximum occurs at the time $t = t_m$, the first throw deflection is given by

$$\theta_m = \frac{N\phi_m q}{2G\sqrt{\alpha^2 - \omega_1^2}}(e^{\mathcal{D}_1 t_m} - e^{\mathcal{D}_2 t_m}). \tag{7–28}$$

If two experiments are performed with the same galvanometer and the same circuit with the same resistances, but with the two different values, q_1 and q_2, for q, then the expressions for the first throws, θ_{m1} and θ_{m2} respectively, will have identical values for N, ϕ_m, α, ω_0, $\mathcal{D}_1$, $\mathcal{D}_2$, and t_m. Therefore, in such experiments,

$$\frac{\theta_{m1}}{\theta_{m2}} = \frac{q_1}{q_2}$$

regardless of which type of damping is used. When sufficient sensitivity is available, the critically damped (or just slightly underdamped) galvanometer is the most convenient.

Galvanometer sensitivities. The sensitivity of a galvanometer to steady currents is described best in terms of the ratio θ/I, where θ is the final steady angular deflection produced by a steady current I in the coil. *The sensitivity of a galvanometer to a charge pulse* is described by the ratio θ_i/q, where θ_i is the maximum angular deflection produced by the passage of a pulse of total magnitude q through the coil. Pulse sensitivity, so defined, has meaning only if the source of the pulse has infinite resistance and is connected directly across the coil; e.g., the source may be a capacitor. Otherwise pulse sensitivity should be stated for the galvanometer combined with its circuit. It is left as an exercise to show that current sensitivity and pulse sensitivity are closely related.

Fluxmeter. A fluxmeter consists of a galvanometer in which $k = G\omega_0^2 = 0$ and in which $\alpha' \simeq 0$. It can be used to read the change in the magnetic flux that threads through a calibrated search coil, even when this change requires a relatively long time. In operation, the indicator may be first set at zero and the flux change then causes a permanent deflection that is proportional to this flux change. The instrument is equivalent to a thoroughly overdamped galvanometer. In fact, a galvanometer that has a very high current sensitvity can be converted to an excellent fluxmeter by connecting a low-resistance shunt across its terminals.

With the foregoing assumptions, the differential equation can be written as

$$\frac{d^2\theta}{dt^2} + 2\alpha \frac{d\theta}{dt} = SN_s \frac{d\phi}{dt},$$

where N_s is the number of turns in the search coil and $d\phi/dt$ is the rate of change of flux through it. If we let $d^2\theta/dt^2 = d\omega/dt$, this equation can be multiplied by dt and integrated directly; that is

$$\int_0^\infty \frac{d\omega}{dt}\, dt + 2\alpha \int_0^\infty \frac{d\theta}{dt}\, dt = SN_s \int_0^\infty \frac{d\phi}{dt}\, dt,$$

which becomes

$$(\omega_f - \omega_i) + 2\alpha(\theta_f - \theta_i) = SN_s(\phi_f - \phi_i),$$

where the subscripts i and f identify respectively the initial and final values of the quantities. At the start of the experiment ω_i was set equal to zero, and the final angular velocity ω_f must be zero to satisfy the law of conservation of energy. Therefore, from the definitions of symbols given for Eq. (7–18),

$$\phi_f - \phi_i = \frac{2\alpha(\theta_f - \theta_i)}{SN_s} = \frac{R_e N}{R_e N_s} \phi_m(\theta_f - \theta_i),$$

and the total deflection is proportional to the total change in flux through the search coil. It is instructive to note that the value ϕ_f for the final flux through the search coil is usually established before θ and ω reach their final values of θ_f and $\omega_f = 0$, since any residual energy, $\frac{1}{2}G\omega^2$, of the swinging coil must be converted to heat in the resistors.

PROBLEMS

7–1 Figure 7–12 shows a triangular loop of wire in which there is a current of 6 A. The magnetic intensity $\mathbf{B} = 1.1$ Wb m^{-2} is uniform over the triangle and parallel to the side AC, as shown.

a) Find the magnitude and direction of the force acting on the sides AB, BC, and AC.

b) Calculate the dipole moment of the loop and the magnitude and direction of the torque acting on it.

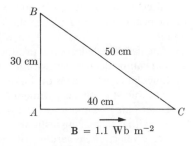

$\mathbf{B} = 1.1$ Wb m^{-2}

Figure 7–12

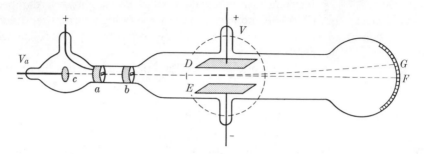

Figure 7–13

7–2 In Fig. 7–13 an electron is projected along the axis midway between the plates of a cathode-ray tube with an initial velocity of 2×10^7 m/s. Assume that there is a downwardly directed electric field, having the uniform magnitude 20,000 N/C over 4 cm of the path, but zero over the remaining 12 cm of the path, to the fluorescent screen at F.

a) How far above the axis has the electron moved when it reaches the end of the plates?

b) At what angle with the axis is it moving as it leaves the plates?

c) How far above the axis will it strike the fluorescent screen F?

7–3 In 1897 Sir J. J. Thomson proved that the "cathode glow" in evacuated tubes like the one shown in Fig. 7–13 was due to negatively charged "corpuscles" with a constant charge-to-mass ratio e/m. The figure shows his equipment, in which the corpuscles (now called electrons) were accelerated from the cathode c by a large potential difference V_a, between c and a, and collimated into a narrow beam by passing them through slits at a and b. When there were no other fields in the tube, the electrons traveled straight to F, where they caused the glass to fluoresce. To deviate this beam so that it would impinge on a point G, which was a distance d above F, an electric field was produced by applying a potential difference V between the parallel metal plates D and E, whose distance of separation was s. The beam was then brought back to F by applying just the right strength of magnetic field B, perpendicular to both the beam and the electric field. The velocity v of the electrons passing through the slit at b was never as large as one-tenth the velocity of light, so that Newton's laws of mechanics were sufficiently accurate. To simplify calculations it will be assumed that the E and B fields are uniform over the length l of the path of the beam from b to F. The following quantities could be measured: V_a, V, B, l, s, and d.

a) Calculate the deflection d produced by the electric field alone and state whether e/m can be determined without applying the B field.

b) Calculate e/m in terms of the measurable quantities. Can e and m be separately determined by this experiment?

c) Determine maximum values for V_a and V in terms of l, s, and d.

d) Show that by placing a slit at F this device can be used to select particles with a single velocity v, regardless of the magnitudes of the charges or masses of these particles.

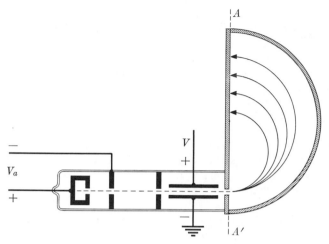

Figure 7–14

7–4 Figure 7–14 shows the principle of the Bainbridge mass spectrograph, which he used to determine the masses of the ionized nuclei of the isotopes of several elements. By means of a device similar to that in Fig. 7–13 a narrow beam of isotopes with different masses but with equal charges is projected into the evacuated chamber shown on the right in the figure, in the shape of a half circle. As pointed out in Problem 7–3(d), the isotopes in such a beam all have the same velocity. With a uniform magnetic field B, perpendicular to the paper in the semicircular chamber of the diagram, the isotopes travel different circles and blacken a photographic plate or film fastened to the plane AA'. Assuming that there are uniform electric and magnetic fields E_v and B_v respectively in the velocity selector:

a) Calculate the masses of the isotopes in terms of the distances, $d_1, d_2, \ldots$, from the slit of the blackened regions on the photographic plate.

b) Calculate the separations between the traces on the photographic plate due to the tin isotopes with mass numbers 116 and 120, from the trace due to that with mass number 118, which is 75 cm from the slit.

c) Assuming that both B and B_v are approximately 1 Wb m^{-2}, calculate the approximate velocity of the isotopes and the approximate potential difference V applied across the plates of the velocity selector, if these plates are 2 cm apart.

7–5 In Aston's mass spectrograph the beam of isotopes may have any distribution of velocities, so that no velocity selector is required. This instrument employs electric and magnetic fields, E and B, that are perpendicular to the original beam but are parallel to each other. Set up coordinate axes with a photographic plate in the yz-plane and a beam of positively charged isotopes projected toward this plane parallel to the negative x-axis. If the fields are uniform and parallel to the positive z-axis, the photographic plate will show parabolic traces.

a) Assuming that the beam enters the uniform fields at $x = x_0$, calculate the equation for these parabolas, for isotopes of mass m and charge e. In this calculation assume that the deflections are so small that the forces on the particles due to the magnetic field can be calculated from their original velocities.

b) Determine the errors, if any, produced by ignoring velocity components in the y- and z-directions in the expression for the forces on the particles due to the magnetic field.

7–6 An electron is moving along the positive y-axis at the constant velocity 220 m s^{-1} when it enters a uniform magnetic field $B_x = 0.4$ Wb/m^2 parallel to the x-axis. Find the magnitude and direction of the electric field required to keep the electron moving at the same velocity along the y-axis.

7–7 An electron with the charge e and the mass m moves at the velocity **v**, where $v \ll c$ in a rectangular coordinate system.
a) Write the general equation governing its motion in the uniform fields **E** and **B**.
b) Assuming that $\mathbf{E} = \mathbf{k}E$ and $\mathbf{B} = \mathbf{k}B$ and also that $v_z = 0$ at $t = 0$, calculate v_z as a function of E, B, e, m and t.
c) Assuming further that $\mathbf{v} = \mathbf{j}v_0$ at $t = 0$, find v_x and v_y as functions of v_0 and the quantities mentioned in (b).

7–8 A cyclotron like that in Fig. 7–1 has a magnetic field of 1.2 Wb/m^2 and a maximum radius of 0.4 m. Assume this is used to accelerate particles to velocities far below the velocity of light.
a) Derive the general formula for the frequency of rotation of these particles.
b) Determine the frequencies needed to accelerate (1) protons, (2) deuterons, and (3) α-particles.
c) Determine the maximum energies that these particles can achieve.

7–9 As pointed out in Section 7–1, an ordinary cyclotron operating with a constant frequency on the dees cannot accelerate protons to very high energies, because the protons fall out of step at this frequency, when their mass increases as they approach the velocity of light. Higher-energy protons can be obtained by periodically decreasing the frequency on the dees to keep in step with the rotational frequency of groups of protons. With this modification the device becomes a synchrocyclotron. At the synchrocyclotron of the Carnegie-Mellon University, the maximum radius available to the protons is 1.74 m, where the magnetic field is nearly uniform at approximately $B = 2$ Wb m^{-2}. The relativistic mass of a particle is given by $m = m_0/\sqrt{1 - \beta^2}$, where $\beta = v/c$, the ratio of the velocity of the particle to the velocity of light, and m_0 is the rest mass of the particle; i.e., m_0 is its mass measured at zero velocity. The energy required to accelerate the particle to the velocity v from rest is $(m_0c^2/\sqrt{1 - \beta^2}) - m_0c^2$.
a) Calculate the frequency of the alternating voltage required between the dees to achieve the initial accelerations.
b) Calculate the velocity of the protons in their outermost circle.
c) What must be the frequency on the dees to keep in step with the protons in this outer circle?
d) Calculate the total increase in the energy of the protons in electron volts.
e) What modification in the frequency pattern would be required to use deuterons instead of protons?

7–10 The fact that elementary particles like the electron have magnetic dipole moments and mechanical angular momentum provides qualitative confirmation for the concept that magnetic dipoles may be produced only by electric charges moving in closed paths. However, attempts to construct detailed models that are completely consistent have met with little success.

a) Calculate the dipole moment of a circular disk of radius a and thickness t rotating about its axis with angular velocity ω, assuming a uniform charge density ρ.

b) Calculate the dipole moment of a uniformly charged sphere rotating about a diameter. Assume that the sphere has a uniform charge density ρ, a radius a, and an angular velocity ω.

7-11 In the linear dynamo of Fig. 7-15, assume $B = 1.5$ Wb m^{-2}, $l = 1.2$ m, $R = 12\ \Omega$ and $\mathscr{E}_B = 36$ V. A mechanical device moves the rod to the right along the rails at the velocity v. Calculate the current and the mechanical power being converted into chemical energy in the storage battery when (a) $v = 30$ m s^{-1}, (b) $v = 20$ m s^{-1}, and (c) $v = 10$ m s^{-1}.

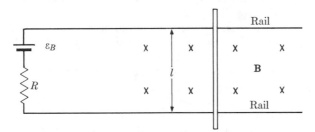

Figure 7-15

7-12 Assume that the linear generator is to produce a 3-A charging current through the battery, where $\mathscr{E}_B = 24$ V, $R = 4\ \Omega$, $l = 0.8$ m and $B = 1.5$ Wb m^{-2}.

a) How fast and in what direction must the rod be pulled along the rails?

b) How much force is required to produce this motion?

7-13 The general principles of a dc dynamo are simplified in Fig. 7-15. Let $R = 3\ \Omega$, $\mathscr{E}_B = 36$ V, $l = 0.9$ m, and $\mathbf{B} = 1.5$ Wb m^{-2}, into and normal to this paper. The mass of the rod is 0.3 kg.

a) With no friction and no load on the rod, calculate the magnitudes and the directions of its initial acceleration, and its final velocity.

b) If the rod must pull a load requiring a force of 6.2 N, what will be its final velocity?

c) If this dynamo is to act as a generator to produce a 3-A charging current through the battery, in what direction and how fast must the rod be pulled along the rails?

d) How much force is required to maintain this motion?

e) Calculate the electric power delivered by the rod and compare this with the mechanical power delivered to the rod.

7-14 Suppose that the rod of the linear dynamo shown in Fig. 7-15 is pulled to the right along the rails at a constant velocity of $v = 2V_B/Bl$.

a) Calculate the magnitude and direction of the current through the battery.

b) Calculate the mechanical power delivered to the rod and show in detail what happens to this power.

7-15 Suppose that the rod of the linear dynamo shown in Fig. 7-15 is connected to pull a load whose motion is resisted by the force Gv.

a) Calculate the terminal velocity of the rod and the mechanical power it delivers at this terminal velocity.

b) Calculate the power delivered by the battery. What happens to this power?

7–16 The individual loops of a 20-turn search coil encloses an average area of 1.5 cm². If this coil is rotated about a loop diameter at 60 cps, what is the smallest magnetic field that can be detected by it? Assume that the smallest amplitude that can be detected is 2×10^{-3} V.

7–17 A single circular loop of wire of radius a and resistance R lies in the xy-plane with its center at the origin. A magnetic field parallel to the z-axis reduces to zero according to the relation $\mathbf{B} = kB_0 e^{-\alpha t}$.

a) Assuming that R is large, so that the magnetic field of the induced current can be ignored, calculate the current, the charge that has passed any given point on the loop, and the power dissipated as heat.

b) Show that the total charge passing any given point depends only on the total change in flux, and is independent of $f(t)$, where $B = f(t)$, even when R is small.

7–18 Suppose that the magnetic field in Problem 7–17 is constant at $\mathbf{B} = kB_0$ but the loop rotates about the y-axis at the constant angular velocity ω rad/s. Using the assumption that R is large, calculate (a) the current, (b) the torque on the loop, and (c) the mechanical power required to rotate the loop. Compare the mechanical power with the rate of heat generation.

7–19 A thin conducting disk of radius a and thickness h is in a uniform magnetic field that is perpendicular to its surface. The resistivity ρ of the disk is sufficiently large that the magnetic fields due to the induced currents (which are called eddy currents) can be ignored. If the field is given by $B = B_0 \cos \omega t$, (a) find the current density as a function of r and t, and (b) find the power dissipated by the eddy currents as a function of time.

7–20 One of the most successful devices for producing high-energy electrons is the betatron invented by D. W. Kerst in 1940. The principle can be understood from Problem 7–17, where the change in flux through a circular loop of wire caused the conduction electrons in the wire to flow around it. If the wire is replaced with an evacuated hollow toroid of larger cross section, free electrons injected into this space can be accelerated to rather large energies by flux changes through the loop that are quite easily achieved. For the device to be successful the electrons must be constrained to move in the evacuated space of the toroid, preferably along the central circle. This is accomplished by maintaining a magnetic field of the required magnitude in the toroid along this central circle. Call the average field inside this circle B_a and the field at the circle itself B_0. Fortunately, if the electrons are at rest when these two magnetic fields start increasing from zero, the electrons can be held to their circular orbit by maintaining the correct constant ratio between B_0 and B_a.

The force on a particle is always given by $\mathbf{F} = d(m\mathbf{v})/dt$, where $m = m_0/\sqrt{1 - \beta^2}$ and $\beta = v/c$.

a) Calculate the velocity achieved by the electron when the average field inside an orbit of radius R has increased from zero to B_a.

b) Calculate the radius R' of the circle an electron will travel when it is moving with velocity v perpendicular to a field B_0.

c) Show that the electron will move in a stable orbit of constant radius $R' = R$ provided that, as the two fields are increased from zero, the ratio $B_a/B_0 = 2$ is maintained.

d) Suppose that such a machine could be built to maintain the above relations for an orbit of radius 1 m and a final average field within the orbit of 1.6 Wb m^{-2}. What would be the increase in energy of the electrons in electron volts?

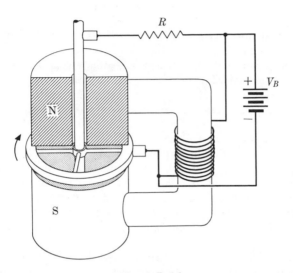

Figure 7–16

7–21 Figure 7–16 shows a homopolar motor which consists of a four-spoked metal wheel free to revolve in a magnetic field which is constant at $B = 2.0$ Wb m^{-2} over the spokes and zero over the rim. The spokes sweep out a circle of radius 30 cm and each carries the same outward radial current at all times, because the rim has a much lower resistance than the spokes even though these are negligible compared with R.

a) In which direction will the wheel rotate, in the direction indicated or in the opposite direction?

b) If the friction were truly negligible, what final speed, in radians per second, would the wheel achieve when connected to a storage battery having $\varepsilon_B = 90$ V and negligible resistance?

c) Calculate the speed of the wheel when the friction and the mechanical load on the motor amounts to a torque of 0·45 N · m and $R = 6$ Ω.

7–22 Integrate the power, Eq. (7–16), over the space occupied by the rotating disk in the homopolar generator of Fig. 7–7. Perform the same integration for the homopolar motor and compare both results with the power relations obtained in the text.

7–23 According to the definition of an emf given in Section 6–5, the Hall effect sample shown in Fig. 7–2 produces an emf ε_H in the circuit connected to the potentiometer. Power will be dissipated in this circuit whenever there is a current I_p through the potentiometer. Show that this dissipated power is $\mathcal{E}_H' I_p$ and that this is just equal to the power $\mathcal{E}_H' I$ that must be absorbed from the main circuit by a Hall-effect-induced back emf $\mathcal{E}_H'$.

7-24 Suppose the potential difference between the spherical terminal of a Van de Graaff generator and the point at which charges are sprayed onto the upward moving belt is 2 million volts. If the belt delivers negative charge to the sphere at the rate of $2 \times 10^{-3} \, \text{C s}^{-1}$, what horsepower must be expended to drive the belt against electrical forces?

7-25 Derive the relations given in Eqs. (7–17) and (7–18) for the effective resistances in the galvanometer circuit of Fig. 7–10.

7-26 Bohr's original model for the hydrogen atom assumed that an electron of charge e and mass m traveled in a circle of radius a with angular velocity ω around a much heavier nucleus that had a positive charge of the same magnitude e. He assumed that the angular momentum of the electon was quantized, so that $ma^2\omega = nh/2\pi$, where $h/2\pi = 1.05 \times 10^{-27} \, \text{erg·s} = 1.05 \times 10^{-34} \, \text{J·s}$ and $n = 1, 2, 3, \ldots$

a) Find a in terms of n, h, m, and e.

b) Find the total energy, W_n (kinetic plus potential), in terms of the same quantities.

c) According to Bohr's hypothesis, light is emitted by a hydrogen atom whenever the electron falls from an orbit having a large n to one with a smaller n. The frequency f of the light emitted was given by $hf = W_n - W_n'$. Determine the frequency and wavelength of the light emitted when the elcctron falls from the orbit where $n = 4$ to that where $n = 2$. Is this frequency visible to the human eye?

7-27 In the differential equation (7–20), the constant ω_0 is fixed by the construction of the galvanometer, but α depends upon the resistance of the external circuit. When $\alpha^2 < \omega_0^2$, Eq. (7–21) can be written as $\mathcal{D}_1 = -\alpha + j\beta$ and $\mathcal{D}_2 = -\alpha - j\beta$, where β is real. For this case, show that Eq. (7–23) is equivalent to Eq. (7–22) and that A and δ are real if $\mathcal{B} = \mathcal{A}^*$, the complex conjugate of $\mathcal{A}$.

7-28

a) Show that Eq. (7–23) approaches Eq. (7–25) in the limit as $\beta \to 0$. Calculate C and D in terms of A and δ.

b) Show that Eq. (7–24) approaches Eq. (7–25) in the limit as $\gamma \to 0$. Calculate C and D in terms of B and δ.

7-29 A galvanometer is connected into the circuit shown in Fig. 7–10. The resistances R, R_s, and R_g are adjusted so that the α of Eq. (7–20) equals $3\omega_0/5$. The source of emf is a dry cell, which produces the final steady deflection θ_p in this circuit.

a) Obtain an expression for the transient deflections, $\theta = f(t)$ assuming the dry cell is connected to the circuit at $t = 0$. The expression should contain only θ_p, ω_0, e, t, and some numerical constants.

b) Sketch the general shape of the θ versus t curve.

7-30 The steady current sensitivity of a galvanometer, θ/I, is closely related to its charge-pulse sensitivity, θ_i/q. When a certain sensitive galvanometer had no circuit connected to its terminals, its coil was observed to make ν complete oscillations per second with negligible damping. Its θ_i/q then was determined by discharging a capacitor through its coil. Calculate (a) the steady current sensitivity, and (b) the steady voltage sensitivity, if the coil resistance is R_g.

7-31 The current sensitivity θ/I of a sensitive galvanometer has been determined, where θ is the final steady deflection produced by a steady current through the coil of I. This galvanometer is to be used in the circuit of Fig. 7–10 for measuring the charge from a capacitor as it is discharged through the circuit. The magnitude of the shunt resistance is adjusted to $R_s = R_x$, the critical damping resistance. Calculate the charge sensitivity of the galvanometer circuit to obtain an expression for θ_i/q, where θ_i is the maximum deflection produced when the capacitor discharges the charge q through the circuit.

7-32 A certain very sensitive galvanometer encounters so little friction that its α' may be neglected.
a) Show that its ω_0 may be obtained by timing the oscillations of its coil on an open circuit.
b) Show that when ω_0 is known the important ratio $N\phi_m/G$ can be obtained by observing the steady deflection produced by a known steady current.
c) With the results obtained in (a) and (b), show that both $N\phi_m$ and G can be obtained by determining the critical damping resistance.

7-33 The galvanometer in Problem 7–32 can be used as a fluxmeter by making R_s, in the circuit of Fig. 7–10, very small compared with its critical damping resistance.
a) Prove this statement.
b) Show that the measurements outlined in (a), (b), and (c) of Problem 7–32 are sufficient for calibrating this fluxmeter, provided all resistances are known. A precise fluxmeter can be obtained by using a galvanometer with very high current sensitivity which is reduced and adjusted with a low-resistance shunt. The deflection of such a galvanometer will take many hours to drift to its rest position.

7-34 In Fig. 7–15 replace the battery with a capacitor having the capacitance C and the original charge Q. Connect this capacitor to the rails at the instant when $t = 0$. Derive (a) the equation for the current i versus t, and (b) the equation for the velocity v versus t. (c) What is the initial acceleration?

Chapter 8 MAGNETIC FIELDS OF ELECTRIC CURRENTS

In 1820 Hans C. Oersted made the discovery that an electric current would exert a force on a magnetized needle. Investigations of this phenomenon were quickly started by several scientists, including Dominique Arago and André M. Ampere, who independently magnetized needles by wrapping them with coils of wire carrying electric currents. Ampere's investigations, which were both experimental and mathematical, led him to the concept of magnetic sheets which is discussed in Section 9–10. Wilhelm Weber had proposed that the magnetization of iron resulted from the orientation of molecular magnets, and Ampere proposed that the magnetic moment of the molecules might originate in circular electric currents. Consequently, the circulating charges of the atoms and molecules are often referred to as "amperian currents." The mathematical relation between steady currents and the magnetic fields induced by them was first formulated by Biot and Savart.

8–1 The Biot-Savart law

The law formulated by Biot and Savart and listed in Section 2–14 is usually written in vector notation as

$$dB = \mu_0 \frac{I \, dl \times a_r}{4\pi r^2}, \tag{8-1}$$

where $d\mathbf{B}$ is the contribution to the field, at point P in Fig. 8–1(a), that is due to a current I in the wire element dl. The vector distance of P from this element is $\mathbf{r} = r a_r$. In this form the law cannot be verified experimentally, because it is impossible to isolate an element of wire carrying a steady current to determine its effect upon the field. Steady currents are found only in closed circuits; therefore the magnetic field at any point due to a steady

254

current is the vector sum of the contributions from all parts of the closed circuit. The field at the point P in Fig. 8–1(a) is given by

$$\mathbf{B} = \mu_0 \oint \frac{I\, d\mathbf{l} \times \mathbf{a}_r}{4\pi r^2} \tag{8–2}$$

integrated over the closed circuit, provided the wire carrying the current I is so fine that its dimensions can be neglected. With finite conductors of various shapes, like those in the circuit of Fig. 8–1(b), the expression is

$$\mathbf{B} = \mu_0 \int \frac{\mathbf{J} \times \mathbf{a}_r\, d\tau}{4\pi r^2} \tag{8–3}$$

integrated throughout the region where there are currents. The magnetic fields due to many shapes of closed circuits have been calculated by means of Eq. (8–3) and the results have been verified experimentally. The calculated values have agreed so exactly with experiment that the correctness of the integral form of the law for steady currents is no longer questioned.

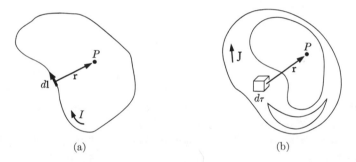

Fig. 8–1 (a) Current-carrying wire giving rise to a **B** field at P, in accordance with the law of Biot-Savart. (b) Current-carrying loop of finite dimensions.

Fields of circuits with straight wire sections, It is often necessary to calculate magnetic fields due to circuits composed of several straight sections of wire. An expression for the field due to a single section of wire can be used to obtain the total field of a more complex figure by taking the vector sum of the fields produced by the individual straight segments. In Fig. 8–2(a), for example, the field **B** at P due to the current I in the rectangular wire loop is the vector sum of the fields at P due to the four straight sides.

The method for calculating the field caused by a wire segment at an arbitrary point P is illustrated in Fig. 8–2(b). The term $d\mathbf{l} \times \mathbf{a}_r$ in Eq. (8–2) is seen to be directed out of the paper for all values of ϕ and is numerically equal to $dl \sin \theta = dl \cos \phi$. Also from the figure, $r = a \sec \phi$, $l = a \tan \phi$,

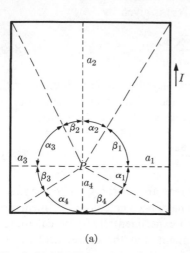
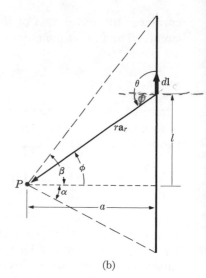

(a)   (b)

Fig. 8–2 (a) Illustration showing angles and lengths needed to calculate **B** at *P* within a current-carrying rectangular wire loop. (b) Diagram for calculating **B** at *P* due to the current *I* in a finite segment of a straight wire.

and $dl = a \sec^2 \phi \, d\phi$, so that the magnitude of **B** at *P* becomes

$$B = \int_{-\alpha}^{\beta} \frac{\mu_0 I \cos \phi \, d\phi}{4\pi a} = \frac{\mu_0 I}{4\pi a} (\sin \beta + \sin \alpha), \qquad (8\text{–}4)$$

where **B** is a vector directed out of the paper. For the special case of an infinite wire, $\sin \beta = \sin \alpha = 1$ and $B = \mu_0 I / 2\pi a$. Beginning with Eq. (8–4), we can readily verify that the magnitude of **B** at point *P* in the rectangle is given by

$$B = \frac{\mu_0 I}{4\pi} \sum_{n=1}^{4} \frac{1}{a_n} (\sin \beta_n + \sin \alpha_n),$$

where the a_n's represent the perpendicular distances from *P* to the respective sides and the α_n's and β_n's are the angles indicated in Fig. 8–2(a). For the special case of a point in the center of a square with sides $2a$ in length, this becomes

$$B = \frac{\mu_0 \sqrt{2}\, I}{\pi a}.$$

Axial field of a solenoid. Fairly strong fields may be produced by winding wires in the form of a solenoid, so that the fields due to successive turns are additive. A common type of solenoid, shown in Fig. 8–3, consists of fine insulated wire wound in a tight spiral of one layer. By making the axis of the solenoid coincide with the *z*-axis and by placing its lower circle an unspecified distance *Z* above the *xy*-plane, the calculation of the field at the origin provides an expression for the field at a general point on the axis.

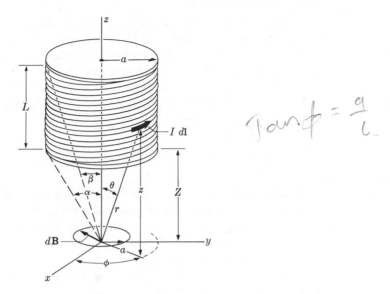

$\tan \phi = \dfrac{a}{L}$

Fig. 8–3 Single-layer cylindrical solenoid. The magnetic field at a general point on the axis is calculated from the Biot-Savart law, by integrating along the wire winding.

If the winding, starting at the point $(a, 0, Z)$ in rectangular coordinates, proceeds in the direction of increasing ϕ which is also the direction of the current I, the element of length of the wire for the Biot-Savart relation can be expressed in spherical coordinates as $d\mathbf{l} = \mathbf{a}_\phi a\, d\phi$. In these spherical coordinates $\mathbf{a}_r$ points away from the origin, while in the Biot-Savart law it points away from the element. Therefore the equation must be written as

$$d\mathbf{B} = -\mu_0 I \frac{d\mathbf{l} \times \mathbf{a}_r}{4\pi r^2} = \mathbf{a}_r \times \mathbf{a}_\phi \frac{\mu_0 I a}{4\pi r^2}\, d\phi = -\mathbf{a}_\theta \frac{\mu_0 a I\, d\phi}{4\pi r^2}. \qquad (8\text{–}5)$$

The integration must progress along the full length of the wire winding from $\phi = 0$ to $\phi = 2\pi N$, where N is the total number of turns in the length L. If $n = N/L$, the diagram shows that

$$z = Z + \frac{\phi}{2\pi n} = a \operatorname{ctn} \theta,$$

which can be differentiated to give

$$d\phi = -2\pi n a \csc^2 \theta\, d\theta.$$

Since $r = a \csc \theta$, Eq. (8–5) becomes

$$d\mathbf{B} = \mathbf{a}_\theta \frac{\mu_0 n I\, d\theta}{2} \qquad (8\text{–}6)$$

or

$$d\mathbf{B} = (\mathbf{a}_r \cos \theta - \mathbf{a}_z \sin \theta)\frac{\mu_0 n I\, d\theta}{2}.$$

Elements of wire on the two ends of any diameter that is perpendicular through the axis produce radial components of field that cancel* each other, although their axial components are all directed upward parallel to the z-axis. Thus each complete turn of wire produces no net radial component. This can be demonstrated by writing $\mathbf{a}_r = \mathbf{i} \cos \phi + \mathbf{j} \sin \phi$ in Eq. (8–5) and integrating. Ignoring the radial components, the field at the origin is then given by

$$\mathbf{B} = -\mathbf{a}_z \frac{\mu_0 nI}{2} \int_\alpha^\beta \sin \theta \, d\theta = \mathbf{a}_z \frac{\mu_0 nI}{2} (\cos \beta - \cos \alpha). \tag{8–7}$$

The $\mathbf{B}$ field along the axis of a solenoid is shown in Fig. 8–4. The expression for this field has the same mathematical form as the $\mathbf{B}$ field along the axis of the uniformly magnetized cylindrical rod described in Section 9–4, provided the $\mu_0 nI$ is replaced by $\mathcal{M}$. The field on the axis near the middle of a long solenoid is $B = \mu_0 nI$ because $\alpha = \pi$ and $\beta = 0$. Later in this chapter it is shown that $B = \mu_0 nI$ is the magnitude of the field everywhere inside an infinitely long solenoid.

Maxwell's second relation. The relation $\nabla \cdot \mathbf{B} = 0$ was introduced in Section 7–4 as having been proved experimentally. It is easily shown to be a necessary consequence of the Biot-Savart law when all magnetic fields, including those produced by magnetic polarizations, are presumed to be the result of the motions of elementary charges in free space. Then, from Eq. (8–3), the field at point P in Fig. 8–5 due to steady currents in the volume τ is

$$\mathbf{B} = \mu_0 \int_\tau \frac{\mathbf{J} \times \mathbf{a}_r \, d\tau}{4\pi r^2}.$$

In performing this integration, the coordinates X, Y, and Z are varied to include all of the volume τ, but the position of P remains fixed, with x, y, and z constant. The integration, however, provides an expression for $\mathbf{B}$ at P in terms of x, y, and z, which can be differentiated with respect to these coordinates. Thus, in calculating $\nabla \cdot \mathbf{B}$ the coordinates x, y, and z are treated as the variables. The divergence can be written as

$$\nabla \cdot \mathbf{B} = \nabla \cdot \int_\tau \frac{\mu_0 \mathbf{J} \times \mathbf{a}_r \, d\tau}{4\pi r^2} = \frac{\mu_0}{4\pi} \int_\tau \nabla \cdot \left(\frac{\mathbf{J} \times \mathbf{a}_r}{r^2} \right) d\tau,$$

where the divergence of the integrand is to be evaluated with $\mathbf{J}$, X, Y, and Z held constant, whereas the integration over τ is performed with x, y,

---

* This ignores a small component due to the pitch of the windings, which adds up to the field produced by the current I directed upward in a hollow conducting cylinder of radius a.

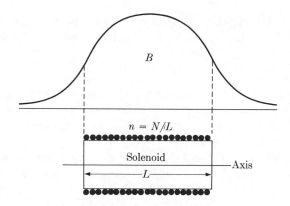

Fig. 8–4 Magnetic fields along the axis of a short solenoid.

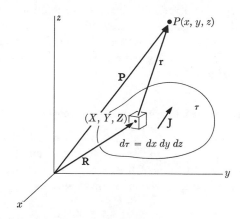

Fig. 8–5 Diagram for using the Biot-Savart law to calculate the **B** field at P due to current densities **J** distributed throughout τ.

and z held constant. Taking the divergence of the integrand, we find

$$\nabla \cdot \left(\frac{\mathbf{J} \times \mathbf{a}_r}{r^2}\right) = \frac{\mathbf{a}_r}{r^2} \cdot \nabla \times \mathbf{J} - \mathbf{J} \cdot \nabla \times \left(\frac{\mathbf{a}_r}{r^2}\right),$$

and since **J** is constant, $\nabla \times \mathbf{J} = 0$. The term $\nabla \times (\mathbf{a}_r/r^2)$ may be evaluated by using

$$\frac{\mathbf{a}_r}{r^2} = \mathbf{i}\left(\frac{x - X}{r^3}\right) + \mathbf{j}\left(\frac{y - Y}{r^3}\right) + \mathbf{k}\left(\frac{z - Z}{r^3}\right),$$

where $r^2 = (x - X)^2 + (y - Y)^2 + (z - Z)^2$. The x-component of $\nabla \times (\mathbf{a}_r/r^2)$ is

$$-\frac{3(y - Y)(z - Z)}{r^5} + \frac{3(z - Z)(y - Y)}{r^5} = 0.$$

The y- and z-components can be obtained from the x-component by cyclic substitution. Since they are also zero, the integrand of the volume integral is zero:

$$\mathbf{\nabla} \cdot \mathbf{B} = 0.$$

The Biot-Savart law of Eq. (8–3) that is used in this demonstration is valid only for currents in closed circuits. It will be shown in Section 8–3 that it can be made valid for fluctuating currents in circuits that are not closed by replacing $\mathbf{J}$ with $\mathbf{J}' = \mathbf{J} + \dot{\mathbf{D}}$.

The substitution of $\mathbf{J}'$ for $\mathbf{J}$ in the foregoing development does not alter the treatment or the conclusion. While Maxwell's second relation $\mathbf{\nabla} \cdot \mathbf{B} = 0$ has been developed here from the Biot-Savart law, it would be somewhat more logical to assume all of the Maxwell field relations, summarized in Section 8–4, and derive the more elementary relations from them. Maxwell's relations are accepted universally, because predictions based upon them always agree with the results of carefully performed experiments. The procedure of demonstrating Maxwell's relations from elementary concepts is adopted here only because the great majority of students find this procedure to be the more satisfactory.

8–2 Ampere's law for all charges in steady motion

Ampere's law for total current is given by $\oint \mathbf{B} \cdot d\mathbf{l}$ integrated around any closed path. With no loss in generality we can calculate this integral for a steady current in a closed loop of fine wire as is shown in Fig. 8–6. From Eq. (8–1) the field at P due to the element of wire $d\mathbf{g}$ is

$$d\mathbf{B} = \frac{\mu_0 I}{4\pi} \frac{d\mathbf{g} \times \mathbf{a}_r}{r} \tag{8–8}$$

where the element is designated $d\mathbf{g}$ to avoid confusion with $d\mathbf{l}$, the element of path length in the line integral.

While the direct integration of $d\mathbf{B}$ around the loop can in principle be accomplished, it is more convenient mathematically to integrate $d\mathbf{B} \cdot d\mathbf{l}$

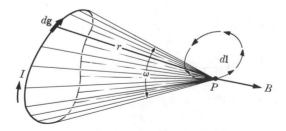

Fig. 8–6 Current I in a closed loop of wire produces the magnetic field $\mathbf{B}$ at the general point P. The value of $\oint \mathbf{B} \cdot d\mathbf{l}$ around the closed path indicated by arrows can be obtained from a general expression for $\mathbf{B}$.

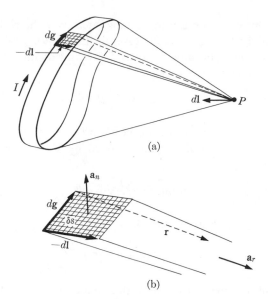

(a)

(b)

Fig. 8–7 Method for evaluating $\oint \mathbf{B} \cdot d\mathbf{l}$ about a current loop. Greater clarity is achieved by displacing the loop by $-d\mathbf{l}$ rather than displacing P by $d\mathbf{l}$.

around this loop, with $d\mathbf{l}$ constant. Integration around the current loop therefore yields

$$\oint d\mathbf{B} \cdot d\mathbf{l} = d\mathbf{l} \cdot \oint d\mathbf{B} = d\mathbf{l} \cdot \mathbf{B}. \tag{8-9}$$

In the diagram of Fig. 8–7(a), every point on the current loop has been displaced a distance $-d\mathbf{l}$ instead of moving P a distance $d\mathbf{l}$. This is done because only relative displacements are important, and because displacement of the loop rather than the point simplifies the analysis. The displacement of the loop produces the ribbonlike surface shown in Fig. 8–7(a) which subtends at P a small solid angle designated by $d\omega$. The crosshatched portion of the ribbon subtends a smaller solid angle $\delta\omega$, and from the magnified drawing of this crosshatched surface in Fig. 8–7(b) its area δs can be seen to be given by

$$-d\mathbf{l} \times d\mathbf{g} = \delta s \mathbf{a}_n,$$

where $\mathbf{a}_n$ is a unit vector directed outward and normal to the surface. From the definition of solid angles,

$$\delta\omega = \frac{-\delta s \mathbf{a}_n \cdot \mathbf{a}_r}{r^2} = \frac{d\mathbf{l} \times d\mathbf{g} \cdot \mathbf{a}_r}{r^2} = \frac{d\mathbf{g} \times \mathbf{a}_r \cdot d\mathbf{l}}{r^2}.$$

Combining this with Eq. (8–8), we find that

$$d\mathbf{B} \cdot d\mathbf{l} = \frac{\mu_0 I}{4\pi} \frac{d\mathbf{g} \times \mathbf{a}_r \cdot d\mathbf{l}}{r^2} = \frac{\mu_0 I}{4\pi} \delta\omega.$$

Both sides of this equation are to be integrated over the ribbonlike surface. The integral of the left-hand side is given in Eq. (8–9), while the integral of $\delta\omega$ merely sums the solid angles subtended by all of the area elements like δs. Therefore the integration yields

$$\mathbf{B} \cdot d\mathbf{l} = \frac{\mu_0 I}{4\pi} \, d\omega. \tag{8–10}$$

Since ω depends upon the location of P, it is a function of the coordinates of P. With the gradient of ω being defined by $\boldsymbol{\nabla}\omega \cdot d\mathbf{l} = d\omega$, Eq. (8–10) may be written as

$$\mathbf{B} \cdot d\mathbf{l} = \frac{\mu_0 I}{4\pi} \, \boldsymbol{\nabla}\omega \cdot d\mathbf{l},$$

which makes it possible to express $\mathbf{B}$ as

$$\mathbf{B} = \frac{\mu_0 I}{4\pi} \, \boldsymbol{\nabla}\omega = -\mu_0 \boldsymbol{\nabla}\left(\frac{-I\omega}{4\pi}\right).$$

From this result it appears that the $\mathbf{B}$ due to a current loop can be expressed as μ_0 times the negative gradient of a scalar magnetic potential, given by

$$\Omega = \frac{-I\omega}{4\pi}. \tag{8–11}$$

The potential Ω, however, is single-valued only where $\oint \mathbf{B} \cdot d\mathbf{l} = 0$, which is true only so long as the path of integration does not pass through real or amperian current loops.

Conservative field limitation. Integration of $\mathbf{B} \cdot d\mathbf{l}$ around the path represented by the string of arrows in Fig. 8–8 can be readily accomplished by using Eq. (8–10), and the result of integrating $d\omega$ about this path is graphically shown in the figure. The solid angle ω subtended by point P at any location in space is equal to the heavily shaded area on the unit sphere whose center is at P. As the point is moved counterclockwise along the path shown in this figure the shaded area increases from $\omega = 0$ to $\omega = 4\pi$. Since the integration of $d\omega$ around such a path gives 4π, Eq. (8–10) gives

$$\oint \mathbf{B} \cdot d\mathbf{l} = \mu_0 I. \tag{8–12}$$

Thus the $\mathbf{B}$ magnetic field is not conservative in regions where there are currents of any kind. For these potentials to provide correct values for $\mathbf{B}$, the currents must include the amperian currents along with the drifting charges of observable currents, i.e., all moving charges.

Amperian currents. The relations developed at the beginning of this section made use of currents produced by charges drifting through wires or other conductors, but they are valid for currents due to any electric charges in

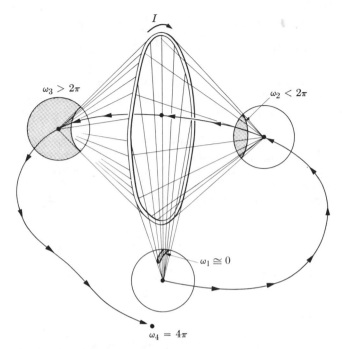

$\omega_3 > 2\pi$

$\omega_2 < 2\pi$

$\omega_1 \cong 0$

$\omega_4 = 4\pi$

Fig. 8–8 The line integral $\int d\omega$ along the path marked by arrows gives 4π for each time this path threads through the current loop.

motion. All materials exhibit magnetic properties and many experiments indicate that these are produced primarily, if not entirely, by motions of their internal charges. The electrons orbit around positive nuclei and also spin about their own axes. Such motions produce closed current loops which result in dipole-like fields. Many of these loops are so oriented that the external magnetic fields they produce nearly cancel each other. In only a few materials, such as iron, cobalt, nickel, and a number of alloys, can any of these loops be oriented so as to produce strong external magnetic fields. These materials are said to be ferromagnetic. The closed-loop currents that are responsible for the magnetic properties of materials are called *amperian currents*, and are practically impossible to measure. Currents due to charges drifting through materials, however, are called "real" currents and are relatively easy to measure. Knowledge of amperian currents is usually obtained from measurements of the external fields they produce.

8–3 Maxwell's fourth relation

In Section 7–2 it is shown that current loops in magnetic fields experience the same torques as should be experienced by magnetic dipoles; i.e., $\mathbf{L} = \mathbf{p}'_m \times \mathbf{B}$, Eq. (7–5). Furthermore, from Eq. (8–11) the scalar magnetic potential of a current loop is

$$\Omega = -\frac{IA \cos \theta}{4\pi r^2}$$

since $\omega r^2 = A \cos \theta$. Thus each amperian current loop inside a material acts just like a magnetic dipole of magnitude $p_m = \mu_0 IA = \mu_0 p'_m$, where A is the area enclosed by the loop.

Magnetization. We now define magnetization at a physical point in any material as $\mathcal{M} = \Sigma p_m / \tau_0$, where τ_0 is the volume of the physical point and Σp_m is the vector sum of the dipole moments within that volume. This quantity can be determined.

The obvious extension of Eq. (8–12) from discrete currents to current densities is

$$\oint \mathbf{B} \cdot d\mathbf{l} = \mu_0 \int_s \mathbf{J}_t \cdot d\mathbf{s} = \mu_0 \int_s (\mathbf{J}_m + \mathbf{J}) \cdot d\mathbf{s}, \qquad (8\text{–}12a)$$

where $\mathbf{J}_t$, the total current density, equals the sum of $\mathbf{J}_m$ and $\mathbf{J}$, the current densities of amperian and nonamperian currents respectively. By use of Stokes's theorem this becomes

$$\int_s \nabla \times \mathbf{B} \cdot d\mathbf{s} = \int_s (\mathbf{J}_m + \mathbf{J}) \cdot d\mathbf{s}$$

or

$$\nabla \times \mathbf{B} = \mu_0 (\mathbf{J}_m + \mathbf{J}). \qquad (8\text{–}13)$$

There is considerable evidence that the amperian currents circulating in closed loops cancel out within uniformly magnetized materials. Adjacent box elements each having the dimensions Δx, Δy, Δz and each having one face in the xz-plane are shown in Fig. 8–9. Identical box elements each having one face in the xy-plane are also shown in this figure. It may be simplest to first assume amperian currents confined to the surfaces of these boxes. If $\mathcal{M}$ is the same in all boxes, the currents in the corresponding sides are all the same. Since these currents must circulate about each box, the currents in all adjoining sides cancel. There are then no net amperian currents within a uniformly magnetized object. Net amperian currents would be found only on the outer surfaces where there are no adjacent boxes to produce cancellation.

Now assume there is a net amperian current density J_{mx} directed parallel to the x-axis. In each adjoining surface between the boxes the net current now must be in the positive x-direction; the currents circulating about the elementary volumes decrease in the plus z-direction, but increase in the plus y-direction as indicated in Fig. 8–9. If $\mathcal{M}_y$ is the y-component of $\mathcal{M}$ in the left-hand box in the xz-plane, the y-component of its magnetic moment is $\mathcal{M}_y \Delta x \Delta y \Delta z$ and the y-component of its magnetic moment per unit area is $\mathcal{M}_y \Delta y$. In the middle block this y-component of magnetic

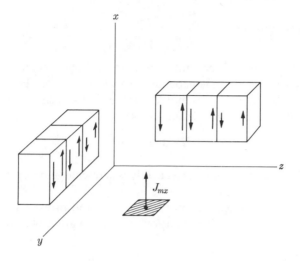

Fig. 8–9 Changes in magnetization required when there are net amperian currents.

moment per unit area is then decreased by $(\partial \mathcal{M}_y/\partial z)\, \Delta y\, \Delta z$. Similar analysis with the other three boxes shows that $\mathcal{M}_z\, \Delta z$ is increased by $(\partial \mathcal{M}_z/\partial y)\, \Delta y\, \Delta z$ as y increases by the amount Δy. Hence a positive value for $\mathbf{J}_{mx}$ requires that

$$\mu_0 \mathbf{J}_{mx}\, \Delta y\, \Delta z = (\partial \mathcal{M}_z/\partial y - \partial \mathcal{M}_y/\partial z)\, \Delta y\, \Delta z.$$

Dividing both sides by $\Delta y\, \Delta z$ reduces this to the x-component of

$$\mu_0 \mathbf{J}_m = \nabla \times \mathcal{M}. \qquad (8\text{–}14)$$

The other two components are easily obtained by cyclical substitutions; i.e., substituting y for x, z for y, and x for z produces the y-components. Repeating these substitutions produces the z-components. Equation (8–14) is then valid, given the assumptions used in the derivation, which are similar to those used in deriving Maxwell's first relation in Section 5–3. Neither charge distributions nor amperian current distributions can truly be uniformly distributed in atomic dimensions. Experiments, however, show our conclusions to be valid.

The H field. Substituting the $\mathbf{J}_m$ from Eq. (8-14) into Eq. (8-13) gives

$$\nabla \times (\mathbf{B} - \mathcal{M}) = \mu_0 \mathbf{J}$$

and provides a much more useful relation than did Eq. (8–13), since $\mathbf{B}$, $\mathcal{M}$ and $\mathbf{J}$ can be measured directly while $\mathbf{J}_m$ cannot. It is customary to replace the difference between the two vectors in the parentheses by a single vector,

$$\mu_0 \mathbf{H} \equiv \mathbf{B} - \mathcal{M}.$$

The resultant equation,

$$\nabla \times \mathbf{H} = \mathbf{J}, \tag{8–15}$$

is valid as long as there is no time dependence. That it cannot be valid when charge distributions are time-dependent can be seen from Eq. (3–23), which demands that $\nabla \cdot \nabla \times \mathbf{H} = 0$, but $\nabla \cdot \mathbf{J}$ does not vanish in general, because the conservation of charge relation gives

$$\nabla \cdot \mathbf{J} = -\frac{\partial \rho}{\partial t}.$$

Therefore Eq. (8–15) can be correct only where charge densities do not vary with time.

Maxwell first pointed out this discrepancy and proposed a modification based upon Eq. (5–4). Differentiation of this equation with respect to time yields

$$\frac{\partial}{\partial t}\nabla \cdot \mathbf{D} = \nabla \cdot \dot{\mathbf{D}} = \frac{\partial \rho}{\partial t}$$

and substitution of this into the charge conservation equation gives

$$\nabla \cdot \mathbf{J} + \nabla \cdot \dot{\mathbf{D}} = 0.$$

Equation (8–15) may be made consistent mathematically by the addition of $\dot{\mathbf{D}}$ to the right-hand side, since the divergences of both sides of the equation then become zero. The equation is written as

$$\nabla \times \mathbf{H} = \mathbf{J} + \dot{\mathbf{D}}, \tag{8–16}$$

which is the general differential form of Maxwell's fourth relation. If both sides of the equation are integrated over a surface s and the left side is transformed into a line integral by Stokes's theorem, the general integral form of Maxwell's fourth relation is obtained as

$$\oint \mathbf{H} \cdot d\mathbf{l} = \int_s (\mathbf{J} + \dot{\mathbf{D}}) \cdot d\mathbf{s}. \tag{8–17}$$

The displacement vector. Integration of Biot-Savart's equation over completely closed circuits in which there are steady currents gives the correct value for the magnetic field at every point in space. However, it does not give the correct magnetic fields due to fluctuating currents because, according to Eqs. (8–16) and (8–17), not only currents but also time variations in $\mathbf{D}$ produce magnetic fields. Since $\mathbf{J}$ and $\dot{\mathbf{D}}$ perform the same function in this equation and necessarily have the same units, Maxwell presumed that $\dot{\mathbf{D}}$ represented a type of current density. He named this current the "displacement current" and called $\mathbf{D}$ the *displacement vector.*

The nature of this displacement current can be partially visualized by considering what happens in a dielectric material between two capacitor plates. As the capacitor is charged, the electric field between the plates increases, thus increasing the polarization of the dielectric. The shift of positive charges one way and negative charges the other way is equivalent to an electric current, even though the polarization charges never break away from their localized positions within the dielectric. This current density, given by $\partial \mathbf{P}/\partial t = \dot{\mathbf{P}}$, is only part of the displacement current; the total displacement current is defined by

$$\dot{\mathbf{D}} = \epsilon_0 \dot{\mathbf{E}} + \dot{\mathbf{P}}.$$

Therefore, even in a vacuum where no polarization charges exist, there is a displacement current represented by $\epsilon_0 \dot{\mathbf{E}}$, and there is no satisfactory method for visualizing this portion of the displacement current. The best verification of the concept is furnished by the propagation of electromagnetic waves in space as discussed in Chapter 12. It was, in fact, the concept of the displacement vector which first led Maxwell to his prediction of the existence of these waves before they had been discovered experimentally. The fact that oscillating electric fields in a vacuum do produce magnetic fields has been well established.

The vector field $\mathbf{H}$ appears to have been introduced here primarily to simplify this fourth fundamental relation. We shall see that it has many other useful properties for dealing with real materials. For example, in static cases with $\mathbf{J} = 0 = \dot{\mathbf{D}}$, $\mathbf{H}$ is conservative while $\mathbf{B}$ is not, since then $\nabla \times \mathbf{H} = 0$ while $\nabla \times \mathbf{B} \neq 0$. Furthermore, in the next chapter we will find that $\mathbf{H}$ can be considered the magnetizing field for magnetizable materials. It will be remembered that $\mathbf{D}$ was introduced with Eq. (5–4) for simplifying Maxwell's first relation. Both $\mathbf{D}$ and $\mathbf{H}$ have many useful properties for dealing with materials. If we dealt only with free space, neither one would be needed. In free space $\mathbf{D} = \epsilon_0 \mathbf{E}$ and $\mathbf{B} = \mu_0 \mathbf{H}$ so that only $\mathbf{E}$ and $\mathbf{B}$ would be needed.

In Chapter 7 it is shown that a closed current loop in a magnetic field in free space experiences the torque, Eq. (7–5),

$$\mathbf{L} = \mathbf{p}'_m \times \mathbf{B},$$

where $\mathbf{p}'_m = \mathbf{n}IA$ and $\mathbf{n}$ is a unit vector normal to the loop. We now see that

$$\mathbf{L} = \mathbf{p}'_m \times \mu_0\mathbf{H} = \mathbf{p}_m \times \mathbf{H}, \qquad (8\text{–}18)$$

where $\mathbf{p}_m = \mu_0\mathbf{p}'_m$ is in agreement with the magnetic-dipole definition used in this section. Equation (8–18) is similar in form to the torque on an electric dipole in an electric field in free space,

$$\mathbf{L} = \mathbf{p} \times \mathbf{E} \qquad (8\text{–}19)$$

given in Section 5–12.

8–4 Summary of Maxwell's field relations

Four very general relations, known as Maxwell's relations, have been obtained in the foregoing pages as extensions of empirically determined laws. These equations are collected here in their differential and integral forms and renumbered for easy reference:

$$\nabla \cdot \mathbf{D} = \rho, \qquad \int_s \mathbf{D} \cdot d\mathbf{s} = \int_\tau \rho \, d\tau, \qquad \text{(M–1)}$$

$$\nabla \cdot \mathbf{B} = 0, \qquad \int_s \mathbf{B} \cdot d\mathbf{s} = 0, \qquad \text{(M–2)}$$

$$\nabla \times \mathbf{E} = -\dot{\mathbf{B}}, \qquad \oint \mathbf{E} \cdot d\mathbf{l} = -\int_s \dot{\mathbf{B}} \cdot d\mathbf{s}, \qquad \text{(M–3)}$$

$$\nabla \times \mathbf{H} = \mathbf{J} + \dot{\mathbf{D}}, \qquad \oint \mathbf{H} \cdot d\mathbf{l} = \int_s (\mathbf{J} + \dot{\mathbf{D}}) \cdot d\mathbf{s}. \qquad \text{(M–4)}$$

Equation (M–1) is obtained directly from the inverse square law for the force one point charge exerts upon another. The vector **D** appears here instead of **E** because of the decision to separate the effects due to polarization charges from the effects due to "real" charges, with ρ designating the volume density of "real" charges. Since there are no "real" poles but materials do contain magnetic dipoles that can be aligned in much the same way that electric dipoles are aligned, Eq. (M–2) is similar to Eq. (M–1). The zero on the right in this relation proclaims that isolated magnetic poles do not exist. Equation (M–3) is obtained directly from Faraday's law of induction. The steady-current form of Eq. (M–4) is obtained from Biot-Savart's law, while the addition of the $\dot{\mathbf{D}}$ term is a modification necessitated by the mathematical requirement that $\nabla \cdot (\nabla \times \mathbf{H}) = 0$. These four relations are valid for space containing materials of all kinds, dielectric, conducting, or magnetic, where ρ is the density of the "real" or nonpolarization charges and **J** is that part of the total current density that is not amperian.

Maxwell's equations in their differential and integral forms combined with the force-on-a-charge equation constitute a very concise expression of electromagnetic theory. The law for the conservation of charge also is essential, although it need not be stated explicitly, since it can be obtained by combining (M–1) with (M–4). All problems in classical electricity and magnetism can be solved with these equations. This does not, however, destroy the usefulness of the older laws; in fact, one who knew only Maxwell's relations would find it desirable to use them to derive less general relations, such as Biot-Savart's law, for the solution of certain problems. *Each of the quantities in these five relations must be measured in the observer's coordinate system.* It is shown in Chapter 13, Relativity, that conclusions obtained from measurements in one system of coordinates are the same as those obtained from measurements all made in another system of coordinates moving with respect to the first.

While the validity of Maxwell's relations for problems already solved by the older laws was readily established, the usefulness of this more general formulation lies in its ability to solve unsolved problems or to predict phenomena as yet unobserved. A crucial test of Maxwell's relations was their prediction of the existence of electromagnetic waves before they had been observed, and the beauty of their formulation cannot be appreciated fully without a study of these phenomena.

Although the five fundamental relations given here apply to macroscopic problems with real materials, only minor changes are required to make them applicable to interatomic spaces. Equations (M–2), (M–3) and (7–1) are directly applicable to mathematical points, since they contain **E** and **B** only. Equations (M–1) and (M–4) are made applicable to interatomic spaces by writing them as

$$\mathbf{\nabla} \cdot \mathbf{E} = \rho_t/\epsilon_0 \tag{MA–1}$$

and

$$\mathbf{\nabla} \times \mathbf{B} = \mu_0 \mathbf{J}_t + \mu_0 \epsilon_0 \dot{\mathbf{E}}, \tag{MA–4}$$

where $\rho_t = \rho + \rho_p$, $\mathbf{J}_t = \mathbf{J} + \mathbf{J}_m$, and $\epsilon = \epsilon_0$.

8–5 General scalar and vector potentials

The differential equations of Laplace and Poisson have proved to be very useful for solving problems concerning static and steady-state phenomena. However, the general forms for Maxwell's relations show that, in general, neither of these equations can be valid for time-dependent phenomena because neither $\mathbf{\nabla} \times \mathbf{E}$ nor $\mathbf{\nabla} \times \mathbf{H}$ equals zero. It is desirable, therefore, to investigate the possibility of developing differential equations that are valid in general. The fact that $\mathbf{\nabla} \cdot \mathbf{B} = 0$ everywhere suggests that **B** can be expressed in terms of a vector potential function **A** such that

$$\mathbf{B} = \mathbf{\nabla} \times \mathbf{A}, \tag{8–20}$$

since according to Eq. (3–23) the divergence of the curl of a vector is zero. Equation (M–2) then becomes

$$\mathbf{\nabla} \cdot \mathbf{B} = \mathbf{\nabla} \cdot (\mathbf{\nabla} \times \mathbf{A}) = 0$$

and substitution of the time derivative of Eq. (8–20) into Eq. (M–3) yields

$$\mathbf{\nabla} \times (\mathbf{E} + \dot{\mathbf{A}}) = 0.$$

This defines a new vector $\mathbf{E} + \dot{\mathbf{A}}$ whose curl is zero, so that it can be expressed as the gradient of a scalar potential:

$$\mathbf{E} + \dot{\mathbf{A}} = -\mathbf{\nabla} V. \tag{8–21}$$

It should be noted that the definition of this new scalar potential is consistent with the earlier definition of the scalar potential in which $\mathbf{E} = -\mathbf{\nabla} V$ for the steady-state condition in which $\dot{\mathbf{B}} = 0$.

The potential functions contained in Eqs. (8–20) and (8–21) will not be useful for electromagnetic problems unless they can be shown to satisfy the other two Maxwell relations. To check this, Eqs. (M–1) and (M–4) can be written as

$$\mathbf{\nabla} \cdot \mathbf{E} = \frac{\rho}{\epsilon} \quad \text{and} \quad \mathbf{\nabla} \times \mathbf{B} = \mu \mathbf{J} + \mu\epsilon \dot{\mathbf{E}}$$

respectively. For the relations in this section it is assumed that $\mathbf{D} = \epsilon\mathbf{E}$ and $\mathbf{B} = \mu\mathbf{H}$, where ϵ and μ are true constants. *The expressions are then valid in free space or in materials with linear constants.* The introduction of the new potentials into these two equations gives

$$\nabla^2 V + \mathbf{\nabla} \cdot \dot{\mathbf{A}} = -\frac{\rho}{\epsilon}$$

$$\nabla^2 \mathbf{A} - \mu\epsilon \ddot{\mathbf{A}} = -\mu\mathbf{J} + \mathbf{\nabla}(\mathbf{\nabla} \cdot \mathbf{A} + \mu\epsilon \dot{V}).$$

Lorentz recognized that these could be reduced to two symmetrical equations, in which V and $\mathbf{A}$ would be separated, by introducing the relation

$$\mathbf{\nabla} \cdot \mathbf{A} + \mu\epsilon \dot{V} = 0, \tag{8–22}$$

which is called the Lorentz "gauge." Such a relation is possible because $\mathbf{B} = \mathbf{\nabla} \times \mathbf{A}$ does not define $\mathbf{A}$ uniquely. The resulting equations:

$$\nabla^2 V - \mu\epsilon \ddot{V} = -\frac{\rho}{\epsilon}, \tag{8–23}$$

$$*\nabla^2 \mathbf{A} - \mu\epsilon \ddot{\mathbf{A}} = -\mu\mathbf{J}, \tag{8–24}$$

provide general differential equations for the potentials V and $\mathbf{A}$ respectively. While these appear to be independent equations, they must be related by the equation of charge conservation. Substitution of these equations into $\mathbf{\nabla} \cdot \mathbf{J} + \partial\rho/\partial t = 0$ yields the wave equation

$$\nabla^2(\mathbf{\nabla} \cdot \mathbf{A} + \mu\epsilon \dot{V}) - \mu\epsilon \frac{\partial^2}{\partial t^2}(\mathbf{\nabla} \cdot \mathbf{A} + \mu\epsilon \dot{V}) = 0,$$

which obviously is satisfied by the Lorentz "gauge" of Eq. (8–22).

Equation (8–24) can be written as three separate scalar equations, one for each of the three rectangular components of $\mathbf{A}$, each of which is identical in form to Eq. (8–23). Any general solution for Eq. (8–23) in rectangular coordinates then automatically provides a general solution for Eq. (8–24). See the footnote at the end of Section 3–8 if other coordinates are needed.

8–6 Low-frequency solutions of the potential equations

The differential equation for scalar potential functions, Eq. (8–23), is identical with Poisson's equation whenever $\ddot{V} = 0$, which indicates that it merely gives a more general expression for the potential function that was

---

* Except in rectangular coordinates $\nabla^2\mathbf{A}$ is a complicated expression.

first introduced for solving static problems. At low frequencies, in fact, the $\mu\epsilon\ddot{V}$ term in Eq. (8–23) and the $\mu\epsilon\ddot{A}$ term in Eq. (8–24) are so small that they can be neglected. This might be expected from the fact that $\mu_0\epsilon_0 = 1/c^2 \simeq 10^{-17}$ s^2 m^{-2}, which suggests that $\ddot{V}$ and $\ddot{A}$ must be quite large before these terms significantly affect the solution of any problem. Since such arguments are by no means conclusive, it is desirable to accept the statement here, with reservations that may be resolved when a particular problem is solved both with the general relations of Eqs. (8–23) and (8–24) and with these relations simplified by setting $\ddot{V} = 0$ and $\ddot{A} = 0$.

In Section 12–7, the electric and magnetic fields of an oscillating electric dipole are calculated by means of the general differential equations, but the terms are separated in such a manner that those terms that would be obtained from the simplified relations can be identified. The amplitudes of the oscillating electric and magnetic fields of the dipole can be estimated from the g's and h's, respectively, given in that section: g_3 and h_2 give the amplitudes of the important terms obtained by ignoring the $\ddot{V}$ and $\ddot{A}$ terms, and g_1, g_2, and h_1 give the amplitudes of the terms added by including the time-derivative terms in the differential equations. Since at low frequencies in any laboratory g_3 is much larger than g_2, which is much larger than g_1, a conservative estimate of the errors caused by the simplifications are obtained by calculating g_2/g_3 and h_1/h_2. When observations are to be made at 60 cps within 5 m of the oscillating currents, the fractional errors do not exceed

$$\frac{g_2}{g_3} = \frac{h_1}{h_2} = \frac{2\pi f r}{v} = 6 \cdot 10^{-6}$$

or 6 parts per million. For the errors produced by neglecting $\ddot{V}$ and $\ddot{A}$ to exceed 1% in such a laboratory, the frequencies must exceed 10^5 cps. Since the amplitudes of the oscillating energies depend upon the squares of the field amplitudes, the errors in the energy amplitudes are much lower. They are approximately

$$\left(\frac{2\pi f r}{v}\right)^2.$$

For low frequencies, then, these two equations can be written as $\nabla^2 V = -\rho/\epsilon$ and $\nabla^2 A = -\mu J$, respectively.

The first of these is Poisson's equation, which is strictly valid for static densities of real charges in media where ϵ is a true constant. When the right side is written as $-\rho_t/\epsilon_0$, where $\rho_t = \rho + \rho_p$ is the density of all charges, including those due to polarizations, this equation is valid for all static problems. Such problems are treated in some detail in Chapters 4 and 5. A general solution for the potential at P in Fig. 8–5 is given by

$$V = \int_\tau \frac{\rho \, d\tau}{\epsilon_0 4\pi r},$$

provided all the real charges lie within τ and all space is filled with a medium whose dielectric constant is ϵ. If in this same figure the real current densities are all confined within τ and all space is filled with a medium where the permeability is μ, a general solution of $\nabla^2 \mathbf{A} = -\mu \mathbf{J}$ is

$$\mathbf{A} = \mathbf{i} \int_\tau \frac{\mu J_x \, d\tau}{4\pi r} + \mathbf{j} \int_\tau \frac{\mu J_y \, d\tau}{4\pi r} + \mathbf{k} \int_\tau \frac{\mu J_z \, d\tau}{4\pi r}$$

or

$$\mathbf{A} = \int_\tau \frac{\mu \mathbf{J} \, d\tau}{4\pi r}. \tag{8–25}$$

For most of the applications the medium will be air, where $\epsilon \cong \epsilon_0$ and $\mu \cong \mu_0$.

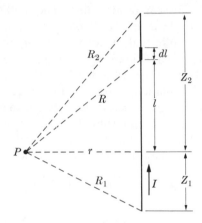

Fig. 8–10 Section of a straight wire carrying a current I parallel to the z-axis, with $Z_1 \gg r$ and $Z_2 \gg r$. The **B** field at P can be obtained from the vector potential.

A vector potential calculation. The use of the vector potential **A** for calculating magnetic fields can be demonstrated with the section of straight wire shown in Fig. 8–10. From Eq. (8–25),

$$\mathbf{A} = \int_{l=-Z_1}^{l=Z_2} \frac{\mu_0 \mathbf{k} I \, dl}{4\pi R},$$

and since r is constant for this integration,

$$\mathbf{A} = \frac{\mu_0 \mathbf{k} I}{4\pi} \int_{l=-Z_1}^{l=Z_2} \frac{dl}{\sqrt{r^2 + l^2}} = \frac{\mu_0 \mathbf{k} I}{4\pi} \{\ln (Z_2 + R_2) - \ln (-Z_1 + R_1)\}.$$

Since

$$(-Z_1 + R_1) = \frac{R_1^2 - Z_1^2}{R_1 + Z_1} = \frac{r^2}{R_1 + Z_1},$$

the vector potential of the current I in this straight wire is

$$\mathbf{A} = \mathbf{k}\frac{\mu_0 I}{4\pi} \{\ln (Z_1 + R_1)(Z_2 + R_2) - 2 \ln r\},$$

which is a vector function parallel to the z-axis. If the wire is very long, so that $Z_1 \gg r$ and $Z_2 \gg r$, then $(Z_1 + R_1)(Z_2 + R_2) \cong 4Z_1 Z_2$ is constant and the **B** field is easily obtained in cylindrical coordinates from the expression

$$\mathbf{B} = \nabla \times \mathbf{A} = -\mathbf{a}_\phi \frac{\partial A_z}{\partial r} = \mathbf{a}_\phi \frac{\mu_0 I}{2\pi r}.$$

This field is the same as that obtained for an infinitely long wire by direct integration of the Biot-Savart law in Section 8–1. In specific problems the use of the vector potential seldom simplifies the calculations, but in general discussions it frequently leads to important simplifications.

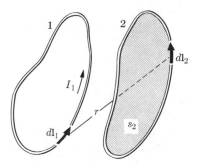

Fig. 8–11 Changes in the primary current I_1 produce emf's $\mathcal{E}_2$ in the secondary circuit, given by $\mathcal{E}_2 = M\, dI_1/dt$, where M is the mutual inductance.

8–7 Mutual and self-inductance

When there is a current I_1 in the primary circuit No. 1 of Fig. 8–11, some of the magnetic flux lines that are produced thread through the secondary circuit No. 2. Time changes in I_1 then produce induced voltages in this secondary circuit, the magnitude of which can be written as

$$\mathcal{E}_{21} = -\frac{d\phi}{dt} = -M_{12}\frac{dI_1}{dt},$$

where ϕ is the magnetic flux threading through the imaginary surface s_2 shown in the diagram and M_{12} is a constant that depends upon the shapes and the relative locations of the two circuits. If the roles of the two circuits are reversed, changes in the current in No. 2 produce induced voltages in No. 1, given by

$$\mathcal{E}_{12} = -M_{21}\frac{dI_2}{dt}.$$

The mks unit for mutual inductance is the henry (H). The mutual inductance between two circuits is one henry when an emf of one volt is induced in one circuit by current changing in the other circuit at the rate of one ampere per second.

The two constants M_{12} and M_{21} are not equal in general, as, for example, when ferromagnetic materials are nearby. However, in the important special cases where μ is constant, the relation

$$M_{12} = M_{21}$$

can be shown to be valid.

In the case where changes in the current I_1 in Fig. 8–11 induce voltages in the second circuit, the magnetic flux through this circuit is given by

$$\phi = \int_{s_2} \mathbf{B} \cdot d\mathbf{s}_2 = \int_{s_2} \nabla \times \mathbf{A} \cdot d\mathbf{s}_2 = \oint_2 \mathbf{A} \cdot d\mathbf{l}_2 \qquad (8\text{–}26)$$

where Stokes's theorem is used to obtain the line integral. According to Eq. (8–25), the vector potential at the element $d\mathbf{l}_2$ is given by

$$\mathbf{A} = \frac{\mu I_1}{4\pi} \oint_1 \frac{d\mathbf{l}_1}{r}, \qquad (8\text{–}27)$$

which must be integrated over the first circuit for each line element $d\mathbf{l}_2$ of the second. Substituting (8–27) in (8–26) gives

$$\phi = \frac{\mu I_1}{4\pi} \oint_2 \oint_1 \frac{d\mathbf{l}_1 \cdot d\mathbf{l}_2}{r},$$

from which

$$\mathcal{E}_2 = -\frac{d\phi}{dt} = -\frac{dI_1}{dt} \frac{\mu}{4\pi} \oint_2 \oint_1 \frac{d\mathbf{l}_1 \cdot d\mathbf{l}_2}{r}$$

and

$$M_{12} = \frac{\mu}{4\pi} \oint_2 \oint_1 \frac{d\mathbf{l}_1 \cdot d\mathbf{l}_2}{r}. \qquad (8\text{–}28)$$

Interchanging the roles of the two circuits merely interchanges the subscripts, which does not alter the value of the double integral. Therefore

$$M_{21} = M_{12} = M$$

and the subscripts can be omitted when μ is constant throughout space.

Circular circuits on the same axis. The mutual inductance of the two circular circuits on the same axis shown in Fig. 8–12 can be calculated by means of Eq. (8–28). The double integral can be obtained by first integrating $d\mathbf{l}_2$ around the second circuit while $d\mathbf{l}_1$ remains fixed, and then integrating $d\mathbf{l}_1$ around the first circuit. Now $d\mathbf{l}_1 \cdot d\mathbf{l}_2 = dl_1\, dl_2 \cos\theta$, $r^2 = x^2 + (2a\sin\theta/2)^2$,

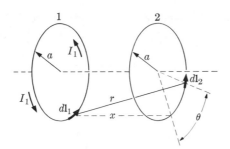

Fig. 8–12 Two circular circuits which are coaxial and have equal radii. When $x \gg a$, the mutual inductance is $M \cong \mu\pi a^4/2x^3$.

and $dl_2 = a\, d\theta$. The double integral can then be written as

$$\oint_1 dl_1 \int_0^{2\pi} \frac{a\cos\theta\, d\theta}{r} \; ,$$

which will give a general relation for any value of the constants. Since the integration is not simple for small values of x, only the case where x is considerably larger than $2a$ will be evaluated. With this simplification, the θ integral becomes

$$\int_0^{2\pi} \frac{a}{x}\left(1 - \frac{2a^2}{x^2}\sin^2\frac{\theta}{2} + \cdots\right)\cos\theta\, d\theta,$$

which can be evaluated easily in terms of the variable $\theta/2$. The value of this integral is $\pi a^3/x^3$, and the mutual inductance is then given by

$$M = \frac{\mu}{4\pi}\left(\frac{\pi a^3}{x^3}\right)\oint_1 dl_1 = \frac{\mu\pi a^4}{2x^3} \; ,$$

since $\oint_1 dl_1 = 2\pi a$. This relation for M is much more simply obtained directly from Eq. (8–23) by assuming that when x is large the magnetic field over the second circuit has the uniform value $B = \mu a^2 I_1/2x^3$ as calculated for its center. Therefore

$$-\frac{d\phi}{dt} = -(\pi a^2)\frac{\mu a^2}{2x^3}\frac{dI_1}{dt} = -M_{12}\frac{dI_1}{dt}$$

or

$$M_{12} = \frac{\pi\mu a^4}{2x^3} \; .$$

Equation (8–26) is useful for making accurate calculations of mutual inductances and for demonstrating the equivalence of the coefficients M_{12} and M_{21}. However, the actual integrations with this equation are usually so complex that simpler methods based directly upon flux changes should be used whenever possible.

Self-inductance. When the current I_1 changes in circuit No. 1 of Fig. 8–11, the magnetic flux changes through this circuit as well as through the second circuit. The voltages induced in the first circuit then are said to be self-induced, and the coefficient L_1 in the relation

$$\mathcal{E}_1 = -L_1 \frac{dI_1}{dt}$$

is called its coefficient of self-inductance or merely its self-inductance. A circuit is said to have one henry of self-inductance when an emf of one volt is induced in this circuit by its current changing at the rate of one A s^{-1}.

It might be expected that L_1 could be calculated from Eq. (8–26) by integrating both $d\mathbf{l}_1$ and $d\mathbf{l}_2$ over the first circuit. This procedure is valid in principle only when the wire is infinitesimally fine and hence the integral becomes infinite. Actually the self-inductance of any circuit like the ones in Figs. 8–10 and 8–11 is strongly dependent upon the diameter of the wire and would be infinite if the wire could be made infinitesimal.

It is instructive to calculate *the self-inductance per meter of length of a long and straight coaxial cable*, consisting of a solid inner cylinder of copper of radius a and a thin and hollow copper cylinder of radius $b > a$. From Eq. (8–12), the magnetic field between cylinders is $B = \mu_0 I/2\pi r$, and the total flux per meter of length is

$$\phi_l = \int_a^b B_0 \, dr = \frac{\mu_0 I}{2\pi} \ln \frac{b}{a},$$

from which

$$L_l = \frac{\mu_0}{2\pi} \ln \frac{b}{a}. \tag{8–29}$$

The self-inductance is strongly dependent upon this radius of the inner cylinder, approaching infinity as a approaches zero. In this calculation the magnetic fields within the cylinders have been ignored. This is valid for frequencies that are high enough to confine the currents to the outer surface of the conductors. This phenomenon is discussed in Section 12–5 and is called the *skin effect*. At lower frequencies currents and magnetic fields will be found inside the conductors. It is then common to calculate the self-inductance of a circuit from the magnetic energy stored by a given current. From the equations of Section 8–8, twice the stored energy divided by the square of the current in the system will give the self-inductance. This calculation is done with dc and the result can be valid only for low frequencies. If high accuracy is needed at intermediate frequencies, a thorough investigation with Maxwell's equations should be undertaken.

Parallel cylinders or wires. We shall later need inductances for transmission lines having different geometries, which will generally be used to transmit high frequencies. Therefore, we will now calculate the *inductance per unit length of a long transmission line consisting of two parallel and conducting cylinders each of radius a, having the distance 2d ≫ 2a between their axes.* This can best be accomplished with the complex variable solutions of Section 4–11. The flux per unit length between the cylinders is

$$\phi_l = \mu(G_2 - G_1) = \mu A \ln m_2/m_1$$

and the current in either cylinder is

$$I = H_2 - H_1 = 2\pi A.$$

Hence the inductance per unit length is

$$L_l = \phi_l/I = \frac{\mu \ln m_2/m_1}{2\pi} = \frac{\mu \ln m_2}{\pi}, \tag{8–30}$$

since $m_1 = 1/m_2$. Here $m_2 = d/a + \sqrt{(d/a)^2 - 1}$.

8–8 Energy stored in linear inductive circuits

Work must be done in setting up currents in the circuits shown in Fig. 8–11, in addition to the $\int I^2 R \, dt$ energy loss that is dissipated in heat. This additional work is stored as energy in the field, and whenever μ is constant it may be recovered as the currents decrease to zero. To calculate this magnetic field energy it is desirable to assume that the circuit resistances are negligible. The current in the first circuit then can be increased from zero to I_1 and that in the second circuit can be increased from zero to I_2 by applying the emf's $\mathcal{E}_1'$ and $\mathcal{E}_2'$ to the two circuits respectively, where these emf's are equal and opposite to the emf's induced by the changing flux. The power delivered to circuit No. 1 is $p_1 = \mathcal{E}_1' i_1$ and the energy delivered to it in the time dt is

$$du_1 = p_1 \, dt = \mathcal{E}_1' i_1 \, dt$$

or

$$du_1 = i_1 \left(L_1 \frac{di_1}{dt} + M \frac{di_2}{dt} \right) dt.$$

Similarly, the energy delivered to circuit No. 2 is

$$du_2 = L_2 i_2 di_2 + M i_2 di_1.$$

The total energy delivered to both circuits is the integral of the sum of these two differential energies.

Since $d(i_1 i_2) = i_1 \, di_2 + i_2 \, di_1$ and $M(i_1 \, di_2 + i_2 \, di_1) = M \, d(i_1 i_2)$,

$$U = \int (du_1 + du_2) = L_1 \int_0^{I_1} i_1 \, di_1 + L_2 \int_0^{I_2} i_2 \, di_2 + M \int_0^{I_1 I_2} d(i_1 i_2)$$

and

$$U = \tfrac{1}{2} L_1 I_1^2 + \tfrac{1}{2} L_2 I_2^2 + M I_1 I_2.$$

When the second circuit current I_2 is zero or when the first circuit is alone, this reduces to

$$U = \tfrac{1}{2}L_1 I_1^2$$

and the energy in the field due to a single circuit is $\tfrac{1}{2}LI^2$. When many closed circuits are involved, the energy expression can be expanded into the form

$$U = \tfrac{1}{2}\{L_1 I_1^2 + M_{12}I_1I_2 + M_{13}I_1I_3 + \cdots$$
$$+ M_{12}I_2I_1 + L_2 I_2^2 + M_{23}I_2I_3 + \cdots$$
$$+ M_{13}I_3I_1 + M_{23}I_3I_2 + L_3 I_3^2 + \cdots\}.$$

8–9 Forces and torques between current circuits

The foregoing energy relation is useful for calculating forces and torques between current circuits. For example, when there are two rigidly shaped circuits in which constant currents are maintained, only the mutual inductance M_{12} in this equation is changed by relative motions between the circuits. An infinitesimal displacement $\delta\mathbf{l}$ of either circuit with respect to the other results in a change of energy which is given by

$$\delta U = I_1 I_2 \nabla M_{12} \cdot \delta\mathbf{l} = I_1 I_2\, \delta M_{12},$$

and this is positive when I_1 and I_2 have the same sign and δM is positive. However, in those cases where I_1 and I_2 do have the same sign, the force acting on either circuit tends to pull it in the direction in which M increases most rapidly. For example, the circular circuits of Fig. 8–12 attract each other whenever their currents are in the same direction, and their mutual inductance obviously would increase if they were to move closer together. When such internal forces move either circuit, external work is performed and at the same time the energy of the system increases. This apparent anomaly can mean only that some source or sources of energy have been ignored. In this case it is the energy that must be supplied to maintain the currents constant.

The flux through the second circuit is

$$\phi_2 = L_2 I_2 + M_{12}I_1$$

and that through the first circuit is

$$\phi_1 = L_1 I_1 + M_{12}I_2.$$

If the second circuit is allowed to move an infinitesimal distance $\delta\mathbf{l}$ closer to the first, it will do an amount of external work given by

$$\delta W = \mathbf{F} \cdot \delta\mathbf{l},$$

where $\mathbf{F}$ is the force on the second circuit due to the first. At the same time the flux changes through both circuits produce induced emf's that must be counteracted if the currents are to remain constant. The emf's required are $\mathcal{E}_1' = d\phi_1/dt$ and $\mathcal{E}_2' = d\phi_2/dt$, and the work done by these emf's is

$$\delta W = (\mathcal{E}_1' I_1 + \mathcal{E}_2' I_2)\, \delta t = 2I_1 I_2\, \delta M_{12},$$

which must equal the sum of the mechanical work done and the increase in energy. Hence

$$2I_1 I_2\, \delta M_{12} = \mathbf{F} \cdot \delta \mathbf{l} + I_1 I_2\, \delta M_{12}$$

or

$$\mathbf{F} \cdot \delta \mathbf{l} = I_1 I_2\, \delta M_{12},$$

and the force can be written as

$$\mathbf{F} = \nabla U = I_1 I_2 \nabla M_{12}. \tag{8–31}$$

Similar arguments show that the torque on either of the two rigid circuits, produced by the presence of the other circuit, is

$$\text{Torque} = \frac{\partial U}{\partial \theta} = I_1 I_2 \frac{\partial M_{12}}{\partial \theta}, \tag{8–32}$$

where the partial derivative has the positive sign because, again, energy must be supplied to keep the currents constant.

8–10 The standard ampere

In Chapter 1 it was pointed out that the "practical" units, ampere, volt, ohm, etc., were defined by the arbitrary relations 1 ampere = 0.1 abampere, 1 volt = 10^8 abvolts, 1 ohm = 10^9 abohms, etc., where the prefix *ab* designates units from the cgs-emu system which had been called the "absolute system." The cgs-emu system, was originally developed from Coulomb's law for forces between isolated magnetic poles. This artificial approach was abandoned when it was realized that the current unit could be established directly from mechanical units by using the forces between current-carrying circuits of known geometry. When the mks system was adopted to include the "practical" units, it was realized that the ampere itself could be established directly from mechanical units. The mks system is now developed from the *standard* ampere, which is defined in terms of forces between current-carrying circuits, and this amounts to defining μ_0 as exactly $4\pi 10^{-7}$ H m^{-1}. It is possible to construct and position solenoids so accurately that their self-inductances and their mutual inductances can be calculated with no larger error than one part per million. According to Eqs. (8–29) and (8–30), the magnitude of a steady current I can be determined by passing this current through two such solenoids connected in series and measuring either the force or torque on one while the other is held stationary.

Ampere balance. The balance shown in Fig. 8–13 was used by the U.S. Bureau of Standards in 1957 to recheck the value of the standard ampere that had been established with a similar balance in 1942. A small coil, shown above the fixed coil in Fig. 8–14, is suspended from one arm of the precision balance at the center of the larger fixed coil. A small break at the center of the fixed coil allows the current in the upper half to be directed opposite to that in the lower. This provides a gradient in the magnetic field of the fixed coil that can be accurately calculated. Since the coils are connected in series, the current has the same magnitude in the suspended coil and in each half of the fixed coil. The weights required to keep the suspended coil balanced at the center determine the magnitude of this current.

The 1957 experiments at the Bureau of Standards also determined the standard ampere by measuring the torque on a small coil suspended in the relatively uniform field of a long solenoid that carried the same current as

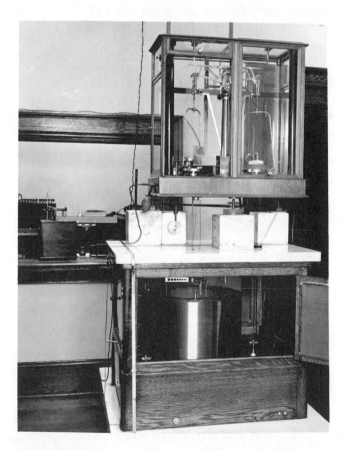

Fig. 8–13 Photograph of the ampere balance used in the 1957 establishment of the ampere. (Courtesy of the U.S. Bureau of Standards.)

the small coil. The two measurements agreed to within 5 parts per million, although their average differed from the 1942 results by 10 parts per million. Since it was impossible to preserve a standard ampere, this last comparison had to be made with the substandards of resistance and voltage that had been established by the earlier experiments and preserved at the Bureau. These substandards, the standard cell for constant voltage and the standard resistor for constant resistance, had apparently varied little during the intervening years.

It is not possible to establish substandards for both resistance and voltage with the single experiment that determines the ampere. Either the ohm or the volt must be established independently.

Fig. 8–14 Photograph showing the construction of the two coils used in the ampere balance shown in Fig. 8–13. Here the smaller coil is shown suspended above the larger coil, although during a standardizing operation it is suspended in the middle of the larger coil. (Courtesy of the **U.S. Bureau of Standards.**)

Standard resistors. In principle, the ohm could be established directly from measurements of the heat losses in a resistor when it carries a known current, but calorimetric measurements are far too inaccurate for this purpose. As will be shown in Chapter 11, a sinusoidal current $i = I_m \sin \omega t$ in a pure inductor results in a sinusoidal potential difference

$$v = L \frac{di}{dt} = \omega L I_m \cos \omega t = V_m \cos \omega t,$$

where $V_m/I_m = \omega L$ is called the reactance, in ohms, of the inductor. Since the L for a precisely constructed inductor can be calculated to six significant figures and the frequency $f = \omega/2\pi$ can be determined to even better precision, the reactance of a precision inductor provides an ideal standard for calibrating standard resistors. Several ingenious bridge-type circuits have been devised for comparing such inductive reactances with resistances.

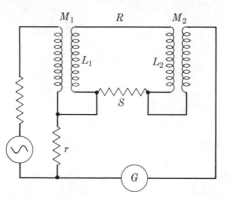

Fig. 8–15 Campbell bridge-type circuit for calibrating the standard resistance r by comparing it with the geometrical average of the precisely determined mutual inductances M_1 and M_2.

One of the simplest of the bridge-type circuits used for standardizing resistors is the Campbell circuit, which uses the two precision mutual inductances, M_1 and M_2 in Fig. 8–15, instead of a single self-inductance. To obtain a balance, the alternating current through the very sensitive indicating instrument G is reduced to zero by varying both S and the frequency f of the ac generator that is shown at the left of the diagram. When a balance has been achieved, the unknown resistance r is given by

$$r = 2\pi f \sqrt{\frac{M_1 M_2}{\mathcal{R}}}, \tag{8–33}$$

where the ratio $\mathcal{R} = R/r$ can be accurately determined in terms of lengths with a Wheatstone bridge. The resistance R is the total resistance of the middle upper loop, including S, and the resistances of the secondary of M_1

and the primary of M_2. The inductances M_1 and M_2 are obtained from bridge-type comparisons with a standard inductor whose mutual inductance has been calculated accurately from its precisely determined dimensions. Analysis of this Campbell bridge shows that it cannot be balanced without ensuring that both

$$(L_1 + L_2)r = M_1S$$

and the relation given in Eq. (8–33) are satisfied. The proof of these relations is left as an exercise.

Once a circuit of this type has been set up, its operation is sufficiently simple so that it can be used frequently to check the constancy of standard resistors. The Campbell circuit was first used by the British National Physical Laboratory, but somewhat more complex modifications of this circuit are now in use at other standardizing laboratories.

PROBLEMS

8–1 A long straight wire carries a current of 10 A along the z-axis, as shown in Fig. 8–16. A uniform magnetic field whose flux density is 10^{-6} Wb m^{-2} is directed parallel to the x-axis. What is the resultant magnetic field at the following points: (a) $x = 0$, $y = 2$ m, (b) $x = 2$ m, $y = 0$, (c) $x = 0$, $y = -0.5$ m?

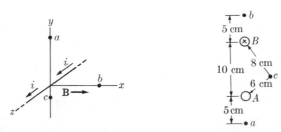

Figure 8–16   Figure 8–17

8–2 Two long straight wires, A and B are arranged in a vertical plane 10 cm apart, as in Fig. 8–17. B carries a current of 6 A into the plane of the paper.
a) What must be the magnitude and direction of the current in A for the resultant field at point a, 5 cm under A, to be zero?
b) What is then the resultant field at b, 5 cm above B, and at c, 6 cm from A and 8 cm from B?

8–3 A long straight wire carries a current of 1.5 A. An electron travels with a velocity of 5×10^6 cm s^{-1} parallel to the wire, 10 cm from it, and in the same direction as the current. What force does the magnetic field of the current exert on the moving electron?

8–4 A closely wound coil has a diameter of 40 cm and carries a current of 2.5 A. How many turns does it have if the magnetic induction at the center of the coil is 1.26×10^{-4} Wb m^{-2}?

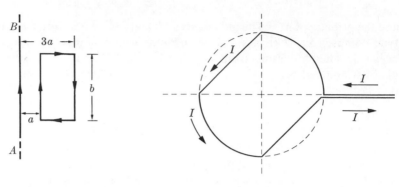

Figure 8–18   **Figure 8–19**

8–5 A solenoid is 10 cm long and is wound with two layers of wire, having an average diameter of 3 cm. The inner layer consists of 50 turns, the outer layer of 40 turns. The current is 3 A, in the same direction in both layers. What is the magnetic induction at a point near the center of the solenoid?

8–6 The long straight wire AB in Fig. 8–18 carries a current of I_1. The rectangular loop whose long edges are parallel to the wire carries a current of I_2. Find the magnitude and direction of the resultant force exerted on the loop by the magnetic field of the wire.

8–7 A solenoid 50 cm long and 8 cm in diameter is wound with 500 turns. A closely wound coil of 20 turns of insulated wire surrounds the solenoid at its midpoint, and the terminals of the coil are connected to a ballistic galvanometer. The combined resistance of coil, galvanometer, and leads is 25 Ω. Find the quantity of charge displaced through the galvanometer when the current in the solenoid is quickly decreased from 3 A to 1 A.

8–8 The wire circuit shown in Fig. 8–19 follows along arcs and chords of a circle whose radius in meters is a. Find the direction and magnitude of the magnetic field produced at the center of this circle by a current of I (amperes) in this circuit.

8–9

a) Directly from the Biot-Savart law, show that the B field at the center of a solenoid, whose length L is very small compared with its radius a, is $B = \mu_0 NI/2a$.

b) The B field inside a very long solenoid is uniform and is given by $B = \mu_0 NI/L$. Show that the general relation of Eq. (8–7) gives these expressions for the B fields at the centers of the very short and the very long solenoids, respectively.

8–10 Figure 8–20 shows a wire carrying a current I in a logarithmic spiral that is given by $r = ae^{-\phi/\pi}$. Find **B** at the origin.

8–11 In Fig. 8–21 there is a current I upward in the wire from A to B and downward in the wire from C to D. The two wires have symmetrical shapes about the z-axis, AB being in the yz-plane and CD being in the xz-plane. The distance r from the origin of any element dl of either wire is given by $r = a/(1 + \sin^2\alpha) \cos \alpha$ and each wire extends from $\alpha = -\pi/2$ to $\alpha = \pi/2$. Find **B** at the origin.

8–12 A regular tetrahedron consisting of four equilateral triangles with sides of length a rests on a table. A current I is directed counterclockwise in a wire that coincides with the periphery of the base. Calculate the field **B** at the apex.

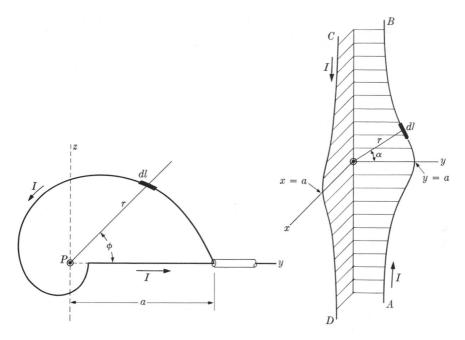

Figure 8–20   Figure 8–21

8–13 There are two circles of wire in the yz-plane, both of which have centers at the origin. A current I is directed clockwise in the circle whose radius is a and the same current I is directed counterclockwise in the circle having the radius $2a$. Find the magnetic field **B** due to this current distribution at a point along the x-axis, which is x meters from the origin.

8–14 An infinitely long and solid cylindrical conductor of radius a carries a uniform current density J_z, which is parallel to the cylindrical axis. It is surrounded concentrically by a hollow cylindrical shell with inner and outer radii b and c, respectively. This shell is also a conductor and carries the uniform current density $-J_z$. Find the field **B** as a function of r for the following points:

a) $a > r > 0$ b) $b > r > a$
c) $c > r > b$ d) $r > c$.

8–15 An infinitely long wire of radius a carries a current I. If r is the perpendicular distance of a point P from the center of the wire, the field at P is independent of the size of the wire so long as $a < r$.

a) Prove this statement directly from Eq. (8–3).
b) Prove this statement by means of Eq. (8–12a).
c) Find the field inside the wire where $r < a$.

8–16 Two very long, straight, and parallel filaments of wire carry currents I_1 and I_2 in the same direction. Suppose that they are a distance d apart and calculate the force exerted by the first wire on each meter of length of the second. Show from Problem 8–15 that the result will not be changed if the first wire has a finite radius $a_1 \cong d/4$. Would the result be changed if both wires had finite radii, $a_1 \cong a_2 \cong d/4$?

8–17 In Fig. 6–4, the two concentric cylinders of Cu and Ni (the latter being solid) provide electrodes for the electrolytic cell on the left, whose emf is V_e, and also provide a transmission line for carrying current to the carbon washer on the right. Assume that the resistivity ρ_c of carbon is so much larger than that of any of the rest of the circuit that all the other resistances can be neglected.

a) Find the current I as a function of V_e, ρ_c, and the dimensions of the cylinders.
b) Find **E** as a function of r for $0 < r < a$; $a < r < b$; $b < r < c$; and $r > c$.
c) Find **H** as a function of r for $0 < r < a$; $a < r < b$; $b < r < c$; and $r > c$, where **H** is induced by a current I.

8–18 Two long, straight wires are parallel and are separated by the distance $2s$. They carry a current I in opposite directions. Draw a diagram, similar to that of Fig. 4–16, of a cross section normal to the wires locating a field point in terms of the coordinates r_1, r_2, θ_1, and θ_2, where r_2 is the distance in this plane of the field point from the center of the wire carrying current into the paper and θ_2 is the angle between r_2 and a line through the centers. The coordinates r_1 and θ_1 are corresponding coordinates measured from the center of the wire carrying current out of the paper.

a) Write an equation in these coordinates for the magnetic potential Ω.
b) Using this potential, calculate **H** in the plane of symmetry as a function of y.

8–19

a) Show that a general solution can be obtained for Problem 8–18 from the complex-variable solution for eccentric cylinders found in Section 4–11.
b) If each wire has a finite diameter compared with the distance between them, determine whether the periphery of each wire will coincide with a circle in Fig. 4–17 having a constant m or will have its center coincide with a pole where either $m = 0$ or $m = \infty$. If the latter should be true, how would you treat those circles surrounding the poles that are partly inside and partly outside the wires?

8–20 A fine wire is bent to form a circle of radius a, which is centered at the origin about the z-axis. It carries a current I, directed past the x- toward the y-axis.

a) Calculate **H** along the z-axis in terms of the rectangular coordinates and its unit vectors.
b) Calculate **H** throughout space in spherical coordinates for the special case in which $r \gg a$.
c) From the **H** obtained in (a), calculate $\int \mathbf{H} \cdot d\mathbf{l}$ from $z = -Z$ to $z = +Z$, where $Z \gg a$. Ignore all terms less than $(a/Z)^2$.
d) From Eq. (8–11) calculate the scalar magnetic potential of this current loop in spherical coordinates for distant points where $r \gg a$.
e) Calculate $\oint \mathbf{H} \cdot d\mathbf{l}$ for a closed path consisting of the straight line along the z-axis from $z = -Z$ to $z = +Z$ combined with a semicircle on which $r = Z \gg a$.

8–21 In the immediate neighborhood of the origin in rectangular coordinates a certain static magnetic field with no moving charges is described by the equations,

$$B_y = m(x + y + z) \quad \text{and} \quad \frac{\partial B_z}{\partial z} = \frac{\partial B_x}{\partial z} = 0.$$

Use Maxwell's equations to find expressions for B_x and B_z.

8-22 Assume a magnetic field that is derivable from the vector potential $\mathbf{A} = kC \ln r$. Find the magnetic flux through a rectangle, whose four corners are at $(r, 0, z_1)$, $(r_2, 0, z_1)$, $(r_2, 0, z_2)$, and $(r_1, 0, z_2)$, respectively.

8-23 A wire extends along the z-axis from $z = 0$ to $z = Z$, where Z is a constant much larger than r, in the cylindrical coordinates of this problem. This wire carries the current I in the positive z-direction.

a) Find the vector potential $\mathbf{A}$ at a general point (r, ϕ, z) in these cylindrical coordinates where $z \ll Z$.

b) Calculate $\mathbf{B}$ at this point from $\mathbf{A}$.

8-24 A transmission line consists of two long straight wires a distance d apart that carry the current I in opposite directions.

a) Find the expression for $\mathbf{A}$ outside these wires.

b) Calculate $\mathbf{B}$ from $\mathbf{A}$.

8-25 Currents in a thin spherical shell of radius a produce magnetic fields whose value throughout the interior of the shell is given by $\mathbf{H} = k\mathbf{G}$.

a) Find the distribution of currents in this spherical shell.

b) Find the field outside the shell.

8-26 A spherical shell of radius a has a surface density of charge σ and is rotating about the z-axis with the angular velocity ω.

a) Show that the vector potential $\mathbf{A}$ has no components in either the $\mathbf{a}_r$ or $\mathbf{a}_\theta$ directions in spherical coordinates.

b) Show that if $\mathbf{A} = \mathbf{a}_\phi A_\phi$, where $A_\phi = (K \sin \theta)/r^2$ outside and $A_\phi = Nr \sin \theta$ inside the shell, then $\nabla \times \mathbf{A}$ gives magnetic fields like those found in Problem 8-25. What are K and N?

8-27

a) Show that in spherical coordinates

$$(\nabla \cdot \nabla)\mathbf{a}_\phi A_\phi = \mathbf{a}_\phi(\nabla^2 A_\phi - A_\phi/r^2 \sin^2 \theta).$$

b) Do the expressions for A_ϕ given in Problem 8-25(b), satisfy the differential equation $\nabla^2\mathbf{A} = 0$ as it is expressed in the (a) part of this problem?

8-28 Calculate the mutual inductance between the long straight wire and the rectangular coil shown in Fig. 8-18.

8-29

a) Calculate the approximate self-inductance of the Bureau of Standards' large solenoid described in Problem 8-33, assuming that the uniform field found in very long solenoids exists throughout this solenoid.

b) A better approximation is obtained by assuming the axial component of the internal field to vary as it does along the axis, but having this constant value in each plane normal to the axis. Calculate the self-inductance on this assumption and compare it with the Bureau of Standards' calculation of 103 mH.

8-30 Two toroidal windings, as shown in Fig. 8-22, are each tightly wound in single layers of wire. They each have the same centerline radius r_c, but the smaller solenoid is wound with an internal radius a and is inside the larger solenoid, which is wound with an internal radius $b > a$. The numbers of the turns on the inner and outer solenoids are N_a and N_b respectively. The inductance of such a device is easily calculated with considerable accuracy, although the device might be difficult to construct accurately. Each of the calculations that follow should be done (1) assuming $r_c \gg b$, and (2) assuming r_c only slightly larger than b.

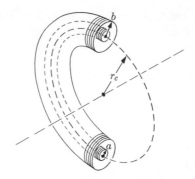

Figure 8–22

a) Calculate the self-inductances L_a and L_b for the two solenoids.

b) Calculate the mutual inductance M between the two solenoids.

8–31 Two rigidly mounted circuits have self-inductances L_1 and L_2 and mutual inductance M. Calculate the self-inductances of this system for the circuits connected in series (a) when they are connected so that the magnetic field from the first adds to the field of the second, and (b) when these fields are opposed to each other.

8–32 Suppose the circular circuit on the right in Fig. 8–12 is free to rotate about an axis through one diameter.

a) Calculate M as a function of the angle α through which this circle has been rotated, assuming $x \gg a$.

b) Calculate the force and the torque on the circle at the right if each circle carries the current I in the same direction.

8–33 To check the standard ampere determined by the balance in Figs. 8–13 and 8–14, the Bureau of Standards measured the torque on a small coil mounted in the center of a long solenoid, where both coil and solenoid carried the current to be checked. The solenoid had 1000 turns 35 cm in diameter and 1 m long, and the small coil had 110 turns 15 cm in diameter and 11 cm long. Calculate the torque on the coil when it is oriented for maximum torque, with a current of 1 A.

8–34 In standardizing the ampere in 1957, Driscoll and Cutkosky of the National Bureau of Standards used the balance shown in Figs. 8–13 and 8–14. The small coil was supported by the balance at the middle of the large solenoid, shown mounted below the balance. Since the force on this small coil depends upon the gradient of the field in the solenoid, center taps were provided to direct the current in opposite directions in the upper and lower halves. The exact fields of this solenoid were calculated from formulas developed by Snow. These formulas are too complex to be used here but the magnitude of the force on the small coil can be obtained by assuming that (1) the solenoid is infinitely long with n turns/m and with the current I in opposite directions in the upper and lower halves, and (2) in the space occupied by the small coil, the axial component of the solenoid's magnetic field equals that along its axis.

a) Calculate $\mathbf{B}$ and ∇B at the center of the solenoid.

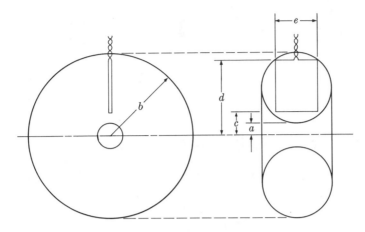

Figure 8–23

b) Calculate the electromagnetic force on a very short coil of N_2 turns whose centerlines follow paths described by circles of radius a about the axis. Assume that the small coil is exactly centered at the middle of the solenoid, with the two axes coinciding, and that the small coil carries the same current I as is carried by each half of the solenoid.

c) Determine the magnitude of this force and the absolute accuracy that would be required in the weighing to establish 1 A to 5 parts per million.

Assume the following dimensions, which were used in the Bureau of Standards' experiment: n of the solenoid $= 1000$ turns/m, $N_2 = 31$ turns in 3.1 cm of length, and $a = 14$ cm.

8–35 A toroid in the shape of a large doughnut with a small hole is shown in Fig. 8–23. It is uniformly wound with N turns of fine wire, which carry the current I. The magnetic field H inside this toroid varies with the distance r measured from the toroid axis.

a) Calculate H at the centerline of this toroid, where $r = (a + b)/2$.

b) Calculate the maximum and minimum values for H inside this toroid.

8–36 A single rectangular loop of wire is inserted into the toroid shown in Fig. 8–23. Calculate the mutual inductance between this loop and the toroid.

8–37

a) Calculate the torque on the rectangular loop in Fig. 8–18 with the current I_1 in the long straight wire and the current I_2 in the loop.

b) Consider the loop in Fig. 8–18 to be rotated 90° about an axis that is parallel to the long straight wire and is at the distance $2a$ from it. Calculate both the force and the torque on the rectangular loop in this position.

Chapter 9 MAGNETIC MATERIALS

All real materials exhibit dielectric and magnetic properties as well as electrical conductivity. The relative magnitudes of these effects differ greatly from one material to another, making it desirable to classify materials according to these magnitudes. The fact that the relative resistivities of materials may differ by a factor as great as 10^{23} at room temperature has created the division of materials into three general classes—conductors, semiconductors, and insulators—as described in Chapter 5.

Dielectric properties of materials vary over a much narrower range. Although magnetic properties differ much more from one material to another than do the dielectric properties, the two kinds of properties can be treated with similar mathematical expressions.

In Chapter 5 the relation

$$\mathbf{D} = \epsilon_0 \mathbf{E} + \mathbf{P}$$

is introduced to define $\mathbf{D}$ for the electrostatic fields in materials. Whenever $\mathbf{P}$ is proportional to $\mathbf{E}$ this is simplified to

$$\mathbf{D} = \epsilon \mathbf{E},$$

where $\epsilon = K\epsilon_0$. In Chapter 8 a somewhat similar relation,

$$\mu_0 \mathbf{H} = \mathbf{B} - \boldsymbol{\mathcal{M}},$$

is introduced to define $\mathbf{H}$ for the magnetostatic fields in materials. This can be made very similar to the electrostatic field relation by transposing to give

$$\mathbf{B} = \mu_0 \mathbf{H} + \boldsymbol{\mathcal{M}}. \tag{9-1}$$

A number of authors write this equation $\mathbf{B} = \mu_0(\mathbf{H} + \mathbf{M})$ because they define magnetization as $\mathbf{M} = \Sigma p'_m / \tau_0$. However, this eliminates the symmetry between magnetic and electric quantities that is clearly displayed in Table 9-1. This symmetry is important in clarifying the roles of the various vector fields and is even more important in the solution of specific problems. Equation (9-1) could also be written as $\mathbf{B} = \boldsymbol{\mathcal{H}} + \boldsymbol{\mathcal{M}}$ with $\boldsymbol{\mathcal{H}} = \mu_0 \mathbf{H}$,

Table 9–1

MAGNETOSTATIC SYMBOLS THAT CORRESPOND MATHEMATICALLY TO
ELECTROSTATIC SYMBOLS

Electrostatic symbols and units		Corresponding magnetostatic symbols and units	
Electrostatic field	$\mathbf{E}$ V m^{-1}	Magnetizing field	$\mathbf{H}$ A m^{-1}
Electric displacement	$\mathbf{D}$ C m^{-2}	Magnetic induction	$\mathbf{B}$ Wb m^{-2}
Electric dipole moment	p $\text{C} \cdot \text{m}$	Magnetic dipole moment	p_m Wb m^{-1}
Polarization	$\mathbf{P}$ C m^{-2}	Magnetization	$\mathcal{M}$ Wb m^{-2}
Density of "real" charges	ρ C m^{-3}	No real poles	
Surface density of "real" charges	σ C m^{-2}	No real poles	
Volume density of polarization charges	ρ_p C m^{-3}	Volume density of magnetic poles	ρ_m Wb m^{-3}
Surface density of polarization charges	σ_p C m^{-2}	Surface density of magnetic poles	σ_m Wb m^{-2}
Electrostatic potential	V V	Magnetostatic potential	Ω A
Electric constant, $\epsilon_0 = 1/c^2 \mu_0$	ϵ_0 F m^{-1}	Magnetic constant, chosen as $4\pi/10^7$	μ_0 H m^{-1}
Dielectric constant, ϵ/ϵ_0	K	Relative permeability, μ/μ_0	K_m

which is useful for some purposes. For isotropic materials in which the magnetization is proportional to $\mathbf{H}$, Eq. (9–1) is usually written as

$$\mathbf{B} = \mu_0 \mathbf{H}(1 + \chi_m) = K_m \mu_0 \mathbf{H} = \mu \mathbf{H},$$

which defines the *magnetic susceptibility as* χ_m *and the relative permeability as* K_m.

Paramagnetic, diamagnetic, and ferromagnetic materials. Materials are classified magnetically according to their magnetic susceptibilities. These classifications are discussed in more detail in Section 9–12, but they can be briefly defined as follows: In paramagnetic materials $0 < \chi_m < 1$, in diamagnetic materials $0 > \chi_m > -1$, and in ferromagnetic materials χ_m is a multiple-valued function that is usually much greater than one. In attempting to line up the magnetic dipoles in a ferromagnetic material, non-conservative forces are encountered that can result in materials remaining magnetized when either **B** or **H** is zero.

Comparison of magnetic and electric fields. While the fields **E** and **B** can be given definite values at mathematical points in space by making the test charge in the defining Eq. (7–2) a point charge of infinitesimal magnitude, the fields **P** and $\mathcal{M}$ must be considered to be averages over small volumes. In defining **D** and **H**, with $\mathbf{D} = \epsilon_0 \mathbf{E} + \mathbf{P}$ and $\mathbf{H} = (\mathbf{B} - \mathcal{M})/\mu_0$, the symbols **E** and **B** should be taken to represent the average values of these fields at physical points. The vectors **D** and **H** also represent averages over the volumes of physical points.

9–1 Calculation of magnetic fields

There are two different but equally valid procedures for calculating magnetic fields in and around magnetic materials. One uses the amperian current approach and the other uses the magnetic pole approach. Both procedures have their proponents. The first procedure is generally preferred by those who believe the amperian current concept to be more realistic. The second is generally preferred by those who need to make actual calculations on real problems. Whenever a problem can be solved by either method, both procedures yield the same results. This will be demonstrated with the short cylinder problem discussed in Section 9–2.

The first method makes use of the relation in Eq. (8–14),

$$\mu_0 \mathbf{J}_m = \nabla \times \mathcal{M}.$$

The second method makes use of the fact that a magnetic pole density can be defined by

$$\rho_m = -\nabla \cdot \mathcal{M},$$

which will be demonstrated. One important reason for preferring this latter method in actual calculations is that it is generally much easier to work with the scalar quantities of the second method than with the vector quantities of the first method. Furthermore, and most important, the second method retains the parallel mathematical forms between the electrostatic and the magnetostatic relations. This parallelism makes it possible to lift electrostatic solutions bodily and use them directly in magnetostatic problems that have the same geometry.

To demonstrate the parallelism we take the divergence of Eq. (9–1),

$$\mathbf{\nabla} \cdot \mathbf{B} = \mathbf{\nabla} \cdot \mu_0 \mathbf{H} + \mathbf{\nabla} \cdot \mathcal{M} = 0, \qquad (9\text{–}2)$$

since $\mathbf{\nabla} \cdot \mathbf{B} = 0$ always. Thus $\mathbf{\nabla} \cdot \mu_0 \mathbf{H} = -\mathbf{\nabla} \cdot \mathcal{M}$ is frequently not zero. Recalling that according to Eq. (5–3) $-\mathbf{\nabla} \cdot \mathbf{P} = \rho_p$, we now define

$$-\mathbf{\nabla} \cdot \mathcal{M} = \rho_m, \qquad (9\text{–}3)$$

where ρ_m can be thought of as a volume density of magnetic poles due to magnetization. Equations (9–2) and (9–3) can now be combined and integrated over an arbitrary volume τ,

$$\int_\tau \mathbf{\nabla} \cdot \mu_0 \mathbf{H} \, d\tau = \int_\tau \rho_m \, d\tau.$$

By application of Gauss's theorem this can be written

$$\int_s \mu_0 \mathbf{H} \cdot d\mathbf{s} = \int_\tau \rho_m \, d\tau, \qquad (9\text{–}4)$$

where s is the surface surrounding τ. This is identical in form to the Gauss's law that was derived from the inverse-square law for fields due to static electric charges. Conversely, since Eq. (9–4) is valid, the fields due to the poles in a small volume $d\tau$ must obey the inverse-square law. The field due to the magnetization of materials can be calculated from

$$\mathbf{H}_m = \int \frac{\mathbf{a}_r \rho_m \, d\tau}{4\pi \mu_0 r^2}. \qquad (9\text{–}5)$$

9–2 Axial fields of a magnetized cylinder

When a short cylinder, like that shown in Fig. 9–1, is uniformly magnetized parallel to its axis, its fields can be calculated throughout space from the principles developed in Section 9–1. The calculation of the fields throughout space is long and complex, but the calculation of the fields along the axis is relatively simple. Since the general principles for such calculations can be illustrated by calculating only the axial fields, they alone will be calculated here.

This is one of the few problems that can be solved as easily with amperian currents as with magnetic pole distributions. We shall use the amperian current approach first.

Calculations with amperian currents. We start with $\mu_0 \mathbf{J}_m = \mathbf{\nabla} \times \mathcal{M}$ and note that $\mathbf{\nabla} \times \mathcal{M} = 0$ everywhere but at those points of the cylinder that lie on its cylindrical surface. Here it is perpendicular to the plane that includes both the axis and the radius that passes through the point in question.

To determine the magnitude of these amperian currents, we write

$$\int_s \mu_0 \mathbf{J}_m \cdot d\mathbf{s} = \int_s \mathbf{\nabla} \times \mathcal{M} \cdot d\mathbf{s} = \oint \mathcal{M} \cdot d\mathbf{l},$$

which can be integrated around a closed rectangular loop lying in the plane previously mentioned. This loop consists of two relatively long lines of length l parallel to the axis—one just inside and the other just outside the cylinder. The loop is completed by two very short lines, perpendicular to the surface, which connect the ends of the longer lines.

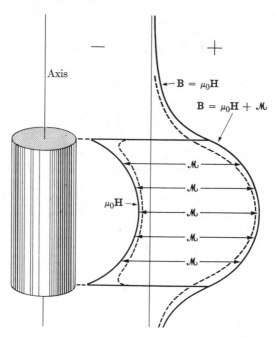

Fig. 9–1 Solid lines indicate the **B** and $\mu_0\mathbf{H}$ fields on the axis of a uniformly magnetized permanent magnet of the cylindrical shape shown. Broken lines indicate the same fields for a real magnet.

The right-hand integral gives just $\mathcal{M}l$, since $\mathcal{M} = 0$ outside the cylinder. The left-hand integral gives μ_0 times the total amperian current along the length l. Referring now to Fig. 8–4 and Eq. (8–7), we find that our problem has already been solved. The amperian currents flow in the same direction as do the real currents in this short solenoid. Since nI is the current per unit length, we can replace $\mu_0 nI$ with $\mathcal{M}$ in Eq. (8–7). This gives the value of **B** on the axis, throughout its length both inside and outside the cylinder. The curve showing the magnitude of **B** is plotted in both Fig. 8–4 and Fig. 9–1.

The curve of $\mu_0\mathbf{H}$ is easily obtained directly from its definition,

$$\mu_0\mathbf{H} = \mathbf{B} - \mathcal{M}.$$

This curve is plotted in Fig. 9–1 and shows that the **H** field is negative through-out the interior of such a cylinder. In fact negative **H** fields are characteristic of all permanent magnets.

Using the amperian current procedure, we have calculated **B** first and have obtained $\mu_0\mathbf{H}$ by subtracting $\mathcal{M}$ from this **B**. For the pole-density procedure we shall calculate $\mu_0\mathbf{H}$ first and then obtain **B** by adding $\mathcal{M}$. For the amperian current calculation we were fortunate to have the short-solenoid solution available. No such solution is available for the pole-density calcu-lation. It will be presented in this section. However, we shall obtain a solution that can be used for the fields of a uniformly polarized electret.

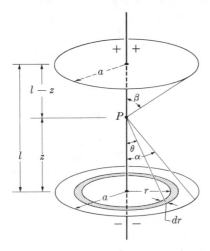

Fig. 9–2 Calculation of the **H** field along the axis of a uniformly magnetized bar of cylindrical shape. This **H** field is due to uniform distributions of poles on the end surfaces.

Calculations with the pole-density concept. The pole density $\rho_m = -\nabla \cdot \mathcal{M}$ is zero everywhere except on the two disk-shaped end surfaces, where a Gauss' pill-box reveals that the surface density of poles is $\sigma_m = -\mathcal{M}$. Figure 9–2 shows the top and bottom surfaces of a magnetized cylinder, each of radius a. The field **H** at the general point P, which is located on the axis a distance z above the base, may be calculated by integrating over the field contributions arising from the excess magnetic poles residing on the bottom and top surfaces. Since the pole density on the bottom surface is $\sigma_m = -\mathcal{M}$, the expression for the poles enclosed in a ring of radius r and width dr is given by $dq_m = \sigma_m 2\pi r\, dr = -\mathcal{M}2\pi r\, dr$. Because of the symmetry of the ring element, all the radial components of the field at P cancel, leaving only the axial components, given by

$$dH_- = \frac{dq_m \cos\theta}{\mu_0 4\pi(z^2 + r^2)} = \frac{-z\mathcal{M}r\, dr}{2\mu_0(z^2 + r^2)^{3/2}}$$

which is negative because it is directed downward. Integration gives

$$H_- = -\int_0^a \frac{z\mathcal{M}}{2\mu_0} \frac{r\,dr}{(z^2 + r^2)^{3/2}} = \frac{\mathcal{M}}{2\mu_0}\left[\frac{z}{(z^2 + r^2)^{1/2}}\right]_0^a$$

or

$$H_- = \frac{\mathcal{M}}{2\mu_0}(\cos\alpha - 1).$$

The total field at P is the sum of the fields due to the positive poles on the top surface and the negative poles on the bottom surface; that is,

$$H = H_+ + H_- = \frac{-\mathcal{M}}{2\mu_0}(1 - \cos\alpha) - \frac{\mathcal{M}}{2\mu_0}(1 - \cos\beta)$$

or

$$\mu_0 H = -\mathcal{M}\left(1 - \frac{\cos\alpha + \cos\beta}{2}\right),$$

which applies only to points on the axis inside the cylinder, where H is always negative. On the axis above the top surface the positive poles produce a field that is directed upward, although the field due to the bottom surface is still directed downward. *The field on the axis above the cylinder is*

$$\mu_0 H = \mu_0(H_- + H_+) = \frac{-\mathcal{M}}{2}(1 - \cos\alpha) + \frac{\mathcal{M}}{2}(1 - \cos\beta)$$

or

$$\mu_0 H = \mathcal{M}\left(\frac{\cos\alpha - \cos\beta}{2}\right) = B,$$

which is positive, since it is directed upward. Since in free space $\mathbf{B} = \mu_0\mathbf{H}$, this also gives the expression for $\mathbf{B}$ on the axis above the cylinder.

The fields along the axis of a uniformly magnetized cylinder whose length l is four times its radius a are plotted in Fig. 9–1. The $\mathbf{B}$ field is continuous, with a smooth maximum at the center of the cylinder and with no sudden change in value on passing out into empty space. The $\mu_0\mathbf{H}$ field, on the other hand, changes sharply from its negative values within the cylinder to positive values in empty space. The magnitude of this sharp change is just $\mathcal{M}$, since $\mu_0\mathbf{H} = \mathbf{B} - \mathcal{M}$ inside and $\mu_0\mathbf{H} = \mathbf{B}$ outside the bar.

As should be expected, the amperian current and the pole-density approaches yield identical results for this idealized cylinder. The modifications required for real magnetized objects are more easily visualized from the pole-density approach.

In practice it is very hard to produce a uniformly magnetized cylinder like the one assumed in this example, because of the very large demagnetizing fields at the ends of such cylinders. Usually the strong negative $\mathbf{H}$ fields reduce the magnetization near the ends, thus causing the poles on the end surfaces to spread around onto the cylindrical surface that is adjacent to them. A body polarization also may appear near the ends, because $\nabla \cdot \mathcal{M}$ may not be zero inside the cylinder. The broken lines in Fig. 9–1 show the kinds of fields that may be produced by a real permanent magnet. The fact that uniform magnetic fields cannot be produced in magnetizing a cylindrical bar makes it difficult to use them for determining magnetic properties. No exact solution has been obtained for the magnetic fields of a ferromagnetic cylinder of finite length; however, exact solutions can be obtained for ferromagnetic materials formed into spheres or into ellipsoids of revolution.

9–3 Fields of permanent magnets and electrets

To clarify the relations obtained in the previous section, it is desirable to obtain solutions for problems in magnetostatics and electrostatics that have the same geometries. Two right circular cylinders with identical dimensions, one uniformly polarized parallel to the axis and the other uniformly magnetized parallel to the axis, provide an excellent example. The nature of each field can be demonstrated by drawing flux lines having densities proportional to the magnitudes of the fields at each point in space. This requires four diagrams: one each for the **D** and the **E** field of the electret, and one each for the **B** and the **H** field of the permanent magnet. Since the geometries are identical and since the **D** and the **B** fields follow the same mathematical relations, it is possible to choose proportionality factors such that these two fields may be represented by a single set of flux lines, as shown in Fig. 9–3(a). With this choice for the densities of the **D** and **B** lines, the $\epsilon_0\mathbf{E}$ and the $\mu_0\mathbf{H}$ fields can be correctly represented by another set of flux lines, as shown in Fig. 9–3(b).

In this figure the fields can be considered to be due to the polarization charges with the electret and to the poles with the permanent magnet. On the top surface of the electret there is a uniform distribution of charges given by $\sigma_p = P$ and on the top surface of the permanent magnet there is a uniform distribution of poles given by $\sigma_m = \mathcal{M}$. The charges and poles on the bottom surfaces are equal but opposite in sign. Flux lines of $\epsilon_0\mathbf{E}$ and $\mu_0\mathbf{H}$ emerge from the positive surfaces and terminate on the negative surfaces. These fields are conservative, but the divergences are not zero on the flat end surfaces. The **D** and **B** fields are obtained by adding **P** and $\mathcal{M}$ respectively to the fields shown in (b). These **D** and **B** lines of flux are continuous throughout space, since the divergences are zero everywhere. However, these fields are not conservative, as is obvious from the fact that $\oint \mathbf{D} \cdot d\mathbf{l} > 0$ and $\oint \mathbf{B} \cdot d\mathbf{l} > 0$ for paths of integration that follow any of the flux lines in the positive direction around the complete circuit. Outside the cylinders, the flux patterns of the (a) and (b) diagrams are identical, since in empty space $\mathbf{D} = \epsilon_0\mathbf{E}$ and $\mathbf{B} = \mu_0\mathbf{H}$.

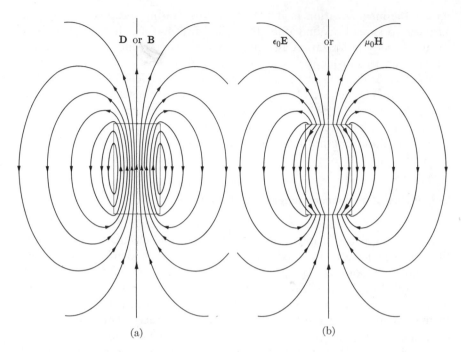

Fig. 9–3 The **D** and **B** fields of a cylindrical electret and permanent magnet are shown in (a). To provide direct comparison, the fields $\epsilon_0\mathbf{E}$ and $\mu_0\mathbf{H}$ of these devices are shown in (b).

D *versus* **E** *and* **H** *versus* **B**. In free space **D** and $\epsilon_0\mathbf{E}$ are completely equivalent concepts, as are also **B** and $\mu_0\mathbf{H}$. From the atomic point of view, where fields are created in free space by charges at rest or in motion, there is no need for the **D** and **H** field concepts because all phenomena can, in principle, be accounted for by the **E** and **B** fields alone. The **D** and **H** fields have been invented solely to simplify the analyses of phenomena observed on a macroscopic scale with polarizable and magnetizable materials.

While it has been known for a long time that only one electric and one magnetic field are needed for analyses on the atomic scale, there have been disagreements as to which two of the four fields should be considered fundamental. Actually there has been little or no disagreement with the conclusion that **E** represents a more basic concept than **D**. However, arguments as to whether the **B** or the **H** field is more basic persisted for a long time before experiments of a rather sophisticated nature were able to resolve the issue.

Obviously, magnetic fields vary greatly from point to point within magnetized materials. To decide between **B** and **H**, it is sufficient to determine which of the two best represents the magnetic field averaged over a sufficiently large volume. The difficulty arises from the fact that different types of probes see different average fields. For example, conduction electrons see a much different average magnetic field than do neutrons projected through the material. The problem has been studied with theoretical methods (Wannier*) with measurement of deflections of cosmic rays passing through iron (Rasetti†) and with measurement of diffraction of neutrons in magnetic materials (Hughes‡ and others). The motions of conduction electrons are strongly affected by the positively charged nuclei and the tightly bound electrons that surround them. Therefore conduction electrons do not move at random through magnetized material and no averages obtained with them should be taken too seriously. However, neutrons and high-speed cosmic-ray particles should pass randomly through these materials, and these experiments do indicate that **B** rather than **H** represents the average field. From the analysis presented in Chapter 8, these particles could see fields that averaged to **B** only by passing in a strictly random fashion through all current loops, including electron orbits and spinning electrons.

The **H** *field.* Since it has been experimentally established that the average magnetic field in a magnetic material is **B**, it is reasonable to question why **H** should be introduced at all. The answer is simply that **H** strongly affects the alignment of the current loops (magnetic dipoles) within the medium. For example, the permanent magnet shown in Fig. 9–3 tends to be demagnetized by its **H** field, which is directed nearly opposite to its magnetization. The **H** field, which is designated as *the magnetizing field*, will cause a decrease of magnetization if the magnet is jarred or heated. Note that if **B** were responsible for changing the magnetization, the magnetization described by Fig. 9–3 should increase with jarring rather than decrease. When the magnetic properties of materials are measured, the results are usually plotted as B versus H or $\mathcal{M}$ versus H.

Probably the simplest methods for distinguishing **H** from **B** in actual experiments are based upon the boundary conditions that are presented in Section 9–4.

9–4 Boundary-value problems

The field $\mathbf{H}_m$ due to magnetized materials has been shown to be conservative. Therefore it can always be expressed as $\mathbf{H}_m = -\nabla\Omega$, where Ω is a scalar potential function. Since $\nabla \cdot \mathbf{B} = 0$ everywhere, Laplace's equation

$$\nabla^2\Omega = 0$$

* G. H. Wannier, *Phys. Rev.* **72**, 305 (1947).
† F. Rasetti, *Phys. Rev.* **66**, 1 (1944).
‡ D. J. Hughes, *Pile Neutron Research*, Addison-Wesley, Reading, Mass., 1954, especially Sections 11–4 and 10–6.

is valid in those regions where $\mathbf{B} = \mu_0 K_m \mathbf{H}_m$, with K_m a true constant. This relation is also valid in other regions, for example in regions where the fields and the magnetization are known to be constant. It does not hold, in general, in regions that include surfaces of separation between materials having different magnetic properties. When problems are to be solved for a region containing such surfaces, the region is divided into two or more regions in each of which Laplace's equation can be satisfied. The solutions for Ω for these separate regions must satisfy the boundary conditions along common boundaries. The procedures are essentially the same as were used in Chapter 5 with dielectric materials.

Boundary conditions. The correspondence between $\mathbf{B}$ and $\mathbf{D}$ and between $\mathbf{H}$ and $\mathbf{E}$ makes it possible to write boundary conditions immediately for $\mathbf{B}$ and $\mathbf{H}$ at any surface separating two magnetic materials. From Eqs. (5–8) and (5–9) these are

$$B_{1n} = B_{2n} \tag{9–6}$$

and

$$H_{1t} = H_{2t}, \tag{9–7}$$

where B_{1n} and B_{2n} are the components of $\mathbf{B}$ normal to the surface between medium 1 and 2, and H_{1t} and H_{2t} are the components of $\mathbf{H}$ tangent to this surface. The proofs of these conditions follow the patterns set in Section 5–5. Applying the integral $\int_s \mathbf{B} \cdot d\mathbf{s} = 0$ to a pillbox like that in Fig. 5–5 gives the first condition, Eq. (9–6), and applying $\oint \mathbf{H} \cdot d\mathbf{l} = 0$ to a closed path like that in Fig. 5–6 gives the second condition, Eq. (9–7). If there should be real currents in the surface, H_{1t} may differ from H_{2t} by a surface density of real current. Its magnitude may be determined from Ampere's law, by evaluating the line integral over the appropriate closed path.

Spheres and cylinders in uniform fields. While uniform magnetizations cannot be produced in most geometrical shapes, they can be produced rather easily in a few geometries, such as spheres, ellipsoids of revolution, and infinitely long cylinders fashioned of homogeneous and isotropic materials. The axes of the ellipsoids must be parallel to the field, while the axis of the cylinder should be perpendicular to the field.

   As an example, assume a sphere of radius a, in which $\mathbf{B} = \mu \mathbf{H}$ with μ a true constant, placed in a uniform field of strength $\mathbf{B}_0 = \mu_0 \mathbf{H}_0$ parallel to the z-axis. This problem is mathematically identical to the problem of the dielectric sphere of radius a in which $\mathbf{D} = \epsilon \mathbf{E}$ with ϵ a true constant. The solution for the magnetic problem is obtained directly from the solution for the electric problem found in Section 5–7 by substituting $\mathbf{B}$ for $\mathbf{D}$, $\mathbf{H}$ for $\mathbf{E}$, μ for ϵ, K_m for K, and Ω for V. The internal and external potentials, from Eq. (5–17), are

$$\Omega_i = - \frac{3}{K_m + 2} H_0 z \quad \text{and} \quad \Omega_o = - \left(1 - \frac{K_m - 1}{K_m + 2} \frac{a^3}{r^3} \right) H_0 r \cos \theta.$$

The internal and external fields can be obtained from these potentials or directly from the electrostatic fields given in Section 5–7 for the corresponding problem. The internal fields are

$$\mathbf{H}_i = \frac{3}{K_m + 2}\, \mathbf{H}_0 = \frac{3}{\chi_m + 3}\, \mathbf{H}_0, \qquad (9\text{--}8)$$

$$\mathbf{B}_i = \frac{3K_m}{K_m + 2}\, \mathbf{B}_0 = \frac{3(\chi_m + 1)}{(\chi_m + 3)}\, \mathbf{B}_0, \qquad (9\text{--}9)$$

and

$$\mathcal{M} = \mathbf{B}_i - \mu_0 \mathbf{H}_i = \frac{3(K_m - 1)}{(K_m + 2)}\, \mathbf{B}_0 = \frac{3\chi_m}{\chi_m + 3}\, \mathbf{B}_0, \qquad (9\text{--}10)$$

and it is evident that the magnetization is uniform throughout the sphere. In the diamagnetic materials discussed in Section 9–12, $0 > \chi_m > -1$, and the magnetization $\mathcal{M}$ is opposite in direction to the original field $\mathbf{B}_0$, as can be seen by making χ_m negative in the above equations.

The solution for an infinitely long cylinder made from the same kinds of material and placed with its axis normal to the uniform field $\mathbf{B}_0 = \mu_0 \mathbf{H}_0$ can be obtained directly from the equations in Section 5–8. The potentials are

$$\Omega_i = -\frac{2}{K_m + 1}\, H_0 z$$

and

$$\Omega_0 = -\left(1 - \frac{K_m - 1}{K_m + 1}\frac{a^2}{r^2}\right) H_0 r \cos \theta.$$

The internal fields are given by

$$\mathbf{H}_i = \frac{2}{K_m + 1}\, \mathbf{H}_0 = \frac{2}{\chi_m + 2}\, \mathbf{H}_0,$$

$$\mathbf{B}_i = \frac{2K_m}{K_m + 1}\, \mathbf{B}_0 = \frac{2(\chi_m + 1)}{\chi_m + 2}\, \mathbf{B}_0,$$

and

$$\mathcal{M} = \frac{2(K_m - 1)}{(K_m + 1)}\, \mathbf{B}_0 = \frac{2\chi_m}{\chi_m + 2}\, \mathbf{B}_0,$$

which show that the magnetization is also uniform throughout such a cylinder.

The foregoing solutions are applicable only to paramagnetic, diamagnetic, and a few ferromagnetic materials in weak fields. In the next section it is demonstrated that with some modifications these solutions can be applied to any isotropic and homogeneous ferromagnetic material formed into these same shapes.

9–5 Ferromagnetic sphere in a uniform field

Solutions of boundary-value problems for ferromagnetic materials are at least as difficult and usually more difficult than solutions for nonferromagnetic materials. This follows logically from the nonlinear, multivalued nature of the relation between $\mathcal{M}$ and H, as shown by the typical curve for ferromagnetic materials in Fig. 9–4. However, exact solutions can be obtained in special cases where $\mathcal{M}$ is uniform throughout the material, as can be achieved with infinite cylinders, spheres, and ellipsoids of revolution.

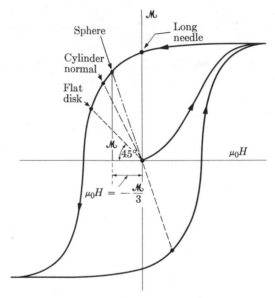

Fig. 9–4 Magnetizations and demagnetizing fields of permanent magnets formed in various shapes from a given material. In each $\mathcal{M}$ must reduce in value along the second quadrant of the hysteresis loop until the demagnetizing field $\mu_0 H$ is (a) $-\mathcal{M}/3$ in a sphere, (b) $-\mathcal{M}/2$ in a cylinder (magnetized normal to its axis), (c) $-\mathcal{M}$ in a flat plate, but (d) zero in a very long needle.

It should be noted, however, that when nonferromagnetic materials of these shapes are properly oriented in a uniform field, as was done in the last section, the internal magnetization is necessarily uniform, but that for ferromagnetic materials the past history is important. For example, if a sphere is placed in a rather weak uniform field and a permanent magnet is then placed near it to induce nonuniform magnetization, some of the nonuniform magnetization will remain even after the magnet is removed. This is caused by the multivalued nature of most magnetization curves. Similarly, a sphere with nonuniform residual magnetization will retain some of this nonuniform magnetization in the presence of a uniform applied field until the field becomes strong enough to saturate the sample. When $\mu_0 H$ is increased well beyond the values plotted in Fig. 9–4, $\mathcal{M}$ remains constant at the largest value shown in that figure and the material is said to be saturated. All of the internal dipoles that can be aligned parallel to the applied field are so aligned.

The very special case of a homogeneous, isotropic, ferromagnetic sphere which initially had no magnetization will now be considered. When a uniform magnetic field is superimposed upon this sphere and its magnitude is increased from zero to some final value, the internal magnetization increases but remains uniform throughout the sphere. While the functional relation between H and $\mathcal{M}$ is not known, it is evident that for a given H the magnetization can be expressed by $\mathcal{M} = \chi'_m \mu_0 H$, where χ'_m is not a constant susceptibility but rather a field-dependent proportionality factor having a fixed value for the conditions specified. Since for any given field value χ'_m is merely a constant of proportionality, as is χ_m, the solution for a ferromagnetic sphere can be obtained directly from the solution for a nonferromagnetic sphere by substituting χ'_m for χ_m.

Making this substitution in Eqs. (9–12) and (9–13) and multiplying the first equation by μ_0, we obtain

$$\mu_0 \mathbf{H}_i = \mu_0 \mathbf{H}_0 - \frac{\chi'_m}{3}\mu_0 \mathbf{H}_i = \mu_0 \mathbf{H}_0 - \tfrac{1}{3}\mathcal{M} \qquad (9\text{–}11)$$

and

$$\mathbf{B}_i = \mathbf{B}_0 + \frac{2\chi'_m}{\chi'_m + 3}\mu_0 \mathbf{H}_0 = \mathbf{B}_0 + \tfrac{2}{3}\mathcal{M} \qquad (9\text{–}12)$$

for the internal fields. Note that the internal $\mu_0 \mathbf{H}_i$ is less than the applied field by $\mathcal{M}/3$, while the internal $\mathbf{B}_i$ is greater than the applied field by $2\mathcal{M}/3$. The correctness of these results can be checked by subtracting Eq. (9–11) from (9–12) to obtain

$$\mathbf{B}_i - \mu_0 \mathbf{H}_i = \mathbf{B}_0 - \mu_0 \mathbf{H}_0 + \mathcal{M},$$

and since $\mathbf{B}_0 = \mu_0 \mathbf{H}_0$, this becomes

$$\mathbf{B}_i = \mu_0 \mathbf{H}_i + \mathcal{M},$$

which is obviously correct.

Spherical permanent magnet. An interesting case arises when a sphere is first saturated in a strong external field which is then reduced uniformly to zero. This case can be described by setting $\mu_0 \mathbf{H}_0 = \mathbf{B}_0 = 0$ in Eqs. (9–11) and (9–12) to give

$$\mu_0 \mathbf{H}_i = -\tfrac{1}{3}\mathcal{M}$$

and

$$\mathbf{B}_i = \tfrac{2}{3}\mathcal{M}.$$

For this case in which the applied field is zero, the ratio between $\mathcal{M}$ and $\mu_0 H_i$ represented by χ'_m is

$$\chi'_m = \frac{\mathcal{M}}{\mu_0 H_i} = -3.$$

The two equilibrium positions for the permanent magnetization of such a sphere in a zero field are represented by the intersections of the line of slope -3 with the hysteresis loop in Fig. 9–4. In the more general case of a uniformly magnetized sphere that was not originally saturated, the ratio between $\mathcal{M}$ and $\mu_0 H_i$ would still be -3, but their magnitudes would be less. It is easily verified that χ'_m is -2 for the case of an infinite cylinder permanently magnetized perpendicular to its axis.

It is evident that good materials for permanent magnets should have very broad, open hysteresis loops like the one in Fig. 9–4, to give as large a residual magnetization as possible. It is also evident that the shape of the magnet is important, since the value of χ'_m is shape-dependent. In general, a long thin rod magnetized *parallel* to its axis maintains its magnetization better than a short stubby one, since $|\chi'_m|$ is larger in the first case.

Demagnetizing factors. For the special geometries studied in this section, with $\mathcal{M}$ uniform through the sample, the internal field may be expressed by

$$\mu_0 H_i = \mu_0 H_0 - \mu_0 H_d,$$

where H_0 is the uniform applied field and H_d, directed opposite to the applied field, is the field generated by the magnetization of the sample. This quantity $\mu_0 H_d$, which is known as the *demagnetizing field*, is proportional to $\mathcal{M}$, and may be expressed as

$$\mu_0 H_d = -D\mathcal{M},$$

where D is known as the *demagnetizing factor*.

Examination of Eq. (9–11) reveals that $D = 1/3$ for a sphere, and the reader may verify that $D = 1/2$ for an infinite cylinder whose axis is perpendicular to the direction of magnetization. The demagnetizing factor D for four different geometries is given by the absolute magnitude of the slopes of the broken lines in Fig. 9–4. The value of D for a flat sheet magnetized perpendicular to its surface is approximately unity, since most of the lines of **H** passing from the north poles on one face to the south poles on the other must pass through the sheet. For a long thin rod magnetized parallel to its axis $D \cong 0$, since the north and south poles are relatively weak and relatively far from the center of the rod.

The demagnetizing factor is not an exact concept for shapes in which $\mathcal{M}$ is not uniform throughout. However, it is convenient in practice to ascribe some average D to various shapes in order to indicate the magnitude of the internal field.

9–6 Measurements of B versus $\mu_0 H$

The demonstration that uniform magnetizations and uniform fields can be produced with particular shapes suggests how such curves can be obtained in the laboratory. Figure 9–5 shows a cross section through the equator of the sphere studied in Section 9–5, in which the fields are perpendicular to the plane of the diagram. Two search coils are shown wound around the sphere, one as close to the equator as possible and the other of somewhat larger diameter. The inner coil is connected to a fluxmeter to determine changes in the total flux through the sphere, which are expressed by $\delta\phi = \pi a^2\,\delta B_i$.

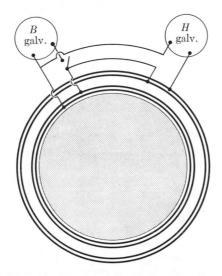

Fig. 9–5 Cross section through the equator of the ferromagnetic sphere of Section 9–6, showing search coils wound to determine changes in B and H. The inner coil alone determines δB, while the inner and outer coils connected in series-opposition determine δH.

Successive values of δB_i can be added to determine B_i as the external field is built up in steps from zero. If the outer coil has the same number of turns and is connected in series-opposition with the inner coil, the fluxmeter then reads only the changes in the flux threading between the two coils. The fluxmeter reading then determines changes in $\mu_0 H_i$ because in this annular region the external and internal H fields are equal. Here at the equator, where the fields are tangential to the spherical surface, the boundary conditions require that $H_{ot} = H_{it}$. With one fluxmeter connected to the inner coil and another to the two coils in opposition, B_i versus $\mu_0 H_i$ curves can be determined by combining the results obtained with the incremental changes.

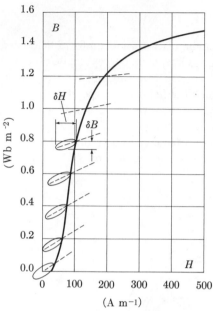

Fig. 9–6 Normal magnetization curve for iron. Minor hysteresis loops for small changes δH in H at various points are shown. Values for the incremental permeability $\mu_i = \delta B / \delta H$ at these points are also indicated.

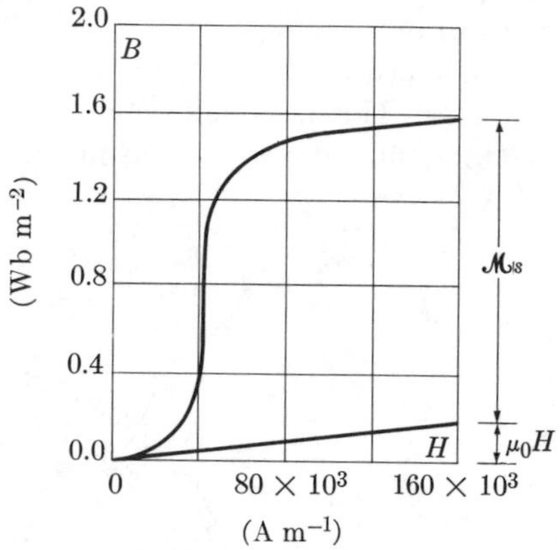

Fig. 9–7 Normal magnetization curve for alnico 5. Since $\mathcal{M}$ becomes a constant $\mathcal{M}_s$, the slope at high fields becomes μ_0.

This method of measurement is particularly easy to apply to samples shaped into prolate ellipsoids, because the location of the coils at the equator is less critical than for a sphere. In fact, for this purpose a long cylindrical rod can be used, because it makes an excellent approximation to a prolate ellipsoid and is much easier to fabricate. The errors caused by the lack of uniform magnetization near the ends of the bar can be neglected in sufficiently long rods.

When calculations are to be made with ferromagnetic materials it is essential to have experimentally determined B versus H curves available. The curves should be ones that have been determined under essentially the same conditions as will be experienced in the work being calculated. While there are many different B versus H curves that can be determined for a given ferromagnetic material, there are only three or four types that are of general importance: (1) the virgin curve, (2) the normal magnetization curve, (3) the hysteresis loop approaching saturation on each half-cycle, and (4) some δB versus δH curves at different fixed-field biases. The virgin curve (1) is obtained when the material is first magnetized after having been cooled

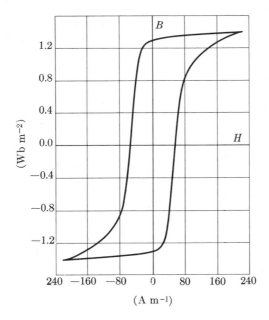

Fig. 9–8 Hysteresis loop for iron.

from above its Curie temperature in a zero magnetic field. This is not nearly as important as the normal magnetization curve (2) that is obtained by a cycling process. In the cycling process the material is first demagnetized by cyclically reversing the applied magnetic field and slowly reducing it. After demagnetization a small field H is applied and reversed several times. One-half the change in magnetic induction produced by reversing H is taken as the value of B to be plotted versus H. When B versus H curves are given without specifying the experimental procedure they are nearly always "normal magnetization" curves. Normal curves for iron and alnico 5 are shown* in Figs. 9–6 and 9–7 respectively. Figure 9–6 also illustrates a type (4) curve from which incremental permeabilities can be obtained. At each point on the normal curve a given reduction δH in H produces different changes δB in B. The incremental permeability is defined as $\mu_i = \delta B/\delta H$. In obtaining hysteresis loops like the one for iron shown in Fig. 9–8, the material is usually first magnetized to saturation by some maximum field $H_{\max}$. Changes δB in B are then observed when the field is suddenly reduced from $H_{\max}$ to H. Values of B corresponding to H are then obtained from $B = B_{\max} - \delta B$, where $B_{\max}$ is the value of B when $H = H_{\max}$.

---

* Figures 9–6, 9–7 and 9–8 are reproduced by permission from R. M. Bozorth, *Ferromagnetism*, D. Van Nostrand, Princeton, N.J., 1951.

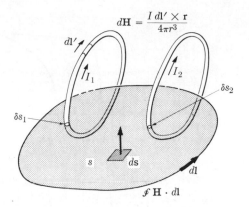

$$dH = \frac{I\,dl' \times \mathbf{r}}{4\pi r^3}$$

Fig. 9–9 Integral of $\mathbf{H} \cdot d\mathbf{l}$ about the path indicated yields a value $I_1 + I_2$.

9–7 Magnetizing fields of real currents

In Section 8–2, where the Biot-Savart law was used for obtaining Maxwell's fourth relation, all magnetic fields are designated by $\mathbf{B}$. The currents, therefore, must include amperian currents as well as real currents. When the amperian currents are excluded, the $\mathbf{B}$ in Eq. (8–12) must be replaced by $\mu_0\mathbf{H}$. Dividing both sides by μ_0 gives

$$\oint \mathbf{H} \cdot d\mathbf{l} = I.$$

A region where $\oint \mathbf{H} \cdot d\mathbf{l} = 0$ and $\mathbf{H}$ is conservative can be mapped out by stretching a sheet of any shape but having the current loop like that in Fig. 8–8 as its periphery. Such a sheet must prevent the closed path of integration from looping the current.

Ampere's law for real currents only. When the path of integration loops two currents as shown in Fig. 9–9

$$\oint \mathbf{H} \cdot d\mathbf{l} = I_1 + I_2$$

and, in general,

$$\oint \mathbf{H} \cdot d\mathbf{l} = \sum_i I_i,$$

where the I_i's represent all the currents looped. The obvious extension from discrete currents to current densities is

$$\oint \mathbf{H} \cdot d\mathbf{l} = \int_s \mathbf{J} \cdot d\mathbf{s}, \tag{9–13}$$

which is the integral form of Maxwell's fourth relation for steady currents, where $\mathbf{J} \cdot d\mathbf{s}$ must be integrated over a surface bounded by the path of inte-

Axis of symmetry

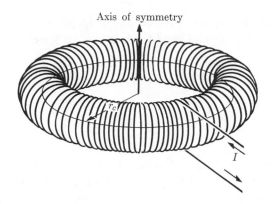

Fig. 9–10 Nonmagnetic toroid surrounded by a wire-wound solenoid.

gration for $\oint \mathbf{H} \cdot d\mathbf{l}$. It is often called Ampere's law. The left-hand side of Eq. (9–13) may be transformed into an area integration by Stokes's theorem, Eq. (3–17), with the result

$$\oint \mathbf{H} \cdot d\mathbf{l} = \int_s \mathbf{\nabla} \times \mathbf{H} \cdot d\mathbf{s} = \int_s \mathbf{J} \cdot d\mathbf{s}.$$

Since the last two integrals are always taken over the same general surface, the integrands must be equal, so that

$$\mathbf{\nabla} \times \mathbf{H} = \mathbf{J}. \tag{9–14}$$

This is the differential form of Maxwell's fourth relation for steady currents.

Equations (9–13) and (9–14) are useful for calculating **H** in many situations where the necessary symmetry exists. For example, the field near a long straight wire carrying a current I can be calculated by integrating along the circumference of a circle, perpendicular to the wire and centered upon it. On such a circle **H** must be tangent to the circle and constant in magnitude. Equation (9–13) gives $H2\pi r = I$ or $H = I/2\pi r$, if r is the radius of the circle.

Symmetrical toroid. One of the simplest and most useful examples is that of calculating H in a symmetrical toroid such as a hoop made of nonmagnetic material formed into a shape similar to that of a bicycle tire and uniformly wound with insulated wire in a tight spiral, as shown in Fig. 9–10. Since this figure is symmetrical about the central axis corresponding to the axle of the bicycle wheel, any circle drawn about this central axis defines a path on which the field must be constant and directed tangent to the circle unless the circle passes through or too close to the individual turns of wire. On circles passing close to the wires, H fluctuates from turn to turn, but these fluctuations can be reduced by using finer wires more closely wound. All other circles that can be drawn about the central axis either are wholly within the solenoid and loop all of the N turns, or are wholly outside of the solenoid and loop no turns. If the current in the wire is I, then

$$\oint \mathbf{H} \cdot d\mathbf{l} = H2\pi r = NI$$

or

$$H = \frac{NI}{2\pi r}$$

for all points inside the solenoid, where r is the radial distance of the point from the axis of symmetry. Axially symmetric circles outside the solenoid do not loop any turns, so that $H_\phi = 0$ everywhere outside the solenoid. There is an external field, a function of r and z, which is just equal to that of a single wire carrying the current I along the center line of the wire loops making up the turns of the solenoid. This can be seen by making the path of integration for Eq. (9–13) loop the solenoid.

If the solenoid is shaped like a fat doughnut with a small hole, the magnitude of $H = NI/2\pi r$ varies considerably within the doughnut. However, when the internal radius of the solenoidal winding is small compared with the radius r_c of its centerline, the field is practically constant and is given by $H \cong NI/2\pi r_c$. The length of this latter solenoid is $L = 2\pi r_c$, and hence

$$H = \frac{NI}{L} = nI,$$

which was given as the expression for the field on the axis of an infinitely long solenoid. Here the expression is shown to be true for all points inside the winding, and is not limited to the axis. Since an infinitely long solenoid cannot be constructed, this constitutes one practical approximation.

9–8 Ampere's law

In Section 9–3 it is pointed out that $\mathbf{H}$ fields due only to magnetized materials are conservative. This is obvious also from Eq. (9–14), $\nabla \times \mathbf{H} = \mathbf{J}$, since in such fields $\mathbf{J} = 0$. Then for fields due to magnetized materials

$$\oint \mathbf{H}_m \cdot d\mathbf{l} = 0$$

always, and in Section 9–7 it was shown that for fields due to steady currents

$$\oint \mathbf{H}_c \cdot d\mathbf{l} = \int_s \mathbf{J} \cdot d\mathbf{s} = \sum_i I_i,$$

where $\sum_i I_i$ is the sum of all currents looped by the path of integration. Since the total field is the vector sum $\mathbf{H} = \mathbf{H}_c + \mathbf{H}_m$, due both to currents and to magnetized materials, the equation

$$\oint \mathbf{H} \cdot d\mathbf{l} = \int_s \mathbf{J} \cdot d\mathbf{s} = \sum_i I_i \tag{9–15}$$

is Ampere's law for steady-state conditions. This means that the presence of any ferromagnetic material in the field of a solenoid or other current-carrying circuit does not alter the magnitude of the line integral along any given closed path even though it will usually alter the field at all locations.

Ferromagnetic toroid. It the solenoid shown in Fig. 9–10 is wound around a toroid of homogeneous and isotropic ferromagnetic material, then

$$\oint \mathbf{H} \cdot d\mathbf{l} = NI$$

for every circular path about the axis of symmetry that threads through the N turns of wire each carrying the current I. Symmetry requires that $\mathbf{H}$ be tangential to the path and that its magnitude be the same at all points. Therefore, $H2\pi r = NI$ and

$$H = \frac{NI}{2\pi r},$$

which is independent of the properties of the ferromagnetic material. This result stems from the fact that there are no free magnetic poles. This geometry has been used for some of the most precise determinations of B versus H or $\mathcal{M}$ versus H curves of materials. The incremental method of the previous section was employed but only one search coil and one fluxmeter were needed, since H was obtained directly from the magnitude of the current and from the geometry. Changes in total flux were easily determined by winding a search coil around the toroid and connecting it to a fluxmeter. Subtracting the change in flux due to the change in current determined the change in flux due to the material. Thus δB or $\delta \mathcal{M}$ was easily determined.

Toroid with thin saw cut. Rather remarkable changes are produced by merely making a saw cut through a toroid normal to the centerline. Integration along the centerline still gives

$$\oint \mathbf{H} \cdot d\mathbf{l} = NI,$$

but the symmetry is destroyed, so that H is no longer constant along this path. To determine the field as a function of current it is now necessary to invoke Maxwell's second relation, which guarantees the continuity of the B lines of flux. If the width t of the saw cut is small, most of these flux lines will go across the air gap normal to the surfaces of the gap, although there

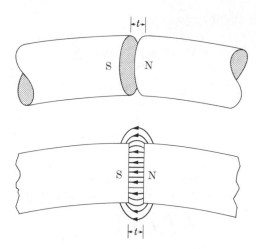

Fig. 9–11 Portion of a ferromagnetic toroid showing a narrow saw cut of width t. The passage of "fringing flux" outside the air gap is shown in the lower cross section.

will be some fringing at the edges, as shown in Fig. 9–11. Aside from this fringing, which may be neglected to a good approximation only when t is very small compared with the internal diameter of the toroid, the rest of the field is still confined to the region inside the solenoid. Thus B along the centerline circle will be practically constant. From previously determined B versus H curves applicable to this material, the value of the internal magnetizing field H_i required to produce a given internal flux B_i along the centerline in the iron can be determined. In the air gap, $K_m \cong 1$, so that $H \cong B_i/\mu_0$ and hence the integral along the centerline is

$$NI = \oint \mathbf{H} \cdot d\mathbf{l} = H_i(2\pi r_c - t) + \frac{B_i t}{\mu_0}.$$

This can be written as

$$NI = H_i 2\pi r_c + \frac{(B_i - \mu_0 H_i)t}{\mu_0}$$

or

$$N(I - I_0) = \frac{(B_i - \mu_0 H_i)t}{\mu_0},$$

where $H_i 2\pi r_i = NI_0$ is the number of ampere-turns required to produce B_i in the original uncut toroid. Although t was made small, B_i is generally so much larger than $\mu_0 H_i$ that I may need to be considerably larger than I_0. The difference $N(I - I_0)$ expresses the number of ampere-turns required for the air gap, and may be thought of as arising from the need to overcome the demagnetizing field produced by the poles induced on the surface of the cut.

9–9 Magnetic "Kirchhoff's laws"

The foregoing calculations suggest that magnetic flux in ferromagnetic materials follows laws that are mathematically similar to Kirchhoff's laws for electric current circuits, especially if the great majority of the flux is confined within the ferromagnetic material. From the continuity principle, the flux into any junction of these materials equals the flux out of this junction, or

$$\phi_1 + \phi_2 + \phi_3 + \cdots = 0,$$

which is like the junction relation for currents. Cyclotron magnets and many other electromagnets are often constructed like the one shown in Fig. 9–12. The flux passes upward through the air gap in the center and divides into two equal parts, half returning to the bottom pole through the iron on the right and the other half through that on the left. These two paths for the flux remind one of parallel electric circuits each carrying half the current.

The line integral for any path that threads through the coils wound around the two center poles is

$$\oint \mathbf{H} \cdot d\mathbf{l} = NI,$$

where NI gives the total ampere-turns in these coils. A convenient choice of path follows a line of flux that passes through the center of the bar of iron on the right, as shown by the broken line in Fig. 9–12. Wherever the flux is constant along the length of any iron bar, B is constant and the H required to produce that B is constant. While such ideal conditions are seldom encountered in practice, it is frequently desirable to assume that they do exist, because more accurate assumptions usually lead to very long and laborious calculations.

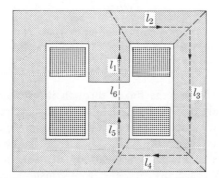

Fig. 9–12 Typical design for an electromagnet in which dotted rectangles represent the coil windings. A path l for magnetic flux is divided into approximate lengths of l_1, l_2, l_3, l_4, and l_5 through the iron and l_6 across the air gap.

It can be assumed that the closed flux circuit indicated by the broken line is made up of six sections of lengths l_1, l_2, l_3, l_4, l_5 and l_6 having cross-sectional areas A_1, A_2, A_3, A_4, A_5 and A_6 respectively, and that all the flux ϕ passes through sections 1, 5 and 6, while only half, or $\phi/2$, passes through sections 2, 3 and 4. The magnetic induction in each section is then $B_1 = \phi/A_1$, $B_2 = \phi/2A_2$, $B_3 = \phi/2A_3$, $B_4 = \phi/2A_4$, $B_5 = \phi/A_5$ and $B_6 = \phi/A_6$. Often, by construction, sections 1 and 5 are identical and sections 2 and 4 are identical, which simplifies the calculations. The area A_6 of the air gap is not well defined, but it is frequently assumed to be given by $A_6 \cong A_1 = A_5$. The narrower the gap between the pole faces, the better is this approximation. In designing a magnet the value of B that is desired in the air gap is usually known, so that the total flux $\phi = B_6 A_6$ can be determined. This establishes the values for the remaining B's. Curves of B versus H which have been determined experimentally for the materials used in constructing each section should then be consulted to determine the value of H_1 required to produce B_1, of H_2 required to produce B_2, etc. In the air gap, $H_6 = B_6/\mu_0$. The line integral now gives

$$\oint \mathbf{H} \cdot d\mathbf{l} \cong H_6 l_6 + H_1(l_1 + l_5) + H_2(l_2 + l_4) + H_3 l_3 \cong NI, \quad (9\text{--}16)$$

from which the required number of ampere-turns can be determined.

If the magnetic inductance in these materials were proportional to the magnetizing fields, so that B/H were a true constant, Eq. (9–18) could be simplified and made more useful. The quantity

$$Hl = \frac{Bl}{\mu} = \frac{\phi l}{\mu A}$$

could then be written for each bar as

$$Hl = \phi \mathfrak{R},$$

where $\mathfrak{R} = l/\mu A$ is called the *reluctance* of the bar. Furthermore, NI is usually called the *magnetomotive force* or mmf, and Eq. (9–16) is written

$$\phi_1 \mathfrak{R}_1 + \phi_2 \mathfrak{R}_2 + \phi_3 \mathfrak{R}_3 + \cdots = \text{mmf},$$

where ϕ_1, ϕ_2, ϕ_3, etc. are the fluxes through the respective sections. This relation corresponds to the circuital law for closed electric current circuits given by Kirchhoff. Note that the reluctance equation $\mathfrak{R} = l/\mu A$ is similar to the resistance equation $R = l/\sigma A$, where σ is the conductivity. This suggests why μ has been called permeability. Unfortunately for this type of analysis, $\mu = B/H$ is usually very far from constant, and the procedure is therefore limited to obtaining crude estimates. In general, the procedure outlined in leading up to Eq. (9–16) gives much more accurate results.

9–10 Magnetized sheet and its equivalent current

The external fields produced by magnetized materials can always be simulated by current circuits. For example, it is easily demonstrated that the current circuit shown in Fig. 9–13(a) is equivalent to the uniformly magnetized sheet in part (b) of that figure. The potential at P due to the element $d\mathbf{s}$ of the magnetized sheet can be obtained from the expression for the potential of an ideal dipole, provided the linear dimensions of this element are small compared with r. This requires that $r \gg t$ and that $r^2 \gg ds$, which

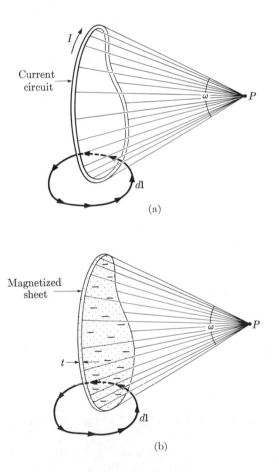

(a)

(b)

Fig. 9–13 The magnetic scalar potential at P due to the current loop in (a) is identical to that at P due to the uniformly magnetized sheet in (b) and is proportional to the solid angle subtended by the loop area at P. The integral $\oint \mathbf{H} \cdot d\mathbf{l}$ is equal to I for the current loop but is zero for the magnetized sheet.

can be satisfied for any value of r provided t and ds are made arbitrarily small. If the magnetic moment per unit area of the magnetized sheet in Fig. 9–14 is p_a, then the moment of the element is $p_a\, ds$, and the potential $d\Omega$ at P due to this element is

$$d\Omega = \frac{-p_a\, d\mathbf{s} \cdot \mathbf{a}_r}{\mu_0 4\pi r^2} = \frac{-p_a}{4\pi\mu_0}\, d\omega,$$

where the minus sign appears because the magnetization is directed away from P in Fig. 9–14, with the south poles on the side of the sheet nearest P. With p_a constant, integration of this equation over the surface gives

$$\Omega = \frac{-p_a\omega}{4\pi\mu_0}, \tag{9–17}$$

where ω is the solid angle subtended by the whole surface s at P.

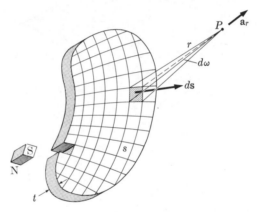

Fig. 9–14 A small segment ds of a magnetized sheet gives rise to a magnetic dipole field at P.

The magnetic scalar potential of current loops like the one in Fig. 9–13(a) is given in Eq. (8–11) as

$$\Omega = -\frac{I\omega}{4\pi}.$$

Comparison of these potentials shows that (a) and (b) in Fig. 9–13 are equivalent provided

$$p_a = \mu_0 I. \tag{9–18}$$

This equivalence of the magnetic sheet and the current loop means that they produce identical magnetic fields everywhere except inside the magnetic sheet. Even inside the sheet the **B** fields are identical, but the **H** fields are very different. The line integral $\oint \mathbf{H} \cdot d\mathbf{l}$ can be evaluated over each of the identically shaped paths indicated by arrows in (a) and (b) of Fig. 9–13.

Its magnitude is I for the path in (a) and zero for the path in (b). Because the external fields are identical, the two integrals must have the same value over all the path outside the magnetized sheet. Therefore, inside this sheet,

$$\int \mathbf{H} \cdot d\mathbf{l} = -I = H_i t,$$

where $H_i = -I/t$ gives the value of $\mathbf{H}$ inside the sheet.

Since fields add vectorially, the fields due to any uniformly magnetized object can be analyzed by imagining it to be made up of a stack of many thin sheets, each uniformly magnetized. Since each magnetized sheet in the object can be replaced with a current loop, it is evident that the external field of any uniformly magnetized object can be duplicated by suitably chosen current loops. For example, a current has been chosen in Section 9–2 for the short solenoid of Fig. 8–4 such as to make it nearly equivalent to the uniformly magnetized cylinder of Fig. 9–1. It could be made exactly equivalent if the finite number of turns could be replaced by an infinite number of turns of infinitesimal diameter but carrying the same number of ampere-turns per meter.

The interchangeability of the fields of these two devices can be used for making accurate measurements of susceptibility.* A cylindrical specimen in a uniform magnetic field is oscillated in and out of a many-turn search coil connected to a sensitive voltage detector. The magnetic moment of the oscillating sample results in a time variation in flux linkage in the search coil and a corresponding voltage signal. If a fine-wired solenoid is wrapped around the cylinder and the correct current sent through it, then the external field of the sample can be exactly canceled. The voltage detector will read zero when this null condition is achieved, and the magnetization of the sample is calculated from the equivalent current. A major advantage of this system is that the flux linkages between the sample and search coil need not be known.

9–11 Work required to produce magnetic fields

It was shown in Chapter 5 that the work required to produce electro-static fields is the sum of all the energies needed to produce the fields at each of the physical points involved. To determine the work required to produce the final field $\mathbf{D}_f$ at any point in a dielectric having arbitrary properties, the charging of a parallel-plate capacitor homogeneously filled with such a dielectric was studied. This geometry has the advantage that with no restriction placed upon the properties of the dielectric the fields within it are homogeneous at all times and the required work is the same for every physical

---

* A. Arrott, "Principle for Null Determination of Magnetization," *Rev. Sci. Inst.* **28**, 99–102 (1957).

point. The work per unit volume is

$$w = \int_0^{D_f} \mathbf{E} \cdot d\mathbf{D},$$

and the power per unit volume is

$$p = \mathbf{E} \cdot \dot{\mathbf{D}}.$$

A similar procedure can be adopted for determining the work and the power per unit volume required to produce the final field $\mathbf{B}_f$ in materials having arbitrary magnetic properties. An infinitely long solenoid homogeneously filled with such a magnetic material has the same advantage as had the parallel-plate capacitor. The $\mathbf{B}$ and $\mathbf{H}$ fields are both confined to the inside of such a solenoid and these fields are uniform throughout. To avoid complications with joulean heat in the windings, it is best to ignore their resistance, although in real situations the energy required to produce the joulean heating must be added. Starting with both fields and the current all equal to zero, fields can be produced by increasing the current. If I is the current and n the number of turns per meter, $H = nI$. If the material fills the solenoid and has the cross-sectional area A, a back emf is produced in the winding by the increase in magnetic induction. The emf per meter required to overcome this back emf is

$$\mathcal{E} = n \frac{d\phi}{dt} = nA\dot{B}_a,$$

where B_a is the axial component of $\mathbf{B}$. The power per meter of length is

$$P = \mathcal{E}I = \frac{nA\dot{B}_a H}{n} = A\mathbf{H} \cdot \dot{\mathbf{B}},$$

and the power per unit volume required to produce these fields is

$$p = \mathbf{H} \cdot \dot{\mathbf{B}}. \tag{9–19}$$

The work per unit volume required to produce the final field $\mathbf{B}_f$ is

$$w = \int_0^{B_f} \mathbf{H} \cdot d\mathbf{B}.$$

A numerical value for the work required to produce a given field can be determined only if the relationship between $\mathbf{B}$ and $\mathbf{H}$ is known. This energy is usually not recovered when the magnetization disappears. This is especially true with ferromagnetic materials, for which some B versus H relationships are shown in Figs. 9–6 through 9–8. When a material is taken through a hysteresis loop like that shown in Fig. 9–8, energy is converted into heat, and the amount of heat per unit volume produced each time the cycle is completed is given by the area of the loop shown in the figure. As should be expected, a unit of area on any of these four figures is

$$\frac{\text{amp}}{\text{m}} \cdot \frac{\text{weber}}{\text{m}^2} = \frac{\text{joule}}{\text{m}^3}.$$

Equation (9–19) has been justified here only for the special case of a homogeneously filled solenoid, but more advanced theory indicates that it is a general relationship. Further evidence for its general validity is presented in Chapter 10 in the discussion of the Poynting vector. For static electromagnetic fields that are linear, where $\mathbf{D} = \epsilon\mathbf{E}$ and $\mathbf{B} = \mu\mathbf{H}$, the energy density is given by

$$U = (\epsilon\mathbf{E} \cdot \mathbf{E} + \mu\mathbf{H} \cdot \mathbf{H})/2. \qquad (9\text{–}20)$$

9-12 The atomic origin of magnetism

According to the Bohr-Sommerfeld model of the atom, negatively charged electrons circulate about the positively charged nucleus in closed orbits which may be thought of as closed current loops. Because a closed current loop gives rise to a magnetic dipole moment, each orbiting electron should have associated with its motion about the nucleus not only an angular momentum but also a magnetic moment. Each electron also spins rapidly about its own axis (electron spin), giving rise to more angular momentum and another source of magnetic dipole moment. It is well known, however, that there is no net orbital or spin angular momentum associated with completely filled atomic shells. This is because quantum-mechanical rules require that there be just as many electrons rotating clockwise as counterclockwise about the nucleus in a closed shell. The electron spins in a filled shell are similarly required to be oriented in space so that they give no net angular momentum. Just as the angular momentum averages to zero in closed shells, so the current in these closed loops also averages to zero. The angular momentum and magnetic dipole moment of each atom must therefore come only from incompletely filled shells. This has been well verified by experiment.

The gyromagnetic ratio. A very simple relation exists between the dipole moment and the angular momentum of electrons. The dipole moment is $\mu_0 I A$, which reduces to $\mu_0 e\omega r^2/2$, since $A = \pi r^2$ and $I = e(\omega/2\pi)$ for an electron traveling in a circular orbit about a nucleus. Since the angular momentum is $m\omega r^2$, the ratio of magnetic dipole moment to the angular momentum is $\mu_0 e/2m$, which is known as the *gyromagnetic ratio*. (This ratio is more commonly written in gaussian units as $e/2mc$.) The values of angular momentum which an orbital electron may have are $0, h/2\pi, 2h/2\pi, 3h/2\pi$, etc., where h is Planck's constant and equal to $6.62 \cdot 10^{-34}$ J·s or $6.62 \cdot 10^{-27}$ erg·s. The smallest nonzero magnetic moment associated with the orbital motion of an electron is therefore $(\mu_0 e/2m)(h/2\pi)$ in mks units and $(e/2mc)$ $(h/2\pi)$ in gaussian units, and it is known as one *Bohr magneton*.

If the electron were a sphere of negatively charged matter spinning rapidly about its axis, by a similar calculation the same gyromagnetic ratio would be obtained for electron spin as for electron orbital motion. Experimentally, however, the ratio between spin magnetic moment and spin angular momentum is just twice this large, being $\mu_0 e/m$ in mks units and e/mc in gaussian units. This apparent discrepancy has been resolved by relativistic considerations. Since the angular momentum of an electron associated with its spin may have values of plus or minus $h/4\pi$, the magnetic momentum associated with one electron spin is just one Bohr magneton. A general expression for the gyromagnetic ratio can be written as $g\mu_0 e/m$ in mks units and $g(e/2mc)$ in gaussian units, where g, the *magnetomechanical factor*, is unity for orbital motion and two for electron spin. Thus $1 < g < 2$ when both orbital motion and electron spin contribute to the magnetization.

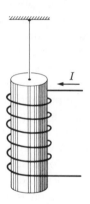

Fig. 9–15 The Einstein-de Haas method of measuring gyromagnetic ratios. Current reversal in the magnetizing coil reverses the magnetization and the associated angular momentum in the cylindrical specimen, thus causing the sample to rotate.

Experimental verification of these relations was first achieved by Einstein and de Haas in 1915. They suspended a ferromagnetic cylinder in a magnetizing coil by a fine wire, as illustrated in Fig. 9–15. The magnetization of the sample was reversed by reversing the magnetizing field of the coil, and thus a corresponding reversal of angular momentum was produced. The magnetization reversal can be measured by the methods discussed in Section 9–6, and the change in angular momentum can be determined by the resulting rotation of the sample about the vertical axis. Some values of g recently determined by more precise magnetic-resonance techniques are indicated in Table 9–2.

Table 9–2

EXPERIMENTAL VALUES FOR THE
MAGNETOMECHANICAL FACTOR

Substance	g
Iron	1.93
Cobalt	1.85
Nickel	1.88
Permalloy (FeNi)	1.90
Magnetite (Fe_3O_4)	1.93

The relatively large values of g indicate that the orbital contribution to the total magnetic moment is quite small. This is explained by the fact that the motion of electrons in the unfilled outer shells is sufficiently perturbed by the proximity of the electrons of other atoms so that their angular momentum is reduced almost to zero. Because of the reduction in orbital angular momentum occurring when an atom is situated in a solid crystal, the magnetic moment of many materials is caused almost entirely by electron spin.

Paramagnetism. In paramagnetic materials the magnetic dipoles associated with the atoms are randomly oriented in the absence of an applied magnetic field. The tendency of these dipoles to align themselves parallel to the applied field is proportional to the strength of the field and may be expressed by

$$\mathcal{M} = \mu_0 \chi_m \mathbf{H}.$$

The quantity $\mathcal{M}$ may be written in terms of the dipole moments $\mathbf{p}_i$ of individual atoms as

$$\mathcal{M} = \frac{1}{\tau} \sum_i \mathbf{p}_i,$$

summed over all atoms in the volume τ.

The tendency of the dipole moments to have random orientation in the absence of an applied field is caused by thermal vibrations. It is therefore not surprising that the ease with which these dipoles can be aligned is inversely proportional to the temperature in most materials, obeying the Curie law

$$\chi_m = \frac{C}{T},$$

where C is the Curie constant and T is the temperature in degrees Kelvin.

Diamagnetism. In many materials the magnetization tends to align itself opposite to the applied field, obeying the same law as do the paramagnetic materials:

$$\mathcal{M} = \mu_0 \chi_m \mathbf{H},$$

except that χ_m is negative. To understand the negative polarizations of these diamagnetic materials, consider an attempt to set up a magnetic field inside a loop of material with zero resistance. From Lenz' law, the current induced in such a loop will be in the direction to prevent the establishment of the field. The resulting current loop would constitute a magnetic dipole pointing opposite to the applied field, and an array of such loops would behave as a diamagnetic medium. The orbital electrons in completely filled atomic shells behave very much like nonresistive current loops. In the absence of an applied field they have no net magnetic moment. Application of a magnetic field alters their orbits slightly so as to oppose the applied field, and the new orbital motion remains indefinitely until the applied field is again changed. This diamagnetic susceptibility is virtually temperature-independent.

Conduction electrons. Thus far only the diamagnetic contribution of electrons in closed shells and the paramagnetic contribution resulting from the orbital and spin angular momentum of electrons in unfilled shells have been treated. The free electrons in a conduction band, however, make a paramagnetic contribution from their spins and a diamagnetic contribution from their motion through the lattice. While the reader can no doubt think of classical arguments to make these contributions plausible, this is a case where classical considerations fail to give even an approximately correct numerical result. In this case the statistical treatment of quantum mechanics shows the paramagnetic effect to be about twice the diamagnetic effect, and both effects to be almost temperature-independent.

Some magnetic susceptibilities. Most materials make both paramagnetic and diamagnetic contributions, and nonferromagnetic materials are classed as *dia-* or *para*magnetic, depending upon which of these contributions is the larger. The magnetic susceptibilities of some typical materials are listed in Table 9–3. While there is considerable variation in the values of susceptibility from one material to another, all values are quite small. In ferromagnetic materials these contributions to the total magnetization can usually be neglected, since they are of the order of one thousand to one million times smaller than the ferromagnetic contribution even in relatively large fields.

9–13 Ferromagnetic materials

Materials that have large magnetizations even in the presence of very weak magnetic fields are called *ferromagnetic*. Only the three elements iron, nickel, and cobalt are ferromagnetic at room temperature and above. Almost all ferromagnetic alloys and compounds contain one or more of these three elements or manganese, which belongs to the same group of transition elements in the periodic table. The ease of magnetization of these

Table 9–3

SOME TYPICAL MAGNETIC SUSCEPTIBILITIES AT
ROOM TEMPERATURE

Material	χ_m
Aluminum	$+1.75 \times 10^{-6}$
Copper	-0.76
Lithium	$+1.90$
Sodium	$+0.68$
Uranium	$+4.86$
Graphite	-7.9
Nickel chloride	$+150.$
Sodium chloride	-1.09

materials results from quantum-mechanical forces which tend to align neighboring atomic spins parallel to each other even in the absence of an applied magnetic field. The quantum-mechanical energy reduction associated with the complete alignment of neighboring atomic spins is called the *exchange energy*. The large magnetizations exhibited by ferromagnetic materials have led to a variety of commercial uses for them. It is therefore desirable to study their properties in some detail.

Temperature dependence. The magnetic properties of ferromagnetic materials are strongly temperature-dependent. In the absence of an applied field they exhibit spontaneous magnetization below a temperature T_C known as the Curie temperature, and are strongly paramagnetic above that temperature, with susceptibilities that decrease with increasing temperature. The ferromagnetic moment is temperature-dependent, starting with zero at the Curie temperature and approaching a fixed value asymptotically as the temperature is lowered toward absolute zero. Above the Curie temperature, the susceptibility may be described rather well by the Curie-Weiss law

$$\chi_m = \frac{C}{T - T_C},$$

but there is no simple expression for the variation of spontaneous magnetization below T_C. A typical plot of C/χ_m above T_C is given in Fig. 9–16, as well as a curve showing the spontaneous magnetization $\mathcal{M}_s$ divided by the spontaneous magnetization at absolute zero, $\mathcal{M}_0$, for temperatures below T_C. The Curie temperatures for iron, cobalt, nickel, and permalloy are listed in Table 9–4, and from the Curie temperatures, it is evident that these materials have almost as much magnetic moment at room temperature as they have at the low temperatures approaching absolute zero.

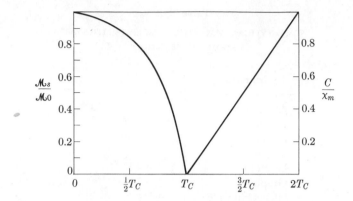

Fig. 9–16 Temperature dependence of some properties of ferromagnetic materials. The dependence of the saturation magnetization is shown up to the Curie point (from $T = 0$ to $T = T_c$) and the dependence of the reciprocal of the paramagnetic susceptibility is shown from $T = T_c$ to $T = 2T_c$. Properties of real materials vary somewhat from these idealized curves. (By permission, from R. M. Bozorth, *Ferromagnetism*, D. Van Nostrand Co., 1951.)

Magnetization of metals. The magnetization $\mathcal{M}_s$ of Co, Ni, and many of their alloys can be understood from a very simple model. All these metals acquire most of their ferromagnetism from the unpaired spins in the partially filled 3d shell. According to Table 9–2 the orbital contribution can be virtually neglected, so that one expects a magnetization contribution of one Bohr magneton for each vacancy in the almost filled 3d shell.

Examination of the periodic table indicates that after all the lower shells are filled, Co has nine remaining electrons in the outer 3d shell, which has room for a maximum of 10 electrons. Thus there appears to be only one

Table 9–4

PROPERTIES OF SOME FERROMAGNETIC METALS

	Fe	Co	Ni	Permalloy (80% Ni 20% Fe)
Measured Curie temperature (°K)	1043	1400	631	780
Electrons available for outer shell	8	9	10	9.6
Average number of conduction electrons	0.6	0.6	0.6	0.6
Vacancies in outer shell	2.6	1.6	0.6	1.0
Measured Bohr magnetons per atom	2.2	1.7	0.6	1.0

vacancy in this shell; however, it is known experimentally that Co, Ni, and many alloys of these elements have, on the average, 0.6 conduction electron per atom. By placing an average of 0.6 electron per atom into the nonferromagnetic conduction band, there are left 1.6 vacancies on the average in the outer shell of Co, which might be expected to cause ferromagnetism. When the experimentally measured magnetization is converted into magnetization per atom in Bohr magnetons, the result is 1.7. This is quite good agreement for such a simple model. A similar calculation for Ni gives 0.6 vacancy per atom, which corresponds very closely to the 0.6 Bohr magneton per atom determined from its saturation magnetization at low temperatures. While the explanation for the 2.2 Bohr magnetons per atom found in Fe must follow a similar pattern, the details are less certain. However, alloys of Co-Ni and Fe-Ni, which have more electrons per atom than pure Co, do fit the simple picture presented here. For example, the permalloy shown in Table 9–4 has $(8 \times 0.2 + 10 \times 0.8) = 9.6$ electrons per atom available for the outer shells. With 0.6 electron per atom in the conduction band, it has one vacancy per atom, in agreement with the one Bohr magneton per atom found by experiment.

9–14 Ferrimagnetic materials

A number of oxides containing iron, nickel, or cobalt exhibit a magnetic behavior described as ferrimagnetic. The most common of these is Fe_3O_4 and the family of ferrites described by the chemical formula $MO \cdot Fe_2O_3 = Fe^{+++}(Fe^{+++}M^{++})O_4^{--}$, where M is any divalent metal such as iron, cobalt, nickel, manganese, magnesium, copper, etc., or a mixture of these. The magnetic moment of these molecules has been found to be substantially less than would be expected if the magnetic spins of the three magnetic ions were lined up parallel to each other. It was proposed by Néel[*] that the magnetic moment of the trivalent ferric ion Fe^{+++} located on a so-called B site of the crystal was aligned antiparallel to the other ferric ions and to M^{++} which were both located on so-called A sites of the crystal. Thus in the case of

$$Fe^{+++}(Fe^{+++}Ni^{++})O_4^{--}$$

the moments of the two Fe^{+++} ions would exactly cancel, leaving the moment of Ni^{++} or two Bohr magnetons per molecule. This has indeed been shown experimentally to be correct. A special case of ferrimagnetism occurs when the antiparallel moments are exactly equal, giving a net magnetic moment of zero. This is called *antiferromagnetism*. As in the case of ferromagnetism, there is a temperature below which ferrimagnetism is exhibited and above which paramagnetism occurs. This is known as the Néel temperature.

---

[*] L. Néel, *Ann. Phys.* **3**, 137 (1948).

Magnetic oxide materials have become increasingly important commerically. They are fabricated with high coercivity for use in magnetic tapes and disks for information storage or with low coercivity and formed into tiny doughnuts for electronically addressable information storage. Other shapes and types have been developed for various other electronic devices. The major advantage of oxides over metals is a low electrical conductivity which permits their use at high frequencies where eddy currents in ferromagnetic metals would tend to prevent the movement of magnetic lines of flux into or out of the metal.*

9–15 Ferromagnetic domains

The preceding discussion suggests that below its Curie or Néel temperature a piece of ferromagnetic material should exhibit a strong magnetic moment even in the absence of an applied magnetic field. However, ferromagnetic and ferrimagnetic materials usually show little or no magnetization until they have been subjected to at least a small magnetic field. This can be understood by considering the classical forces acting between magnetic dipoles. Just a little experience with permanent bar magnets is needed to show that it is difficult to hold a group together with all the north poles pointing in the same direction. Strong forces tend to rotate the magnets so that they lie next to each other in head-to-tail positions. In a ferri- or ferromagnetic material, the quantum-mechanical forces tending to align the individual atomic dipoles parallel to each other are in competition with these classical forces trying to align them antiparallel. The net result is that up to a certain size bundle of dipoles, the quantum forces hold all the dipoles parallel. Larger pieces of material are divided by classical magnetostatic forces into regions within which all atomic dipoles are parallel to each other while the direction of alignment differs from one region to the next. A region within which all atomic dipoles are aligned parallel is known as a magnetic domain, and the separation between two domains is known as a domain wall. Magnetic domains may have volumes from 10^{-6} cm³ to 10^{-2} cm³, therefore bulk materials usually contain millions of domains.

Atomic dipoles have a tendency to align themselves parallel to certain crystalline directions. In the body-centered cubic structure of iron, for example, the dipoles tend to lie parallel to one of the three perpendicular directions defined by the cube edges. These are known as easy directions of magnetization. This anisotropy in the magnetic properties of metallic crystals is caused by a quantum-mechanical coupling between the electron spins and the electrostatic crystalline field. In normal bulk materials, which contain many tiny crystallites joined together in a complex crystal structure,

---

* For further reading see R. F. Soohoo, *Theory and Applications of Ferrites*, Prentice-Hall, Englewood Cliffs, N.J., 1960.

there is some tendency for each crystallite to be a single domain, with its magnetic moment lying parallel to an easy axis. This one-to-one correspondence between domain walls and crystallite boundaries is strongly perturbed, however, by the effects of other energy contributions.

The actual shapes and sizes of magnetic domains can be calculated, in general, only by a process of minimization of the total energy contributed from all causes. Such a calculation would demonstrate the experimentally observed fact that some crystallites contain several domains while some domains contain several crystallites. A detailed study of this problem will not be attempted here. A considerable understanding of the problem can be obtained, however, from the special case of domain walls within a large single crystal.

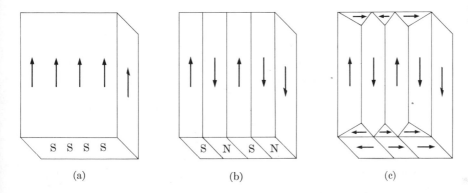

(a)   (b)   (c)

Fig. 9–17 Domains in a ferromagnetic single crystal. The configuration in (a) produces a large external field and therefore a large magnetostatic energy. The external field and the magnetostatic energy are progressively reduced from (a) to (b) to (c).

In the single crystal illustrated in Fig. 9–17(a) all the dipoles are aligned parallel to one easy axis. In Fig. 9–17(b) the magnetostatic energy has been greatly reduced by dividing the crystal into four domains lying in a head-to-tail configuration. The total reduction in energy is not as great as magnetostatic considerations alone would predict, since there is a large energy increase, proportional to the domain wall area, resulting from the quantum-mechanical exchange forces attempting to prevent disalignment between dipoles along the wall.

Domain walls. A domain wall actually consists of a transition region of finite width between two differently directed magnetized domains, with the dipoles becoming more nearly aligned parallel to the second domain as the distance from the first to the second domain is traversed. The change of alignment of the dipoles in a 180° domain wall is illustrated in Fig. 9–18.

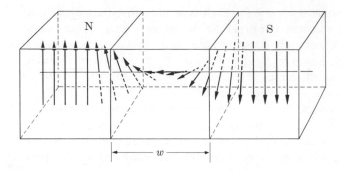

Fig. 9–18 Schematic illustration of variation of magnetic dipole alignment through domain wall of width *w*.

The quantum exchange energy between two dipoles is proportional to the square of the angle between them. For a 180° wall the energy between two dipoles is approximately $JS^2(\pi/N)^2$, where N is the total number of dipoles lying in a line extending across the width of the wall, J is the exchange energy term, and S is the dipole spin. The total energy of this string of N dipoles is therefore proportional to $1/N$. Minimization of exchange energy would yield an infinitely wide wall; however, this neglects anisotropy and magnetostatic and magnetoelastic energy terms. When these terms are included, a finite wall is obtained, which for iron is about 10^{-4} mm. Using $N \simeq 300$, the total energy of the wall is then rather well determined from the dominant exchange energy term

$$W = \frac{JS^2\pi^2}{N}.$$

It is found to be about 1 erg cm^{-2} or 10^{-3} J m^{-2}. A minimization of the sum of the total wall energy and the magnetostatic energy gives a good estimate of the number of walls in a given volume.

If lines of magnetic flux were drawn in Fig. 9–17(a) and (b), it would be apparent that the external field is far more extensive in (a) than in (b), and this necessarily increases the total energy. A further reduction in energy can, therefore, be accomplished by the addition of domains of closure as illustrated in Fig. 9–17(c), where the external field has been virtually eliminated. A low total energy is further maintained by this pattern, since all domains are magnetized parallel to an easy axis.

Actual photographs of a domain pattern similar to that described are shown in Fig. 9–19. The Bitter technique used for this observation, developed by Francis Bitter in 1931, consists in observing finely divided ferromagnetic particles in a colloidal suspension placed on the polished surface of the crystal under observation. A microscope cover-glass placed in contact with the colloidal suspension flattens out any surface irregularities and permits

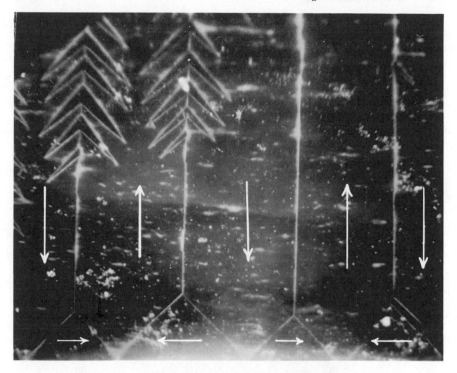

Fig. 9–19 Domains of closure in Si–Fe single crystal photographed by Bitter technique. Arrows illustrate direction of magnetization within individual domains. The complex treelike domain structure observed in the upper left corner is beyond the scope of this discussion, but further reduces the total energy under some conditions. (Courtesy of H. J. Williams, Bell Telephone Laboratories, Murray Hill, New Jersey.)

good optical observation with a microscope. The large field gradients in the vicinity of the domain walls are seen to attract the particles in the suspension and thus reveal the domain pattern. A great variety of domain patterns in many materials have been observed in this manner.*

9–16 Magnetization curves

Interesting changes in domain patterns caused by changes in the applied field have been observed in iron whiskers by De Blois and Graham. These whiskers consist of long, hairlike single crystals whose growth results from the tendency of metallic vapor to condense on the end of a solidified strand

---

* Some of these patterns as well as an excellent review article on magnetic domains may be found in an article by C. Kittel and J. K. Galt, "Solid State Physics," Vol. 3, Academic Press, New York, 437–564, 1956.

rather than on the sides. Since such whiskers are nearly perfect crystals and often show optically perfect faces, they are ideally suited for the study of domain patterns by the Bitter technique. Figure 9–20 shows a sequence of four domain patterns taken for different values of the applied field. A gradual and continuous wall motion occurs as the field is reduced from (a) to (b) and then increased in the reverse direction from (b) to (c). However, a further increase in H from (c) to (d) results in a discontinuous displacement of the walls and a complete change in the domain pattern.

These changes can be correlated with the typical magnetization curve shown in Fig. 9–21. The initial portion of the curve results from reversible wall motion, as from (b) to (c) in Fig. 9–20, while the steep portion corresponds to discontinuous changes such as occurred from (c) to (d) in Fig.

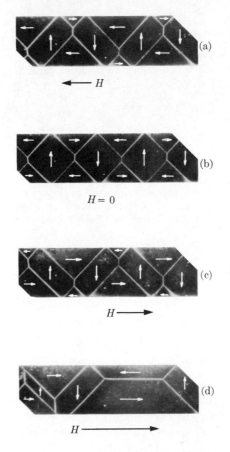

Fig. 9–20 Reversible and irreversible domain wall motion. Walls move smoothly from (a) to (b) to (c) and back, but jump suddenly from (c) to (d) when fields are applied as indicated. Whisker is 50 μ on a side. (Photographs by R. W. DeBlois and C. D. Graham, Jr., *Journal of Applied Physics*, **29**, 936, 1958.)

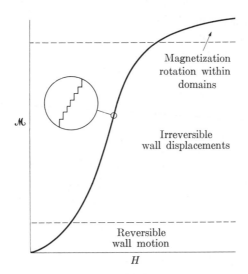

Fig. 9–21 Typical magnetization curve, showing regions in which reversible and irreversible effects dominate.

9–20. As early as 1919, Barkhausen observed discontinuous steps in the steep portion of magnetization curves of small iron wires. These steps, illustrated in the magnified inset of Fig. 9–21, are now referred to as Barkhausen jumps and may be related to discontinuous domain wall displacements. The leveling-off region at the top of the magnetization curve occurs after all domains in the bulk material are magnetized nearly parallel to the field. It is due to a gradual rotation of the magnetization within the domains, tending to arrange it so as to lie more nearly parallel to the magnetizing field.

It should not be surprising that most of the energy loss of the magnetization reversal occurs in the steep portion of the curve where predominantly irreversible wall displacements occur. These irreversible displacements are responsible for the wide area enclosed by a hysteresis loop which is proportional to the energy dissipated in the cycle. In real materials the magnetization curves cannot be broken down into purely irreversible and purely reversible portions; however, the division of the magnetization curve in Fig. 9–21 does illustrate the regions where these processes dominate.

9–17 The extraordinary Hall effect in ferromagnetic materials

The Hall effect in nonferromagnetic materials is discussed in Section 7–1, where it is shown that the Hall electric field per unit current density in the sample is given by

$$\rho_H = RB,$$

where ρ_H is called the Hall resistivity.

Hall effect measurements in ferromagnetic materials pose other problems. From 1880,* when Hall first reported this effect, until 1930, when an extraordinary Hall effect was rediscovered,† the coefficients reported for ferromagnetic materials were of orders of magnitude too large for theoretical understanding. Kundt‡ had reported in 1893 that Hall emf's measured in ferromagnetic materials at low fields were proportional to M rather than to either B or H. This fact was completely ignored by the hundreds of researchers who made many measurements on these materials before this extraordinary effect was rediscovered. During the decade following 1930 many precise measurements§ demonstrated how constant the ratio ρ_H/M is for ferromagnetic materials in low magnetic fields.

The failure of early researchers to recognize this important relation was due to their failure to distinguish the internal field H_i from the H_a field which they applied and measured. This is not too surprising since no distinction is needed with nonferromagnetic materials, where internal fields in typical samples like that in Fig. 7–2 are practically equal to the applied fields. The situation is very different when the kind of sample shown in Fig. 7–2 is made of ferromagnetic materials. The demagnetizing field in such a sample is so large that in general $H_i \ll H_a$. In fact, the boundary conditions developed in Section 9–4 guarantee that, if the sample is very thin, $B_a = B_i = H_i + 4\pi M$ in the gaussian units in vogue at that time or $B_a = B_i = \mu_0 H_i + \mathcal{M}$ in our mks units. To observe the dependence of Hall emf's on the magnetization in the ferromagnetic materials it was necessary to use a bar of the material in which both B_i and H_i could be accurately measured, together with simultaneous measurements of the Hall emf. Furthermore, accurate Hall measurements could be obtained only with a large current density perpendicular to the magnetic field. Naturally, it was important to be able to accurately determine the magnitude of this current density. Figures 6–5 and 6–6 show how this was accomplished and Fig. 9–5 illustrates the principles used in determining H_i and B_i.

When it was discovered that low field Hall emf's were proportional to M, a relation equivalent to

$$\rho_H = R_0 H + R_1 4\pi M$$

was proposed since it seemed certain that there must also be an effect pro-

---

* E. H. Hall, *Phil. Mag.*, **9**, 225 and **10**, 301 (1880).

† Emerson M. Pugh, *Phys. Rev.* **36**, 1503 (1930).

‡ A. Kundt, *Wied. Ann.* **49**, 257 (1893).

§ Emerson M. Pugh and T. Lippert, "Hall emf and Intensity of Magnetization" *Phys. Rev.* **42**, 709 (1932); E. M. Pugh, N. Rostoker and A. I. Schindler, *Phys. Rev.* **80**, 688 (1950).

portional to the magnetic field. In mks units this can be written as

$$\rho_H = R_o\mu_0 H_i + R_1\mathcal{M}$$

or

$$\rho_H = R_o B_i + R_e\mathcal{M}, \tag{9-20}$$

where $R_e = R_1 - R_o$ is the extraordinary and R_o is the ordinary coefficient. The latter form is now preferred for this relation. Since in general $\mu_0 H_i \ll \mathcal{M}$ and $R_o \ll R_1$, $R_o\mu_0 H_i$ is so small compared to $R_1\mathcal{M}$ that the ratio $\rho_H/\mathcal{M}$ usually appears constant at low fields even with accurate measurements. The situation changes at high fields.

Differentiation of Eq. (9–20) with respect to B_i gives

$$\partial\rho_H/\partial B_i = R_0 + R_e\,\partial\mathcal{M}/\partial B_i,$$

which gives the slope of an ρ_H versus B_i curve at any value of B_i. At temperatures well below the Curie point, $\mathcal{M}$ becomes constant at high fields so that $R_e\partial\mathcal{M}/\partial B_i = 0$ and R_o is determined by the slope of the straight line. As a rule, this slope is quite small, so that R_o was not measured* until the 1950's, when higher fields with better measuring techniques became available. After these measurements were reported, a number of laboratories carried out similar experiments and in general confirmed these findings. A word of caution is needed here to emphasize that reliable values for R_o can be obtained only with accurate measurements at high fields. One laboratory reported positive values for R_o in two Co-Ni alloys that had been determined as negative by other laboratories. In these alloys the ρ_H versus B_i curve rises at low fields, because R_e is positive, but reaches a maximum and takes on a negative slope at the higher fields. Their positive values resulted from failure to use sufficiently high fields. If their measurements had been more accurate they would have seen that the curve had not yet become straight at their highest fields.

As the temperature approaches close to the Curie point the term $R_e\partial\mathcal{M}/\partial B_i$ approaches a constant, which is not equal to zero at high fields. While the slope of the ρ_H versus B_i still approaches a straight line at large values of B_i, this slope no longer determines R_o. Instead it gives $R_o + R_e\,\partial\mathcal{M}/\partial B_i$ and creates an apparent anomaly† in the high-field slope versus temperature curve. This anomaly continues rather far above the Curie point, because the material is strongly paramagnetic in this region and R_e increases rapidly with temperature.

Because R_e is generally very large, Hall emf measurements provide a sensitive means for determining the manner in which ferromagnetic materials approach saturation‡ in increasing fields.

---

* A. I. Schindler and E. M. Pugh, *Phys. Rev.* **89**, 295 (1953).
† F. E. Allison and Emerson M. Pugh, *Phys. Rev.* **102**, 1281 (1965).
‡ S. Foner, *Phys. Rev.* **101**, 1648 (1956).

PROBLEMS

9–1 A solid sphere of radius a is inserted into a uniform field, $\mathbf{D}_a = k\epsilon_0\mathbf{E}_a$, whose magnitude can be varied. The sphere consists of a homogeneous and isotropic but nonlinear dielectric material, whose polarization P versus E can be obtained from experimental curves like those in Figs. 5–13 or 5–14. Suppose that, when this unpolarized sphere is inserted, the field is increased from zero to E_a. From a curve like that in Fig. 5–13 one finds a D_i corresponding to E_i. The polarization is then $P = D_i - \epsilon_0 E_i$. Show that the internal fields are given by $\mathbf{E}_i = k(E_a - P/3\epsilon_0)$ and $\mathbf{D}_i = k(\epsilon_0 E_a + 2P/3)$ and the external field is given by

$$\mathbf{E}_e = kE_a + d_f Pa^3/3\epsilon_0 r^3.$$

9–2 A solid sphere of radius a is a permanently polarized electret with the uniform polarization $\mathbf{P} = kP$. Show that the internal field is given by $\mathbf{E}_i = -kP/3\epsilon_0$ and the external field is given by

$$\mathbf{E}_e = d_f Pa^3/3\epsilon_0 r^3.$$

9–3 A solid sphere of radius a is inserted into a uniform field, $\mathbf{B}_a = k\mu_0\mathbf{H}_a$, whose magnitude can be varied. The sphere consists of a homogeneous and isotropic but nonlinear ferromagnetic material, whose "normal magnetization" of $\mathcal{M}$ versus H can be obtained from experimental curves like those in Figs. 9–6 or 9–7. Assume that the cycling process described in Section 9–6 has been carried out with the sphere to produce "normal magnetization" within it. If B_a is then the applied field, the magnetization $\mathcal{M}$ can be obtained from these curves,

$$\mathcal{M} = B_i - \mu_0 H_i.$$

a) Show that the internal fields are given by $\mu_0 \mathbf{H}_i = k(\mu_0 H_a - \mathcal{M}/3)$ and $\mathbf{B}_i = \mu_0 H_i + \mathcal{M} = k(\mu_0 H_a + 2\mathcal{M}/3) = k(B_a + 2\mathcal{M}/3)$.
b) Show also that the external field is given by

$$\mathbf{B}_e = \mu_0 \mathbf{H}_e = kB_a + d_f \mathcal{M}a^3/3r^3.$$

9–4 Show that the internal and external fields of a sphere of ferromagnetic material, which has the uniform permanent magnetization $\mathcal{M}$ parallel to the z-axis, is given by $\mathbf{B}_i = k2\mathcal{M}/3$, $\mu_0 H_i = -k\mathcal{M}/3$, and $\mathbf{B}_e = \mu_0 H_e = d_f \mathcal{M}a^3/3r^3$, respectively.

9–5 Assume that Fig. 9–4 represents the hysteresis loop of the material in the sphere of Problem 9–4. Assume also that the coordinates of the tips of these loops are at (2 Wb m^{-2}, 2 Wb m^{-2}) and (-2 Wb m^{-2}, -2 Wb m^{-2}). Calculate approximately the $\mathbf{B}_i$, $\mu_0 H_i$, and $\mathbf{B}_e$.

9–6 Right circular cylinders of infinite length, which are oriented perpendicular to the uniform fields, have properties similar to those of the sphere in Problem 9–4. Such cylinders of ferromagnetic materials can be used profitably for some Hall measurements. With a current parallel to the cylindrical axis and an applied field perpendicular to this axis, Hall probes can be located on a diameter that is perpendicular to both of these directions.
a) Show that, if the applied field is $\mathbf{B}_a = i\mu_0 H_a$, the internal fields are given by

$$\mathbf{B}_i = i(B_a + \mathcal{M}/2) \quad \text{and} \quad \mu_0 H_i = i(B_a - \mathcal{M}/2).$$

b) Assume the cylinder is made of the iron, whose normal magnetization curve is shown in Fig. 9–6. Show that when $B_i = 0.8$ Wb m^{-2} (8000 G) the $\mu_0 H_i \ll B_i$ and hence $B_i \cong \mathcal{M}$ while $B_a \cong \mathcal{M}/2$. Show also that, since the Hall field is given by $\mathbf{E}_H = \mathbf{J} \times (R_0 \mathbf{B}_i + R_e \mathcal{M})$, $E_H \cong (R_0 + R_e)\mathcal{M}$; i.e., the Hall emf is nearly proportional to the applied fields at low fields. This cylinder has an advantage over the flat sheet shown in Fig. 7–2. Since $\mathcal{M} = 2B_a$, saturation is reached at smaller applied fields. A disadvantage stems from the need for a large current through the cylinder to produce large current densities required for measurements. This can be overcome by sawing the cylinder in two through its axis. The material removed by the saw should then be replaced in the form of a thin sheet of the material insulated from the rest of the cylinder by thin films. Current is then sent only through this thin sheet.

9–7 Find the fields of the permanently magnetized sphere in Problem 9–4 using amperian currents. First determine the distribution of these currents and then calculate the vector potential. Problems 8–25, 8–26 and 8–27 will be found useful.

9–8 Assuming the sphere of Problem 9–3 is made of the iron whose hysteresis loop is shown in Fig. 9–8, calculate the electromagnetic energy absorbed each time the sphere is taken through a complete hysteresis cycle.

9–9 An Alnico magnet of U-shape rests with its flat and circular pole faces, which are 4 cm in diameter, on a flat surface of a large block of soft iron. This magnet has a uniform and permanent magnetization of 1.2 Wb m^{-2} perpendicular to its pole faces. It can be assumed that all the flux out of and into these pole faces passes through the iron.

a) Calculate the work required to lift both poles of this magnet simultaneously, a distance δ above the flat surface. Ignore its weight.

b) Calculate the force required for this lifting.

9–10 A copper wire of radius a carries a steady current I.

a) What is the flux density at a point within the material of the wire, at a distance r from its axis?

b) Find the magnetic energy density at points within a thin cyclindrical shell of radius r, thickness dr, and length l.

c) Find the total magnetic energy within a portion of the wire of length l.

9–11 Suppose a cylindrical bar magnet of alnico 5, 1 cm in diameter and 10 cm long, is permanently and uniformly magnetized to an intensity of magnetization of 1.2 Wb m^{-2}.

a) What is the strength of the poles at the ends of the magnet?

b) What is the magnetic moment of the magnet?

c) What is the torque exerted on the magnet when it is suspended in air at right angles to a magnetic field of flux density 1 mWb m^{-2}?

d) Compare the magnitude of H in the above field with the coercive force of alnico 5.

9–12 A solenoidal winding of 100 turns of wire is wound on a wooden rod of the same dimensions as the bar magnet in Problem 9–11. What current must be sent through the wire in order that the torque acting on the solenoid in an external field shall equal the torque on the bar magnet?

9–13 Sketch the lines of induction of a diamagnetic sphere in a uniform external field.

9–14 In Fig. 9–22, a long straight conductor perpendicular to the plane of the paper carries a current *I*. A bar magnet having point poles of strength q_m at its ends lies in the plane of the paper.

Let the distance between the poles of the magnet be 8 cm and its pole strength $q_m = 10^{-5}$ Wb. Let $a = 6$ cm, $b = 5$ cm, $c = 5$ cm, and let $I = 100$ A. Find the magnitude of the magnetic intensity **H** at point *P*.

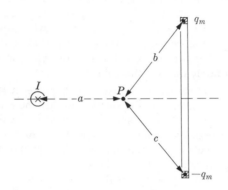

Figure 9–22

9–15 A permanent magnet consists of a right circular cylinder that is uniformly magnetized like the cylinder in Fig. 9–1, but whose length *l* is large compared with its radius *a*. Calculate **B** and **H** for the following points as a function of $\mathcal{M}$:
a) On the axis outside but very close to the surface that is positively magnetized.
b) On the axis inside but very close to the surface that is positively magnetized.
c) At the center of the bar on the axis.
d) Near the center of the bar but just outside the lateral surface.
e) Use the relations governing fields on the two sides of a bounding surface to check your results.

9–16 The similarity of the magnetostatic relations with those for electrostatic phenomena suggests that a magnetic pole q_m should experience a force of $q_m\mathbf{H}$ in a magnetic field. Assume the cylinder of Fig. 9–1 placed in a uniform field with the positive direction of its axis at an angle θ to the field. Calculate the torque on the cylinder (a) from the force on the poles q_m and $-q_m$ on the end surfaces, assuming that they can be considered as concentrated at their respective centers, and (b) from the sum of the torques on the individual dipoles of the cylinder.

9–17 The magnetic field near a long straight wire coinciding with the *z*-axis can be written in cylindrical coordinates as $\mathbf{H} = \mathbf{a}_\phi I/2\pi r$. Place the center of the magnetized cylinder of Fig. 9–1 a distance *r* from the wire, where *r* is quite large compared with the length of the cylinder, with its axis parallel to $\mathbf{a}_\phi$. Calculate the net force on the cylinder.

9–18 A ferromagnetic sphere of radius *a* is permanently magnetized uniformly in the *z*-direction to a magnetization $\mathcal{M}$. Suppose this sphere to be surrounded by a mixture of very heavy oil and powdered iron whose relative permeability is $K_m \cong 5$. Further, suppose that this mixture extends to very large *r*'s and does not affect the value of $\mathcal{M}$ in the sphere.

a) Find the magnetic potentials and the fields inside and outside the sphere in terms of the total dipole moment $p_m = 4\pi a^3 \mathcal{M}/3$, the K_m, and the spherical coordinates.

b) Actually, if the internal fields are altered by the presence of the permeable mixture, the value of $\mathcal{M}$ will change. Compare the values of $\mathcal{M}$ when the sphere is in the mixture with the $\mathcal{M}$ when the sphere is in a vacuum, assuming that in each case the magnetization is produced by following the hysteresis loop of Fig. 9–8.

9–19 Place a small magnetized cylinder, of the kind illustrated in Fig. 9–1 but with $l = 50a$, in the permeable mixture described in Problem 9–18. Calculate potentials and fields at points where $r \gg l$, in terms of the total dipole moment $p_m = \pi a^2 l \mathcal{M}$, the K_m, and the spherical coordinates. Can you give a physical explanation for the fields in this case having a different dependence upon K_m than they did in the case of the magnetized sphere of Problem 9–18?

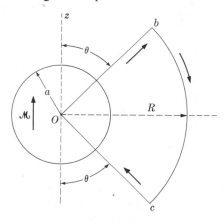

Figure 9–23

9–20 A permanently magnetized sphere is shown in Fig. 9–23. Express the **B** and **H** fields inside and outside this sphere and use these expressions in calculating

$$\int \mathbf{H} \cdot d\mathbf{l} \quad \text{and} \quad \int \mathbf{B} \cdot d\mathbf{l}$$

for each portion of the closed path $ObcO$. From these parts determine the values of the closed-path line integrals

$$\oint \mathbf{H} \cdot d\mathbf{l} \quad \text{and} \quad \oint \mathbf{B} \cdot d\mathbf{l}.$$

9–21 Suppose that small cavities have been cut out of a very large ferromagnetic material in which there is a uniform magnetization $\mathcal{M}$ and a uniform magnetizing field **H** that is parallel to $\mathcal{M}$. Find the fields $B_c = \mu_0 H_c$ inside cavities of the following shapes: (a) a spherical cavity of radius a, (b) a very long and thin cylindrical cavity parallel to $\mathcal{M}$, (c) a thin disklike cavity perpendicular to $\mathcal{M}$, and (d) a long cylindrical cavity of radius a with its axis perpendicular to $\mathcal{M}$. [*Note:* The cavities (b) and (c) have been imagined by many authors to illustrate the difference between the **H** and **B** fields respectively in a ferromagnetic medium.]

9–22

a) A Rowland ring, shown in Fig. 9–24, has a cross section of 2 cm², a mean length of 30 cm, and is wound with 400 turns. Find the current in the winding that is required to set up a flux density of 0.1 Wb m⁻² in the ring, (1) if the ring is of annealed iron (Table 9–5), (2) if the ring is of silicon steel (Table 9–6). (3) Repeat the computations above if a flux density of 1.2 Wb m⁻² is desired.

b) Find the current required to establish a flux of 4×10^{-4} Wb in the ring if it is of silicon steel.

c) Compute the flux in the ring if the current in the windings is 6 A.

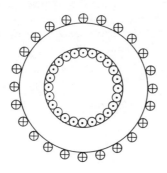

Figure 9–24

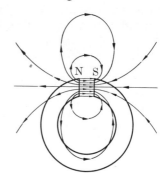

Figure 9–25

9–23 The hysteresis loop of the iron that is formed into a toroid is shown in Fig. 9–8. The surface of the iron has the shape that would be swept out by a circle of small radius a as it moved perpendicular to its plane with its center on a larger circle of radius $r_c > 20a$. An air gap like that shown in Fig. 9–25 is produced in this toroid by sawing out a nearly cylindrical rod of iron of radius a and length $l = 2a$. Assume that the iron is given a uniform magnetization $\mathcal{M}_i$ parallel to the large circle. Although the iron removed from the toroid cannot be a true cylinder, it may be assumed to be cylindrical for the calculations that follow. Calculate the values of $\mu_0 H$ and B in terms of $\mathcal{M}_i$ and a at the following points on the large circle: (a) in the middle of the air gap, (b) at one end of the air gap in the air, (c) at the same end of the air gap but just inside the iron, and (d) at a point diametrically opposite the center of the air gap.

9–24 The air gap in the iron toroid described in Problem 9–23 is filled with nickel that saturates in relatively weak fields at $\mathcal{M}_n = 0.6$ Wb m⁻². Assuming that the iron is uniformly magnetized, as in Problem 9–23, to $\mathcal{M}_i = 1.3$ Wb m⁻², calculate the fields B and $\mu_0 H$ at the points on the large circle specified in (a), (b), (c), and (d) of that problem.

9–25 A region can be partially shielded from external magnetic fields by surrounding it with a sufficient thickness of high-permeability ferromagnetic material. To determine the effectiveness of such shielding, calculate the field B_i inside a hollow ferromagnetic cylinder when there is an external field B_0 perpendicular to the cylindrical axis. The internal and external radii of the cylinder are a and b, respectively, and the length of the cylinder is very large compared with either a or b. The relative permeability of the shield is K_m, which may be only 10^3 but can be as large as 10^5 in some new materials, especially in weak fields. This problem can be

Table 9–5

MAGNETIC PROPERTIES OF ANNEALED IRON

Magnetic intensity	Flux density	Permeability	Magnetization	Relative permeability
H, A m^{-1}	B, Wb m^{-2}	$\mu = B/H$, H m^{-1}	$\mathcal{M} = B - \mu_0 H$, Wb m^{-}	$K_m = \mu/\mu_0$
0	0	$3{,}100 \times 10^{-7}$	0	250
10	0.0042	4,200	0.0042	330
20	0.010	5,000	0.010	400
40	0.028	7,000	0.028	560
50	0.043	8,600	0.043	680
60	0.095	16,000	0.095	1270
80	0.45	56,000	0.45	4500
100	0.67	67,000	0.67	5300
150	1.01	67,500	1.01	5350
200	1.18	59,000	1.18	4700
500	1.44	28,800	1.44	2300
1,000	1.58	15,800	1.58	1250
10,000	1.72	1,720	1.71	137
100,000	2.26	226	2.13	18
800,000	3.15	39	2.15	3.1

Table 9–6

MAGNETIC PROPERTIES OF SILICON STEEL

Magnetic intensity H, A m^{-1}	Flux density B, Wb m^{-2}
0	0
10	0.050
20	0.15
40	0.43
50	0.54
60	0.62
80	0.74
100	0.83
150	0.98
200	1.07
500	1.27
1,000	1.34
10,000	1.65
100,000	2.02
800,000	2.92

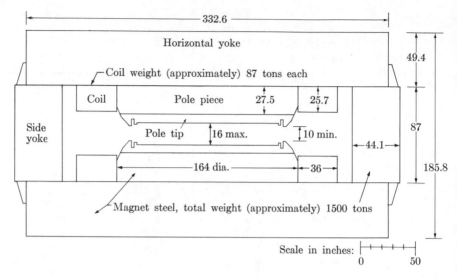

Figure 9–26

modified by assuming a hollow sphere instead of a hollow cylinder. The resulting fields have values that do not differ very much from those in the cylinder.

9–26 The shape and dimensions of the large magnet constructed for the Carnegie-Mellon cyclotron are shown in Fig. 9–26. The two pole pieces are round and both have the maximum diameter of 164 in. The remainder of the magnetic circuit is made up of rectangular sections whose dimensions perpendicular to the diagram are each 208.7 in. The purest iron available in these sizes, whose magnetic properties are close to those given in Table 9–5, was used. During operation the field at the center is 2.050 Wb m^{-2} and this falls off linearly with radius to 1.970 Wb m^{-2} at the radial distance 68.4 in., whence it falls rapidly to zero at 72 in. The coils, which are made of oil-cooled copper straps, are connected in series and carry 21,000 A.

a) Estimate the total ampere-turns and the number of turns required for this magnet.

b) Assuming that only 10% of the space in the coils is required for insulation and cooling oil, estimate the cross section of the copper straps.

c) What voltage and what power are required for the magnetization?

9–27 The two toroids, sections of which are shown in Fig. 9–27, are single-layer compact windings that share the same circle of centers, with the radius r_c. Assume that the small solenoid is wound on a core of the special ferromagnetic alloy *perminvar*, for which $\mu \cong 300 \, \mu_0$. Further assume that the radii a and b of the two coils are related by $b^2 = 2a^2$ and that the circle passing through their centers has the radius $r_c \gg b$.

a) Calculate the self-inductances of the two coils, L_a and L_b, and their mutual inductance, M.

b) Calculate L_a, L_b, and M for a situation in which the perminvar fills all the space inside the large as well as the small solenoid.

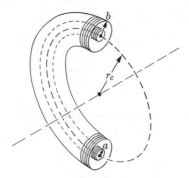

Figure 9–27

9–28 Suppose that a current I is sent through the two solenoids of Problem 9–27, which are connected in series aiding. Calculate the electromagnetic energy (a) with the perminvar inside the small solenoid only, and (b) with the perminvar filling all the space inside both the large and small solenoids.

Chapter 10 THE POYNTING VECTOR

Most of our discussion thus far has concerned electromagnetic fields, culminating in the four Maxwell field relations of Section 8–4 and the Lorentz force relation of Section 7–1. We have pointed out that these five equations solve all electromagnetic problems as long as all quantities are measured in the observer's coordinate system. Only in Chapters 6 and 7 have we discussed energy conversions and we have not previously mentioned energy transfers. Electromagnetism would create little or no interest if energy transfer and energy conversion were not involved. A general expression for the flow of energy into an arbitrary volume surrounded by a closed surface s is needed. Such energy conversions involve a vector that has been named for its discoverer, J. H. Poynting.*

10–1 Energy transfer and conversion

Any electromagnetic energy that flows into τ through s increases the electric field energy, increases the magnetic field energy, or is converted into other forms of energy. According to Section 9–11, the power per unit volume expended in changing the electric and the magnetic fields is $\mathbf{E} \cdot \dot{\mathbf{D}}$ and $\mathbf{H} \cdot \dot{\mathbf{B}}$, respectively, and according to Section 7–7 the power per unit volume being converted to other forms is $\mathbf{E} \cdot \mathbf{J}$. The power flowing into τ then is obtained by integrating the sum of these three terms over the volume τ; that is,

$$P = \int_\tau (\mathbf{E} \cdot \dot{\mathbf{D}} + \mathbf{H} \cdot \dot{\mathbf{B}} + \mathbf{E} \cdot \mathbf{J}) \, d\tau. \tag{10–1}$$

This integrand can be simplified and written in terms of a single vector by

---

* J. H. Poynting, *Collected Scientific Papers*, Art. 10, pp. 175–193, Cambridge University Press (1920) and *Phil. Trans. Roy. Soc.* **178**, 343 (1884).

introducing Maxwell's third and fourth relations, $\dot{\mathbf{B}} = -\nabla \times \mathbf{E}$ and $\dot{\mathbf{D}} + \mathbf{J} = \nabla \times \mathbf{H}$ respectively. The integrand then becomes

$$\mathbf{E} \cdot \nabla \times \mathbf{H} - \mathbf{H} \cdot \nabla \times \mathbf{E} = \nabla \cdot (\mathbf{H} \times \mathbf{E}) = -\nabla \cdot \mathbf{S},$$

where the single vector $\mathbf{S}$, which is called the *Poynting vector* in honor of its inventor, is defined by

$$\mathbf{S} = \mathbf{E} \times \mathbf{H}. \tag{10–2}$$

Significantly, $\mathbf{S}$ has the unit W m^{-2}, since $\mathbf{E}$ has the unit V m^{-1} and $\mathbf{H}$ has the unit A m^{-1}. The electromagnetic power entering the volume τ through s is

$$-\int_\tau \nabla \cdot \mathbf{S} \, d\tau = -\int_s \mathbf{S} \cdot d\mathbf{s},$$

where the surface integral is obtained by applying Gauss' mathematical theorem. Equation (10–1) then can be written as

$$\int_\tau (\mathbf{E} \cdot \dot{\mathbf{D}} + \mathbf{H} \cdot \dot{\mathbf{B}} + \mathbf{E} \cdot \mathbf{J}) \, d\tau + \int_s \mathbf{S} \cdot d\mathbf{s} = 0. \tag{10–3}$$

The volume integral gives the total electromagnetic power expended within the volume to change the electric and magnetic fields and to convert electromagnetic energy to other forms. A negative sign for any of these terms indicates that that particular energy conversion is progressing in the opposite direction. The surface integral gives the flux of $\mathbf{S}$ out of the volume.

This vector $\mathbf{S} = \mathbf{E} \times \mathbf{H}$ has been adopted for determining the rate of flow of energy in electromagnetic waves by practically all authors of that subject. It determines the power being broadcast by radio, television and radar antennas. It is used for analysis of light intensities and for many other related phenomena. Careful experiments have verified the predictions made by the use of this Poynting vector with oscillating electromagnetic fields.

10–2 Momentum density in electromagnetic fields

In mechanics there is an important theorem which states that, whenever there is energy flowing through a unit area per unit time given by Σ, there must be an associated volume density of momentum given by Σ/c^2, where c is the velocity of light. Thus the momentum density in electromagnetic fields must be given by

$$\mathbf{p} = \mathbf{S}/c^2 = \mathbf{E} \times \mathbf{H}/c^2. \tag{10–4}$$

In other words, when an electromagnetic wave strikes a perfectly reflecting surface normally, it should be expected to produce a force per unit area on that surface equal to $-2\mathbf{p}c$. At such a surface the direction of the wave is reversed and this expression gives the rate of change of momentum per unit area. This has been observed experimentally.

In fact the predictions obtained from these relations for rate of energy flow and momentum density have been experimentally verified* for both reflecting and absorbing surfaces at various angles of incidence. The momentum density relation also has been obtained theoretically with our five basic equations by analyzing the impingement of electromagnetic waves on such surfaces. A simple procedure for obtaining this relation involves the analysis of electromagnetic waves impinging on and being reflected from a perfectly conducting surface. The electric field of the wave induces currents in the surface that interact with the magnetic field. The details are given in Section 12–5, where the correctness of our momentum density expression is verified.

Field energy correlations. In Section 12–3 we will see that in sinusoidal and plane electromagnetic waves the time average of the Poynting vector is given by

$$\bar{\mathbf{S}} = \mathbf{v}\bar{U},$$

where $\bar{U}$ is the average value of the electromagnetic energy density in these waves. One can think of the energy in the wave being propagated at the velocity $\mathbf{v}$ of the wave. It should not be concluded, however, that the magnitudes of $\mathbf{S}$ and $\mathbf{S}/c^2$ at all points in space are always proportional to the energy densities at such points. Static electric or magnetic fields can be superimposed upon electromagnetic waves without affecting these magnitudes. Nevertheless, these static fields will add to the energy densities.

This warning is especially important when Poynting vectors in static fields are being considered. There are many static fields in which $\mathbf{S}$ is not equal to zero and in most of these fields S is not proportional to U. In this chapter we consider all fields to be in free space where $\epsilon = \epsilon_0$ and $\mu = \mu_0$.

Energy flow in dc networks. In dc networks there is no time dependence and outside of the conductors there are no currents. The equations (M–3) and (M–4) give $\nabla \times \mathbf{E} = 0$ and $\nabla \times \mathbf{H} = 0$, respectively, and these fields are static and conservative outside the conductors. However, power is generally being transmitted by these networks, e.g., from generator to motor, storage battery, or resistance. In these networks mechanical power is converted to electromagnetic power by the generator and transmitted to devices which convert the electromagnetic power to other forms. In referring to such networks we will call the conversion devices electromagnetic sources or sinks, depending on the direction of the conversion. In later sections we show how $\mathbf{S}$ can be used to chart the flow of energy from any source to any sink.

The use of $\mathbf{S}$ in static fields has been objected to by some, because it often has magnitude in static fields where no flow of energy can be detected. However, in all such cases it can be shown that $\nabla \cdot \mathbf{S} = 0$ everywhere. To

---

* R. A. Houston, *A Treatise on Light*, 5th Ed., Longman's, Green and Co., Ltd., London, 1927, pp. 453–454, provides excellent references for these demonstrations.

detect energy flow one must be able to absorb some of it. This can be done only where $\nabla \cdot \mathbf{S} \neq 0$. In sinks, for example, $\nabla \cdot \mathbf{S}$ is primarily negative, in sources it is primarily positive, and in between sources and sinks $\nabla \cdot \mathbf{S} = 0$.

Above the earth's surface magnetic field lines pass between the north and south poles while electric field lines are radial. This results in an east-west Poynting vector, suggesting that electromagnetic energy flows endlessly around the earth. Some persons have been so troubled by this that they have recommended discarding the Poynting vector concept in static fields. Feynman,* however, has proposed a gedanken experiment (one performed in thought only) which demonstrates that static electromagnetic fields do carry the angular momentum that is predicted by the momentum density expression, $\mathbf{p} = \mathbf{S}/c^2$. He concludes that in static fields, "There really is a momentum flow. It is needed to maintain the conservation of angular momentum in the whole world."

Feynman's experiment is qualitative. Before proceeding to use the $\mathbf{S}$ concept in static fields, we will present in the next section a gedanken experiment in which Feynman's statement can be proven quantitatively.†

10–3 Angular momentum conservation in static electromagnetic fields

To investigate whether static electromagnetic fields have angular momentum, it is desirable to analyze the changes in momentum that take place in the establishment of these static fields. For this investigation we will start with a permanent magnet inside a capacitor consisting of concentric spherical shells (Fig. 10–1). We consider the permanently magnetized inner sphere to be a nonconductor covered with a thin conducting film, so that any current flow is confined to the surface of this sphere, whose outer radius is a. It has a uniform magnetization $\mathcal{M}_0$ parallel to the z-axis and it is free to rotate without friction about that axis. The outer shell is a nonmagnetic conductor of inner radius b that is also free to rotate independently and without friction about the z-axis.

When this capacitor is charged with $+Q$ on the inner sphere and $-Q$ on the outer sphere, there is an electric field only in the space between the spheres that is given by

$$\mathbf{E} = \mathbf{a}_r Q/4\pi\varepsilon_0 r^2. \tag{10–5}$$

This is easily verified with Gauss' law of Eq. (3–5). The magnetic field due to the permanent magnet is given by

$$\mathbf{B} = \mu_0 \mathbf{H} = \mathcal{M}_0 a^3 \mathbf{d}_f/3r^3, \tag{10–6}$$

---

* R. P. Feynman, R. B. Leighton, and M. L. Sands, *The Feynman Lectures on Physics*, Vol. II, p. 27–11, Addison-Wesley, Reading, Mass., 1964.
† Emerson M. Pugh and George E. Pugh, "Physical Significance of the Poynting Vector in Static Fields," *Am. Journ. Phys.* **35**, pp. 153–156, 1967.

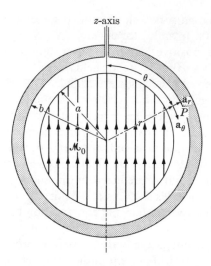

Fig. 10–1 Nonconducting sphere surrounded by two conducting spherical shells of radii a and b. The inner sphere is solid and magnetized.

where $\mathbf{d}_f = (\mathbf{a}_r\, 2 \cos \theta + \mathbf{a}_\theta \sin \theta)$ from Eq. (2–25a). Here **B** extends from $r = 0$ to $r = \infty$. This equation can be obtained from Eq. (2–25) by replacing the electric dipole moment **p** with the magnetic dipole moment $\mathbf{p}_m$, the electric field **E** with the magnetic field **H** and the ϵ_0 with the μ_0. It is shown in Section 9–5 that the magnetic field outside a uniformly magnetized sphere is the same as if all of the sphere's magnetic moment were concentrated at its center, which gives

$$p_m = \mathcal{M}_0 \frac{4\pi a^3}{3}.$$

Now, $\mathbf{a}_r \times \mathbf{a}_r = 0$ and $\mathbf{a}_r \times \mathbf{d}_f = \mathbf{a}_r \times \mathbf{a}_\theta \sin \theta = \mathbf{a}_\phi \sin \theta$ so that in the space from $r = a$ to $r = b$,

$$\mathbf{S} = \mathbf{E} \times \mathbf{H} = \mathbf{a}_\phi\, Q\mathcal{M}_0 a^3 \sin \theta / 12\pi\epsilon_0\mu_0 r^5. \tag{10–7}$$

This suggests that, in this space, energy flows in circles about the z-axis. Since the two spheres can only rotate about the z-axis, any momentum density function must be directed along these circles to satisfy the law of conservation of angular momentum.

To determine what happens as these fields are established, we start with the capacitor uncharged and with the spheres at rest. The magnetic field is already established but there is no Poynting vector because there are no **E** fields. To charge the capacitor we cause positive and negative charges to flow down a concentric cable whose axes coincide with the z-axis, as shown in Fig. 10–1, giving the inner sphere a positive and the outer sphere a negative charge. Assume that these charges flow so slowly that they can always be considered uniformly distributed over the spherical surfaces. The resulting electric currents in the sphere interact with the magnetic field to produce opposite but not equal angular momenta in the two spheres. The combined angular momentum of the two spheres is not zero. Thus, a system that originally has no angular momentum and into which no angular momentum is introduced is given a net mechanical angular momentum. Detailed calculations in the next section show, in fact, that this mechanical angular momentum is exactly equal and opposite to the angular momentum stored in the Poynting vector between the two spheres.

10–4 Angular momentum calculations*

The current densities on the spherical surfaces need to be calculated. At any time t, the magnitude of the uniform charge on either sphere is $q = q(t)$, where

$$\int_0^\infty q(t)\, dt = Q.$$

The current past a point P on the outer sphere can be derived from the rate of change of the negative charge below the zone of P. This charge is $-q(1 + \cos\theta)/2$ and its rate of change is $\frac{1}{2}(1 + \cos\theta)\, dq/dt$. Hence the total current $\mathbf{J}$ flowing past the zone of P is given by

$$\mathbf{J} = -\mathbf{a}_\theta[\tfrac{1}{2}(1 + \cos\theta)](dq/dt)$$

and the rotational force per unit zone width at P is given by

$$\mathbf{F} = \mathbf{J} \times \mathbf{B} = \mathbf{J} \times (\mathcal{M}_0 a^3/3b^3)(2\mathbf{a}_r \cos\theta + \mathbf{a}_\theta \sin\theta)$$
$$= \tfrac{1}{2}(1 + \cos\theta)(dq/dt)(\mathcal{M}_0 a^3/3b^3)\mathbf{a}_\phi 2\cos\theta.$$

The force on the zone width $b\,d\theta$ at P is

$$d\mathbf{F} = \mathbf{a}_\phi(\mathcal{M}_0 a^3/3b^2)(1 + \cos\theta)\cos\theta\, d\theta(dq/dt),$$

and the resulting torque about the axis is

$$dL = (\mathcal{M}_0 a^3/3b)(1 + \cos\theta)\sin\theta\cos\theta\, d\theta(dq/dt).$$

Now the angular impulse delivered to the zone of the outer sphere between θ and $\theta + d\theta$ is

$$dI = \int_0^\infty dL\, dt,$$

or

$$dI = \frac{\mathcal{M}_0 a^3}{3b}(1 + \cos\theta)\cos\theta\sin\theta\, d\theta \int_0^\infty \frac{dq}{dt}\, dt,$$

$$dI = (Q\mathcal{M}_0 a^3/3b)(1 + \cos\theta)\cos\theta\sin\theta\, d\theta$$

---

* Ibid.

and the total angular impulse delivered to the outer sphere is

$$I_b = \int_0^\pi dI,$$

which gives

$$I_b = \frac{Q\mathcal{M}_0 a^3}{3b} \int_0^\pi (1 + \cos\theta)\cos\theta\sin\theta\, d\theta,$$

$$I_b = (2/9)(Q\mathcal{M}_0 a^3/b).$$

The same type of analysis shows that the total angular impulse delivered to the inner sphere is $I_a = -(2/9)(Q\mathcal{M}_0 a^3/a)$. Thus the net impulse is

$$I = I_a + I_b = -(2/9)(Q\mathcal{M}_0 a^3)[(1/a) - (1/b)], \qquad (10\text{–}8)$$

which must give the net mechanical angular momentum of the system if the spheres rotate without friction.

Thus, a system having no angular momentum has been taken and, without adding any angular momentum, has been transformed into a system containing a net mechanical angular momentum. This might appear to violate the law of conservation of angular momentum. Since we do believe that angular momentum in any isolated system must be conserved, we assume that this missing angular momentum will be found in the electromagnetic fields.

The momentum density in the field from Eqs. (10–4) and (10–7) is $\mathbf{p} = \mathbf{S}/c^2 = \mathbf{a}_\phi(\mathcal{M}_0 Q a^3 \sin\theta/12\pi r^5)$, since $c^2\mu_0\epsilon_0 = 1$. The total angular momentum of the Poynting vector is given by

$$P = \int pr\sin\theta\, d\tau$$

$$= \int_0^{2\pi}\int_0^\pi\int_a^b \frac{\mathcal{M}_0 Q a^3}{12\pi r^4} r^2 \sin^3\theta\, d\phi\, d\theta\, dr$$

$$P = (2/9)(\mathcal{M}_0 Q a^3)[(1/a) - (1/b)]. \qquad (10\text{–}9)$$

and the total angular momentum of this system is, from Eqs. (10–8) and (10–9),

$$I + P = 0.$$

Thus by assigning the momentum density $\mathbf{p} = \mathbf{E} \times \mathbf{H}/c^2$ to the field, the conservation law has been satisfied. Obviously the same results will apply in the limit where $b \to \infty$. Thus, for example, the earth might be thought of as a charged magnetic sphere, which would have a significant circulating Poynting-vector flux.

In the preceding calculations, we have dealt only with the ϕ component of the Poynting vector (which remains after the charging of the spheres is complete). During the actual charging process, there is also a small θ component which results from the interaction of the static electric field with the magnetic field produced by the charging current itself (this field was ignored in the previous calculation). As guaranteed by the divergence derivation of the Poynting vector, this θ component is exactly the size required to account for the spatial distribution of the electrostatic and circulating energy accumulated in the field between the two spheres.

Ambiguity in the definition of **S**. The fact that **S** is defined only in terms of $\mathbf{\nabla} \cdot \mathbf{S}$ gives it a certain ambiguity. In fact an infinite number of expressions can satisfy this definition. For example, $\mathbf{S}' = \mathbf{S} + \mathbf{f}$ with any function for which $\mathbf{\nabla} \cdot \mathbf{f} = 0$ will satisfy it. We have been able to choose an **f** that makes the magnitude of $\mathbf{S}'$ proportional or even equal to $\dot{U}$ at all points in the experiment treated here. However, **f** must be modified for every different value of Q or $\mathcal{M}$. Furthermore the function **f** would require drastic revision for every different geometry; e.g., any of those in the following sections.

The function $\mathbf{S} = \mathbf{E} \times \mathbf{H}$ is the only single function we have found that is at all reasonable for all geometries and all situations. In the following sections we treat **S** and **p** as unique for determining rates of energy flow and momentum densities.

10–5 The Poynting flux from dc sources of emf

Since power is transmitted by dc circuits, the Poynting vector must have both sources and sinks. In such circuits the sources are the sources of emf and the sinks are any device that uses electromagnetic power, e.g., resistances, motors, or storage batteries being charged. By applying Gauss's theorem to the last integral in Eq. (10–3), the integrands can be separated to give

$$\mathbf{E} \cdot \dot{\mathbf{D}} + \mathbf{H} \cdot \dot{\mathbf{B}} + \mathbf{E} \cdot \mathbf{J} = -\mathbf{\nabla} \cdot \mathbf{S}. \qquad (10\text{–}10)$$

In dc circuits the first two terms on the left are zero because the time derivatives are zero. The sign of $\mathbf{\nabla} \cdot \mathbf{S}$ then depends on the sign of $\mathbf{E} \cdot \mathbf{J}$. Within sources of emf $\mathbf{\nabla} \cdot \mathbf{S}$ must be predominantly positive, since electromagnetic power does emerge from them. This means that $\mathbf{E} \cdot \mathbf{J}$ must be predominantly negative in all dc sources of emf, many of which are discussed in Chapters 6 and 7.

A brief analysis with two of these sources should be illuminating. The Van de Graaf generator in Fig. 6–3 has a belt that carries positive charges up to charge the metal sphere. The positive charges being carried up by that portion of the belt that is on the left constitute an electric current directed vertically upwards. The electric field is directed downwards in this region and $\mathbf{E} \cdot \mathbf{J}$ is negative. Examination of the magnetic field near this part of the belt due to the charges it carries upward reveals that, as should be expected, the $\mathbf{E} \times \mathbf{H}$ is directed away from the belt. The downward moving portion of the belt carries a small fraction of the positive charges which did not flow onto the sphere. Here $\mathbf{E} \cdot \mathbf{J}$ is positive and $\mathbf{E} \times \mathbf{H}$ is directed into this portion of the belt. The magnitudes of $\mathbf{E} \cdot \mathbf{J}$ and $\mathbf{E} \times \mathbf{H}$ around the downward moving belt are much smaller than those around the upward moving section. Thus the term $\mathbf{E} \cdot \mathbf{J}$ is predominantly negative in this emf source. The charged particles accelerating down the tube on the right constitute the sink, where electromagnetic energy is being converted to the kinetic energy of the particles.

The linear generator of Fig. 7–5 furnishes another example. The movement of the rod through the magnetic field causes a current in the general direction of $\mathbf{v} \times \mathbf{B}$ but the electric field between the rails is $\mathbf{E} = -\mathbf{v} \times \mathbf{B}$, if the resistance of the rod is negligible. Hence $\mathbf{E} \cdot \mathbf{J}$ is negative.

As discussed in Section 6–5, the term $\mathbf{E} \cdot \mathbf{J}$ is predominantly negative in both thermocouples and chemical cells when they are delivering electromagnetic power to dc circuits. When conditions are such that these devices are receiving electromagnetic power from dc circuits and converting it into other forms, the $\mathbf{E} \cdot \mathbf{J}$ becomes predominantly positive.

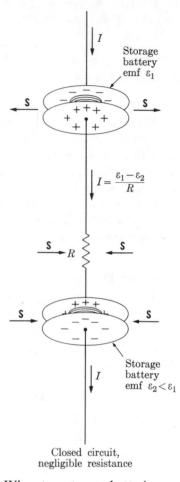

Fig. 10–2 One disk-shaped storage battery charging another disk-shaped storage battery through a total resistance R. Each battery is between two metal disks of larger diameter. The return circuit is not shown.

Closed circuit,
negligible resistance

Storage batteries charging storage batteries. When two storage batteries are connected in opposition to each other in a simple resistive circuit, the discharging of the one with the larger emf will charge the one with the smaller emf. Such a situation is illustrated in Fig. 10–2, where the disk-shaped storage battery at the top is connected to charge the other disk-shaped storage battery at the bottom through a resistance R. The return circuit is not shown. The resistances in the batteries are presumed negligible and the total resistance in the rest of the circuit is just R. The disk-shaped storage batteries are each placed between metal disks of larger diameter to more clearly determine the nature of the *Poynting flux near these batteries.* The current $I = (\mathcal{E}_1 - \mathcal{E}_2)/R$ is directed downwards in both batteries. The magnetic field $\mathbf{H}$ within the space between the metal disks circles to the left near the front edges of the disks enclosing each battery. The electric field $\mathbf{E}$ is vertically upward between the top disks and vertically downward between the bottom disks. Thus **S** is directed radially outward from the upper battery and radially inward toward the lower battery. It is also directed inward toward the resistor. The detailed manner in which **S** is absorbed by resistors is treated in Section 10–6.

Between the metal disks outside the battery at the top $\mathbf{E} = \mathbf{a}_z\, \mathcal{E}_1/d$ and from Ampere's law $\mathbf{H} = -\mathbf{a}_\phi I/2\pi r$ so that $\mathbf{S} = \mathbf{a}_r \mathcal{E}_1\, I/2\pi rd$, where d is the distance between the metal plates. Consider this upper battery in a volume between the metal plates which is also surrounded by a cylindrical surface of radius r. The Poynting flux $\mathcal{E}_1 I$ passes out through this cylindrical surface, indicating that this upper battery is contributing that much electromagnetic power to the system. The same type of analysis can be used inside the battery. Although the electric field in the battery varies greatly from one metal plate to the other, the average is still given by $\mathbf{E} = \mathbf{a}_z \mathcal{E}_1/d$. To determine **H** we again use a circular path of radius r about the z-axis. If a is the radius of the cylindrical battery, this circle only encloses a current given by $I\,r^2/a^2$, so that here $\mathbf{H} = -\mathbf{a}_\phi I\,r/2\pi a^2$. The $\mathbf{S} = \mathbf{a}_r \mathcal{E}_1\, I\,r/2\pi a^2 d$ and the Poynting flux through the cylindrical surface of radius r is $\mathcal{E}_1\, I\,r^2/a^2$, which builds up from zero at the center of the battery to $\mathcal{E}_1\, I$ at $r = a$.

In the space between the metal plates enclosing the lower battery, $\mathbf{E} = -\mathbf{a}_z \mathcal{E}_2/d$ while **H** has the same value as in the upper battery. Thus $\mathbf{S} = -\mathbf{a}_r \mathcal{E}_2\, I/2\pi rd$ outside the battery and $\mathbf{S} = -\mathbf{a}_r \mathcal{E}_2 I\,r/2\pi a^2 d$ inside. The corresponding inwardly directed Poynting fluxes are $\mathcal{E}_2 I$ and $\mathcal{E}_2 I\,r^2/a^2$, respectively. In other words this inwardly directed flux has the magnitude $\mathcal{E}_2 I$ until it reaches $r = a$, where it begins being absorbed in the charging of the battery. At $r = 0$ it has all been absorbed. Since

$$IR = \mathcal{E}_1 - \mathcal{E}_2,$$

then

$$\mathcal{E}_1 I = \mathcal{E}_2 I + I^2 R,$$

which shows that all of the Poynting flux $\mathcal{E}_1 I$ that leaves the upper battery finds its way back to the battery being charged and the resistor being heated. In the diagram of Fig. 10–2, it is not easy to follow the paths taken by the Poynting flux. These paths can be followed in detail if the two batteries and the resistor are placed between concentric spherical shells that are conducting. The solution of this problem is left as an exercise.

Magnetic dynamos, thermocouples, and some other sources of emf can be used either as sources or sinks in somewhat the same manner as these storage batteries. When used as sinks the dynamo becomes a motor and the thermocouple becomes a refrigerator extracting heat from its surroundings. When a load is put on one of these motors it generally slows down and its back emf decreases to allow a larger current through it. The net result is to increase the Poynting flux into the motor. Of course, there must then be an increase in the flux out of the source.

10–6 Direct-current energy propagation along a coaxial cable

The simply constructed coaxial transmission line shown in Fig. 10–3 is chosen to demonstrate the role of the Poynting vector in the propagation of energy. The solid inner cylinder of nickel and the hollow outer cylinder of copper provide electrodes for the electrolytic cell on the left and also provide a transmission line, conducting current to the carbon washer on the right.

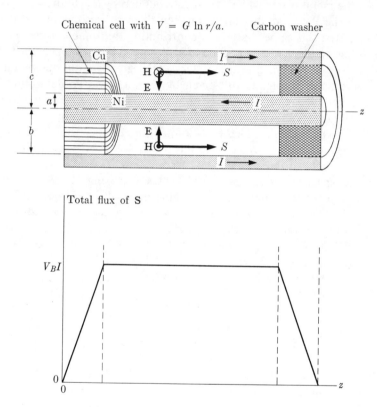

Fig. 10–3 Poynting-vector flux along a dc transmission line consisting of coaxial cylinders. The Poynting flux builds up linearly in the chemical cell, remains constant in the intervening space, and is absorbed linearly in the carbon washer.

If the cylinders are long but have sufficient cross section that their resistances are negligible, the field **E** in the air space between cylinders will have the same radial dependence as it has in the washer. This is not true of the **E** field in the electrolyte, where chemical forces cause the potential to vary rapidly near each conductor. However, a bimetallic hollow cylinder, with Cu inside and Ni outside, inserted into the electrolyte concentric with the original cylinders would change this potential distribution. In principle it would be possible to insert many such bimetallic cylinders of different diameters to produce any desired potential distribution. In particular, it could be made to approach the potential distribution in the carbon washer. Such a distribution will be assumed because it greatly simplifies the treatment of the problem. The radial potential distribution throughout the device is then given, in cylindrical coordinates centered on the z-axis, by $V = G \ln r/a$, where G is some constant. Chemical energy is converted to electric energy in the cell and electric energy is converted to thermal energy in the carbon washer. In order to maintain the steady-state conditions assumed for the device, this heat must be removed as rapidly as it is produced.

The electric field at all points between cylinders is $\mathbf{E} = -\boldsymbol{\nabla}V = -\mathbf{a}_r(G/r)$. The steady current I that is directed to the right in the Cu and to the left in the Ni produces a magnetic field that can be calculated from the integral relation of Eq. (9–13)

$$\oint \mathbf{H} \cdot d\mathbf{l} = \int_s \mathbf{J} \cdot d\mathbf{s}.$$

Convenient surfaces for evaluating these integrals are flat circular disks, each of radius r, centered on and perpendicular to the axis of the device. The line integral then can be evaluated around the periphery and the surface integral can be evaluated over the flat surface. For each of these disks the left-hand side of Eq. (9–13) gives $H2\pi r$, since H is constant everywhere on the periphery. For disks having radii greater than a and less than b and lying between the cell and washer, the surface integral gives I. In this region,

$$\mathbf{H} = -\mathbf{a}_\phi \frac{I}{2\pi r}$$

and

$$\mathbf{S} = \mathbf{a}_r \times \mathbf{a}_\phi \frac{IG}{2\pi r^2} = \mathbf{k} \frac{IG}{2\pi r^2}.$$

It should be noted that $\mathbf{S} = 0$ inside the metal conductors because $\mathbf{E} = 0$, and for $r > c$ it is also zero since $\mathbf{H} = 0$. The expression for the Poynting vector indicates that energy is being propagated parallel to the z-axis and directed from the cell to the washer. The areal density of this propagated power varies with r, but the total flow into the washer can be obtained by integrating over the flat inside surface of the washer; that is,

$$P = \int_s \mathbf{S} \cdot d\mathbf{s} = \int_a^b \frac{IG}{2\pi r^2} 2\pi r \, dr = IG \ln \frac{b}{a},$$

where G must be determined from the boundary conditions. If the potential of the Ni cylinder is zero and that of the Cu cylinder is V_B, the boundary conditions are satisfied by $V_B = G \ln b/a$. The power entering the washer is then

$$P = IV_B,$$

which is also the power leaving the flat inner surface of the electrolytic cell.

Starting from the outside surface of the cell and progressing parallel to the axis, the Poynting vector at each radial distance builds up linearly within the cell and reaches its maximum values at the inner surface. Likewise, each of these Poynting vectors decreases linearly to zero on passing through the carbon washer from its inner to its outer surface. This behavior stems from the fact that **H** varies linearly with z inside both the cell and the washer, while **E** is independent of z throughout the device. The variation in **H** can be shown by applying Eq. (9-13) to one of the imaginary flat disks as it is moved parallel to the axis. The line integral always gives $H2\pi r$, but the values obtained from the surface integrations vary linearly from 0 to I on passing from the outer to the inner surface of the cell or washer. Since the **S** for each value of r has the type of z-dependence shown in Fig. 10-3, the total flux of **S** also has this z-dependence. It is the curve for the total **S** flux that is plotted in Fig. 10-3, since it has the shape that is characteristic of the magnitude of **S** at each value of r.

10-7 Source of joulean heat in wires

These gedanken experiments indicate that in dc circuits electromagnetic energy is propagated through space from the source of emf to the regions where it is converted to other forms of energy. We are all familiar with the fact that the energy needed to actuate our radio and television sets is propagated through space from the broadcasting station to our sets. We should not then be surprised to find that even in dc and low-frequency ac the electromagnetic energy is propagated through space from the source to the sink just as it is with the higher-frequency ac circuits.

For completeness we should investigate how energy is transferred to wires having resistance. This can be done conveniently by referring to Fig. 10-3 and assuming that the coaxial cable shown there is very long compared to the radius c. The carbon washer should be replaced with a copper one and the central cylinder (Ni in the diagram) should have so much resistance that the resistances in the rest of the circuit can be neglected. All of the drop in potential now occurs between the two ends of the central cylinder or wire. Lines representing the Poynting vector flux now emerge from the chemical cell and follow curved lines into the surface of this cylinder.

Let us calculate the direction and magnitude of the Poynting vector inside this central cylinder. Use an imaginary cylindrical surface of radius $r = a$ and of length one meter, whose axis is concentric with that of the wire as is shown in Fig. 10-4. In this imaginary surface the electric field is

$$\mathbf{E} = \rho\mathbf{J} = -\mathbf{a}_z \rho I/\pi a^2$$

and the magnetic field is $\mathbf{H} = -\mathbf{a}_\phi J\pi r^2/2\pi r$, from Ampere's law, where I is the total current in the wire. Since $J = I/\pi a^2$, **H** can be written

$$\mathbf{H} = \mathbf{a}_\phi I\, r/2\pi a^2.$$

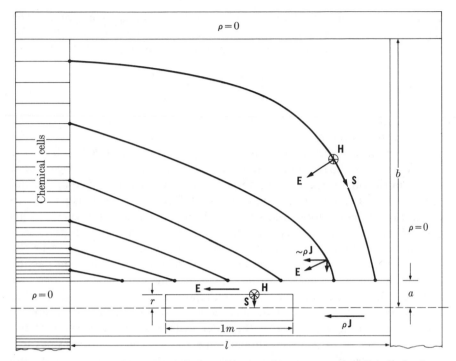

Fig. 10–4 Section of coaxial cylinder with greatly exaggerated radial dimensions. Central cylinder of length l has $\rho \neq 0$ but elsewhere $\rho = 0$.

Then $\mathbf{S} = \mathbf{E} \times \mathbf{H} = \mathbf{a}_z \times \mathbf{a}_\phi \rho I^2 r / 2\pi^2 a^4$ or $\mathbf{S} = -\mathbf{a}_r \rho I^2 r / 2\pi^2 a^4$. Thus $\mathbf{S}$ is directed radially inward. The total flux of $\mathbf{S}$ through the imaginary cylindrical surface is $P = S\, 2\pi r = \rho I^2 r^2 / \pi a^4$, or $P = I^2 R_l r^2 / a^2$, where R_l is the resistance per meter of length of our inner cylinder or wire. The total flux of $\mathbf{S}$ through the outer surface of our cylinder or wire is

$$I^2 \rho l / \pi a^2 = I^2 R_l l,$$

where $R_l l$ is the resistance of this section of the wire. This represents the rate of joulean heating. This rate of heating is distributed uniformly throughout the volume of the wire, since at any radius r the total flux inward is proportional to $r^2 l$, i.e., to the volume remaining inside this radius. The foregoing analysis applies to all wires having ohmic resistance and carrying steady currents. Apparently the electromagnetic energy flows into such wires through their cylindrical surfaces.

    This geometry provides one more example for which the flow pattern of Poynting vector flux from source to sink can be calculated in detail with little difficulty. Laplace's equation $\nabla^2 V = 0$ holds for the empty space between the concentric cylinders of Fig. 10–4. From Section 4–5 this equation can be written for the cylindrical coordinates of r and z, independent of ϕ. The relevant solutions are z, $\ln r$ and z, $\ln r$. Using the boundary conditions for this empty space we find the potential distribution to be given by

$$V = IR_t \left[z + (l - z) \frac{\ln r/a}{\ln b/a} \right].$$

With this relation we have plotted in Fig. 10–4 the traces, in any plane passing through the axis, of six equipotential surfaces. It will be seen that these traces also represent lines of Poynting vector flux proceeding from the chemical cells to the resistive central cylinder, where they are absorbed. As is shown in the figure, the $\mathbf{E}$ in the empty space is normal to the equipotential surfaces and $\mathbf{H}$ is normal to the plane. Therefore $\mathbf{S} = \mathbf{E} \times \mathbf{H}$ is parallel to the traces of the equipotential surfaces throughout the empty space. This statement is true also for the paths $\mathbf{S}$ takes in the other coaxial cable problems. Although the $\mathbf{S}$ lines that follow the equipotential traces do not strike the surface of the inner cylinder normally, they do become normal on passing into the interior through the surface. Just outside the surface of the inner cylinders the $\mathbf{E}$ vector does have the z-component $\rho\mathbf{J}$ but it also has a radial component due to the capacitive effect between the two cylinders. There are negative and positive charges on the outer surface of the inner cylinder and the inner surface of the outer cylinder respectively. The magnitudes of these surface densities of charge are the same as on coaxial cylindrical capacitors with the same geometry and the same potential distribution. Thus in the situation we have assumed, the Poynting vector energy emerges from the chemical cell and disappears within the wire that is assumed to have all of the resistance of the circuit.

The emergence of Poynting flux from each of the sources of emf in the foregoing gedanken experiments stems from making the integral of the $\mathbf{E} \cdot \mathbf{J}$ term in Eq. (10–1) negative. This term can be negative only if $\mathbf{E}$ and $\mathbf{J}$ are oppositely directed within these sources. This requires that electric charges be caused to move against the electric fields within the sources. Some other examples of sources of emf in which charges move against electric fields are given in Section 6–5.

10–8 Poynting vector in oscillating fields

If the chemical cells in either of the Figs. 10–3 or 10–4 are reversed in polarity, the current will change direction but the flow of energy will remain the same in both direction and magnitude. When the emf changes direction, both the current and the electric field change direction. Thus $\mathbf{S} = \mathbf{E} \times \mathbf{H}$ retains the same magnitude and the same direction, namely, from the source toward the sink, which is the washer in Fig. 10–3 and the inner cylinder in Fig. 10–4. In other words, the direction of the current can be alternated without changing the direction of the energy flow.

Let the chemical cell in Fig. 10–3 be replaced by a generator producing an emf that alternates at low frequency, which is given by $\mathcal{E} = V_0 \cos \omega t$. The Poynting vector is confined to the space between the cylinders. Its flux integrated over this space gives

$$P = \mathcal{E}I = \frac{V_0^2 \cos^2 \omega t}{R},$$

where R is the resistance of the washer. This is the power delivered to and absorbed by the washer at any instant of time t. Since the time average over a single cycle of $\cos^2 \omega t$ is $\frac{1}{2}$, the power delivered to the washer averaged over many cycles is $V_0^2/2R$. In a similar fashion electromagnetic power can be delivered to a resistor with alternating currents in many other transmission lines. A very common form of transmission line consists of two straight wires in parallel. If we can ignore the resistances of these parallel wires as we ignored the resistance of the coaxial cable, the power delivered at low frequencies to the resistor will be practically the same.

The main difference between these two transmission lines is that with the parallel wires the Poynting vector flux extends throughout space instead of being confined to the space between cylinders as it is with the coaxial cable. While most of this flux does find its way back to the resistor, where it is absorbed, a minute portion of it continues out into space and is lost to the circuit. At low frequencies this loss should be ignored. However, it increases so rapidly with increasing frequencies that parallel wires are poor transmission lines for high frequencies. The power that does not return to its circuit, but goes on out into space, is the same as that used in radio and television broadcasting. The mechanisms producing these waves are treated in Sections 12–7 and 12–8, where their frequency dependence is determined.

Coaxial cables make much better transmission lines for high frequencies because they confine the Poynting vector flux to the space between their cylinders.

High-frequency oscillations. With sufficiently long lines and sufficiently high frequencies, currents and potential differences travel in waves at the velocity of light along these transmission lines. These waves are accompanied by waves of Poynting flux that carry energy away from their generator. It has long been recognized that waves of Poynting flux do determine the power that is being transmitted by the transmission lines. It has also been recognized that these waves carry momentum that is transferred to anything that absorbs their energy. In fact, any substance that reflects such waves back on themselves may receive up to twice as much momentum as does a perfect absorber. This should be expected, because the law of conservation of momentum requires that when any momentum is completely reversed by impingement upon a massive object, that object must receive twice the original momentum. This principle is used in Section 12–5 to obtain the expression for momentum density in electromagnetic fields.

Resonance. An interesting variation can be produced with the coaxial cable of Fig. 10–3 by replacing the high resistance washer with a perfectly conducting one. This ensures that the potential difference between the two line conductors remains zero at the surface of the washer upon which the wave impinges. This in turn requires the superposition of another wave whose electric vector will always cancel that of the incoming electric vector at the surface of this washer. In other words, a wave is being reflected at the surface of this washer. By changing either the length of the line or the frequency of the oscillations, a pattern of standing waves can be established in the intra-cylindrical space. The established condition is called *resonance*. It can be shown that the Poynting vector flux oscillates back and forth within these standing waves. This condition of resonance could be maintained by the generator without adding any power were it not for the nonnegligible resistivity of the cylindrical walls.

Further study of this interesting subject must be deferred until the mathematical relations for alternating currents and electromagnetic waves are formulated in Chapters 11 and 12, respectively.

PROBLEMS

10–1 Assume the quantities $\mathcal{M}$ and $\dot{U}$ are each well defined.

a) Is ρ_m uniquely defined by $\nabla \cdot \mathcal{M} = \rho_m$?

b) Is **S** uniquely defined by $\nabla \cdot \mathbf{S} = \dot{U}$?

10–2 A solid sphere of copper of radius a is surrounded concentrically with a shell of copper having internal and external radii b and c, respectively. With the origin of spherical coordinates centered at the center of these spheres, the z-axis coincides with the axes of two circular cylinders that connect the solid sphere and its surrounding shell in two regions. Along the positive z-axis the connection consists of a storage battery (emf $\mathcal{E}_1$) oriented to make the inner shell positive. It extends from $\theta = 0$ to $\theta = \theta_0$. Along the negative z-axis the connection consists of two parts in series, both of which extend from $\theta = \pi$ to $\theta = \pi - \theta_0$. In direct contact with the inner sphere is a second storage battery (emf $\mathcal{E}_2 < \mathcal{E}_1$), oriented to make the inner sphere positive. An ohmic semiconductor of resistance R completes the circuit.

a) Calculate **S** as a function of r between the spheres.

b) What are the values of **E** and **H** outside the spheres from $r = c$ to $r = \infty$?

c) Calculate the total **S** flux through a surface on which θ (between θ_0 and $\pi - \theta_0$) is constant.

10–3

a) For the circuit described in Problem 10–2 calculate the **S** flux as a function of θ in the storage battery with the emf $\mathcal{E}_1$.

b) If the results in (a) vary with θ, explain what happens to the corresponding energy.

10-4 Assume that the chemical cells in the coaxial cable of Fig. 10-3 are replaced with a homopolar generator like the one shown in Fig. 7-7. Let the disk rotate at the constant velocity ω_0, without friction, and yet make intimate electrical contact with both the inner and outer cylinders. The disk could be made to rotate about the inner cylinder. Let a uniform magnetic field **B**, parallel the z-axis, extend through both the generator and the carbon disk, which is presumed to have practically all of the resistance of the circuit. Take the emf generated at $\omega = \omega_0$ to be $\mathcal{E}_0 = I_0 R$, where R is the resistance of the washer and I_0 is the steady current through it.
a) Find the expression for **S** close to the rotating disk.
b) What happens to **S** if the resistivity of the washer $\rho = \infty$? Qualitatively what then happens to **S** as the disk is speeded up from $\omega = 0$ to $\omega = \omega_0$? Explain these results.
c) Calculate the **S** flux emerging from the generator disk for a washer with the resistance $R = R$ and also for one with the resistance $R = \infty$.

10-5 Replace the carbon washer of Problem 10-4 with a homopolar motor like the one shown in Fig. 7-8. Assume that this motor can rotate with negligible friction, that its resistance is negligible, and that it is not loaded. Assume that the rate of rotation of the generator disk is controlled at $\omega = \omega_0$ regardless of any current in the circuit.
a) Describe qualitatively what happens to the motor disk after it is suddenly connected with its disk originally at rest. Remember we are assuming negligible resistance throughout the circuit.
b) Calculate initial and final value of $\mathbf{S} = \mathbf{f}(r)$ very close to the generator disk.
c) Calculate the initial and final **S** flux emerging from the generator disk, if the initial current is I_i.

10-6 Assume a homopolar generator connected by a coaxial cable to a homopolar motor as is described in Problem 10-5. Consider the cable to be long compared to its radii but still with negligible resistance. Load the motor with a constant torque load such that there is a current I_l through it.
a) What is the steady angular velocity of the motor, assuming the generator velocity is maintained at $\omega = \omega_0$?
b) Calculate the value of **S** very near the motor disk.
c) Calculate **S** at a cross section through the middle of the cable.
d) Calculate the total **S** flux into the motor.
e) Calculate the total **S** flux through the middle cross section that is normal to the z-axis.

10-7 Suppose the coaxial cylinders in Fig. 10-3 have the total length $t_c + l + t_R$, where t_c is the length in contact with the cell and t_R is the length in contact with the washer.
a) Find the magnetic field energy U_m stored in the air space.
b) Find the electric field energy U_e stored in the air space and compute the ratio U_m/U_e.

10-8 Coaxial cylinders like those shown in Fig. 10-3 are made very long and the cell is replaced by a sinusoidal generator with the frequency $f = \omega/2\pi$. The space between cylinders is filled with air, and $(b - a) \ll \lambda$ but $l \gg \lambda$, where $\lambda = c/f$ is the wavelength of the electromagnetic waves.

a) Show that the magnitudes of the electric and magnetic fields are given by the real parts of

$$E_r = \frac{V_m e^{j(\omega t - gz)}}{r \ln b/a} \quad \text{and} \quad H_\phi = \frac{I_m e^{j(\omega t - gz)}}{2\pi r}.$$

b) Calculate the magnitude and direction of the Poynting vector as a function of cylindrical coordinates.

c) Calculate the time average of the total Poynting vector flux.

d) If no power is reflected by the washer, what is the complex impedance of this transmission line?

10-9 An electromagnetic plane wave, given by $E_y = G \cos \omega(t - z/v)$, strikes a barrier and is reflected as $E_y' = -G \cos \omega(t + z/v)$

a) Calculate the expression for the standing waves that are produced.

b) Calculate the Poynting vector **S** as a function of t and z.

c) Calculate the time average of **S**.

Chapter 11 ALTERNATING CURRENTS

Electric currents that oscillate sinusoidally play an important role in modern society. Oscillating currents in homes range from 60 cycles/s for lights, clocks, washers, and similar appliances to 10^8 cycles/s for short-wave radio and television. The range of frequencies utilized in military and commerical applications is even broader.

Because of the time dependence of alternating currents, the simple procedures for solving problems involving steady currents are not directly applicable. For example, in the derivation of Kirchhoff's law for junctions, Eq. (6–16), the time derivative in the conservation of charge relation

$$\int_s \mathbf{J} \cdot d\mathbf{s} = -\int_\tau \frac{\partial \rho}{\partial t} \, d\tau$$

was set equal to zero. This is not permissible with alternating currents. Furthermore, in demonstrating Kirchhoff's law for closed circuits, as stated in Eq. (6–17), the electric field was assumed to be conservative, a condition that applies only where $\nabla \times \mathbf{E} = 0$. However, by Maxwell's third relation, $\nabla \times \mathbf{E} = -\dot{\mathbf{B}}$, and $\mathbf{E}$ can be conservative only when $\mathbf{B}$ is independent of time. This can be true only if the current densities are independent of time.

As should be expected, the difficulties encountered in solving network problems increase with increasing frequencies of the oscillations. The disturbing effect of the $\partial \rho / \partial t$ term depends upon the capacitances in the network, and the disturbing effect of the $\dot{\mathbf{B}}$ term depends considerably upon the inductances. Neither of these terms becomes important in simple resistor circuits until fairly high frequencies are reached. However, most circuits utilizing alternating currents employ devices specifically designed to enhance the effect of the $\partial \rho / \partial t$ and $\dot{\mathbf{B}}$ terms. These are called capacitors and inductors, respectively. When these are included in a network, it is necessary to modify the simple techniques used with steady currents even at low frequencies.

11–1 Lumped linear circuit element approximation

The electromagnetic power flow into any collection of circuit elements can be obtained from the expression

$$P = -\int \nabla \cdot \mathbf{S} \, d\tau = \int (\mathbf{E} \cdot \dot{\mathbf{D}} + \mathbf{H} \cdot \dot{\mathbf{B}} + \mathbf{E} \cdot \mathbf{J}) \, d\tau, \qquad (11\text{–}1)$$

where $\mathbf{S}$ is the Poynting vector as defined in Eq. (10–2) and the integral must be taken over the total volume in which all circuit elements and associated fields are contained. The first two terms on the right represent the storage or release of energy in electric and magnetic fields set up by the total circuit and result in no loss of electromagnetic energy *except* when hysteretic materials are present, or at very high frequencies by electromagnetic radiation discussed in Chapter 12. The third term always represents the dissipation of electromagnetic energy and its associated conversion to other forms of energy. In typical circuit elements, like that shown in Fig. 11–1, these three terms represent, respectively, the capacitive, inductive, and resistive components of power absorption.

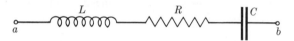

Fig. 11–1 A series element, basic for all ac networks, containing inductance L, resistance R, and capacitance C.

In going from the very general expression of Eq. (11–1) to the concepts frequently used to simplify ac circuit theory, the following approximations are made:

1. The frequency of the alternating current is low enough that

   a) its wave length is very long compared to any device in the circuit and in fact long compared to the longest path through the circuit,

   b) the current travels through the conductors and other circuit devices unlimited by the penetration depth considerations that are discussed in Section 12–5, and

   c) energy loss due to electromagnetic field propagation out from the circuit may be neglected.

2. Each circuit element contains one and only one of the three terms in the right-hand side of Eq. (11–1).

3. There is a unique value for μ, ϵ, and resistivity in each element of the circuit which is independent of the frequency and magnitude of $\mathbf{H}$, $\mathbf{E}$, and $\mathbf{J}$.

The first two constitute the lumped circuit element approximation while the third introduces the simplifying assumption of linearity.

For circuits of length l, condition 1(a) is satisfied when $l \ll v/2\pi f$, where v is only slightly less than the velocity of light in vacuum as discussed in Chapter 12. For 60-cycle current, therefore, wires must be much shorter than 800 km, while for frequencies in the megacycle range used by commercial radio and television, wires would have to be much shorter than 50 m.

The importance of assumption 1(b) can be understood by considering how the resistance of a length of wire is determined, namely $R = \rho l/A$, where ρ is the resistivity, l the length, and A the cross-sectional area through which the current flows. If such high frequency is used that the current is confined to a region near the surface of the wire, then the effective A is less than the actual cross section of the wire and R is larger for this frequency than for low frequencies or dc. Inductance and capacitance are also geometric parameters which become frequency-dependent at high frequencies where penetration depth must be considered.

The limitation imposed by assumption 1(c) will be better understood after our discussion of radiation from an oscillating dipole in Section 12–7. Normally, unless the conductors have large cross sections, conditions 1(b) and 1(c) are automatically satisfied if 1(a) is satisfied, and it is fairly easy to determine when the limits of that approximation are being approached.

The second assumption of the lumped circuit parameter approximation is quite good over a broad range of frequencies with commercially available resistors, capacitors, and inductors. However, as frequency increases, smaller inductors and capacitors are used in the circuits because their effects are proportional to frequency: $\omega L = X_L$ and $\omega C = 1/X_C$. Thus, at higher frequencies the intrinsic inductance or capacitance of the current carrying wires must be considered. Even the inductance of the loop created by soldering a small commercial carbon resistor into a circuit may be important. In this case, however, a solution for the circuit can be obtained by constructing an "equivalent circuit" in which the real resistor and soldered connection are replaced by a pure resistor in series with a pure inductor of appropriate value.

Real circuit elements can, in fact, always be replaced by an equivalent circuit consisting of pure lumped circuit elements so long as the frequency is low enough as defined by assumption 1. The equivalent circuit is then analyzed using simple ac theory. *In this manner assumptions 1 and 2 reduce to the single assumption that the frequency is low.*

The third assumption relates to the properties of the materials from which the various circuit elements are fabricated. In practice ϵ is constant in most circuit elements and surrounding media; however, there are numerous important cases in which the linearity approximation does not hold because either the resistivity or the permeability is not independent of current or magnetic field. Some of these cases are discussed in Sections 11–11 and 11–13.

All dielectric and magnetic materials have natural resonant frequencies, as illustrated for example in Fig. 5–2. These phenomena can introduce non-linear effects, but they normally occur only for frequencies above 10^7 cycles/s or higher.

In spite of the limitations of the *lumped linear circuit element* approximation, it is a very useful concept whose application permits simple solutions to many practical circuit problems, some of which will now be studied. Within the framework of this approximation, a network containing resistors, inductors, and capacitors will be treated as though it contains a number of what are called "series elements," which are defined to consist of a resistor, a capacitor, and an inductor connected in series between two junctions. Since most of the inductance in such an element is in the inductor, all of the inductance will be considered concentrated in this device. Likewise, the capacitance can be considered to be concentrated in the capacitor. The total resistance of the element, including that of the inductor and of the capacitor, can be considered as concentrated in the resistor.

11–2 Differential equation for an ac series element

Development of new laws governing ac networks requires the determination of relationships between potential differences and currents in a single series element like that shown in Fig. 11–1. If all collections of charges are assumed to be negligible except across the capacitor, then currents in R and L at each instant are the same and are equal to $i = dq/dt$. Furthermore, the charge on the left plate of the capacitor is $q = q(t)$ and the charge on the right plate is just $-q$, so that the charge leaving the right plate produces the same current i directed toward the junction b. The time-dependent potential difference across C is $v_C = q/C$ and the potential difference across R is just $v_R = iR = R(dq/dt)$. While it is practically impossible to make an inductor with zero or even negligible resistance, no error is introduced by assuming such a device, provided its actual resistance is included in R. The potential difference across a pure inductor is

$$v_L = L\frac{di}{dt} = L\frac{d^2q}{dt^2}.$$

Assuming that all magnetic field effects due to changing currents are lumped into the inductor, these three potential differences can be added to give the potential difference across the series element,

$$v = v_L + v_R + v_C.$$

This can be written as a differential equation:

$$L\frac{d^2q}{dt^2} + R\frac{dq}{dt} + \frac{q}{C} = v(t). \tag{11–2}$$

Because multiple frequencies associated with nonsinusoidal oscillations tend to introduce undesired power losses in many circuits, alternating currents are usually produced to be as nearly sinusoidal as possible. Also, non-sinusoidal oscillations may be treated as sums of sinusoidal oscillations of different frequencies. For these reasons, the potential difference between a and b in Fig. 11–1 will be assumed to have the form

$$v(t) = V_m \cos \omega t, \tag{11-3}$$

where $\omega = 2\pi f$ and f is the frequency of the oscillations in cycles/s.

Substitution of Eq. (11–3) into Eq. (11–2) gives a simple differential equation which may be solved readily in terms of real quantities. However, later work in ac circuits will be greatly facilitated by the use of complex quantities; in particular, they will aid the development of graphic and analytic solutions discussed in Section 11–7. The shift to complex notation will therefore be made immediately. Equation (11–3) becomes

$$\mathcal{V}(t) = V_m(\cos \omega t + j \sin \omega t) = V_m e^{j\omega t},$$

where $v(t)$ is just the real part of $\mathcal{V}(t)$, and Eq. (11–2) becomes

$$L\frac{d^2\mathcal{Q}}{dt^2} + R\frac{d\mathcal{Q}}{dt} + \frac{\mathcal{Q}}{C} = \mathcal{V}(t), \tag{11-4}$$

where q is the real part of $\mathcal{Q}$.

It can be readily verified that the derivative of a complex number is the sum of the derivatives of the real and imaginary parts, that the integral is the sum of the integrals of the real and imaginary parts, and that the sum of complex numbers is the sum of the real parts plus the sum of the imaginary parts. However, as is shown in Section 4–8, the process of multiplying one complex quantity by another does not provide a simple product of the original real quantities separated from the imaginary quantities. For this reason, complex quantities are not used in Section 11–4 for calculating ac power.

11–3 Solution of the differential equation with complex variables

Equation (11–4) is an ordinary second-order differential equation with constant coefficients. The general solution for it is the sum of two functions: $\mathcal{Q}_c$, the general solution for Eq. (11–4) with $\mathcal{V}(t) = 0$; and $\mathcal{Q}_p$, some particular integral that satisfies Eq. (11–4) as it is written.

Complementary function. It will now be shown that the first function, called the complementary function, determines the transient phenomena in the series element. The nature of the transient and its rate of decay is determined by the ratio of R to the geometric average of the other two constants, L and $1/C$. It is oscillatory when R is small and monotonic when R is large. The complementary function is found by substituting $\mathcal{Q} = e^{\mathcal{D}t}$ into Eq. (11–4), with $\mathcal{V}(t) = 0$, to obtain

$$LD^2 + R\mathcal{D} + \frac{1}{C} = 0.$$

The solutions of this quadratic equation in $\mathcal{D}$ are given by

$$\mathcal{D} = -\frac{R}{2L} \pm \sqrt{\left(\frac{R}{2L}\right)^2 - \frac{1}{LC}}.$$

When $R/2 > \sqrt{L/C}$, the two solutions can be written as

$$D_1 = -\alpha + \gamma$$

and

$$D_2 = -\alpha - \gamma,$$

where

$$\alpha = \frac{R}{2L} \quad \text{and} \quad \gamma = \sqrt{\left(\frac{R}{2L}\right)^2 - \frac{1}{LC}}.$$

Both terms are real, and $\alpha > \gamma$. When, however, $R/2 < \sqrt{L/C}$, the two solutions can be written as

$$\mathcal{D}_1 = -\alpha + j\beta$$

and

$$\mathcal{D}_2 = -\alpha - j\beta,$$

where

$$\beta = \sqrt{\frac{1}{LC} - \left(\frac{R}{2L}\right)^2}$$

is also real.

For large resistances, then, the complementary function is real and can be written as

$$\mathcal{Q}_C = q_C = Ae^{D_1 t} + Be^{D_2 t}, \tag{11–5}$$

which decays monotonically because both D_1 and D_2 are negative real numbers. For small resistances, the complementary function can be written as

$$\mathcal{Q}_C = e^{-\alpha t}(\mathcal{G}_1 e^{-j\beta t} + \mathcal{G}_2 e^{+j\beta t}).$$

By choosing $\mathcal{G}_1 = a + jb$ and $\mathcal{G}_2 = \mathcal{G}_1^* = a - jb$, this can be transformed into an all real function. Remembering that $e^{j\beta t} = \cos \beta t + j \sin \beta t$ and that $\cos(\beta t - \delta) = \cos \delta \cos \beta t + \sin \delta \sin \beta t$, we can write this function as

$$\mathcal{Q}_C = q_c = Ke^{-\alpha t} \cos(\beta t - \delta), \tag{11–6}$$

where $K = 2\sqrt{a^2 + b^2}$ and $\delta = \tan^{-1}(b/a)$ are two new arbitrary constants which are real. Actually, Eq. (11–6) could have been assumed directly and then checked to show that it satisfies Eq. (11–4) when the right-hand side is zero. Equation (11–6) is an oscillatory function whose amplitude decays with time.

The arbitrary constants A and B for large resistances or K and δ for small resistances are determined from the initial conditions, that is, from the values of q and dq/dt at $t = 0$. When $v(t)$ is a sinusoidal potential difference determined by an ac generator, the values of the arbitrary constants are determined by the phase at which the generator is connected or disconnected. If desired, the effects produced by these transients can be minimized by making the changes at predetermined phases. For example, devices have been invented to interrupt large alternating currents at the instant these currents pass through zero, to avoid the disastrous arcs that otherwise would form.

It should be noted that while the solutions obtained thus far apply specifically to the series element shown in Fig. 11–1, the form of the function $v(t)$ has not been specified. These solutions will thus determine the transient currents in such a series element whether the source of potential difference produces sinusoidal potentials, steady potentials, or any other time-dependent potentials.

The rate of decay of these transient terms is important in most applications. With $R/2 > \sqrt{L/C}$, the second term in Eq. (11–5) decays first, since $D_2 = -(\alpha + \gamma)$ is always more negative than $D_1 = -(\alpha - \gamma)$. When $R/2 \gg \sqrt{L/C}$, the first term decays very slowly, but as $R/2 \to \sqrt{L/C}$ the rate of decay approaches $e^{-Rt/2L} = e^{-t/\sqrt{LC}}$, which gives the maximum rate of monotonic decay. With $R/2 < \sqrt{L/C}$, the transient is an oscillation whose amplitude is proportional to $e^{-Rt/2L}$. The transient oscillations continue for a long time with small resistances, but with $R/2 \to \sqrt{L/C}$ the amplitudes approach the decay rate $e^{-t/\sqrt{LC}}$. The most rapid rate of decay is then $e^{-t/\sqrt{LC}}$, when $R/2 = \sqrt{L/C}$, which is called the *critical damping* condition. It is easily shown that, as this relation is approached, the two forms of the solution, Eqs. (11–5) and (11–6), each approach the relation $Q_C = (F + Gt)e^{-\alpha t}$, which is the solution when the two roots of the quadratic equation in $\mathfrak{D}$ are equal. Figure 11–2 shows the transient time variations in the charge on the capacitor of Fig. 11–1 for the three characteristic damping conditions, namely, (1) overdamped with $R/2 > \sqrt{L/C}$, (2) underdamped with $R/2 < \sqrt{L/C}$, and (3) critically damped with $R/2 = \sqrt{L/C}$.

The complete solution of Eq. (11–4) is

$$Q = Q_C + Q',$$

where Q', the particular integral, is any function that satisfies Eq. (11–4) as it is written. In many ac networks the transients decay so rapidly that they can be ignored. The particular integral, which determines the quasi steady-state relations, then becomes the more important term.

Particular integral. Assuming that the potential difference impressed across the series element in Fig. 11–1 is the real part of $V_m e^{j\omega t}$, the particular integral of Eq. (11–4) can be obtained by assuming $Q' = \mathfrak{B}e^{j\omega t}$, where $\mathfrak{B}$ is a constant

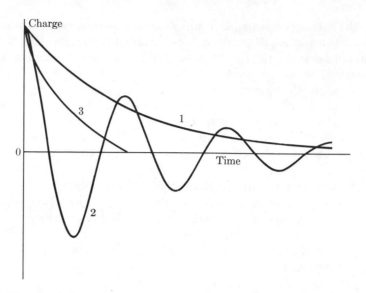

Fig. 11–2 Transient time variations in the charge on a capacitor in a series element for the three characteristic damping conditions: (1) overdamped with $R/2 > \sqrt{L/C}$, (2) underdamped with $R/2 < \sqrt{L/C}$, and (3) critically damped with $R/2 = \sqrt{L/C}$.

that may need to be complex to provide for a possible difference in phase angle between the oscillating charge and the oscillating potential difference. The effect of a complex $\mathcal{B}$ on the phase may be seen by writing $\mathcal{B} = Be^{j\beta}$, so that $\mathcal{B}e^{j\omega t} = Be^{j(\omega t + \beta)}$. Direct substitution into Eq. (11–4) gives

$$\mathcal{B}j\omega e^{j\omega t}\left[R + j\left(\omega L - \frac{1}{\omega C}\right)\right] = V_m e^{j\omega t}. \tag{11–7}$$

Impedance and reactance. The term in brackets can be written as

$$R + j\left(\omega L - \frac{1}{\omega C}\right) = \mathfrak{Z} = Ze^{j\theta}, \tag{11–8}$$

where by definition $\mathfrak{Z}$ is the *complex impedance* and Z is the *real impedance*. The phase angle θ will be more definitely defined soon. By writing time-dependent current in all portions of the series element as the real part of $\mathfrak{I}' = d\mathcal{Q}'/dt = \mathcal{B}j\omega e^{j\omega t}$ and the time-dependent potential difference as the real part of $\mathcal{V}' = V_m e^{j\omega t}$, Eq. (11–7) can be written as

$$\mathfrak{I}'\mathfrak{Z} = \mathcal{V}'. \tag{11–9}$$

Equation (11–9) is like the dc Ohm's law relation $IR = V$, although with ac each term is complex. The symbols $\mathcal{V}'$ and $\mathfrak{I}'$, whose real parts represent steady sinusoidal oscillations, are primed here to save the unprimed symbols

$\mho$ and $\Im$ for some time-independent quantities that will be introduced later for network calculations. The fact that $\Im$ is, in general, complex means that $\Im'$ is seldom in phase with $\mho'$; that is, the time-dependent potential difference and current reach their maxima at different times.

11-4 Steady ac in a series element

Steady alternating currents in networks are so important that the calculation procedures should be simplified as much as possible. The absolute magnitude of the impedance and the phase relations between the current and voltage can be understood more easily if we write

$$\Im = R + jX = Ze^{j\theta},$$

where

$$Z = \sqrt{R^2 + X^2}, \qquad \theta = \tan^{-1}\frac{X}{R}, \qquad X = \omega L - \frac{1}{\omega C}.$$

The relationship among these quantities are graphically presented in the so-called "impedance triangle" of Fig. 11–3. Here Z is the absolute magnitude of the impedance and θ is the phase angle by which the current maximum lags the voltage maximum. The quantity X is defined as the *reactance* of the series element and can be written as $X = X_L - X_C$, where $X_L = \omega L$ and $X_C = 1/\omega C$ are called the *inductive reactance* and the *capacitive reactance* respectively.

Equation (11–7) can be written as

$$\Im' = \frac{V_m}{Z}\, e^{j(\omega t - \theta)},$$

and since the real part of this complex expression gives the time-dependent values of the actual current,

$$i = \frac{V_m}{Z}\cos(\omega t - \theta).$$

The maximum current in this element is $I_m = V_m/Z$, which shows that the maximum values of current and voltage also obey an Ohm's law type of relation.

Power absorption in ac series elements. The power being absorbed at any instant by a series element is the product of the instantaneous current and the instantaneous difference of potential. Since it is the product of the real current and the real voltage that is involved, the product of the complex quantities cannot be used. The time-dependent power absorbed is then

$$p = vi = V_m I_m \cos \omega t \cos(\omega t - \theta),$$

which shows that the element may alternatively absorb and emit power to the rest of the circuit to which it is connected. To determine the net power P absorbed by the element, the time average of p must be obtained. Setting $\beta = \omega t$ and $\omega T = 2\pi$, where T is the period of oscillation, we obtain

$$P = \frac{1}{T}\int_0^T p\, dt = \frac{1}{2\pi}\int_0^{2\pi} p\, d\beta$$

and

$$P = \frac{V_m I_m}{2\pi}\left[\cos\theta\int_0^{2\pi}\cos^2\beta\, d\beta + \sin\theta\int_0^{2\pi}\cos\beta\sin\beta\, d\beta\right].$$

The value of the first integral is π and that of the second is zero. Hence

$$P = \frac{V_m I_m \cos\theta}{2}. \tag{11–10}$$

In its early development, alternating current was used chiefly for lighting, where the incandescent lamp filaments produced elements with nearly pure resistance. In these elements, $X = 0$, $\cos\theta = 1$, and Eq. (11–10) can be written as

$$P_R = \frac{I_m^2 R}{2} \quad\text{or}\quad P_R = \frac{V_m^2}{2R}.$$

When these same filaments were used with dc, the power was given by

$$P = I^2 R \quad\text{or}\quad P = \frac{V^2}{R}.$$

It was therefore decided to define an "effective current" and an "effective voltage" for ac networks as $I = I_m/\sqrt{2}$ and $V = V_m/\sqrt{2}$ respectively. These definitions have proved so useful that ac ammeters and voltmeters are still calibrated to read these "effective" or "root-mean-square" values. It should be noted that this decision made it possible to make filament bulbs whose brightness on 110 V ac was identical with their brightness on 110 V dc. Equation (11–10) can then be written as

$$P = VI\cos\theta, \tag{11–11}$$

where $\cos\theta$ is called the *power* factor.

Alternating-current resistance. From the impedance triangle of Fig. 11–3, it can be seen that Eq. (11–11) can be written as $P = I^2 R$, which shows that all the power absorbed by a simple series element is absorbed in the resistance. However, as will be shown in Section 11–12, nearby circuits may react upon a primary circuit to absorb power from it. In fact, nearby magnetic or conducting materials may absorb power in a similar fashion. This means that circuits measured with ac often exhibit greater resistance than

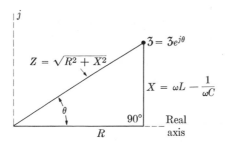

Fig. 11–3 Impedance triangle plotted in a complex plane for a typical series element, in which the magnitudes R, L, and C are lumped into separate devices.

do the same circuits when measured with dc. Resistance to ac is defined by the relation

$$R_{ac} = \frac{P}{I^2},$$

which is the resistance that must be used, instead of the dc resistance, for all ac network problems. The R in ac circuit problems must always be understood to mean R_{ac}, which can be measured with a wattmeter and an ammeter.

11–5 Kirchhoff's laws for ac networks

By considering each series element to have localized inductance, capacitance, and resistance, Kirchhoff's two laws can be used with small modifications for the time-dependent quantities associated with low-frequency alternating currents. The modified laws are remarkably accurate up to and often beyond 1000 cycles/s, depending upon the properties of the circuits. The expression "alternating currents," or just ac, is sufficiently descriptive to distinguish the phenomena in this range of frequencies from those usually designated by "high frequency," where more sophisticated techniques must be employed.

Junctions in ac networks. Kirchhoff's first law states that the algebraic sum of the steady currents out of any junction is zero. So long as the frequencies are not too high, this law can be extended to the time-dependent values of alternating currents out of any junction; that is,

$$\sum_i i_i(t) = 0.$$

When the time-dependencies of these current functions are sinusoidal, complex variables $\mathfrak{J}'_i(t)$ can be chosen whose real parts are equal to these real currents, so that

$$\sum_i \mathfrak{J}'_i(t) = 0. \tag{11–12}$$

The usefulness of such complex expressions can be seen from Fig. 11–4, where the currents out of a grounded junction in a balanced three-phase network are shown. The three resistors are identical, and the potential differences between the three power lines and the ground are identical in magnitude but differ in phase. The currents i_1, i_2, and i_3 in the three resistors are given by the real parts of $\mathfrak{J}'_1 = I_m e^{j\omega t}$, $\mathfrak{J}'_2 = I_m e^{j(\omega t + 2\pi/3)}$, $\mathfrak{J}'_3 = I_m e^{j(\omega t + 4\pi/3)}$, respectively. Now

$$\mathfrak{J}'_1 + \mathfrak{J}'_2 + \mathfrak{J}'_3 = I_m e^{j\omega t}[e^{j0} + e^{j(2\pi/3)} + e^{j(4\pi/3)}],$$

but the terms in brackets can be written as $1 + (-0.5 + j0.866) + (-0.5 - j0.866) = 0$. Now Eq. (11–12) gives $\mathfrak{J}'_1 + \mathfrak{J}'_2 + \mathfrak{J}'_3 + \mathfrak{J}'_g = 0$, if the current in the ground connection is defined as the real part of $\mathfrak{J}'_g$. However, $\mathfrak{J}'_1 + \mathfrak{J}'_2 + \mathfrak{J}'_3 = 0$, so that $\mathfrak{J}'_g = 0$ and there is no current in the ground connection.

In the diagram of Fig. 11–4(b) the three vectors should be visualized as each rotating about the origin with the constant angular velocity ω rad s^{-1} and

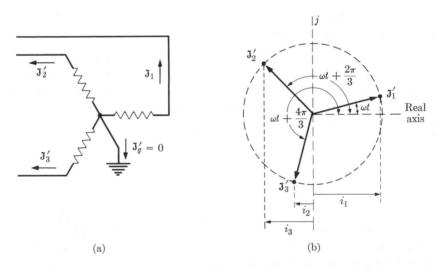

(a) (b)

Fig. 11–4 (a) A three-phase network connected through three identical resistors to a neutral junction which is connected to ground. (b) A time dependent plot in the complex plane of the three currents out of the junction. With the currents in each phase having the same maximum I_m, which is the length of each vector, the algebraic sum of the time-dependent currents $i_1 + i_2 + i_3 = 0$ at every instant and there is no current in the ground connection.

thus maintaining their angular separations of 120°. The projections of these rotating vectors on the horizontal or real axis at any instant gives the currents in the corresponding resistors at that instant. The algebraic sum of these instantaneous currents is always zero because the vector sum of the three vectors is zero.

Provided all the currents in a network have the same frequency, the currents out of any junction can be represented by a vector diagram like that in Fig. 11–4(b), where the vector sum $\sum_i \mathfrak{J}'_i = 0$. The vectors representing currents out of such junctions can be written as $\mathfrak{J}'_i = I_{mi}e^{j(\omega t+\theta_i)}$, but the necessity for visualizing all vectors as rotating tends to confuse the relationships between the vectors. The time-dependence is usually eliminated from these vector diagrams by defining time-independent vectors by

$$\mathfrak{J}_i = \frac{\mathfrak{J}'_i e^{-j\omega t}}{\sqrt{2}} = \frac{I_{mi}}{\sqrt{2}}\, e^{j\theta_i} = I_i e^{j\theta_i},$$

where I_i is the effective value of the ith current. Equation (11–12) then becomes

$$\sum_i \mathfrak{J}_i = 0, \qquad\qquad (11\text{–}13)$$

which is the time-independent form of Kirchhoff's junction law that is generally used in ac circuit problems. With this modification, the vector diagram of Fig. 11–4(b) can be drawn as shown in Fig. 11–5(a), where the absolute magnitudes of the vectors represent the effective currents and the angles between the vectors provide the phase angles. *The absolute orientation*

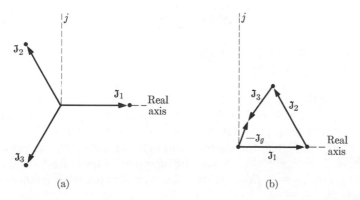

(a)   (b)

Fig. 11–5 (a) Time-independent vector diagram representing the three currents out of the grounded junction of Fig. 11–4(a). Effective or root-mean-square currents are represented by the lengths of the vectors. (b) The same vectors are plotted to more clearly represent the vector sum $\mathfrak{J}_1 + \mathfrak{J}_2 + \mathfrak{J}_3$, which is here shown to equal a finite $\mathfrak{J}_g$, although $\mathfrak{J}_g = 0$ in the problem being illustrated.

of such a diagram is entirely arbitrary, and the horizontal components (the real values) have no physical meaning whatsoever. The vector sum of the vectors shown in Fig. 11–5(a) can be exhibited somewhat more clearly by drawing the vectors head to tail, as shown in Fig. 11–5(b). When such a figure fails to close, the vector sum is the vector distance from the tail of the first to the head of the last. The figure is purposely drawn to show a finite sum which is the negative of $\mathfrak{J}_g$, although $\mathfrak{J}_g = 0$ in the problem illustrated.

Closed loops in ac networks. Kirchhoff's second law for networks of steady currents states that the algebraic sum of the voltage drops around any closed loop equals the algebraic sum of the emf's encountered in traversing that loop. With the inductances concentrated in inductors and with these relatively low frequencies, this law also holds for the time-dependent quantities and can be written as

$$\sum_j v_j(t) = \sum_j e_j(t).$$

Following the procedure used with the current equation, complex functions $\mathfrak{J}'_j \mathfrak{Z}_j$ and $\mathcal{E}'_j$ can be chosen so that their real parts equal v_j and e_j respectively. The foregoing equation then is included in the relation

$$\sum_j \mathfrak{J}'_j \mathfrak{Z}_j = \sum_j \mathcal{E}'_j. \tag{11–14}$$

Still following the procedure used with the current relations, a new complex quantity can be defined as

$$\mathcal{E}_j = \mathcal{E}'_j \frac{e^{-j\omega t}}{\sqrt{2}}.$$

The substitution of this and the current relation into Eq. (11–14) gives

$$\sum_j \mathfrak{J}_j \mathfrak{Z}_j = \sum_j \mathcal{E}_j, \tag{11–15}$$

which is the time-independent form of Kirchhoff's second law for ac circuits that is usually used.

The significance of Eqs. (11–14) and (11–15) can be demonstrated and a better understanding of Eqs. (11–7) and (11–8) can be obtained by considering the series element of Fig. 11–1. The vector diagrams in Fig. 11–6 are drawn on the assumptions that the potential difference across this element is $\mathcal{V}' = V_m e^{j\omega t}$ and that the current through the inductor, capacitor, and resistor is $\mathfrak{J}' = I_m e^{j(\omega t - \theta)}$. Since the complex impedances are $\mathfrak{Z}_R = R$, $\mathfrak{Z}_L = jX_L$, and $\mathfrak{Z}_C = -jX_C$, the time-dependent potential differences are the real parts of

$$\mathcal{V}'_R = \mathfrak{J}'R = I_m R e^{j(\omega t - \theta)},$$
$$\mathcal{V}'_L = \mathfrak{J}'jX_L = I_m X_L j e^{j(\omega t - \theta)},$$

and

$$\mathcal{V}'_C = -\mathfrak{J}'jX_C = -I_m X_C j e^{j(\omega t - \theta)},$$

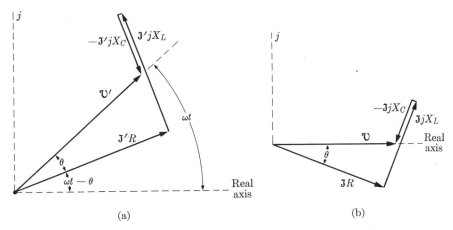

(a)   (b)

Fig. 11–6 Vector diagrams for the potential differences in the series element of Fig. 11–1. In (a) the vectors have lengths corresponding to the maximum potential differences and the projections on the real axis give the true time-dependent values. In (b) the vectors have lengths corresponding to root-mean-square values.

respectively, and these are plotted in Fig. 11–6(a). The corresponding time-independent vectors, which are plotted in Fig. 11–6(b), are

$$\mathcal{V}_R = RIe^{-j\theta}, \qquad \mathcal{V}_L = X_L Ije^{-j\theta}, \qquad \text{and} \qquad \mathcal{V}_C = -X_C Ije^{-j\theta}.$$

They are shorter by the factor $1/\sqrt{2}$ than the time-dependent vectors. In Fig. 11–6(b) all phase angles are related to the phase of the potential difference, which is chosen as the reference vector and is plotted along the real axis, making $\mathcal{V} = V$ and $\mathcal{I} = Ie^{-j\theta}$. If the current had been chosen as the reference vector, the whole diagram would merely have been rotated counterclockwise through the angle θ with $\mathcal{V} = Ve^{j\theta}$ and $\mathcal{I} = I$.

11–6 Admittance, conductance, and susceptance

In solving network problems certain expressions occur so frequently that they have been given special names.

Admittance. The reciprocal of the complex impedance occurs so frequently that it has been named *admittance*. It is defined by

$$\mathcal{Y} = \frac{1}{\mathcal{Z}}.$$

Since $\mathcal{Y}$ is a complex quantity, it can be written

$$\mathcal{Y} = G - jB,$$

from which the quantities G and B have been named the *conductance* and the *susceptance* respectively.

Conductance and susceptance. For the typical series element of Fig. 11–1, the conductance is

$$G = \frac{R}{Z^2}$$

and the susceptance is

$$B = \frac{X}{Z^2}.$$

While admittance is a complex number, whose absolute magnitude is

$$|\mathcal{Y}| = \frac{1}{Z},$$

both conductance and susceptance are real quantities. All three quantities have the unit 1/ohm, which is often called the mho. Note that the conductance is the reciprocal of the resistance only when $X = 0$, and the susceptance is the reciprocal of the reactance only when $R = 0$.

11–7 Analytical and graphical solutions of ac network problems

Equations (11–13) and (11–15) show that ac network problems can be treated analytically by the methods developed for solving dc network problems, provided the complex quantities developed in the previous sections are used. Figures 11–5 and 11–6(b) show that the analytical equations with complex quantities can be represented by diagrams in a complex plane. Solutions can thus be obtained graphically as well as analytically. In graphical solutions, effective currents and effective voltages are treated as simple vector quantities and their complex nature is usually ignored. In both the graphical and the analytical solutions, the lengths of the vectors are determined from the effective rather than the maximum values because effective values are obtained directly from the measuring instruments. Using these simple time-independent complex relations, solutions can be obtained for ac network problems without resorting to the cumbersome time-dependent forms.

Effective impedance. Whenever a group of elements like that in Fig. 11–1 are connected into a network with just two external terminals, the network can be treated as a single series element with some effective impedance. The treatment is essentially the same as that for a dc network of resistors. Networks that can be divided into groups of elements all in series or all in parallel are relatively simple to calculate.

When the complex impedances $\mathfrak{Z}_1$, $\mathfrak{Z}_2$, $\mathfrak{Z}_3$, etc. are all connected in series, the complex current $\mathfrak{J}$ will be the same in all elements, and the complex potential difference across the group will be

$$\mathcal{V}_s = \sum_i \mathfrak{J}\mathfrak{Z}_i = \mathfrak{J}\left(\sum_i R_i + j\sum_i X_i\right).$$

The complex impedance $\mathfrak{Z}_s$ of elements in series is then

$$\mathfrak{Z}_s = \sum_i \mathfrak{Z}_i = \sum_i R_i + j\sum_i X_i. \tag{11–16}$$

When such elements are all connected in parallel, the complex potential difference $\mathcal{V}$ across the group is the potential difference across each one, but the complex current through the group is given by

$$\mathfrak{J}_p = \sum_i \mathfrak{J}_i = \sum_i \frac{\mathcal{V}}{\mathfrak{Z}_i} = \frac{\mathcal{V}}{\mathfrak{Z}_p}.$$

The complex impedance $\mathfrak{Z}_p$ of elements in parallel is given by

$$\frac{1}{\mathfrak{Z}_p} = \sum_i \frac{1}{\mathfrak{Z}_i}. \tag{11–17}$$

11–8 Resonance

A simple series element consisting of some resistance, inductance, and capacitance is resonant (i.e., has minimum impedance) for the case when X_C and X_L are equal. The total series impedance is then given by

$$\mathfrak{Z}_s = R + j(X_L - X_C) = R.$$

Because $X_C = 1/2\pi fC$ and $X_L = 2\pi fL$, a circuit is resonant only at the frequency expressed by

$$f_r = \frac{1}{2\pi\sqrt{LC}}.$$

Such circuits can be tuned to particular frequencies by varying either the capacitance or the inductance.

A different type of resonant circuit is produced by connecting a capacitor in parallel with an inductor, as shown in Fig. 11–7. The resonant frequency for such parallel circuits depends upon the resistance as well as upon the inductance and capacitance. Whenever the resistance is small enough to be neglected, the resonant frequency is given by $f = 1/2\pi\sqrt{LC}$ as in the series circuit. A good capacitor usually has negligible resistance, and inductors for resonant circuits are constructed to have as low resistance as possible. The ratio $Q = X_L/R_L$ is taken as one measure of the quality of an inductor for

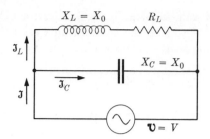

Fig. 11–7 Parallel resonant circuit which acts as a frequency filter.

such purposes. If Q is large compared with unity, the circuit shown in Fig. 11–7 has a sharp maximum impedance very near to the frequency for which $X_L = X_C = X_0$. Under these conditions the circuit can be analyzed in a simple and informative manner by defining $r = R_L/X_0$, which is small compared with unity. While Eq. (11–17) could be used, it is at least as simple to calculate the generator current $\mathfrak{J} = \mathfrak{J}_L + \mathfrak{J}_C$ directly.

The effective emf of the generator is taken to be V and its internal impedance is assumed to be negligible. The capacitor current is

$$\mathfrak{J}_C = \frac{V}{-jX_0} = \frac{jV}{X_0}$$

and the current through the inductor is

$$\mathfrak{J}_L = \frac{V}{R + jX_0}\left(\frac{R - jX_0}{R - jX_0}\right) = \frac{V(R - jX_0)}{R^2 + X_0^2}.$$

Dividing the numerator and the denominator by X_0^2, the expression for $\mathfrak{J}_L$ is obtained:

$$\mathfrak{J}_L = \frac{V}{X_0}(r - j)(1 + r^2)^{-1},$$

which can be expanded in increasing powers of r to give

$$\mathfrak{J}_L = -\frac{jV}{X_0} + \frac{V}{X_0}[r + (j - r)(r^2 - r^4 + \cdots)],$$

where the first term is much larger than the second. The generator current is then

$$\mathfrak{J} = \mathfrak{J}_L + \mathfrak{J}_C = \frac{V}{X_0}[r + (j - r)(r^2 - r^4 + \cdots)],$$

in which the higher powers of r can be neglected, to yield

$$\mathfrak{J} \simeq \frac{Vr}{X_0}.$$

The impedance of this parallel network to current through the generator is

$\mathfrak{Z}_p \cong X_0/r$, which can be very large. In fact, if R_L could be reduced to zero, $\mathfrak{Z}_p$ would become infinite. Such circuits are often used to filter out undesirable frequencies. Compared with the small current in the generator, a relatively large current $I_L \cong I_C = V/X_0$ circulates in the inductor and capacitor.

Early underwater mines of World War II could be detected by low-frequency magnetic fields that would be reflected by their metallic surfaces. Some coils for producing these fields were mounted on rafts, with capacitors designed to produce resonance. When such a raft was dragged behind a boat, the current in the coils could be maintained large enough to produce strong magnetic fields with rather small currents in the wires connecting the generator on the boat to the coils on the raft.

The foregoing problem is an excellent one for illustrating graphical solutions, as shown in Fig. 11–8. Choosing the real axis for the applied voltage V, the currents $\mathfrak{J}_L$, $\mathfrak{J}_C$ and their sum $\mathfrak{J} = \mathfrak{J}_C + \mathfrak{J}_L$ have been drawn as vectors, whose lengths give the effective values of the currents in the corresponding elements. Since it is 90° ahead of the voltage, $\mathfrak{J}_C$ is plotted along the positive j-axis and $\mathfrak{J}_L$, which lags the voltage by the angle $\tan^{-1} X_L/R$, is plotted at this angle below the real axis. The generator current $\mathfrak{J}$ is shown as the vector sum of the other two. The solid-line solution has been drawn on the assumption that $X_L = X_C = 12R_L/5$, and the broken-line solution is drawn on the assumption that the frequency alone is varied until $Z_C = Z_L$,

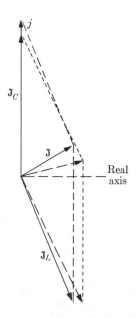

Fig. 11–8 Graphical solutions for the circuit of Fig. 11–7. The solid lines are drawn for $X_L = X_C = 12R_L/5$, while the broken lines are drawn for a variation in frequency that makes $X_C^2 = X_L^2 + R_L^2$.

that is, until $X_C^2 = X_L^2 + R_L^2$. A little study of this diagram reveals that the generator current would be smallest for some solution lying between those represented by the solid and broken lines. The exact solution for minimum generator current can be obtained analytically.

11–9   Routine method for solving networks

The routine method introduced in Section 6–11 for solving dc networks can be applied profitably for obtaining steady-state solutions with ac networks. For example, the relations used for solving the network in Fig. 11–7 could have been obtained by assuming that both of the currents $\mathfrak{J}$ and $\mathfrak{J}_L$ circulate clockwise in the lower and upper loops respectively. A solution for either $\mathfrak{J}$ or $\mathfrak{J}_L$ could be written as the ratio of two determinants. In either case the determinant in the denominator would be the same and would be symmetrical about a diagonal. Furthermore, since both currents are assumed to circulate in the same direction, all the off-diagonal terms must be negative. This provides two routine checks on the correctness of the denominator determinant. The determinant for the numerator is constructed from the denominator determinant. The procedure can be illustrated by setting up the determinant solution for $\mathfrak{J}_L$ as follows.

First, write loop equations for the upper and lower loops, respectively, collecting the impedances multiplying each current, that is,

$$\mathfrak{J}_L(\mathfrak{z}_L + \mathfrak{z}_C) - \mathfrak{J}\mathfrak{z}_C = 0,$$
$$-\mathfrak{J}_L\mathfrak{z}_C + \mathfrak{J}\mathfrak{z}_C = \mathfrak{V},$$

where

$$\mathfrak{z}_L = R_L + jX_L \qquad \text{and} \qquad \mathfrak{z}_C = -jX_C.$$

Second, write the solution of these equations for $\mathfrak{J}_L$ in determinant form, namely,

$$\mathfrak{J}_L = \frac{\begin{vmatrix} 0 & -\mathfrak{z}_C \\ \mathfrak{V} & +\mathfrak{z}_C \end{vmatrix}}{\begin{vmatrix} (\mathfrak{z}_L + \mathfrak{z}_C) & -\mathfrak{z}_C \\ -\mathfrak{z}_C & +\mathfrak{z}_C \end{vmatrix}} = \frac{\mathfrak{V}}{\mathfrak{z}_L}.$$

An outline of this method for solving ac Wheatstone bridge problems is given in one of the problems.

Network theorems. The procedures outlined for solving steady ac network problems are adequate for solving problems involving combinations of series elements with sources of alternating emf's. However, modern circuits are so complex that it is often desirable to use simplifying theorems. Three of the most useful theorems, therefore, will be stated without proof, though they can be proved with the complex forms of Kirchhoff's laws that have just been developed.

To understand network theorems it is desirable to differentiate between ideal and real generators. Ideal generators are either ideal voltage generators or ideal current generators. The ideal voltage generator maintains a constant-amplitude sinusoidal potential difference across its terminals regardless of the current through it. The ideal current generator maintains a constant-amplitude sinusoidal current regardless of the potential difference across its terminals.

In circuit terminology any source of electrical energy may be called a generator. A real ac generator can be replaced either by an ideal voltage generator with an internal impedance in series or by an ideal current generator with this same internal impedance in parallel. When generators are shut down the ideal voltage generator must have zero impedance between its terminals, while the ideal current generator must have infinite impedance, amounting to an open circuit between its terminals. The following theorems are probably the most important of the many theorems that are available. These theorems are stated for ac networks, where they are needed most, but only the obvious modifications are required to make them applicable to dc networks.

1. *Superposition Theorem.* Any network involving several generators may be solved by considering one of them at a time, with all other ideal generators shut off. The actual complex current in any branch is then the sum of the complex currents produced by each generator, taken one at a time.

2. *Thévenin's Theorem.* Any two-terminal network may be replaced by a single ideal voltage generator in series with an impedance. The value of the generator voltage is given by the open-circuit potential across the terminals of the network. The impedance is that determined by shutting off all ideal generators in the circuit and calculating the impedance of the network.

3. *Norton's Theorem.* Any two-terminal network may be replaced by a single ideal current generator in parallel with an impedance. The generator current is equal to the current when the terminals of the actual network are short-circuited. The impedance has the same value as defined in the previous theorem.

These theorems all arise from the linearity of the differential equation, Eq. (11–2). They are not valid when circuit elements contain the nonlinear components discussed in Section 11–11 and 11–13.

11–10 Alternating-current bridges and other instruments

The Wheatstone and other dc bridges have their counterparts in several ac bridges. For example, the bridge shown in Fig. 11–9 is used for the comparison of impedances. When the four impedances have been adjusted so that there is no ac current through the detector *G*, the bridge is balanced.

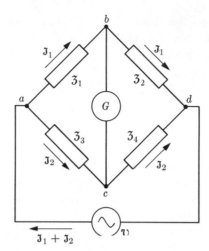

Fig. 11–9 The general form of an ac bridge shown in the balanced condition.

Since there is no current through G, there is a single current $\mathfrak{I}_1$ through both $\mathfrak{Z}_1$ and $\mathfrak{Z}_2$, and another single current $\mathfrak{I}_2$ through $\mathfrak{Z}_3$ and $\mathfrak{Z}_4$.

The fact that there is no current through the finite impedance of the detector G then guarantees that the potential drop from b to c is zero. Therefore the potential drop from a to b equals that from a to c, and the potential drop from b to d equals that from c to d:

$$\mathfrak{I}_1\mathfrak{Z}_1 = \mathfrak{I}_2\mathfrak{Z}_3$$

and

$$\mathfrak{I}_1\mathfrak{Z}_2 = \mathfrak{I}_2\mathfrak{Z}_4.$$

Dividing the first equation by the second to eliminate $\mathfrak{I}_1$ and $\mathfrak{I}_2$, we find

$$\frac{\mathfrak{Z}_1}{\mathfrak{Z}_2} = \frac{\mathfrak{Z}_3}{\mathfrak{Z}_4}, \tag{11–18}$$

which is an equation that is reminiscent of the dc Wheatstone bridge relation among four resistances. However, each of the four impedances may contain three elements, R, L, and C, and these twelve different quantities may make a general analysis difficult. The difficulties inherent in a general analysis of ac bridges has stimulated the design of many different bridges for special purposes. Many of these special bridges use two pure resistance elements to determine unambiguously the ratio between the impedances in the other two arms. While excellent treatments of many bridges can be found in the literature,* the important principles can be learned by studying just a few examples.

---

* L. Page and N. I. Adams, *Principles of Electricity*, 2nd ed., pp. 475–498, Van Nostrand, Princeton, N.J., 1952.

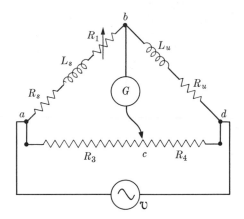

Fig. 11–10 Bridge for comparing an unknown with a standard inductance.

Inductance bridge. A bridge designed for comparing an unknown inductance with a standard inductance is shown in Fig. 11–10, where the $\mathfrak{Z}_3$ and $\mathfrak{Z}_4$ of Fig. 11–9 have been replaced by the purely noninductive resistances R_3 and R_4. The standard inductance L_s with its resistance R_s replaces $\mathfrak{Z}_1$, and the unknown inductance L_u with its resistance R_u replaces the $\mathfrak{Z}_2$ of that figure. It is usually impossible to balance such a bridge by varying only the ratio R_3/R_4. It can be balanced in this way only if $L_u/R_u = L_s/R_s$; that is, if the Q's of the unknown and the standard coils happen to be identical. Ordinarily it will be necessary to add resistance either to the branch containing the unknown or to the branch containing the standard. Assume that a resistance R_1 should be added to the standard's branch, as shown in Fig. 11–10. Equation (11–18) now gives

$$\frac{R_1 + R_s + jX_s}{R_3} = \frac{R_u + jX_u}{R_4},$$

which provides the two conditions

$$\frac{R_1 + R_s}{R_u} = \frac{R_3}{R_4} \quad \text{and} \quad \frac{R_3}{R_4} = \frac{X_s}{X_u} = \frac{L_s}{L_u},$$

both of which must be satisfied to achieve a balance.

Once this bridge is balanced the value of L_u can be obtained from the second of these relations, but to obtain this balance it is necessary to determine not only the correct ratio R_3/R_4, but also the correct value of R_1 to satisfy the first of these relations. A first approximation to the correct value of R_3/R_4 can be obtained by setting $R_1 = 0$ and determining the value of the ratio that produces the minimum reading on G. With the ratio fixed at this approximate value, dc is applied to the bridge to determine the value of R_1 required to obtain a balance. Now R_1 is kept at this approximate value and ac is used to obtain a better value for R_3/R_4. It may be necessary to repeat these procedures several times to achieve a good balance with ac.

A similar bridge can be used to compare capacitors with a standard. However, when dc is being used to determine the ratio of the resistances, the capacitors must either be removed or bypassed with a shorting wire.

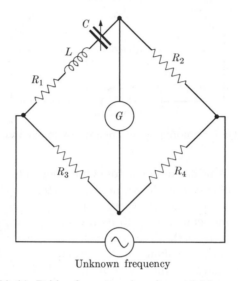

Fig. 11–11 Bridge for measuring sinusoidal frequencies.

Frequency bridge. The fact that a series element containing inductance, capacitance, and resistance resonates when $\omega L = 1/\omega C$ can be useful when a bridge for measuring frequencies is to be made. As shown in Fig. 11–11, only one of the four branches contains inductance and capacitance, while the other three contain purely noninductive resistances. Since three of the branches have only pure resistance, the bridge will balance only when the remaining branch has pure resistance. This can be achieved only by making the reactances cancel, as they will when $X_C = X_L$ or when $f = 1/2\pi\sqrt{LC}$. The balancing of the bridge determines the frequency in terms of L and C. Although the frequency is determined independently of the resistances of the branches, the bridge cannot be balanced unless the relation $R_1/R_2 = R_3/R_4$ is satisfied.

For detecting unbalance in bridge circuits using audible frequencies, headphones may be adequate; any device sensitive enough to detect the small currents resulting from an observable unbalance is satisfactory. However, when a current-indicating device is to be used to determine quantitatively the magnitude of the current through it, the requirements on it are more exacting. Direct-reading meters are usually faster than bridges, but are less accurate.

Alternating-current ammeters and voltmeters. Instruments employing a movable coil suspended in the fixed field of a permanent magnet are not directly useful for ac measurements. The torque on the coil alternates with the current through it and the inertia of the moving system prevents significant deflections. However, by providing a coil assembly with nearly the same mechanical oscillation frequency as the frequency of the alternating current, a sensitive detector for an alternating current of given frequency may be constructed. In such instruments, which are called *vibration* galvanometers, relatively small angular impulses of the right frequency build up resonant vibrations in the coil, whose maximum deflections can then be read.

The fixed permanent magnet design is also used with crystal rectifiers to produce ammeters and voltmeters that are quite satisfactory for alternating currents having wide ranges of amplitude and frequency. These instruments can also be used for measuring voltage or current in dc circuits, *but the markings on the dc scales must be different from those on the ac scales because their calibrations are different.* Figure 11–14 shows a common design for the crystal rectifiers in these instruments in which, however, a moving coil would replace the resistor load. If the ratio of the resistances of these rectifiers for current in the two directions is sufficiently large, the resistance in one direction may be assumed to be zero and in the other infinite. With the design shown in Fig. 11–14 the current through the coil then follows the current *vs.* time graph in that same figure. If the coil assembly has sufficient inertia to maintain a constant deflection, it will be proportional to the time average of the current in the graph. The time-average and the root-mean-square current are the same for dc, but with ac rectified in this fashion, the time-average current is $2I_m/\pi$, while the root-mean-square current is $I_m/\sqrt{2}$, for a current amplitude of I_m. Scale divisions for alternating current must therefore be $2\sqrt{2}/\pi$ times the scale divisions for direct current.

Dynamometers. Devices for measuring alternating current can also be made by replacing the permanent magnet of the galvanometer with a current coil whose magnetic field oscillates with the current. This fixed-field coil is electrically connected in series with the movable coil, so that the instantaneous torque on the latter is proportional to the square of the instantaneous current. The torque is thus unidirectional. The deflections of this instrument, which is called a dynamometer, are proportional to the average of these squares or to the mean square of the current. The chief disadvantage of the dynamometer is that it must have a nonlinear scale, which cannot indicate small currents with accuracy.

Wattmeters. The dynamometer type instrument is ideally suited for measuring the average electric power delivered to any device. With the fixed-field coil of a dynamometer connected like an ammeter, in series with the device, and its movable coil connected like a voltmeter, across the terminals of the device, a torque is produced that is proportional to the product of the instantaneous current through the device and the voltage across it. The instantaneous torque is then proportional to the instantaneous power delivered and, with sufficiently high inertia in the moving coil, *there is a steady deflection that is proportional to the average power delivered.* Since this instrument, which is called a wattmeter, indicates the average power, its deflections are proportional to

$$P = VI \cos \theta$$

but not to the product *VI*. If separate measurements are made of *V* and *I*, the lag angle *θ* can be determined from the wattmeter reading. An important feature of this instrument is that it can be used to measure ac or dc with a single calibration.

Wattmeters must be used with great care. It is very easy to burn out one or the other of the wattmeter coils, because the instrument may read zero with either or both of the coils overloaded. This will occur with lag angles of 90°, for which $\cos \theta = 0$. When $\cos \theta = 1$, the reading may be small because the product *VI* is small, even though either *V* or *I* is much too large for its coil. The values of *V* and *I* should be determined from separate voltmeter and ammeter readings to assure that neither is too large for its wattmeter coil.

11–11 Nonohmic semiconductor elements

In previous discussions it has been assumed that the resistances of all resistor elements are independent of the magnitude or direction of the current through them. Such elements are called ohmic resistors. However, many devices used in ac circuits have resistances that are dependent on the current or voltage, and these are said to be nonohmic.

As examples of nonohmic devices we will discuss the following: thyrite, which exhibits a nonlinear voltage versus current curve symmetrical with current direction, the nonsymmetrical p-n junction device, the n-p-n junction transistor, which provides amplification and control, and the Gunn oscillator, which converts dc into high frequency ac current. Of these devices, the Gunn oscillator represents the most recent development, the thyrite has had the longest commerical use, while the p-n junction and n-p-n transistor devices have the greatest commercial interest, since they figured prominently in the revolution in the electronics industry which followed the invention of transistors and the development of elementary transistor circuits in the late 1940's and early 1950's.

In this revolution vacuum tubes have been largely replaced by transistors for almost all applications from microminiaturized circuitry for space electronics and high-speed circuitry for electronic computers to low-cost circuits for home radio and television. By the 1960's a specially fabricated silicon transistor inside a can less than a quarter of an inch on a side was typically doing the job of a four-inch-high vacuum tube of the 1940's and 1950's. By the end of the 1960's monolithic semiconductor circuits, with transistors, resistors, capacitors, and interconnections fabricated on one silicon wafer, were finding their way into sophisticated electronic equipment. A few very highly integrated units contained silicon chips less than a quarter inch on a side with the equivalent of over 100 transistors and associated passive circuitry, and single silicon chips containing over ten times that many circuits are being developed. The response time of high-performance transistor circuits is measured in nanoseconds (10^{-9} s) as compared to microseconds for the vacuum tube circuitry of the 1950's, and the cost per circuit has also been markedly decreased. The increased function, low power requirements, high packaging density, increased ruggedness and reliability, and low cost of semiconductor devices has opened a panorama of new applications which not only has resulted in major growth of the electronics industry but also has affected how we live, do business, conduct warfare, and explore the universe.

Because of the great significance of these developments, most students of this book have already had some exposure to semiconductor devices and circuits and expect to undertake more detailed study in future courses. The material in this section is, therefore, brief and is designed only to provide a broad introduction and perspective.*

In reading about the simple nonohmic devices presented here one should note that a sinusoidal voltage applied across a nonohmic element usually results in a current with nonsinusoidal waveform, so that the previous linear mathematical treatments of ac problems are not valid. Fortunately, however, nonohmic elements are often used in circuits in such a manner that their output is nearly sinusoidal. Even when this is not true, many physical concepts obtained from the sinusoidal solutions are valid even though the detailed solutions are not.

p-n junctions. The performance of most semiconductor devices is based on the electrical characteristics of junctions between hole-rich semiconductors (called p-type, for positive charge conduction) and electron-rich semiconductors (called n-type, for negative charge conduction). Before treating the junction region, it is well to discuss briefly the properties of the semiconductors themselves. The most commonly used semiconductors are the elements germanium and silicon, and we will limit our discussion of p-n junctions to these.

An individual atom can be visualized as a positively charged nucleus with an equal negative charge carried by the orbital electrons. These elec-

---

* Excellent introductory material on semiconductor devices and circuits is contained in R. M. Warner and J. N. Fordemwalt, Eds., *Integrated Circuits—Design Principles and Fabrication*, McGraw-Hill, New York, 1965 or W. R. Beam, *Electronics in Solids*, Chapter 5, McGraw-Hill, New York, 1965.

trons orbit the nucleus with discrete, well-defined energy levels. The lowest energy levels closest to the nucleus are filled, following the Pauli exclusion principle, leaving four outer electrons in an unfilled energy shell in silicon and germanium.

When these atoms are brought together to form a solid crystal, the electrons in the lower energy levels are only slightly perturbed, while the four outer electrons join with those of neighboring atoms to form the covalent bonds of the crystal. Exactly four electrons from each atom, no more or less, are required for these bonds. The energy of the *bond* or *valence electrons* is no longer a discrete value, but now covers a small range or *band* of energies. Valence electrons are therefore said to be in a *valence band*.

Since all electrons in pure single crystalline silicon or germanium exist in orbit about one nucleus or in a covalent bond between atoms, no conductivity could exist without thermal excitation. Above absolute zero, a small number of covalent bonds are broken, and the freed electrons wander randomly through the crystal until they come to another broken bond where they may *recombine* with the electron remaining at the covalent bond site. While free, these electrons will conduct current under the influence of an applied electric field. There is a broad continuum of energies associated with these conduction electrons which is referred to as the *conduction band*. Between the valence band and the conduction band there is a range of energies which electrons cannot assume, i.e., there are no *electronic states* with these energies. This is known as the forbidden energy gap and is about 1.11 eV in silicon and 0.67 eV in germanium. The number of electrons with sufficient thermal energy to be excited into the conduction band at room temperature is very small but increases rapidly with increasing temperature.

Resistivity in semiconductors decreases with increasing temperature because of the exponentially increasing number of current carriers. This can be contrasted with metal conductors which have on the order of one conduction electron per atom even at absolute zero. The number of current carriers in metals is essentially temperature-independent so that resistivity increases with increasing temperature from thermal scattering.

Electric current in semiconductors results not only from the thermally excited electrons, but also from the resultant *holes* (i.e., missing electrons) in the valence band. Every time a bond is broken, a net positive charge near the bond site results. This is called a *hole*. If a neighboring covalent bond electron jumps into this hole, the effect is the same as if the hole had moved from the atom to that neighboring site. Hole and electron conductivity are both important in semiconductors, although the conductivity of electrons is generally larger due to their higher mobility. Clearly, in pure silicon or germanium, the number of holes must be exactly equal to the number of conduction electrons.

It is possible to add holes or conduction electrons to these semiconductors by adding small amounts of selected impurity atoms. For example, a small addition of material from the fifth column of the periodic table, such as P, As, or Sb, adds one extra electron per impurity atom into an electronic state that is electrostatically coupled to the impurity ion. This electron can be thought of as being in a hydrogen-like orbit about the impurity ion. The orbit is enlarged by the high dielectric constant of the semiconductor so that it encompasses many neighboring atoms. This electron is nearly free and therefore easily ionized into the conduction band; in fact, the energy of the orbital state for the common impurities is so little below the conduction band that the probability of it being ionized is nearly unity at room temperature. These impurities are called *donors* because they donate approximately one conduction electron per impurity atom. Similarly, a small addition of *accepter atoms* from a material in the third column of the periodic table, such as B, Al, or Ga, results in almost one conducting hole per impurity atom. Semiconductors with more donor than accepter atoms are called n-type while the converse are p-type. Impurity concentrations of 10^{-3} to 10^{-7} are typical for most doped devices. Semiconductors with carefully adjusted impurity content are said to be *doped*, while pure semiconductors (less than one impurity atom in 10^9) or those with compensating holes and electrons are said to be *intrinsic* semiconductors.

As one would intuitively expect from our knowledge of gases or liquids placed in contact, when a junction is formed between n- and p-type material, some of the conduction electrons diffuse into the p-region while some holes diffuse into the n-region. In this case, however, diffusion is limited by the electrostatic fields that build up near the junction because the opposite direction of flow by carriers of opposite sign results in a noncancelling net flow of charge. The electrostatic forces that bring this diffusion process to an equilibrium are confined to a very narrow boundary layer, and are of such sign that the normal carriers in both n- and p-regions are repelled from the boundary layer. Obviously a large potential gradient cannot exist where there is a high concentration of current carriers. This is illustrated in Fig. 11–12; the reduction in both positive and negative current carriers near the junction is shown in (a), the net residual charge from the ionized impurity atoms near the junction is shown in (b) and the resultant electrostatic field is shown in (c). This electrostatic field is required to provide a force on the charges to counterbalance the statistical (or quantum mechanical) tendency of the current carriers to diffuse into the neighboring material. The electrostatic energies of electrons in the conduction and valence bands near the junction are shown in (d). The electrostatic potential difference is given by the line integral of the electric field across the junction, i.e., the area under the curve in (c).

The solid straight line labelled φ and running through the forbidden gap between the valence and conduction bands in Fig. 11–12(d) is the Fermi level. It can be defined as the energy above which the probability of an electron state's being filled is less than 0.5 and below which the probability is greater than 0.5.

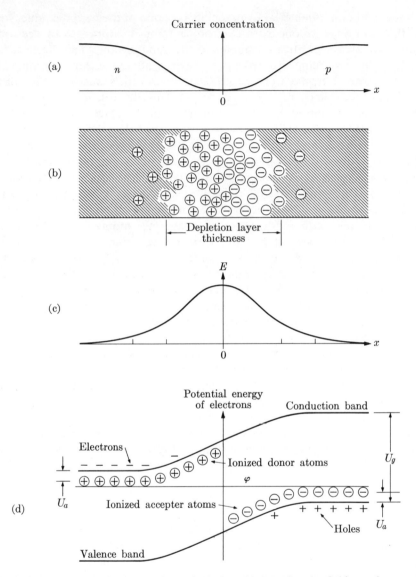

Fig. 11–12 Carrier concentration, depletion layer, electric field, and energy relations across a p-n junction.

In solid-state theory, a description of the p-n junction normally begins with a derivation of the Fermi distribution function followed by a discussion of how the Fermi level must relate to band edges through a p-n junction. It must pass through the center of the energy gap in the neutral region, close to the conduction band of the n-region, and close to the valence band of the p-region. Finally, it is shown that the Fermi level must be the same through all materials in intimate electrical contact. The results from this theory are identical to those of our qualitative treatment. However, the more formal solid-state treatment of the subject has the advantage of providing a quantitative value for the energy difference between the conduction bands in the p- and n-regions, providing one has experimentally determined the energy gap, the ionization energy for p- and n-type impurity atoms, and the impurity density.

Current rectification is an important electrical capability of p-n junctions. When the p-n junction is *forward biased*, that is, when a field is applied from the p- to the n-region, the applied field tends to cancel the electrostatic field at the boundary and the carriers drift toward the junction. This narrows the depletion layer of Fig. 11–12, increases the carrier density in the junction region, and results in increased current through the junction. A *reverse bias*, on the other hand, decreases the carriers in the junction region and results in very little current. Thus the desired diode characteristic of low impedance in one direction and very high impedance in the other is achieved. Using a fairly simple model, one can derive the *rectifier equation* for p-n junctions,

$$J = J_s(e^{qV/kT} - 1),$$

where J is current density, q is the electron charge, V is the applied voltage, T is the absolute temperature, k is Boltzman's constant, and J_s is the saturation current density under reverse bias. The term J_s is dependent on the carrier density and mobility of the particular semiconductor. This derivation may be found in both of the references cited previously.*

Simple ac rectification. Whether the device chosen is a p-n junction, a semiconductor-metal rectifier, or a vacuum tube, the characteristic voltage versus current curve will be essentially of the form illustrated in Fig. 11–13. When a simple sinusoidal voltage, as illustrated below the characteristic curve, is applied to the rectifier, the current through it is as illustrated to the right. The magnitude of the effective current during one half-cycle may be so much smaller than it is during the other half that it can be neglected. In this case the device may be thought of as a simple switch that is on in one direction and off in the other.

For many purposes these unidirectional pulses should be closer together. A circuit that is frequently used to obtain current during both half-cycles is shown in Fig. 11–14. When a three-phase ac circuit is available, unidirectional pulses can be spaced even closer together with the circuit shown in Fig. 11–15. The pulses combine to produce a ripple of higher frequency and of relatively small amplitude superimposed upon a steady dc, as shown in the figure. The amplitude of the ripple can be reduced greatly by passing it

---

* See p. 387.

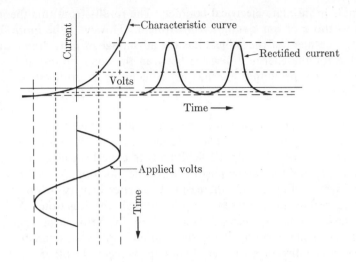

Fig. 11–13 Characteristic curve of a typical rectifier, showing the form of the rectified current curve when it is connected across a sinusoidal voltage.

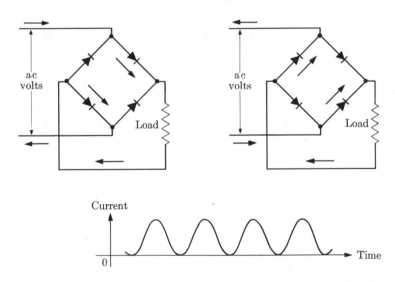

Fig. 11–14 A full-wave rectifying circuit for single-phase ac. Rectifiers are indicated by →⊦ or ⊣←, where the first rectifier allows current to pass to the right, and the second one, to the left. Arrows on the circuit at left indicate the direction of the current in the branches for current entering at the top, while those on the right-hand circuit show directions for current entering at the bottom. The current *vs.* time graph indicates that there is a pulsing, unidirectional current through the load when ac is supplied to the circuit.

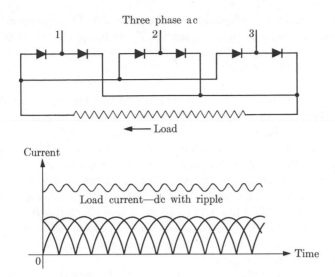

Fig. 11–15 A full-wave rectifying circuit for three-phase ac is shown at the top. The current through the load is dc, with a small ripple as indicated in the current *vs.* time graph. The bottom curves represent the three components whose sum is the load current.

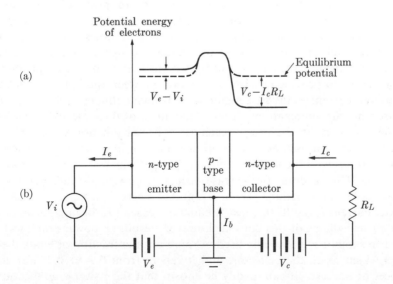

Fig. 11–16 A specific wiring arrangement for an n-p-n amplifier transistor is illustrated in (b) together with the associated electron energy level diagram in (a). The emitter is forward biased by a small voltage V_e modulated by the input voltage V_i. A larger reverse bias of the collector results in an I_c nearly equal to I_e. Because of the large bias voltage, power amplification results in the load R_L.

through a parallel L-C filter like that shown in Fig. 11–7, designed for maximum impedance at the ripple frequency. The inductor coil of this filter should provide very low resistance for the dc and should provide the large Q required for an effective filter.

An n-p-n junction transistor. As a representative transistor, we will consider the n-p-n junction device illustrated in Fig. 11–16(b). It consists of two p-n junctions placed back to back; and, as wired, it provides amplification of voltage and power with little current change. The forward biased junction is called the *emitter* junction because it emits electrons which pass through the base (if it is thin enough) and are collected at the reverse biased *collector junction.* The potential energy of the electrons without bias is shown in Fig. 11–16(a) by the dashed line while the solid line illustrates the shift in these levels resulting from the bias voltages. These lines are essentially level except in the junction regions, since the other regions are heavily enough doped that they contribute little resistance.

From our knowledge of p-n junctions, we know that the forward biased emitter junction offers low impedance to the input V_i while the reverse biased collector junction provides a high output impedance. Thus a small change in V_i will result in a large change in I_e, while a relatively larger change in V_c or R_L will have little effect on I_c. These characteristics are desirable in an amplifying device but are not alone sufficient to provide amplification. Indeed, if the base section of the transistor is made too thick, the electron current carriers from the emitter will *recombine* with holes in the p-type base and never reach the collector junction. The input signal current I_e will then flow around the input circuit (I_e and I_b), with little effect on the output circuit.

In order to provide effective interaction between the input current and the output current I_c, the base region must be made thin enough so that little electron-hole recombination occurs. Then most of these electrons (*minority carriers* in the p-type base) will reach the collector junction where the steep potential gradient assures their continued motion into the collector region. This contrasts with the *majority carriers* (p-type) which are repelled from the junction. Thus a dramatic increase occurs in current through the reverse biased collector junction. With a suitably narrow base, an increase in emitter current I_e results in a nearly equal increase in I_c through the load R_L. The higher voltage of this current, however, results in power amplification.

The symbol α is typically used to represent the fraction of emitted electrons which reach the collector, and it ranges from 0.9 to 0.99 for most devices of interest. It can readily be shown that the power amplification is equal to α^2 times the ratio of the output to the input impedance. Of considerable engineering interest is the β of the transistor which is defined as I_c/I_b and is the current gain of the n-p-n device when wired in the grounded emitter circuit configuration. Simple algebra shows that $\beta = \alpha/(1 - \alpha)$.

Thyrite. One of the best-known materials whose resistance depends on the magnitude but not the direction of the applied voltage is thyrite. It is a black ceramic material formed by heat-treating a mixture of carbon and clay. As the voltage across thyrite is increased, valence electrons are excited into the conduction band so the resistivity decreases with increasing voltage. The characteristic current versus voltage curve for this material is so nonlinear that a tenfold increase in the voltage across it may increase the current through it by a factor of 3000.

This feature makes it very useful for protecting power lines from lightning or, in fact, any electric equipment from unexpected voltage surges. For example, a large electromagnet operating on full current has considerable energy stored in its magnetic fields. An accidental breaking of the circuit may produce a high voltage capable of arcing across and destroying the insulation between the magnet coils. A thyrite block placed across the magnet terminals can provide a closed path for the current in the coils, where the stored energy can be dissipated harmlessly. For this purpose a thyrite resistor with the following characteristics can be chosen: across the normal operating voltage its resistance can be large compared with that of the coils, but this resistance will drop rapidly with moderate increases in the voltage.

Semiconductor junction devices such as the p-n-p-n or thyrector devices have been made with similar characteristics, and they are becoming more widely used for over-voltage protection especially where lower leakage current or well-determined breakover voltage characteristics are required.

Gunn oscillator. The Gunn effect, first reported in 1963,* permits a dc voltage to be transformed directly into high frequency ac with a simple, single-stage device and is unique in that it results from bulk, as opposed to junction, phenomena. Conduction electrons in GaAs have two very different mobilities, depending on their energy relationship within the crystal lattice. Normally, they reside in a low-energy conduction band which happens to have a relatively high mobility. As voltage across the GaAs is increased, electric current increases until the voltage becomes large enough, 1000 to 3000 V cm^{-1}, to raise the electrons to the higher energy level. Further voltage increase actually produces a slight decrease in current due to the lower mobility of electrons in the high-energy band. This condition is somewhat unstable and is accompanied by fluctuations in current and voltage.

---

* *Solid St. Comm.*, Vol. 1, pp. 88–91, Pergamon Press, 1963; J. B. Gunn, *IBM Journal* **8**, 141 (1964); and *IEEE Transactions on Electron Devices*, Vol. Ed-13, No. 1, Jan. 1966, Guest Editor, Rudolf Engelbrecht, especially "Introduction," pp. 1–4, and "Theory of Negative-Conductance Amplification and of Gunn Instabilities in Two-Valley Semiconductors," by D. E. McCumber and A. G. Chynoweth, pp. 4–27.

Gunn observed that the fluctuations in a short specimen of n-type GaAs were periodic and, in fact, that they resulted from the flow of electrons across the crystal in the form of waves. As electrons were accelerated to the front of the wave by the applied voltage they would then enter the lower-mobility, higher-energy conduction band and be slowed down, finally falling back into the high mobility band at the rear of the wave and again be accelerated to the front of the wave. This wave of electrons was called a *domain* and was observed to move across the sample with a velocity of approximately 10^7 cm s^{-1}. When a domain reaches the end of the crystal, a large current pulse is observed, and a new domain is formed at the other end. A pulsed output signal of 10^9 cps would be observed in a sample 10^{-2} cm long. These devices and modifications of them are being studied as possible inexpensive sources of radar or microwave frequencies. A major practical problem to date is that these devices typically exhibit efficiencies of only a few percent.

11-12 Inductive coupling

The existence of a time-dependent current in an ac circuit assures the existence of varying magnetic fields in the neighborhood. If other circuits, magnetic materials, or sizeable conductors are present in these fields, they will interact with the original circuit and change the character of its observable impedance. Usually these interactions absorb some power from the primary circuit. The transfer of power through the secondary winding of a transformer and the heating of a conductor in an induction furnace are familiar examples of such interactions. On the other hand, when a well-designed core of ferromagnetic material is inserted into a choke coil, its inductive reactance can be increased manyfold with little power absorption.

A physical understanding of these phenomena can be obtained by making a careful mathematical study of the inductively coupled circuits shown in Fig. 11-17. The time-dependent currents in the two circuits are the real parts of $\mathfrak{I}_1'$ and $\mathfrak{I}_2'$. The inductive interactions between the circuits take place through the mutual inductance M between the coils, which have self-inductances of L_1 and L_2. A generator maintains a steadily alternating emf in circuit No. 1 that is the real part of $\mathfrak{V}' = V_m e^{j\omega t}$. In the calculations, the total resistance in circuit No. 2 is combined into $R_2 = R_s + R$.

One circuit equation can be written for each closed loop: for the first,

$$R_1 \mathfrak{I}_1' + L_1 \frac{d\mathfrak{I}_1'}{dt} + M \frac{d\mathfrak{I}_2'}{dt} = \mathfrak{V}', \tag{11-19}$$

and for the second,

$$R_2 \mathfrak{I}_2' + L_2 \frac{d\mathfrak{I}_2'}{dt} + M \frac{d\mathfrak{I}_1'}{dt} = 0. \tag{11-20}$$

Since only the quasi-steady relations are desired, these can be reduced to

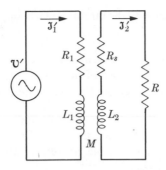

Fig. 11–17 A pair of inductively coupled circuits with the primary circuit No. 1 connected to a source of steadily alternating emf. The total resistance in circuit No. 2 is $R_2 = R_S + R$.

algebraic equations by introducing the concepts from Section 10–4 that the time-dependent complex currents and emf in these equations have the form $\mathcal{B}' = \mathcal{B}e^{j\omega t}$, with the time derivatives $\partial \mathcal{B}'/\partial t = j\omega\mathcal{B}e^{j\omega t}$. The resulting equations can be divided through by $\sqrt{2}\, e^{j\omega t}$ to produce the following simultaneous equations in the two unknown complex currents $\mathfrak{I}_1$ and $\mathfrak{I}_2$:

$$R_1\mathfrak{I}_1 + L_1 j\omega\mathfrak{I}_1 + Mj\omega\mathfrak{I}_2 = \mathcal{V}$$

and

$$R_2\mathfrak{I}_2 + L_2 j\omega\mathfrak{I}_2 + Mj\omega\mathfrak{I}_1 = 0.$$

Solving for $\mathfrak{I}_1$, the complex current in the primary circuit can be written as

$$\mathfrak{I}_1 = \frac{\mathcal{V}}{\mathfrak{Z}_e}, \tag{11–21}$$

where

$$\mathfrak{Z}_e = R_e + jX_e, \tag{11–22}$$

with

$$R_e = R_1 + \frac{R_2 X_m^2}{Z_2^2}, \qquad X_e = X_1 - \frac{X_2 X_m^2}{Z_2^2}, \tag{11–23}$$

and where $X_1 = \omega L_1$, $X_2 = \omega L_2$, $X_m = \omega M$, and $Z_2^2 = R_2^2 + X_2^2$. These are general expressions for the effective resistance, reactance, and impedance of the primary circuit of Fig. 11–17, as modified by the presence of the secondary. Note that if $X_m = 0$, then $R_e = R_1$ and $X_e = X_1$, as should be expected. The effects due to the secondary circuit can be eliminated either by reducing M to zero or by increasing R_2 to infinity, e.g., by opening the secondary circuit.

The power absorbed by the system shown in Fig. 11–17 is

$$P = I_1^2 R_e. \tag{11–24}$$

Study of these relations has proved to be very rewarding in developing physical intuition for work with ac circuits. Two important cases for such a study now will be presented.

Case I. Ideal transformer. The current, voltage, and power relations in an ideal transformer will be studied as a function of the load on the secondary. The ideal transformer will be defined to have the following characteristics:

a) It has a closed flux path of ferromagnetic material with a constant permeability high enough to retain all the magnetic flux within its path. In addition, this material has negligible hysteresis and is laminated to have negligible eddy-current losses.

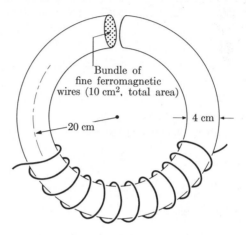

Fig. 11–18 Ferromagnetic core for an ideal transformer, consisting of a bundle of fine insulated wires of material in which μ is large but constant.

b) The primary and secondary coils are each wound around the ferromagnetic material so that all of the flux threads through each coil. For simplicity it will be assumed that the ferromagnetic path has a constant cross section A and length l. For example, if the core of the ideal transformer were the bundle of ferromagnetic wires shown in Fig. 11–18, the length would be $l = 2\pi 20$ cm $= 40\pi$ cm and the cross-sectional area would be $A = 10$ cm^2, the sum of the cross-sectional areas of the wires in the bundle.

c) Since practical considerations limit the size of the coils that can be used, it is convenient to assume that the primary and secondary coils have the same total cross section. Each coil may have either many turns of fine wire or fewer turns of heavier wire, but the relation $N_1 a_1 \simeq N_2 a_2$ is a close approximation, where the primary and secondary coils have N_1 turns of cross section a_1 and N_2 turns of cross section a_2, respectively.

From the characteristics (a) and (b), $L_1 = N_1^2 \mu A/l$, $L_2 = N_2^2 \mu A/l$, and $M = N_1 N_2 \mu A/l$, so that $L_1 L_2 = M^2$ and $X_1 X_2 = X_m^2$. If the wire can be wound so that the average length of the turns in each coil is C, the resistances of the coils are $R_1 = \rho N_1 C/a_1$ and $R_s = \rho N_2 C/a_2$ respectively, and $R_1 \cong R_s N_1^2/N_2^2 = R_s n^2$, where $n = N_1/N_2$. Then for a given size and shape of coil, both R and L are nearly proportional to N^2 and their ratio is nearly constant. Under the conditions specified, which are often closely approximated in real transformers, $X_1/R_1 = Q = X_2/R_s$, but the R_2 of Eq. (11–23) must be written as $R_2 = R_s + R$, where R is the load on the secondary.

In transformers with ferromagnetic cores, $X_2 \gg R_s$ and $Z_2^2 = X_2^2 + (R_s + R)^2 \cong X_2^2$, so that $X_m^2/Z_2^2 \cong n^2$ *when the load resistance R is small.* With ideal transformers and small load resistances Eqs. (11–22) and (11–23) can be written as

$$R_e \cong R_1 + n^2 R_2$$

and

$$X_e \cong X_1 - n^2 X_2.$$

With small load resistances connected to an ideal transformer, $n^2 X_2 \cong X_1$ and $X_e \cong 0$, which makes the power factor approach unity and makes the load appear to be nearly pure resistance. The power absorbed from a generator maintaining a constant amplitude of potential difference is

$$P = \frac{V^2}{R_1 + n^2 R_2},$$

which becomes $V^2/2R_1$ when R is reduced to zero by shorting the secondary.

On the other hand, when the load resistance R is increased to infinity by opening the secondary circuit, $X_e = X_1$ and the power factor becomes R_1/Z_1, which is very small. The power absorbed is then only $P = V^2 R_1/Z_1^2$. Whenever the load resistance R is large compared with the resistance of the secondary winding R_s, the voltage across R is very nearly $V_L = V(N_2/N_1)$. In real transformers, where hysteresis and eddy-current losses may not be neglected, these power losses decrease the effective inductance and increase the effective resistance.

Case II. Inductive effects of nearby conductors. Any conductor near an ac circuit reacts to absorb power from that circuit. Because of the many possible shapes and kinds of conductors, the general problem is complex. However, considerable understanding can be gained by calculating the relatively simple case in which a short-circuited coil absorbs power from a coil connected to an alternator. To further simplify the calculations, the coils can be constructed identically and the primary coil connected directly to a constant-amplitude, constant-frequency alternator with negligible impedance. Under these conditions the quantities in Eq. (11–23) have the relations $X_1 = X_2$

and $R_1 = R_2$. To study effects due to changes in the position of the shorted secondary coil, the mutual inductance M or $X_m = \omega M$ can be treated as variable. The effective ac resistance of the primary is then

$$R_e = R_1\left(1 + \frac{X_m^2}{Z_1^2}\right),$$

and its effective reactance is

$$X_e = X_1\left(1 - \frac{X_m^2}{Z_1^2}\right).$$

When the coils are close together $X_m \to Z_1$, so that $R_e \to 2R_1$, $X_e \to 0$, and the lag angle $\theta \to 0$. The power loss is large, or nearly $V^2/2R_1$, as in the shorted transformer.

When the coils are far apart $X_m \to 0$, so that $R_e \to R_1$, $X_e \to X_1$, and the lag angle $\theta \to \tan^{-1}(X_1/R_1)$. The power loss then is only $P = V^2/R_1(Q^2 + 1)$, which is a relatively small quantity when X_1/R_1 is large compared with unity. It should be noted that X_m does not vary linearly with the separation of the coils. In fact, for a secondary coil displaced without rotation along a radial line through the center of the primary, X_m is proportional to $1/r^3$ for $r \gg a$, where a is the radius of the coils.

These illustrations show that the effective ac resistance of any circuit exceeds its dc resistance when conducting materials are nearby. As frequencies increase, the effective ac resistance increases even when there are no power-absorbing materials in the neighborhood. This increase is due to the so-called "skin effect," which causes the current densities to be larger on the surface of a conductor than within the interior. Inductive effects tending to reduce ac densities have greater magnitudes in the interior than on the surface. Current densities in a solid cylindrical conductor as functions of the frequency and the radial distance from the axis can be calculated with Maxwell's relations, but the results are somewhat complex. It is simpler and quite instructive to show that if a conductor consists of two concentric cylindrical shells that are thin but have the same dc resistance, the inner cylinder carries the smaller current. This problem is left as an exercise.

11–13 Nonlinear magnetic devices

Nonlinear magnetic devices can be classified in two general categories: those utilizing the saturable magnetization or field-dependent permeability, and those utilizing the remanence and threshold in the magnetization hysteresis curve. In the first category are devices such as saturable inductors, saturable transformers, ferroresonant transformers, and magnetic amplifiers. The second category consists of mechanically addressable memory devices such as magnetic tapes and disks and electronically addressable memory devices used in computers and communication switching networks. (The term addressable refers to the ability to store or read information from a specified location.) This second category also contains some rather specialized logic devices which have found limited use in digital computer circuits.*

The use of nonlinear magnetic properties in devices to control or amplify current, voltage, and especially power was proposed as early as 1901. Significant use was not made of these devices, however, until World War II, when they were used especially by the Germans for naval fire control systems. A rapid growth was seen in this area after the war, and numerous sophisticated devices and circuits were devised, frequently employing the capabilities of semiconductor and nonlinear magnetic elements to complement each other. The rapid developments of transistors since the 1950's has overshadowed this area, since many of the promising applications of nonlinear magnetic circuits have been more economically handled by all transistor circuits. There still remain, however, a number of areas, e.g., power supply and distribution systems for electronic circuitry in which magnetic devices provide the most economical solution.

In information storage, nonlinear magnetic devices reign supreme. Magnetic tapes represent the most economical storage media when access time is not critical. They are used for storage of music, speech, video signals, and digital information for data processing systems. Large disks with magnetic oxide on the surface provide access in just a few milliseconds to the stored information anywhere on the disk by moving a magnetic pickup device to the specified track location on the surface of the disk.

Electronically addressable ferrite core memories were introduced in the early 1950's. Information ("ones" or "zeros") was stored by the clockwise or counterclockwise magnetization of each doughnut-shaped magnetic core. In the ferrite core memory introduced on the IBM 701 computer in 1955, for example, each ferrite core was 2 mm o.d. and 1.27 mm i.d. and information could be stored and then read out in 12 microseconds from any core in the memory. Today, many computers use this type of storage element, and several billion ferrite cores are produced annually in the United States. Read-write cycle times of less than 1 microsecond for main memories are commercially available. These increased speeds have resulted from improved core material and reduced core size, as well as better semiconductor drivers and sense amplifiers.

A saturable inductor regulator and a ferrite core memory will now be discussed as examples of devices in the two categories.

---

* For further discussion of subjects in this section see A. J. Meyerhoff *et al.* (Editors) *Digital Applications of Magnetic Devices*, John Wiley & Sons, Inc., New York, 1960; R. K. Richards, *Electronic Digital Components and Circuits*, D. Van Nostrand, Princeton, N.J., 1967, and W. Geyger, *Nonlinear-Magnetic Control Devices*, McGraw-Hill, New York, 1964.

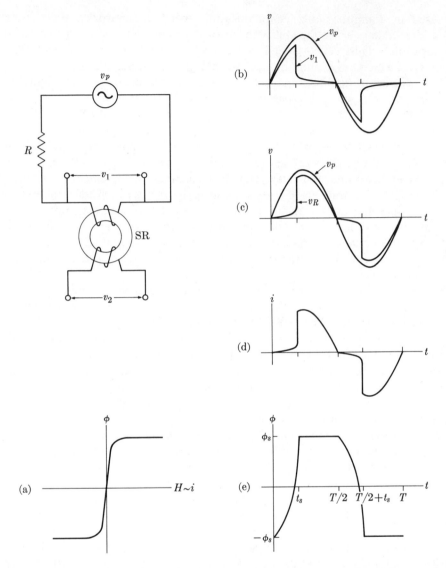

Fig. 11–19 Saturable core (SR for Saturable Reactor) in series with resistor and power supply. Idealized wave forms for v_p, v_1, v_R, i, and $\emptyset$ corresponding to the magnetization curve for the SR as illustrated in (a).

Saturable inductor regulator. The simple circuit in Fig. 11–19 contains a soft (i.e., high permeability) magnetic core inductor in series with an ohmic resistor and an ac power supply. In addition to the N_1 primary windings on the saturable reactor SR there are also N_2 secondary windings. If the sinusoidal power supply voltage v_p provides a volt-time integral

$$\int_0^{T/2} v_p \, dt = N \int_0^{T/2} \frac{d\phi}{dt} \, dt > 2N\phi_s$$

sufficient to saturate the SR *and* the magnetic field generated by the applied current i exceeds the threshold for full magnetization reversal, then it can be shown that the time-average voltages at the input and output of the SR are nearly independent of the wide excursions of the power supply voltage (providing that the lowest excursion is still sufficient to saturate the SR).

The voltage v_1 is given in general by

$$v_1 = N_1 \frac{d\phi}{dt} + iR_c,$$

where R_c is the total resistance in the coil winding N_1, and the time average ac voltage is

$$v_1(\text{avg}) \equiv \frac{2}{T} \int_0^{T/2} v_1 \, dt = \frac{2}{T} \left[\int_0^{t_s} v_1 \, dt + \int_{t_s}^{T/2} v_1 \, dt \right],$$

where T is the period of the power supply voltage and t_s is the time during the first half cycle which is required to saturate the core. An examination of the waveforms in Fig. 11–19 reveals that the current i is very small for $0 \leqslant t < t_s$ and that the flux change $d\phi/dt$ is negligible for $t_s \leqslant t < T/2$. Thus, the time-average ac voltage becomes

$$v_1(\text{avg}) \simeq \frac{2}{T} \left[N_1 2\phi_s + R_c \int_{t_s}^{T/2} i \, dt \right],$$

where in practice the second term is made very small compared to the first term by making R_c small and/or by operating the circuit at a sufficiently high frequency so that T is small. In this case

$$v_1(\text{avg}) \simeq 4fN_1\phi_s,$$

where

$$f = 1/T$$

is the frequency.

The voltage output from the secondary winding is given in general by

$$v_2 = N_2 \frac{d\phi}{dt}$$

so that the time-average ac voltage here is completely independent of R_c, although it will depend on the resistance of the secondary winding if load current is drawn between the output terminals. Thus

$$v_2(\text{avg}) \simeq 4fN_2\phi_s.$$

The following characteristics of the circuit are evident:

1. The time-average voltages v_1 and v_2 at the input and output of the SR are independent of variations in power supply voltage so long as the SR saturates once every half cycle.

2. The time-average output voltage is proportional to frequency and the saturation flux of the SR.

3. The resistor R has no effect on the output voltages but does serve to limit the current in the primary after the core is saturated according to the relation

$$i = v_p/(R + R_c) \qquad \text{for} \qquad t_s < t \leqslant T/2.$$

This circuit can be used as it is for voltage regulation, but more often it is the basis of more complicated circuits for current or voltage reference circuits. The addition of a capacitor in parallel with the secondary winding of such magnitude as to produce resonance will result in a more nearly sinusoidal output voltage. Because of the linearity of output with frequency, the device can also be used as a frequency meter.

It should be noted that the ϕ versus i curve is shown with no hysteresis. In practice some hysteresis is always present in magnetic materials and this results in a fixed hysteretic loss with each cycle.

Ferrite core memory. A hysteresis loop typical of square loop material used in coincident current memories is shown in Fig. 11–20(a), while the wiring scheme for a 3 × 3 array is shown in Fig. 11–20(b). Each core is able to store a "one" or "zero" by being magnetized either clockwise or counterclockwise, and the pattern of "ones" and "zeros" so stored in the array can be coded to represent desired information. Current drivers are situated at the ends of the x-lines and y-lines while a sense amplifier is placed on the z-line. A positive current pulse sent coincidently down the x_1 and y_2 lines, for example, will cause the (1, 2) core to switch into the clockwise direction and hence store a "one," providing that the sum of the magnetic fields from the two currents exceeds the coercive force H_c of the core. The other two cores on the x-line and two cores on the y-line will not be switched if the magnetic fields of a single current pulse is less than the core coercive force. Such pulses are called *half-select pulses.*

The coincidence of two half-select pulses at a core location can also be used to read out the stored information. Assuming the polarity is such as to write a "zero," then the core will switch only if a "one" was previously stored. The resultant $d\phi/dt$ voltage associated with a previously stored "one" is sensed by the sense amplifier. The information in the core is destroyed by the reading process (destructive read-out) and must be written again by a built-in regeneration circuit before subsequent read operations occur. Signal from the core is used, after amplification, to trigger this regeneration circuit as well as to provide data or instructions required else-where in the computer for logical operations.

In practice a three-dimensional selection scheme is usually utilized in order to reduce the number of semiconductor drive circuits for each storage element. A cubic $3 \times 3 \times 3$ memory could be made by placing two 3×3 arrays in planes parallel with and directly over the one shown in Fig. 11–20(b). The x_1-line could then be continued through the two three-core lines above it so that one driver would drive all nine cores corresponding to the coordinate x_1, instead of the original three. The same could be done for all x- and y-lines. Three separate z-lines would then be required, z_1, z_2 and z_3; and each of these would be equipped with a current driver as well as a sense amplifier. In this manner each x-driver drives all cores in a given x-plane, a y-driver drives all cores in a y-plane, and the same holds for the z-drivers and sense amplifiers.

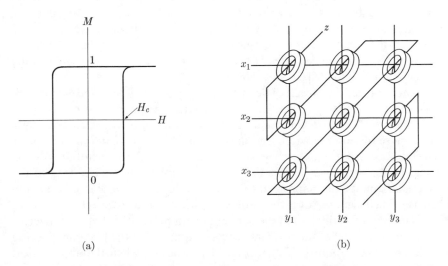

(a)   (b)

Fig. 11–20 Hysteresis loop for "square loop" ferrites used in magnetic core memories (a) and wiring arrangement used for coincident current selection (b). The x, y, and z wires in this figure and the centers of the cores through which they pass all lie in the same plane.

The coincidence of a current pulse in the x_1- and y_2-lines would now occur at location (1, 2) in every z-plane. It is common practice that all the cores on each string along any intersection of an x- and y-plane are written into and interrogated simultaneously. These strings of cores are referred to as *words*. Any desired core(s) in the selected word are prevented from switching by putting a current pulse of opposite polarity down the z-line(s) of the appropriate z-plane(s). This pulse, known as the *inhibit pulse*, makes it possible to store the desired pattern of "ones" and "zeros" in the word during the write operation. The read operation requires no inhibit pulse since all the cores along the selected word line are interrogated simultaneously. The output signal of each core in the word is detected by the sense amplifier at the end of the corresponding z-plane wire.

Commercial memory arrays are, of course, much larger. For example, the basic operating module for the memory used in the IBM 360/65 contains 589,824 cores with 0.35 mm i.d. $\times$ 0.53 mm o.d. which are wired in 36 square planes, each with 128 $\times$ 128 cores. Each x- and y-drive line passes through 128 $\times$ 36 cores, while the job of driving and sensing the 128 $\times$ 128 cores in each plane is actually divided among four drivers and four sense amplifiers. A full read-write cycle requires 0.75 μs.

Faster operation could be accomplished by reducing the number of cores on each drive and sense line and/or by using more sophisticated drive and sense circuitry; however, both of these would increase the cost per storage element. The fastest full-sized experimental ferrite core memory reported to date has a read-write cycle of 110 ns (0.11 μs) and uses cores of 0.18 mm i.d. $\times$ 0.30 mm o.d.* Cycle times under 100 ns have also been obtained in memories which employ a "soft" magnetic metal, permalloy, electroplated onto the surfaces of the drive lines or vacuum deposited as flat magnetic films with strip lines as drivers.†

An important characteristic of ferrite cores is illustrated by the S-curves in Fig. 11–21. A single drive wire is strung through a core; and after the core is written into the "one" state, it is pulsed to the zero state by a single pulse of variable width and magnitude. If all the material in the core was reversed from $-\mathcal{M}_s$ to $+\mathcal{M}_s$, the flux reversal in terms of $\nabla \mathcal{M}/\mathcal{M}_s$ would be 2. Fig. 11–21 shows that 0.7 A is required to switch 75% of the material if applied for 500 ns while 0.9 A is needed if the pulse is shortened to 100 ns. Furthermore, due to inhomogeneities in the material, more than 3 A is required to fully switch the core even with pulses of 500 ns or longer.

A simple model of the reversal process requires that the first contribution to flux reversal (about 0.45 A in Fig. 11–21) occurs when the applied field equals the coercive force H_c at a radius equal to half of the core's inside diameter and that the knee of the curve (about 0.67 A) occurs when the applied field reaches H_c at the outer surface of the core. The reader can quickly verify this concept using Ampere's law, Eq. (9–15), to find that $H_c \cong 400$ A m^{-1}.

---

* G. E. Werner and R. M. Whalen, "A 110-Nanosecond Ferrite Memory," *IEEE Trans. on Magnetics* **2**, 3, p. 584 (1966).

† G. Kohn *et al.*, "A Very-High-Speed, Nondestructive-Read Magnetic Film Memory," *IBM Journal Res. and Develop.* **11**, 162 (1967); and E. W. Pugh *et al.*, "Device and Array Design for a 120-Nanosecond Magnetic Film Main Memory," *IBM Journal Res. and Develop.* **11**, 169 (1967).

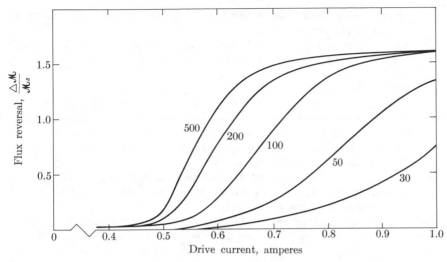

Fig. 11–21 Characteristic S-Curves for ferrite cores with drive current pulses 500 to 30 nsec long. The core is 0.35 mm i.d. by 0.53 mm o.d. and 0.10 mm high and is used in the IBM 360/65 computer. It is made of Fe_3O_4 with a small amount of copper and manganese added to improve performance. $\mathcal{M}_s = 0.375$ Wb m^{-2} (Data courtesy of the IBM Corporation.)

In performing this calculation one may well ponder what happens if the wire is off center or passes through the core at an angle, as indeed it does in practice. A few qualitative considerations quickly reveal that, no matter how far off center the wire is, the core *must* reverse in concentric cylindrical sections otherwise very large demagnetizing fields would be set up in excess of the drive field. It must, therefore, switch from the inside radius to the outside radius and Ampere's law may be applied as though the wire passes through the center of the core.

Since it is desired that the whole core switch in a full-select pulse while none of the core switch with a half-select pulse, it appears that the o.d. should always be made less than twice the i.d. This is indeed so, and this fabrication rule is not violated except in very complex *partial switching* modes of operation which, so far, are of limited use and will not be discussed here.

PROBLEMS

11–1 Show that the first term on the right-hand side of Eq. (11–1) reduces to $P = IV$ in a simple parallel plate capacitor for which $C = \epsilon A/d$.

11–2 Show that the second term on the right-hand side of Eq. (11–1) reduces to $P = IV$ in a simple toroid as illustrated in Fig. 9–10 for which inductance $L = \mu N^2 A/l$.

11–3 Show that the third term on the right-hand side of Eq. (11–1) reduces to $P = I^2R$ in a simple series element.

11–4 An inductor of reactance 10 Ω and a capacitor of reactance 25 Ω (both measured at 60 cps) are connected in series with a 10-Ω resistor across a 100-V, 60-cycle ac line.

a) Find the voltage across each part of the circuit.

b) What are the expressions for the instantaneous line voltage and line current?

11–5

a) A pure resistance and a pure inductance are in series across a 100-V ac line. An ac voltmeter gives the same reading whether connected across the resistance or the inductance. What does it read?

b) The magnitudes of the resistance and inductance in part (a) are altered so that a voltmeter across the inductance reads 50 V. What will the voltmeter read when connected across the resistance?

11–6 The rms terminal voltage of an ac generator is 100 V and the so-called angular frequency $\omega = 2\pi f$ is 500 rad/s. In series across the generator are a 3Ω resistor, a 50 μF capacitor, and an inductor whose inductance can be varied from 10 to 80 mH. The peak voltage ($V_{\max}$) across the capacitor should not exceed 1200 V.

a) What is the maximum allowable current in the series circuit?

b) To what value can the inductance be safely increased?

11–7 The reactances of an inductor and a capacitor are equal at a frequency f_0. Show in the same diagram the reactances of the inductor and the capacitor as a function of frequency, over a frequency range from zero to $2f_0$. Show also the reactance of the two in series.

11–8 An ideal ac voltage generator is connected to terminals a and b of the simple series element in Fig. 11–1. In terms of R, L, and C, at what frequency will the voltage be maximum (a) across the inductor, and (b) across the capacitor?

11–9 A circuit draws 330 W from a 110-V 60-cycle ac line. The power factor is 0.6 and the current lags the voltage.

a) Find the capacitance of the series capacitor that will result in a power factor of unity.

b) What power will then be drawn from the supply line?

11–10 When switch S in Fig. 11–22 is closed, the steady current in the battery is 10 A.

a) Explain how oscillations originate in the L-C circuit when the switch is opened.

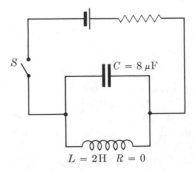

$L = 2\text{H}$ $R = 0$ **Figure 11-22**

b) What is the frequency of the oscillations?

c) What is the maximum potential difference across the capacitor?

11–11 A 0.01 μF capacitor, a 0.1 H inductor whose resistance is 1000 Ω, and a switch are connected in a series circuit. The capacitor is initially charged to a potential difference of 400 V. The switch is then closed.

a) Find the ratio of the oscillation frequency of the circuit to the frequency if the resistance were zero.

b) How much energy is converted to heat in the first complete cycle?

c) How much energy is converted to heat in the complete train of oscillations?

d) What is the maximum additional resistance that can be inserted in the circuit before the discharge becomes nonoscillatory?

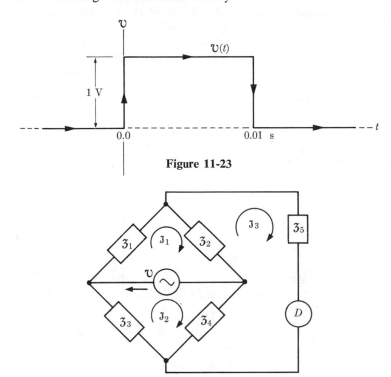

Figure 11-23

Figure 11-24

11–12 A resistance R and a capacitance C in series are connected across the terminals of an ideal voltage generator which produces the rectangular voltage pulse shown in Fig. 11–23. Take $R = 10^6$ Ω and $C = 0.01$ μF, and derive expressions giving $q(t)$ and $i(t)$ (a) during the period $0 < t < 0.01$ s, and (b) when t exceeds 0.01 s. (c) Plot curves of $q(t)$ and of $i(t)$ versus t.

11–13 For the calculation of the detector current in an ac Wheatstone bridge it is desirable to draw the network as it is shown in Fig. 11–24. Since the expression $N - J + 1 = 6 - 4 + 1 = 3$ shows that three unknowns must be considered, the complex currents $\mathfrak{I}_1$, $\mathfrak{I}_2$, and $\mathfrak{I}_3$ can be chosen to circulate in the three loops as shown in the figure.

a) Use the routine method of solution to show that the current through the detector is given by

$$\mathfrak{I}_3 = \mathfrak{V}(\mathfrak{Z}_2\mathfrak{Z}_3 - \mathfrak{Z}_1\mathfrak{Z}_4)/\Delta,$$

where the denominator is

$$\Delta = \mathfrak{Z}_1\mathfrak{Z}_2\mathfrak{Z}_3 + \mathfrak{Z}_1\mathfrak{Z}_2\mathfrak{Z}_4 + \mathfrak{Z}_1\mathfrak{Z}_3\mathfrak{Z}_4 + \mathfrak{Z}_2\mathfrak{Z}_3\mathfrak{Z}_4 + \mathfrak{Z}_1\mathfrak{Z}_3\mathfrak{Z}_5 + \mathfrak{Z}_1\mathfrak{Z}_4\mathfrak{Z}_5 + \mathfrak{Z}_2\mathfrak{Z}_3\mathfrak{Z}_5 + \mathfrak{Z}_2\mathfrak{Z}_4\mathfrak{Z}_5.$$

b) Suppose that $\mathfrak{Z}_4$ is the unknown impedance that is to be determined and that $\mathfrak{Z}_4 = \delta\mathfrak{Z}$, where $\delta \simeq 1$ provides for the inevitable error in the determination of $\mathfrak{Z}_4$ due to the lack of sensitivity in the detector. Obtain an expression for δ in terms of the smallest current $|\mathfrak{I}_m|$ that is observable in the detector, assuming that each of the remaining $\mathfrak{Z}$'s is exactly equal to $\mathfrak{Z}$.

c) Make a plot in the complex plane showing those values of δ that cannot be detected, if $|\mathfrak{I}_m| = 10\ \mu A$, $\mathfrak{Z} = (300 + 400)j\ \Omega$, and $|\mathfrak{V}| = 125$ V.

11–14 Prove the superposition theorem for a complex network containing at least two generators.

11–15 Assume the validity of Thévenin's theorem and prove Norton's theorem. Show that the impedance in parallel with the ideal current generator must be the same as that in series with the ideal voltage generator. What is the relationship between the generator voltage for Thévenin's theorem and the generator current for Norton's theorem?

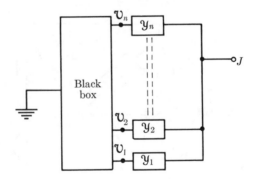

Figure 11-25

11–16 Occasionally another theorem named after Millman is useful. Millman's theorem gives the steady-state voltage at the junction J in Fig. 11–25 as

$$\mathfrak{V}_J = \frac{\sum_1^n \mathfrak{V}_i \mathfrak{Y}_i}{\sum_1^n \mathfrak{Y}_i},$$

where all of the $\mathfrak{V}$'s are measured from ground potential and where the "black box" may contain any combination of linear elements and generators. Prove this theorem.

11–17 The series element shown in Fig. 11–1 is connected across a storage battery that has the emf V_B and a negligible internal resistance. Assuming that a switch provided to complete the circuit is closed at $t = 0$, calculate the charge q on the capacitor as a function of time (a) when $R/2 < \sqrt{L/C}$, and (b) when $R/2 > \sqrt{L/C}$.

11–18 The ac generator in Fig. 11–7 is replaced by a storage battery that has negligible internal resistance and the constant emf $\mathcal{E}_B$. A switch has been placed in series with the battery for completing the circuit. If this switch is closed at $t = 0$, calculate the current i through the inductor as a function of time.

11–19 An ac generator, in series with a switch for completing the circuit, is connected across the series element of Fig. 11–1. Assume that the generator has negligible impedance and produces a sinusoidal emf with the amplitude V_m and the frequency $f = \omega/2\pi$. Assuming also that $R/2 < \sqrt{L/C}$, calculate the transient current in the element as a function of time, when the switch is closed at two different instants: (a) when the generated emf is maximum, and (b) when the generated emf is zero.

11–20 A sinusoidal potential difference of constant amplitude is applied across the series element of Fig. 11–1 at various frequencies from 0 to 500 cps. The magnitudes of L and C are adjusted so that at 250 cps $X_L = X_C$. The effective potential difference is 100 V and $R = 10\ \Omega$. Plot the effective current I versus the frequency for the two different cases in which X_L at 250 cps has values such that (a) $Q = X_L/R = 3$, and (b) $Q = X_L/R = 10$.

11–21 Three wires that are connected to a common junction in an ac network carry the currents $\mathfrak{I}_1$, $\mathfrak{I}_2$, and $\mathfrak{I}_3$ away from this junction.
a) When the effective values of these currents are 7 A, 5 A, and 3 A, respectively, determine graphically the phase angles of $\mathfrak{I}_2$ and $\mathfrak{I}_3$ with respect to $\mathfrak{I}_1$.
b) From your graphical solution draw a diagram like that in Fig. 11–4(b) to indicate the time-dependent values for these currents. Take the instant of time when $\omega t = 70°$ and show clearly the instantaneous values of the currents in each wire at this instant.

11–22 Two ac generators are connected in parallel to feed a single load having the complex impedance $\mathfrak{Z}$. The two generators have different internal impedances $\mathfrak{Z}_1$ and $\mathfrak{Z}_2$, and different emf's $\mathcal{V}_1$ and $\mathcal{V}_2$, which are not in phase, although they have the same frequency.
a) Calculate the current $\mathfrak{I}$ through the load and $\mathfrak{I}_1$ through generator number 1.
b) Determine the numerical values for $\mathfrak{I}$ and $\mathfrak{I}_1$, for $\mathcal{V}_1 = 120$ V, $\mathcal{V}_2 = 120e^{j\pi/2}$ V, $\mathfrak{Z}_1 = 1\ \Omega$, $\mathfrak{Z}_2 = 2\ \Omega$, and $\mathfrak{Z} = (3 + 4j)\ \Omega$.
c) What fraction of the power lost in $\mathfrak{Z}$ is generated by generator number 1?

11–23 Three terminals of a three-phase power line have sinusoidal potentials, with respect to a grounded neutral terminal, of $\mathcal{V}_1$, $\mathcal{V}_2$, and $\mathcal{V}_3$. A junction J is *not grounded*, but is connected to terminal 1 through an impedance $\mathfrak{Z}_1$, to terminal 2 through an impedance $\mathfrak{Z}_2$, and to terminal 3 through an impedance $\mathfrak{Z}_3$.
a) Calculate the current $\mathfrak{I}_1$ through $\mathfrak{Z}_1$, assuming that its positive direction is from terminal 1 toward J, and show that

$$\mathfrak{I}_1 = \frac{(\mathfrak{Z}_2 + \mathfrak{Z}_3)\mathcal{V}_1 - (\mathfrak{Z}_2\mathcal{V}_3 + \mathfrak{Z}_3\mathcal{V}_2)}{\mathfrak{Z}_1\mathfrak{Z}_2 + \mathfrak{Z}_1\mathfrak{Z}_3 + \mathfrak{Z}_2\mathfrak{Z}_3}.$$

b) Calculate the potential $\mathcal{V}_J$ of the junction J.

c) Obtain the numerical value of $\mathfrak{V}_J$ for

$$\mathfrak{Z}_1 = 10 \ \Omega, \ \mathfrak{Z}_2 = 11 \ \Omega, \ \mathfrak{Z}_3 = 9 \ \Omega,$$
$$\mathfrak{V}_1 = 100e^{j2\pi/3} \ \mathrm{V}, \ \mathfrak{V}_2 = 100 \ \mathrm{V}, \ \mathfrak{V}_3 = 100e^{-j2\pi/3} \ \mathrm{V}.$$

11–24 In the parallel circuit shown in Fig. 11–7 $X_L = X_C$ when the frequency is given by $f_0 = 1/2\pi\sqrt{1/LC}$, but this frequency does not make the generator current a minimum. Determine the frequency f_r for which the generator current is a minimum.

11–25 A certain ac electric clock, designed for 110 V and 60 cps, consists of an electromagnet with a toothed steel armature rotating between its poles. While the clock is running, measurements show that $X = \omega L = 3220 \ \Omega$, $R_{\mathrm{ac}} = 1760 \ \Omega$, and $R_{\mathrm{dc}} = 760 \ \Omega$, but while its armature is fastened to prevent rotation, measurements show that $X = \omega L = 3330 \ \Omega$, $R_{\mathrm{ac}} = 1600 \ \Omega$, and $R_{\mathrm{dc}} = 760 \ \Omega$. Assuming that the power losses in the iron core and in the armature are the same whether or not the armature rotates, calculate (a) the power lost in heating the coil, (b) the power lost in heating the core and armature, and (c) the power used in driving the armature.

11–26 Clocks like the one described in Problem 11–25 make excellent laboratory stopwatches when they are provided with suitable on-off switches. Where these clocks are to be used by students and 110 dc as well as 110 ac outlets are available, they should be protected with capacitors connected in series to avoid burning of the wire windings when a dc outlet is accidentally selected. However, these clocks will not run on less than 0.024 A and will burn out if the heat loss increases more than 40% over the normal heat loss.

a) What capacitance, connected in series with the clock, will leave the current unchanged?

b) If the only capacitor available has twice the capacitance calculated in (a), what resistance should be added to the system to keep the current constant?

c) How does the power consumption of the original clock compare with the power consumption of the clock as modified in parts (a) and (b) of this problem?

11–27 The armature of a certain induction motor is a copper "squirrel cage" with no external electric connections. The two poles of the electromagnet have windings connected in series, with a total dc resistance of 0.5 Ω, across a 200-V 60-cps ac line. Running with no load, this motor draws only 5 A. However, when it is delivering 3200 W of mechanical power, it draws 20 A and a wattmeter shows that the electric power being absorbed is 3600 W. Find the effective ac resistance and the effective inductive reactance (a) when the motor is running with no load, and (b) when the motor is driving the 3200-W mechanical load. [*Note.* It can be assumed that all the heat losses in the motor are proportional to I^2.]

11–28 Replace the load R in Fig. 11–17 with a load of $\mathfrak{Z} = R + jX$ and show that the phase angle between the current and voltage of the primary is approximately $\tan^{-1} X/R$, with an ideal transformer in which $M^2 = L_1L_2$, $R \gg R_s$, and $\omega L_1/R_1 = \omega L_2/R_s \gg 1$ and $X_2 \gg |Z|$.

11–29 The full-wave rectifying circuit of Fig. 11–14 is used in a certain experiment to produce a direct current of 2.00 A through a device having a pure resistance of 100 Ω. The maximum potential of each phase to ground is just 120 V.

a) With the device connected as is the load in Fig. 11–14, calculate the maximum, minimum, and average currents through the device.

b) Calculate the resistance required to reduce the average current to just 2.00 A.

c) Assume that half of the resistance calculated in (b) will be in a control rheostat while the other half will be in a circuit like that of Fig. 11–7 for filtering out the ripple. Further, assume that any available inductor will have a Q of 25 on 60 cycles and that the ripple will be sinusoidal with an amplitude that is half the difference between the maximum and the minimum of the curve summing the pulses in Fig. 11–14. Calculate the capacitance needed to make the two reactances of the filter equal for the ripple frequency.

d) Calculate the amplitude of the current ripple that will pass through the filter as designed in (c). Compare the amplitude of the ripple passing through the filter with the magnitude of the steady current by calculating the ratio of the amplitude to the dc magnitude.

11–30 Most transformers have closed paths for the magnetic flux inside iron or other ferromagnetic materials. To simplify construction these paths are made rectangular or some other shape that has relatively sharp angles. Calculations for such shapes must take into account the fact that not all the flux is confined to the iron paths. The toroid ring designed by Rowland for precise measurements of magnetic properties is easier to calculate though harder to construct. A special alloy (30% Fe, 25% Co, 45% Ni) called *perminvar* has been developed to have a constant permeability of $\mu = 300\,\mu_0$, with negligible hysteresis up to $H = 240$ A m^{-1}. If the core in Fig. 11–18 is made up of many fine wires of this material, insulated to prevent eddy currents, an ideal transformer with negligible core losses can be constructed. In winding insulated copper wire around this core the space factor k, defined in Problem 6–19, can be defined as the ratio of the total copper cross section to the minimum cross section of the coil when this is taken perpendicular to the wires. Assuming that k is 0.7,

a) calculate the self-inductance L_1 and the resistance R_1 for a coil of N_1 turns whose minimum cross section lies between the radii 18 cm and 17 cm, and

b) calculate the Q of this coil for a frequency of 60 cps.

c) Calculate the number of turns N_1 that must be used in this coil to avoid exceeding the limitation that $H < 240$ A m^{-1} when the coil is connected directly to a 110-V, 60-cps line.

d) Calculate the Q of a secondary coil of N_2 turns whose minimum cross section lies between the radii 17 cm and 16 cm, using the frequency of 60 cps.

11–31 When the ferromagnetic bundle of wires in Fig. 11–18 is made with the kind or iron that was used in obtaining the hysteresis loop of Fig. 9–8, a sinusoidal potential difference across the coil winding does not produce a truly sinusoidal current in the winding. The nonlinear magnetization curve produces a variable reactance that distorts the current *vs.* time curve. The problem is complicated further by the fact that magnetization curves vary with the frequency at which they are measured and with the maximum current in the coils. In spite of these difficulties, the curve of Fig. 9–8 can be used to obtain fair approximations for special cases, where the current maxima produce maximum fields of $H_{max} \cong 240$ A m^{-1} and the frequency is 60 cps or less. For this approximation a constant permeability should be assumed, and this may be chosen as $\mu = B_{max}/H_{max}$ as taken from the curve of Fig. 9–8.

a) Calculate N_1 for a primary coil that can be connected directly across a 110-V and 60-cps line to produce $H_{max} = 240$ A m^{-1}.

b) Calculate the dc resistance of this coil.

c) Estimate the power loss due to the hysteresis loop plotted in Fig. 9–8 and determine the resistance that this power loss adds to the dc resistance of the primary coil.

d) Determine the ac resistance of the primary coil, assuming there is no secondary circuit.

11–32 A long, straight coaxial cable consists of three thin cylindrical shells of copper separated by air. The two inner cylinders, of radii a and $b > a$, are electrically joined at the two ends to constitute one of the conductors in a two-conductor transmission line. The outer cylinder, of radius $c > b$, constitutes the return path. When direct current is used for transmission the outwardly directed current is equally divided between the two inner cylinders. When alternating current is used for transmission the total effective current $I = I_a + I_b$ is maintained constant, independent of the frequency $f = \omega/2\pi$.

a) Calculate the ratio I_b/I_a as a function of the frequency, where I_a and I_b are the effective currents in the cylinders of radii a and b, respectively. Assume that the resistances per meter of length are R_l, R_l, and $R_l/2$ for the cylinders with the radii a, b, and c, respectively.

b) With dc transmission the power lost in the two inner cylinders is $P_d = (I/2)^2(R_l + R_l) = I^2 R_l/2$. With ac transmission, calculate as a function of frequency the power P_a lost in these two cylinders.

c) Plot I_b/I_a versus f and P_a/P_d versus f, choosing inner cylinders having $a = 1$ cm with 0.04-cm thickness and $b = 2$ cm with 0.02-cm thickness. Use $c = 4$ cm and choose frequencies for plotting that make the inductive effects more important than the dc resistance.

11–33 The ferrite core used in the IBM 360/65 memory is 0.35 mm i.d., 0.53 mm o.d., and 0.10 mm thick. It is made of a copper-manganese ferrite for which $\mathcal{M}_s = 0.375$ Wb m^{-2}.

a) Assuming that 70% of the material participates in the flux reversal process, calculate the change in flux when the core is fully switched from a "zero" to a "one."

b) If the voltage output on the sense line resulting from this reversal looks like an isosceles triangle with the base 200 ns long and the height measured in volts, calculate the output signal in volt ns and peak output signal in mV.

c) Referring to Fig. 11–21, calculate the flux reversal if a 100 ns long pulse is applied of magnitude 0.7 A.

11–34 Assume that an experimental ferrite core of 0.18 mm i.d. and 0.30 mm o.d. has the same material properties as the 0.35 mm × 0.53 mm used in the 360/65 computer.

a) How much steady current must be sent down the line through the center to fully reverse it? (By fully reverse we mean switch the 70% which is easily reversed).

b) How much current will initiate a partial reversal?

11–35 Assume that the 0.35 mm × 0.53 mm core is 70% reversed with a 300 ns long current pulse of 0.7 A.

a) How much energy is absorbed by the hysteretic process of the core?

b) If the average power output of the current source is 80 W, what fraction of the drive current energy would be absorbed by the single core?

11–36 Explain why the ferrite core whose S-curve is illustrated in Fig. 11–21 could not be used in coincident current operation with current pulses shorter than 100 ns.

11–37 Assume that the ferrite core described in Fig. 11–20 was used in the circuit shown in Fig. 11–19 and wired so that $N_1 = N_2 = 2$.

a) Find the output voltage v_2 when a 1000-cps voltage v_p is applied with variations such that the current through the resistor R ranges from 0.7 to 0.8 A.

b) Even though this ferrite core provides regulation its hysteresis properties are not optimum for this application. Why?

Chapter 12 ELECTROMAGNETIC RADIATION

The publication of Maxwell's *A Dynamical Theory of Electromagnetic Field*, in 1865, unified the various laws of electricity then in existence. All known laws then could be obtained from five fundamental relations, namely, Maxwell's four field equations, Eqs. (M–1), (M–2), (M–3), and (M–4), and the relation for the force on a moving charge, Eq. (7–1).

In order to make these five relations consistent mathematically and consistent with the law of conservation of charge, Maxwell had introduced a displacement current. At that time, there were no experiments to justify this concept. However, with the introduction of the displacement current Maxwell's field equations predicted the existence of electromagnetic waves that would propagate through vacuum with a velocity equal to that of light. It was truly amazing that this predicted velocity of propagation, based entirely upon laboratory measurements of electric and magnetic quantities, could be so nearly equal to the measured velocity of light. As measuring accuracies have improved, the two types of measurements have yielded more and more nearly the same results, so that now the two velocities are considered to be identical.* The best value for this velocity is $c = 299{,}792 \pm 2$ km s^{-1}.†

Maxwell's equations also predicted that the velocities of electromagnetic waves through a dielectric would be inversely proportional to the square root of the dielectric constant. Therefore, if light consists of electromagnetic waves, the ratio between refractive indices of two media should equal the square root of the ratio of their dielectric constants. For many materials,

---

* A comparison of these results is given by Jenkins and White, *Fundamentals of Optics*, p. 393, McGraw-Hill, New York, 1959.

† An excellent review article by Mulligan and McDonald, *Am. J. Phys.* **25**, 180 (1957), summarizes measurements by Rank and Bennett, Bennett and Plyler, and Blaine and Conner.

such as melted paraffin and certain gases, this relation was fairly accurate. Subsequent researches have shown, however, that complete agreement with this theory is obtained only when the frequency of the waves is considered, since the index of refraction and the dielectric constant both depend upon the frequency used in the measurement. The validity of Maxwell's relations has been verified by so many thousands of experiments that dielectric constants at high frequencies are now commonly obtained from the refractive index, assuming Maxwell's relations to be exact.

Before the publication of Maxwell's theory, a theory for optical phenomena based upon elastic vibration in a kind of solid "aether" had been developed by Huygens, Fresnel, Young, Green, and others. As the studies progressed, the properties that must be assigned to this aether became more and more fantastic. The phenomena of polarized light could be explained only if the vibrations in the aether were transverse to the direction of propagation. This seemed to require enormous rigidity in a medium which also must have negligible density and must produce no noticeable drag on planetary motion. The theory that identified light waves as an electromagnetic phenomenon provided an ideal solution for these anomalies.

Maxwell recognized that electromagnetic waves longer than those of visible light should exist, but he doubted that they could be detected. However, between 1885 and 1889, Hertz set up oscillating electric circuits for producing and detecting these waves. He measured their velocity and wavelength and showed that they could be refracted, reflected, diffracted, and polarized much like ordinary light. Hertz's experiments verified the theoretical work of Maxwell. Practical applications followed soon. By 1895 Marconi had demonstrated that these electromagnetic waves could be used for commercial telegraphy, to transmit messages across the Atlantic Ocean without wires.

Modern research has detected and used electromagnetic radiation over a wide range of wavelengths. Approximate ranges of some of these are indicated in Fig. 12–1. It is interesting to note the small portion of this range that is visible to the human eye.

12–1 Wave equations for nonconducting isotropic media

Electromagnetic waves start from regions where there are charges and currents, can pass through regions where there are none, and finally end in regions where again there are charges and currents. The phenomena of wave propagation are simplest in those regions where there are no real charges or currents. The perfectly general case would involve the propagation of waves through regions that are neither isotropic nor homogeneous, but such complexities* should be postponed for later consideration. Here the study will

---

* See Chapter XIX in *Theoretical Physics* by Georg Joos, G. E. Stechert, New York, 1934.

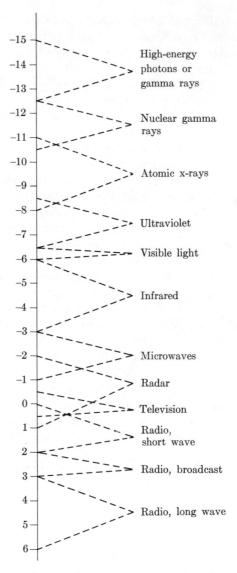

Fig. 12–1 Some names given to important wavelength regions in the continuous spectrum of electromagnetic radiation. The numbers on the vertical axis give the $\log_{10}$ of the wavelength in meters.

be confined to regions where μ and ϵ can be considered constants, which includes all regions in which the media are homogeneous, isotropic, and have no hysteresis effects; the perfect vacuum is simply a special case.

It will be assumed, then, that $\mu = K_m \mu_0$ and $\epsilon = K \epsilon_0$, where K_m and K are true constants for the media, and that $\rho = 0$ and $\mathbf{J} = 0$. Maxwell's field equations then become

$$\mathbf{\nabla} \cdot \mathbf{E} = 0, \tag{12–1}$$

$$\mathbf{\nabla} \cdot \mathbf{H} = 0 \tag{12–2}$$

$$\mathbf{\nabla} \times \mathbf{E} = -\mu\dot{\mathbf{H}}, \tag{12–3}$$

$$\mathbf{\nabla} \times \mathbf{H} = \epsilon\dot{\mathbf{E}}. \tag{12–4}$$

From these four equations in two unknowns separate differential equations for $\mathbf{E}$ and for $\mathbf{H}$ can be obtained. Equations (12–1) and (12–2) provide no relation between $\mathbf{E}$ and $\mathbf{H}$ and no indication of their time-dependence, but Eqs. (12–3) and (12–4) each contain both $\mathbf{E}$ and $\mathbf{H}$ and provide a time-dependent relation. Either $\mathbf{E}$ or $\mathbf{H}$ can be eliminated from these two relations to obtain a single differential equation with time-dependence. By taking the curl of (12–3) and the time derivative of (12–4), $\mathbf{H}$ can be eliminated to provide the separate differential equation for $\mathbf{E}$:

$$\mathbf{\nabla} \times (\mathbf{\nabla} \times \mathbf{E}) = -\mu\mathbf{\nabla} \times \dot{\mathbf{H}} = -\mu\epsilon\ddot{\mathbf{E}},$$

$$\mathbf{\nabla} \times (\mathbf{\nabla} \times \mathbf{E}) = -\mu\epsilon\ddot{\mathbf{E}}.$$

Similarly, $\mathbf{E}$ can be eliminated from (12–3) and (12–4) by taking the curl of (12–4) and the time derivative of (12–3) to obtain $\mathbf{\nabla} \times (\mathbf{\nabla} \times \mathbf{H}) = -\mu\epsilon\ddot{\mathbf{H}}$.

These expressions can be rendered more useful by using the mathematical identity of Eq. (3–24) to obtain

$$\mathbf{\nabla} \times (\mathbf{\nabla} \times \mathbf{E}) = \mathbf{\nabla}(\mathbf{\nabla} \cdot \mathbf{E}) - \nabla^2\mathbf{E} = -\mu\epsilon\ddot{\mathbf{E}},$$

and the introduction of Eq. (12–1) reduces this to $\nabla^2\mathbf{E} = \mu\epsilon\ddot{\mathbf{E}}$, which is a wave equation. Similarly, by using Eq. (12–2) $\nabla^2\mathbf{H} = \mu\epsilon\ddot{\mathbf{H}}$ may be obtained. To see that these are indeed wave equations, consider the special case of a plane wave being propagated along the z-axis. Such a plane wave must be independent of the x- and y-coordinates, so that

$$\frac{\partial \mathbf{E}}{\partial x} = \frac{\partial \mathbf{E}}{\partial y} = 0 \quad \text{and} \quad \frac{\partial \mathbf{H}}{\partial x} = \frac{\partial \mathbf{H}}{\partial y} = 0.$$

The equations then become

$$\frac{\partial^2 \mathbf{E}}{\partial z^2} = \mu\epsilon\ddot{\mathbf{E}} \tag{12–5a}$$

and

$$\frac{\partial^2 \mathbf{H}}{\partial z^2} = \mu\epsilon\ddot{\mathbf{H}}, \tag{12–5b}$$

each of which is a second-order differential equation with two solutions. The two arbitrary vector functions $\mathbf{E} = \mathbf{f}(t - z/v)$ and $\mathbf{E} = \mathbf{F}(t + z/v)$ are both solutions of Eq. (12–5a), as can be demonstrated by direct substitution. Figure 12–2 shows an arbitrarily shaped function of $(t - z/v)$ plotted at

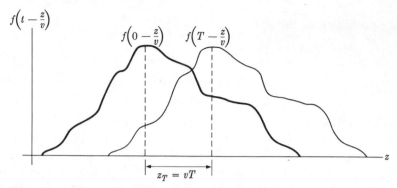

Fig. 12–2 This figure illustrates how a general function of the form $f(t - z/v)$ represents plane waves propagated in the z-direction with velocity v.

time $t = 0$ and again at time $t = T$. It is apparent that in the time T the wave has shifted along the z-axis a distance z_T such that $T = z_T/v$. Such a function, which is independent of x and y, provides a good example of a plane wave of any shape moving with velocity v in the positive z-direction. Similarly, any $\mathbf{F}(t + z/v)$ represents a plane wave moving in the negative z-direction. Since any information that can be obtained from waves propagated in the negative z-direction is as easily obtained from waves propagated in the positive z-direction, the functions represented by $\mathbf{F}(t + z/v)$ will be ignored. For waves moving in the positive direction it is desirable to define

$$\mathbf{f}' = \frac{\partial \mathbf{f}}{\partial(t - z/v)} \quad \text{and} \quad \mathbf{f}'' = \frac{\partial^2 \mathbf{f}}{\partial(t - z/v)^2}.$$

Using these definitions, we can write the following partial derivatives:

$$\frac{\partial \mathbf{E}}{\partial z} = -\frac{1}{v}\mathbf{f}', \qquad \frac{\partial \mathbf{E}}{\partial t} = \mathbf{f}',$$

$$\frac{\partial^2 \mathbf{E}}{\partial z^2} = \frac{1}{v^2}\mathbf{f}'', \qquad \frac{\partial^2 \mathbf{E}}{\partial t^2} = \mathbf{f}'',$$

which satisfy (12–5a) if $\mu\epsilon v^2 = 1$ or if

$$v = \frac{1}{\sqrt{\mu\epsilon}} = \frac{1}{\sqrt{K_m K}\sqrt{\mu_0\epsilon_0}} = \frac{c}{\sqrt{K_m K}},$$

where the constant c satisfies the relation $\mu_0\epsilon_0 c^2 = 1$. An identical treatment for magnetic waves yields these same relations when we use Eq. (12–5b). Very careful experiments have shown that the c thus defined is accurately equal* to the velocity of light in a vacuum. In most media the product

---

* Jenkins and White, *Fundamentals of Optics*, p. 393, McGraw-Hill, New York, 1957.

Table 12–1

INDEX OF REFRACTION OF SOME GASES COMPARED WITH THE
SQUARE ROOT OF THEIR DIELECTRIC CONSTANTS

Gas	$\sqrt{K}$	Index
Hydrogen	1.000132	1.000138–1.000142
Air	1.000295	1.000293
Carbon monoxide	1.000350	1.000335–1.000340
Carbon dioxide	1.000492	1.000448–1.000454
Nitrous oxide	1.000565	1.000516

KK_m is greater than one and the wave propagates at a velocity less than c.
Since experiments show that K_m is approximately one when the frequency
exceeds $c/0.003$ m, visible light should have the velocity $v = c/\sqrt{K}$ and the
index of refraction should be $\sqrt{K}$. Table 12–1 compares, for some gases, the
$\sqrt{K}$ measured at frequencies below 3×10^6 Hz with the index of refraction
for visible light ($\sim 5 \times 10^{10}$ Hz) traveling through these same gases. Since
both of these quantities vary with the frequency* at which they are measured,
the closeness of the agreement is surprising. Similar tables compiled for
solids or liquids do not show such agreements, because the variations with
frequency are more pronounced in these substances.

12–2 Relationships between electric and magnetic fields in plane waves

To investigate relationships for **E** and **H** in plane waves, Maxwell's
first equation can be used in the form

$$\frac{\partial E_x}{\partial x} + \frac{\partial E_y}{\partial y} + \frac{\partial E_z}{\partial z} = 0.$$

Since with plane waves propagating in the z-direction $\partial \mathbf{E}/\partial x = \partial \mathbf{E}/\partial y = 0$,
the derivative of each component of **E** with respect to x or y is separately
equal to zero, and therefore $\partial E_z/\partial z = 0$. This shows that E_z is either constant or zero, but the possibility that E_z may be constant merely means that a
static field may exist, which is of no interest in our study of waves. Therefore,
when plane electromagnetic waves are propagated along the z-axis, $E_z = 0$
and the waves are transverse to the direction of propagation. A similar
argument shows that $H_z = 0$ also, and both of these field vectors are in a
plane perpendicular to the direction of propagation of the wave.

---

* The International Committee on Weights and Measures recommends hertz,
abbreviated Hz, for cycles per second.

The quantitative relationship between **E** and **H** can be obtained by means of (12–3) as follows:

$$\mathbf{\nabla} \times \mathbf{E} = \mathbf{i}\left(\frac{\partial E_z}{\partial y} - \frac{\partial E_y}{\partial z}\right) + \mathbf{j}\left(\frac{\partial E_x}{\partial z} - \frac{\partial E_z}{\partial x}\right) + \mathbf{k}\left(\frac{\partial E_y}{\partial x} - \frac{\partial E_x}{\partial y}\right) = -\mu\dot{\mathbf{H}}$$

and, since $\partial E/\partial x = 0 = \partial E/\partial y$,

$$-\mathbf{i}\frac{\partial E_y}{\partial z} + \mathbf{j}\frac{\partial E_x}{\partial z} = -\mu(\mathbf{i}\dot{H}_x + \mathbf{j}\dot{H}_y + \mathbf{k}\dot{H}_z).$$

Equating components gives

$$- \frac{\partial E_y}{\partial z} = -\mu\dot{H}_x \tag{12–6}$$

and

$$\frac{\partial E_x}{\partial z} = -\mu\dot{H}_y, \tag{12–7}$$

which provide quantitative relationships between **E** and **H**. The following arbitrary functions of $(t - z/v)$ can be chosen for the nonzero components of **E** and **H**:

$$E_x = f\left(t - \frac{z}{v}\right), \qquad E_y = g\left(t - \frac{z}{v}\right),$$

$$H_x = m\left(t - \frac{z}{v}\right), \qquad H_y = n\left(t - \frac{z}{v}\right).$$

Substitution of these functions into Eq. (12–6) gives $(1/v)g' = -\mu m'$ or $g = -\mu v m$, provided that the constant of integration, which has no importance in wave propagation, is neglected. Likewise, Eq. (12–7) gives $-(1/v)f' = -\mu n'$ or $f = \mu v n$. Therefore, since $v = 1/\sqrt{\mu\epsilon}$ and $\mu v = \sqrt{\mu/\epsilon}$, then $g = -\sqrt{\mu/\epsilon}\, m$ and $f = \sqrt{\mu/\epsilon}\, n$, which shows that at any fixed time and fixed location in space the magnetic field is proportional to the electric field and so is in phase with it. Easily remembered expressions for this proportionality are

$$B_y = \frac{E_x}{v} \qquad \text{and} \qquad B_x = -\frac{E_y}{v}. \tag{12–8}$$

Furthermore, since

$$\mathbf{E} \cdot \mathbf{H} = fm + gn = \left(\sqrt{\frac{\mu}{\epsilon}}\, n\right)m - \left(\sqrt{\frac{\mu}{\epsilon}}\, m\right)n = 0$$

the electric and magnetic vectors are at right angles to each other. *The foregoing relations apply only to plane waves.* The fields **E** and **H** are not necessarily perpendicular nor are they necessarily transverse to the direction of propagation in other forms of waves.

To further clarify these phenomena, consider the special case of a sinusoidal plane wave that is polarized in the x-direction. In such a wave, $E_y = 0$ and the x-component can be written as $E_x = E_{0x} \sin \omega(t - z/v)$. From Eq. (12–8), $H_x = 0$ and $H_y = (E_{0x}/\mu v) \sin \omega(t - z/v)$. These expressions are plotted at $t = 0$ in Fig. 12–3.

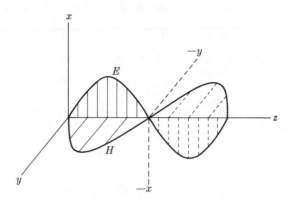

Fig. 12–3 A plane-polarized electromagnetic wave propagating in the z-direction.

12–3 Energy flow in electromagnetic waves

Sinusoidal plane waves have been investigated in the previous section and, for a polarized wave, values of **E** and **H** at $t = 0$ are plotted in Fig. 12–3. The power density propagated by such waves is, according to Section 10–1, given by the Poynting vector **S** = **E** × **H**. Hence

$$\mathbf{S} = EH\mathbf{k} = \sqrt{\frac{\epsilon}{\mu}}\, E_x^2 \mathbf{k} = v\epsilon E_x^2,$$

since $v^2 \epsilon \mu = 1$ and $\mathbf{v} = \mathbf{k}v$.

For the wave shown in Fig. 12–3,

$$\mathbf{S} = v\epsilon E_{0x}^2 \sin^2 \omega \left(t - \frac{z}{v} \right),$$

which pulsates with time at any z, or with z at any time. Its time average at any point in space is given by the integral over a full period from $\beta = 0$ to $\beta = 2\pi$;

$$\bar{\mathbf{S}} = \frac{v\epsilon}{2\pi} \int_0^{2\pi} E_{0x}^2 \sin^2 \beta \; d\beta,$$

where $\beta = \omega(t - z/v)$. Hence

$$\bar{\mathbf{S}} = \frac{\epsilon E_{0x}^2 \mathbf{v}}{2}$$

represents the time-average power transmitted through a unit area by a plane electromagnetic wave and is proportional to the square of the amplitude. While this expression has been derived for plane polarized light, it can readily be extended, since nonpolarized light consists of many plane polarized waves whose directions of polarization are randomly distributed.

It is instructive to compare the $\bar{\mathbf{S}}$ for these waves with the sum of the electric and magnetic energies in them. Since $B = E/v$, this sum is given by

$$U = (\mathbf{H} \cdot \mathbf{B} + \mathbf{E} \cdot \mathbf{D})/2 = \epsilon E_x^2.$$

With a sinusoidal wave the average energy is

$$\bar{U} = \epsilon E_{0x}^2/2.$$

Thus $\bar{\mathbf{S}} = \mathbf{v}\bar{U}$, which means that the energy in these waves travels with the velocity $\mathbf{v}$.

12–4 Laws of optics

If light consists of electromagnetic waves, it must be possible to derive the experimentally determined laws of reflection and refraction from the fundamental laws of electricity and magnetism. First, it will be necessary to determine the boundary conditions for electric and magnetic fields at the interface between two media.

Boundary conditions. It has previously been shown for static conditions that the tangential components of $\mathbf{H}$ and $\mathbf{E}$ and the normal components of $\mathbf{B}$ and $\mathbf{D}$ are continuous across the boundary. These conditions were derived by means of the equations $\nabla \times \mathbf{H} = 0$, $\nabla \times \mathbf{E} = 0$, $\nabla \cdot \mathbf{B} = 0$, and $\nabla \cdot \mathbf{D} = 0$ in Sections 5–5 and 9–4, assuming that no real charges existed on the boundary.

Boundary conditions for electromagnetic waves should be derived from the following more general forms of Maxwell's relations, which only assume ρ to be zero:

$$\nabla \cdot \mathbf{D} = 0, \qquad \nabla \cdot \mathbf{E} = 0,$$
$$\nabla \times \mathbf{E} = -\dot{\mathbf{B}}, \qquad \nabla \times \mathbf{H} = \mathbf{J} + \dot{\mathbf{D}}.$$

The divergence equations for $\mathbf{D}$ and $\mathbf{B}$ are the same as the static ones, so that the normal boundary conditions remain unchanged:

$$D_{1n} = D_{2n} \qquad \text{and} \qquad B_{1n} = B_{2n}. \tag{12–9}$$

The tangential boundary condition for $\mathbf{E}$ may be obtained by applying the integral form of Maxwell's third relation

$$\oint \mathbf{E} \cdot d\mathbf{l} = -\int_s \dot{\mathbf{B}} \cdot d\mathbf{s}$$

to the path illustrated in Fig. 5–6. Starting from the upper left corner and progressing counterclockwise, the left-hand side of the equation becomes

$$-E_{2n}a - E_{1n}a + E_{1t}b + E_{1n}a + E_{2n}a - E_{2t}b$$

and the right-hand side becomes

$$-\bar{\dot{B}}_n 2ab,$$

where $\bar{\dot{B}}_n$ is the average normal component of $\dot{\mathbf{B}}$ and $2ab$ is the area within the path. If the path is chosen very close to the boundary so that b is much greater than a, all terms containing a or ab may be neglected compared with terms containing b only. The left side of the equation then reduces to $E_{1t} - E_{2t}$, while the right side becomes zero. A similar treatment of the equation containing $\mathbf{H}$ yields the same result for H_t, and the tangential boundary conditions on $\mathbf{E}$ and $\mathbf{H}$ become

$$E_{1t} = E_{2t} \quad \text{and} \quad H_{1t} = H_{2t}, \tag{12–10}$$

which are identical with the static conditions.

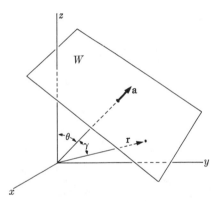

Fig. 12–4 A plane wavefront W, propagating in the **a**-direction. An arbitrary point on the wavefront is located by **r**.

Snell's law. Application of the above boundary conditions to a plane wave at an interface between two media leads to Snell's law. An expression for a plane wave traveling in an arbitrary direction is needed. A plane wavefront is represented in Fig. 12–4 by sheet W, where the propagation vector **a** is a unit vector perpendicular to the sheet. In such a plane wave, the magnitudes of the electric or magnetic fields are the same at every point on sheet W. This condition is satisfied by

$$\mathbf{E} = \mathbf{f}\left(t - \frac{\mathbf{a} \cdot \mathbf{r}}{v}\right) \quad \text{and} \quad \mathbf{H} = \mathbf{g}\left(t - \frac{\mathbf{a} \cdot \mathbf{r}}{v}\right),$$

where $\mathbf{a} \cdot \mathbf{r} = r \cos \gamma$ is just the perpendicular distance from the origin to

the wavefront and is thus independent of the location on the wave as selected
by $\mathbf{r}$. Just as $\mathbf{f}(t - z/v)$ represents a plane wave traveling in the z-direction,
so $\mathbf{f}(t - \mathbf{a} \cdot \mathbf{r}/v)$ represents a plane wave traveling in the $\mathbf{a}$-direction. In
fact, in the special case where $\mathbf{a}$ is chosen to be the unit vector in the z-
direction, $\mathbf{a} \cdot \mathbf{r}$ reduces to

$$\mathbf{k} \cdot (x\mathbf{i} + y\mathbf{j} + z\mathbf{k}) = z$$

and these equations reduce to the familiar

$$\mathbf{E} = \mathbf{f}\left(t - \frac{z}{v}\right) \quad \text{and} \quad \mathbf{H} = \mathbf{g}\left(t - \frac{z}{v}\right).$$

Snell's law may be demonstrated by assuming the space in Fig. 12–5
to consist of two regions containing different transparent dielectrics separated by an interface that coincides with the xy-plane. The velocity of light
in the two dielectrics may be v below the plane where z is negative, and v''
where z is positive. The directions of the incident, reflected, and refracted
rays are designated by the unit propagation vectors $\mathbf{a}$, $\mathbf{a}'$; and $\mathbf{a}''$, respectively.

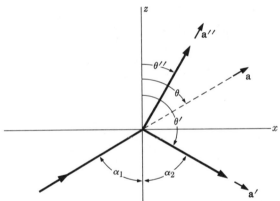

Fig. 12–5 Incident ray lying in the xz-plane impinges on the interface between two
media represented by xy-plane. Directions of incident, reflected, and refracted
waves are determined by the unit vectors $\mathbf{a}$, $\mathbf{a}'$, and $\mathbf{a}''$, respectively.

Since monochromatic, polarized light is composed of sinusoidally oscillating electric and magnetic fields, the electric field in the incident wave may
be represented by

$$\mathbf{E} = \mathbf{E}_0 \sin \omega\left(t - \frac{\mathbf{a} \cdot \mathbf{r}}{v}\right),$$

while in the reflected and refracted waves it will be represented by

$$\mathbf{E}' = \mathbf{E}_0' \sin \left\{\omega'\left(t - \frac{\mathbf{a}' \cdot \mathbf{r}}{v}\right) - \delta'\right\}$$

and

$$\mathbf{E}'' = \mathbf{E}_0'' \sin\left\{\omega''\left(t - \frac{\mathbf{a}'' \cdot \mathbf{r}}{v''}\right) - \delta''\right\}.$$

Primes have been added to the ω's to allow for possible changes in frequency, while δ' and δ'' allow for possible phase shifts. The boundary conditions require that the tangential components of the electric and magnetic vectors on one side of the boundary must equal those on the other side. Therefore, at $z = 0$, Eq. (12–10) yields

$$E_x + E'_x = E''_x, \qquad H_x + H'_x = H''_x,$$

and　　　　　　　　　　　　　　　　　　　　　　　　　　　　　　　(12–11)

$$E_y + E'_y = E''_y, \qquad H_y + H'_y = H''_y,$$

and the boundary conditions on **D** and **B** are not needed.

To apply the boundary conditions of Eq. (12–11), the propagation vectors should be expressed in cartesian coordinates. The general expression for a unit vector is

$$\mathbf{i} \sin\theta \cos\phi + \mathbf{j} \sin\theta \sin\phi + \mathbf{k} \cos\theta,$$

where θ and ϕ are the usual spherical coordinates. Without loss of generality, the unit propagation vector for the incident ray may lie wholly within the xz-plane, provided the reflected and refracted rays are free to make any arbitrary angles with respect to this plane.

The condition on the x-components of **E** thus results in

$$E_{0x} \sin\omega\left[t - \frac{x \sin\theta}{v}\right]$$

$$+ E'_{0x} \sin\left\{\omega'\left[t - \frac{x \sin\theta' \cos\phi' + y \sin\theta' \sin\phi'}{v}\right] - \delta'\right\}$$

$$= E''_{0x} \sin\left\{\omega''\left[t - \frac{x \sin\theta'' \cos\phi'' + y \sin\theta'' \sin\phi''}{v''}\right] - \delta''\right\}.$$

For the equations to be satisfied at all times and for all points in the xy-plane, the arguments of the sine functions must be identical (or different by a factor of π, since this affects only the sign of the amplitude). The δ's may, therefore, be 0 or π. The frequency is unchanged, since $\omega = \omega' = \omega''$. Because the sine function of the incident ray is independent of y, the other sine functions must also be independent of y. This requires that $\sin\phi' = \sin\phi'' = 0$. Thus the propagation vectors of the incident, reflected, and refracted rays all lie in the same plane, as illustrated in Fig. 12–5. (Note that requiring $\sin\theta' = \sin\theta'' = 0$ would also satisfy this condition, but this is not an acceptable choice because application to the x-terms would require $\sin\theta = 0$, which is true only if the incident beam strikes the surface perpendicularly.) The condition that the arguments have the same x-dependency requires that $\sin\theta/v = \sin\theta'/v$, which may be satisfied by $\theta = \theta'$ or $\theta = 180° - \theta'$. The latter expression will be selected, since the choice of $\mathbf{a}'$ as the unit propagation vector of the reflected ray requires θ' to be greater than $90°$. Examination of Fig. 12–5 reveals that $\theta = 180° - \theta'$ is the familiar law of reflection, more commonly written as $\alpha_1 = \alpha_2$, where $\alpha_1 = \theta$. The x-dependencies further require that $\sin\theta/v = \sin\theta''/v''$, where v and v'' are the velocities of light in the two media. This can be written in the familiar form of Snell's law,

$$n \sin\theta = n'' \sin\theta'',$$

where $n = c/v$, with v being the velocity of light in the medium and c the velocity of light in a vacuum.

Fresnel's reflection equations. Application of the boundary conditions to both the $\mathbf{H}$ and $\mathbf{E}$ vectors can result in a determination of the relative intensities of the reflected and refracted rays. In doing this it is convenient to express the electric and magnetic vectors in two components, E_y and H_y pointing in the y-direction and E_i and H_i lying entirely in the incident xz-plane. All $\mathbf{E}$ and $\mathbf{H}$ components must be perpendicular to the direction of propagation of their respective rays. Substitution of the x-component of E_i into Eq. (12–11) gives

$$E_i \cos\theta + E_i' \cos(180° - \theta) = E_i'' \cos\theta'',$$

while the y-component condition is unchanged:

$$E_y + E_y' = E_y''.$$

Substitution of the x-component of H_i into Eq. (12–11) gives the same equations except that E is everywhere replaced by H. These equations for H may be expressed in terms of the electric vectors by use of the proportionality between E and H that exists in plane waves. Since $\sqrt{K}$ is approximately equal to the index of refraction and K_m is approximately one, Eq. (12–8) may be rewritten as

$$H_y = \frac{1}{\mu}\frac{E_x}{v} = \sqrt{\frac{K\epsilon_0}{K_m\mu_0}}\, E_x \cong n\sqrt{\frac{\epsilon_0}{\mu_0}}\, E_x.$$

In the present case, where the direction of propagation is not z but some arbitrary direction $\mathbf{a}$, the proportionality exists between H_i and E_y and between H_y and E_i. Thus the boundary conditions on $\mathbf{H}$ reduce to

$$nE_y \cos\theta + nE_y' \cos(180° - \theta) = n''E_y'' \cos\theta''$$

and

$$nE_i + nE_i' = n''E_i''.$$

Snell's law may be used to express n and n'' in terms of θ and θ''. These four equations may then be used to determine the four quantities E_i', E_i'', E_y', and E_y'' in terms of the incident components E_i and E_y. By liberal application of trigonometric identities, the resulting expressions reduce to the usual form of Fresnel's equations given below:

$$E'_i = E_i \frac{\tan (\theta - \theta'')}{\tan (\theta + \theta'')},$$

$$E'_y = -E_y \frac{\sin (\theta - \theta'')}{\sin (\theta + \theta'')},$$

$$E''_i = 2E_i \frac{\cos \theta \sin \theta''}{\sin (\theta + \theta'') \cos (\theta - \theta'')},$$

$$E''_y = 2E_y \frac{\sin \theta'' \cos \theta}{\sin (\theta + \theta'')}.$$

The energy transmitted by an electromagnetic wave is given by Poynting's vector $\mathbf{S} = \mathbf{E} \times \mathbf{H}$. Using this vector and the proportionality between E and H in plane waves, the relative energy in the reflected and refracted rays for any angle of incidence can be determined from Fresnel's equations.

It is interesting to note that if the angle between the reflected and refracted ray is 90°, that is, $\theta' - \theta'' = \pi/2$, then the denominator of the first equation is infinite and E'_i is zero. Light reflected at this angle would have no component of the electric vector in the incident plane. The angle of incidence required for this effect may be expressed in terms of the index of refraction by substituting $\theta + \theta'' = \pi/2$ into Snell's law. The resulting expression,

$$\tan \theta_B = \frac{n''}{n},$$

is known as *Brewster's law*, where θ_B is Brewster's angle. If ordinary light with random planes of polarization strikes an interface with this angle to the normal, the reflected light will be perfectly polarized, with the electric vector perpendicular to the incident plane. This is a standard laboratory technique for producing plane polarized light.

12–5 Electromagnetic waves with conductors

In dealing with electromagnetic waves impinging on the plane surface of a dielectric, much was learned from assuming an arbitrary direction of impingement and an arbitrary direction of polarization. However, when the dielectric is replaced by a good conductor, it is as useful and much simpler to consider the special case of a plane-polarized wave traveling parallel to the positive z-axis and impinging on the conductor's plane surface, which coincides with the xy-plane through the origin. In the free space below this plane, $\mathbf{H}$ then points along the positive y-axis. We assume that throughout the conductor $\mu = \mu_0$, while ϵ and the conductivity σ are constant at the frequency of the incoming wave. Inside the conductor Maxwell's fourth and third equations can be written respectively as

$$\nabla \times \mathbf{H} = \sigma \mathbf{E} + \epsilon \dot{\mathbf{E}} \qquad \text{and} \qquad \nabla \times \mathbf{E} = -\mu_0 \dot{\mathbf{H}}.$$

Now eliminate $\mathbf{H}$ by differentiating the first with respect to time and taking the curl of the second:

$$\nabla \times (\nabla \times \mathbf{E}) = -\mu_0 \nabla \times \dot{\mathbf{H}} = -\mu_0 \sigma \dot{\mathbf{E}} - \mu_0 \epsilon \ddot{\mathbf{E}}.$$

From Eqs. (3–24) and (M–1), $\nabla \times (\nabla \times \mathbf{E}) = -\nabla^2 \mathbf{E}$, which gives

$$\nabla^2 \mathbf{E} = \mu_0 \sigma \dot{\mathbf{E}} + \mu_0 \epsilon \ddot{\mathbf{E}}.$$

With our assumptions this becomes

$$\frac{\partial^2 E_x}{\partial z^2} = \mu_0 \sigma \dot{E}_x + \mu_0 \epsilon \ddot{E}_x. \tag{12–12}$$

A solution for this equation is

$$E_x = E_0 e^{j\beta}, \qquad \text{where} \qquad \beta = \omega(t - az),$$

and a is a constant to be determined. We will choose E_0 to be the electric amplitude just inside the conductor at $z = 0$. When this solution is substituted into Eq. (12–12) and the result is divided by $e^{j\beta}$ the following relation is obtained:

$$-a^2 \omega^2 = j\mu_0 \sigma \omega - \mu_0 \epsilon \omega^2$$

and since $v^2 \mu_0 \epsilon = 1$, this can be written

$$a^2 v^2 = 1 - \frac{j\sigma}{\omega \epsilon}.$$

Assume $av = V = n - jk$, then $V^2 = n^2 - k^2 - j2nk = 1 - j\sigma/\omega\epsilon$ or $n^2 - k^2 = 1$ and $nk = \sigma/2\omega\epsilon$. From these two equations we find

$$n^2 = \frac{\sqrt{1 + \sigma^2/\omega^2\epsilon^2} + 1}{2} \simeq \frac{\sigma}{2\omega\epsilon}$$

and

$$k^2 = \frac{\sqrt{1 + \sigma^2/\omega^2\epsilon^2} - 1}{2} \simeq \frac{\sigma}{2\omega\epsilon},$$

since $\sigma/2\omega\epsilon \simeq 10^7/10^{-10} f = 10^{17}/f$ for metals. Until f exceeds the very high frequency of ultraviolet light, $10^{17}/f$ is very large compared to one. The electric field in the metal is then given by

$$E_x = E_0 e^{-k\omega z/v} e^{-j\omega(t - nz/v)}.$$

The last term on the right-hand side represents the sinusoidal oscillation traveling into the metal at the velocity v/n, where $\mathbf{v} = 1/\sqrt{\mu_0 \epsilon}$. Then $v = c$ if $\epsilon = \epsilon_0$.

Skin depth. The middle term of this electric field equation gives the reduction in amplitude as the wave progresses inward. Where $z = v/k\omega$, the amplitude is reduced to E_0/e. This value of z is designated by $\delta = 1/\sqrt{\mu\pi f\sigma}$ and is generally called the *skin depth*. It is the depth at which the electric field drops to 0.368 of its value at the surface. This depth is calculated here for a plane surface. However, if the calculated δ is small compared to the diameter of a given wire it is also valid for such a wire. It is somewhat more useful to determine the depth at which the Poynting vector or power flow drops to 0.01 of its value at the surface. Since the H field is proportional to the E field, the Poynting vector is proportional to E^2. The fractional reduction in S from $z = 0$ to $z = z_1$ is given by

$$(e^{-z_1/\delta})^2 = 0.01, \qquad 2z_1 = \delta \ln 100, \qquad z_1 = 2.3/\sqrt{\pi\mu\sigma f}.$$

For copper this becomes

$$z_1 = 0.15/\sqrt{f} \text{ in meters,} \qquad \text{or} \qquad z_1 = 150/\sqrt{f} \text{ in millimeters.}$$

We have seen that the energy for maintaining currents in circuits is carried through the fields by the Poynting vector. Since at 60 Hz $z_1 = 19.4$ mm in copper, with ordinary ac, the currents are uniformly distributed through the cross section of most wires. However, at 2×10^6 Hz, $z_1 = 0.1$ mm and currents oscillating at frequencies at or above this frequency are carried primarily by the outer surface or *skin* of most copper wires. Intermediate frequencies require a more detailed investigation.

Poynting vector momentum. There is momentum as well as energy flow associated with the Poynting vector as is demonstrated in Chapter 10. Probably the simplest way to determine the magnitude of this momentum is to investigate the pressure produced by the reflection of electromagnetic waves from the surface of a perfect conductor. We will assume an electromagnetic wave traveling along the z-axis and impinging on a conducting surface at the origin as is described at the beginning of this section. We will also assume that the conductor is perfect and that $\epsilon = \epsilon_0$ and $\mu = \mu_0$ throughout this space.

Consider three regions: (a) in free space just below the conducting surface, (b) right in this surface, and (c) in the conductor just above this surface. Choose an instant of time when the incoming fields E_{ix} and H_{iy} are positive as is shown in Fig. 12–6. We assume that each of the three regions is very thin and thus E_{ix} and H_{iy} have the same value in each. The E_{ix} field induces a current density J_x in amperes per meter which is confined to the surface. The energy impinging on the surface is carried away by an electromagnetic wave traveling back along the negative z-axis. This reflected wave also has an electric field $-E_{rx}$ in the free space of region (a). Since the perfect conductor cannot support electric fields, $E_{ix} - E_{rx} = 0$ in region (c). This relation is also true in regions (a) and (b), since boundary conditions guarantee equal

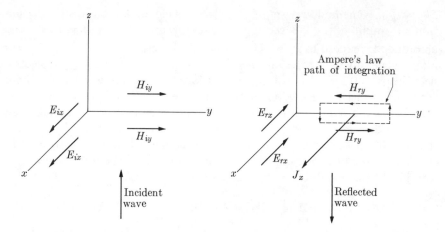

Fig. 12–6 Electromagnetic wave impinging on and being reflected from a perfectly conducting surface, which is in the *xy*-plane through the origin. The single co-ordinate system is shown twice to clarify the fields associated with the incident and reflected waves.

transverse components of electric fields on the two sides of a boundary. While the resultant electric fields are zero at the surface, the same cannot be said of the magnetic fields. The presence of the current density J_x results in the addition of transverse magnetic fields in regions (a) and (c). Symmetry and Ampere's law, Eq. (9–13), require that these two fields be of equal magnitude but opposite in direction. To cancel the magnetic field of the incident radiation within the conductor, the additional field in region (c) must be in an opposite direction from, and of an equal magnitude to that of the incident radiation. Both the magnetic and electric fields must be zero inside this perfect conductor. These considerations determine the magnitudes of the additional magnetic fields and they determine the magnitude of J_x as well; i.e., $H_{ry} - H_{iy} = 0$ and from Ampere's law $J_x = 2H_{ry} = 2H_{iy}$. Clearly the magnetic fields due to J_x do not contribute to the magnetic field in the surface, since they are of equal but opposite sign on each side of it. We are now ready to calculate the force exerted by the electromagnetic wave being reflected from the perfect conductor.

Clearly the magnetic field in region (b) is just H_{iy}. Since J_x is the current density in this region the force per unit area on the surface is, from the Lorentz force equation,

$$J_x \mu_0 H_{iy} \mathbf{a}_z = \mu_0 \mathbf{J} \times \mathbf{H}_i \qquad \text{in N m}^{-2}.$$

Now $J_x = 2H_{iy} = 2E_{ix}/\mu_0 c$ in A m^{-1}, from Eq. (12–8). Therefore, the force per unit area can be written as

$$2\mathbf{E}_i \times \mathbf{H}_i/c.$$

We can think of this pressure as being caused by a reversal in direction of the momentum density $\mathbf{p}$ in the electromagnetic wave by impact with the perfectly reflecting surface. Clearly the rate of change in momentum per second at this surface is

$$-2\mathbf{p}c.$$

We have thus shown that the momentum density in the electromagnetic wave is

$$\mathbf{p} = \mathbf{E} \times \mathbf{H}/c^2, \tag{12–13}$$

which should be considered the momentum density of the Poynting vector. This is the same value that was needed in Section 10–2, with static electromagnetic fields, to satisfy the law for the conservation of angular momentum. In fact this expression for momentum density in electromagnetic fields is found in all imaginable situations.

Reflection from real metals. All real metals have some resistivity ρ. When an electromagnetic plane wave strikes the plane face of such a metal normally, the following boundary conditions apply:

$$E_{ix} - E_{rx} = \rho J_x$$

in the metallic surface (b) and also in regions (a) and (c). Ampere's law gives

$$2H_{ry} = J_x.$$

From Eq. (12–8) we obtain $E_{ix} = c\mu_0 H_{iy}$ and $E_{rx} = c\mu_0 H_{ry}$. Now we substitute the last three relations in the first to obtain

$$c\mu_0(H_{iy} - H_{ry}) = 2\rho H_{ry},$$

which reduces to

$$H_{ry} = \alpha H_{iy}$$

where $\alpha = 1/(1 + 2\rho/\mu_0 c)$. Since both the incident and reflected waves are plane waves in free space,

$$E_{rx} = \alpha E_{ix}.$$

Now $2\rho/\mu_0 c \ll 1$ even for the metal having the highest resistivity in Table 6–1, i.e., Ni with $\rho \simeq 7 \times 10^{-8}$. Hence for Ni

$$2\rho/\mu_0 c = 0.37 \times 10^{-9}$$

and even for carbon $2\rho/\mu_0 c = 2 \times 10^{-5}$. Thus $\alpha \simeq 1$ and the analyses used with perfect conductors is remarkably accurate for all real metals.

12–6 Retarded potentials

In Section 8–5 the general differential equations for scalar and vector potentials,

$$\nabla^2 V - \mu\epsilon\ddot{V} = -\rho/\epsilon \tag{12–14}$$

and

$$\nabla^2 \mathbf{A} - \mu\epsilon\ddot{\mathbf{A}} = -\mu\mathbf{J}, \tag{12–15}$$

were developed. In electrostatic problems or in any problems where the time derivatives in these equations can be ignored, the scalar potentials were shown to be obtained from

$$V = \frac{1}{4\pi\epsilon} \sum_i \frac{q_i}{r_i}$$

with point charges, or from

$$V = \int_{\tau'} \frac{\rho \, d\tau'}{4\pi\epsilon r}$$

with charges whose distribution function $\rho = \rho(x', y', z')$ is known. When the charge distributions are time-dependent and the time derivatives cannot be ignored, these solutions must be modified.

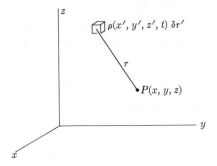

Fig. 12–7 A charge within the volume element $\delta\tau'$ located a distance r from a general point in space.

The time-dependent solution for a distribution of charge will be obtained by calculating the potential δV at the field point P due to the charge density in the volume element $\delta\tau'$, as illustrated in Fig. 12–7. This potential should depend only on charge density ρ, the time t, and the distance r. Using spherical coordinates with the origin at $\delta\tau'$ and with δV independent of θ and ϕ, the Laplacian can be written as

$$\nabla^2 \delta V = \frac{1}{r^2} \frac{\partial}{\partial r} \left(r^2 \frac{\partial \, \delta V}{\partial r} \right).$$

It is desirable to transform this into a function of $r \, \delta V$ by using the relation

$$\nabla^2 \delta V = \frac{1}{r} \frac{\partial^2}{\partial r^2} (r \, \delta V).$$

Equation (12–14) for δV at the field point P then becomes

$$\frac{\partial^2}{\partial r^2} (r \, \delta V) - \mu\epsilon \frac{\partial^2}{\partial t^2} (r \, \delta V) = 0,$$

which is a wave equation in the product $r \, \delta V$. Direct substitution shows that

$r \, \delta V = f \, (t - r/v)$, where $v = 1/\sqrt{\mu \epsilon}$, is a solution that represents a wave traveling outward along the radius r. The other solution, $r \, \delta V = g(t + r/v)$, which represents an oppositely directed wave, is not important here.

The function of $(t - r/v)$ must be chosen to give the static potential in the special cases where there is no time-dependence. Thus if the charge density at $\delta \tau'$ is given by $\rho = \rho(x', y', z', t)$, the potential at P due to the volume element $\delta \tau'$ is

$$\delta V = \frac{\rho(x', y', z', t - r/v)}{4 \pi \epsilon r} \, \delta \tau',$$

where r again is the distance from $\delta \tau'$ to the field point P shown in Fig. 12–7. The potential at P due to any distribution of charges is then

$$V = \int_{\tau'} \frac{\rho(x', y', z', t - r/v)}{4 \pi \epsilon r} \, d\tau', \qquad (12\text{–}16)$$

which gives solutions for static charge distributions when the dependence on time is set equal to zero, as well as the more general solutions for charge distributions that do change with time. The similarity between Eqs. (12–14) and (12–15) makes it possible to obtain the solution for the vector potential by substituting $\mu \mathbf{J}$ for ρ/ϵ in Eq. (12–18). Thus

$$\mathbf{A} = \int_{\tau'} \frac{\mu \mathbf{J}(x', y', z', t - r/v)}{4 \pi r} \, d\tau'. \qquad (12\text{–}17)$$

It should be noted that if a sudden change in the charge or current occurs, its effect will not be noticed at P until the time r/v later. This is physically reasonable if potential is thought of as radiating from the originating charge with a velocity v. Because of the time lag between a given charge distribution in space and the corresponding potential at P, Eqs. (12–16) and (12–17) are said to be equations for *retarded potentials*.

12–7 Radiation from an oscillating dipole

In Section 2–6 the electric field due to a stationary dipole was obtained by calculating the scalar potential V at a general point in space, the electric field being the negative gradient of this potential. However, when the moment of a dipole changes with time, a number of additional complications are introduced. First, since retarded potentials must be used, the phase of the potential depends upon the location of the field point. Second, the electric field is no longer given by the negative gradient of the scalar potential alone, but also involves the vector potential, as indicated by Eq. (8–21),

$$\mathbf{E} = -\nabla V - \dot{\mathbf{A}}.$$

Finally, in addition to the electric field, there is a magnetic field generated in accordance with Eq. (8–20), $\mathbf{B} = \nabla \times \mathbf{A}$.

Oscillating dipole potentials. An oscillating dipole is illustrated in Fig. 12–8. It consists of two small metallic spheres with their centers a distance l apart and joined by a conducting wire. It is desired to calculate the potentials at the general field point P located by the spherical coordinates r and θ, where r is chosen to be large compared with l. The potentials and fields must be independent of ϕ because of the symmetry about the axis of the dipole. Assuming that there is a sinusoidal current between the two spheres that is alternately charging and discharging them, the charge on the top sphere can be written as the real part of $+q = q_0 e^{j\omega t}$ and that on the bottom sphere is the real part of $-q = -q_0 e^{j\omega t}$. The current then is the real part of $I = dq/dt = j\omega q_0 e^{j\omega t}$.

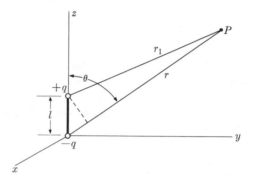

Fig. 12–8 Oscillating electric dipole, where $+q = q_0 \cos \omega t$.

The scalar potential may be determined by substituting an expression for the charge density into Eq. (12–16). Since there is no charge in space, except on the small spheres that are located at the origin and at $z = l$ on the z-axis, the integrand is zero everywhere except over these spheres. If the spheres are very small, r is constant over the sphere at the origin and r_1 is constant over the sphere at $z = l$. Over these spheres the absolute values of $\int \rho \, d\tau' = q_0 e^{j\omega t}$, and therefore the scalar potential is

$$V = \frac{q_0 e^{j\omega(t-r_1/v)}}{4\pi\epsilon r_1} - \frac{q_0 e^{j\omega(t-r/v)}}{4\pi\epsilon r}. \tag{12–18}$$

Because $r \gg l$, $r_1 = r - l \cos \theta$, and Eq. (12–18) may be written as

$$V = \frac{q_0 e^{j\omega(t-r/v)}}{4\pi\epsilon r} \left\{ \frac{e^{j\omega l(\cos \theta)/v}}{1 - l(\cos \theta)/r} - 1 \right\}.$$

The term $\{1 - l(\cos \theta)/r\}^{-1}$ may be written as $\{1 + l(\cos \theta)/r\}$, by using the first two terms of the binomial expansion, and the term $\exp\{j\omega l (\cos \theta)/v\}$ may be expressed by the first two terms of a power series as $1 + j\omega l (\cos \theta)/v$, provided that $\omega l/v < 1$. This last approximation is applicable provided l is much smaller than the wavelength of the radiation, which is true for many dipoles in nature. Equation (12–18) finally reduces to

$$V = \frac{p_0 e^{j\beta} \cos \theta}{4\pi\epsilon r^2}\left(1 + \frac{j\omega r}{v}\right), \tag{12–19}$$

where $\beta = \omega(t - r/v)$ and $p_0 = q_0 l$, the maximum dipole moment. It is interesting to note that Eq. (12–18) gives the static scalar potential of Eq. (2–24) for the special case where the distance r of the point from the dipole is small compared with the wavelength of the radiation, $\lambda = 2\pi v/\omega$, since the second term in the parentheses is then negligible. Before fields can be calculated, the vector potential as well as this scalar potential is needed.

In determining the vector potential of the dipole, Eq. (12–19) reduces to an integral over the length of the wire between the two spheres, since the current density is zero everywhere else. The expression $\int_{\tau'} \mu \mathbf{J}\, d\tau'$ becomes $\mathbf{k} \int_l \mu I\, dl$, where $\mathbf{k}$ is a unit vector parallel to the z-axis, and the vector potential is given by

$$\mathbf{A} = \mathbf{k}\int_0^l \frac{\mu I(t - r/v)}{4\pi r}\, dl.$$

The value of r in this equation actually ranges from r to r_1 over the length of the wire, but since $r \gg l$ and $r_1 = r - l\cos\theta$, r can be treated as constant and taken outside the integral sign. It is proper to treat r as a constant here because, at any instant, the integrand has the same sign over the length of the dipole; it would not have been satisfactory to treat it as constant in Eq. (12–16), where the integrand would cancel out to this approximation because it has opposite signs over the two spheres. The first nonzero term must be maintained. The resulting expression is

$$\mathbf{A} = \mathbf{k}\frac{\mu}{4\pi r}\, lI\left(t - \frac{r}{v}\right) = \mathbf{k}\frac{\mu}{4\pi r}\, lj\omega q_0 e^{j\omega(t-r/v)}$$

or

$$\mathbf{A} = \frac{j\omega\mu p_0 e^{j\beta}}{4\pi r}(\mathbf{a}_r \cos\theta - \mathbf{a}_\theta \sin\theta), \tag{12–20}$$

since $\mathbf{k} = \mathbf{a}_r \cos\theta - \mathbf{a}_\theta \sin\theta$.

Oscillating dipole fields. The electric and magnetic fields of the oscillating dipole can now be calculated directly from these vector and scalar potentials, using the relations $\mathbf{E} = -\nabla V - \dot{\mathbf{A}}$ and $\mu\mathbf{H} = \nabla \times \mathbf{A}$. In performing these differentiations, the following relations are useful: $\beta = \omega(t - r/v)$, $\partial\beta/\partial r = -\omega/v$, and $\partial\beta/\partial t = \omega$. Since the gradient

$$\nabla V = \mathbf{a}_r \frac{\partial V}{\partial r} + \mathbf{a}_\theta \frac{1}{r}\frac{\partial V}{\partial \theta}$$

has many terms, each component will be calculated separately. The r-component is

$$\frac{\partial V}{\partial r} = \left[\frac{1}{r^2}\left(1 + \frac{j\omega r}{v}\right)\left(-\frac{j\omega}{v}\right) + \left(-\frac{2}{r^3} - \frac{j\omega}{vr^2}\right)\right]e^{j\beta}\frac{p_0\cos\theta}{4\pi\epsilon},$$

which, on collecting terms, becomes

$$\frac{\partial V}{\partial r} = -\left(\frac{2}{r^3} + \frac{2j\omega}{vr^2} - \frac{\omega^2}{v^2 r}\right)\frac{p_0\cos\theta}{4\pi\epsilon}e^{j\beta}.$$

The θ-component is

$$\frac{1}{r}\frac{\partial V}{\partial\theta} = -\left(\frac{1}{r^3} + \frac{j\omega}{vr^2}\right)\frac{p_0\sin\theta}{4\pi\epsilon}e^{j\beta}.$$

Direct differentiation of Eq. (12–20) with respect to time gives the simple relation

$$\dot{\mathbf{A}} = -\frac{\omega^2\mu p_0 e^{j\beta}}{4\pi r}(\mathbf{a}_r\cos\theta - \mathbf{a}_\theta\sin\theta).$$

These expressions for ∇V and $\dot{\mathbf{A}}$ can be combined and the terms with like powers of r collected to give the following expression for $\mathbf{E}$:

$$\mathbf{E} = \frac{p_0 e^{j\beta}}{4\pi\epsilon}\left[-\frac{\omega^2\sin\theta}{v^2 r}\mathbf{a}_\theta + \left(\frac{j\omega}{vr^2} + \frac{1}{r^3}\right)(\mathbf{a}_r 2\cos\theta + \mathbf{a}_\theta\sin\theta)\right]. \quad (12\text{–}21)$$

This electric field is made up of terms proportional to $1/r$, $1/r^2$, and $1/r^3$. For static conditions $\omega = 0$ and all but the last terms disappear, with the result

$$\mathbf{E}(\text{static}) = \frac{p_0}{4\pi\epsilon r^3}(\mathbf{a}_r 2\cos\theta + \mathbf{a}_\theta\sin\theta).$$

This is the dipole field that was given in Section 2–13, and it falls off rapidly with the distance r from the dipole. At high frequencies and at large distances the first term dominates, since it falls off only as $1/r$ and is proportional to ω^2.

The magnetic field can be obtained directly from the vector potential by means of the defining equation $\mu\mathbf{H} = \nabla\times\mathbf{A}$, which gives

$$\mathbf{H} = \frac{j\omega p_0}{4\pi r^2\sin\theta}\begin{vmatrix} \mathbf{a}_r & r\mathbf{a}_\theta & r\sin\theta\,\mathbf{a}_\phi \\ \dfrac{\partial}{\partial r} & \dfrac{\partial}{\partial\theta} & \dfrac{\partial}{\partial\phi} \\ \dfrac{e^{j\beta}\cos\theta}{r} & -e^{j\beta}\sin\theta & 0 \end{vmatrix},$$

which may be reduced to

$$\mathbf{H} = \frac{j\omega p_0 e^{j\beta}}{4\pi} \mathbf{a}_\phi \sin\theta \left(\frac{j\omega}{vr} + \frac{1}{r^2} \right). \tag{12–22}$$

As should be expected, this expression for the magnetic field approaches zero when $\omega \to 0$, since a static dipole does not produce a magnetic field. However, when ω is finite and $\omega r/v \ll 1$, the second term predominates and we have

$$\mathbf{H}\,(\text{low frequency}) = \frac{j\omega q_0 l e^{j\omega t}}{4\pi r^2} \sin\theta\, \mathbf{a}_\phi.$$

The time-dependent field, which is the real part of this expression, can be written as

$$\mathbf{H}\,(\text{low frequency}) = \frac{il \sin\theta}{4\pi r^2} \mathbf{a}_\phi,$$

where $i = -q_0 \omega \sin\omega t$ is the time-dependent electric current between the spheres. This last expression could have been obtained directly from the elementary form of the Biot-Savart law. At high frequencies and large distances, the $1/r$ terms are dominant in both the magnetic and electric fields. These terms represent the radiation field in which the $\mathbf{E}$ and $\mathbf{H}$ vectors are in phase and are perpendicular both to each other and to $\mathbf{r}$, the direction of propagation. Thus, at large distances from an oscillating dipole, the radiation field takes on the form of the plane waves discussed in Section 12–2. The lines of electric and magnetic flux shown in Fig. 12–9 are densest in the direction perpendicular to the dipole and drop to zero in the direction of the dipole axis.

It should be remembered that Maxwell's relations can be satisfied exactly at any point in space only by including all of the terms. For example, the lines of electric flux shown in Fig. 12–9 must include radially directed terms if they are to be continuous, and yet must be zero at $\theta = 0$ and $\theta = \pi$. If the number of closed loops of $\mathbf{E}$ flux shown in the figure is kept constant but their length is increased as they are propagated radially outward, the density of the θ-component falls off as $1/r$, while the density of the $\mathbf{r}$-component falls off as $1/r^2$ and becomes negligible as r becomes very large.

12–8 Energy flow from the oscillating dipole

The energy leaving the oscillating dipole per unit of time may be obtained by surrounding the dipole with an imaginary sphere and integrating the normal component of $\mathbf{S} = \mathbf{E} \times \mathbf{H}$ over its surface. As the radius of this imaginary sphere is made arbitrarily large a finite contribution is made only by the $1/r$ terms of $\mathbf{E}$ and $\mathbf{H}$, since the cross product of these terms falls off

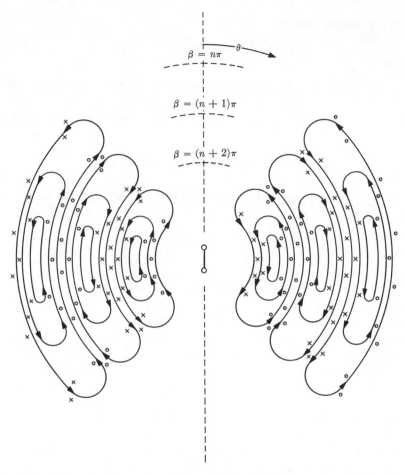

$\beta = n\pi$

$\beta = (n + 1)\pi$

$\beta = (n + 2)\pi$

Fig. 12–9 Schematic illustration of the radiation fields from an oscillating dipole. Solid lines represent electric flux while crosses and circles represent the magnetic flux lines entering and leaving the paper, respectively.

as $1/r^2$ and the area of the sphere increases as r^2. All other terms of $\mathbf{E}$ and $\mathbf{H}$ fall off more rapidly and become negligible at large distances.

Since the real parts of these $1/r$ terms for $\mathbf{E}$ and $\mathbf{H}$ are

$$\mathbf{E} = -\frac{\omega^2 p_0 \cos\beta \sin\theta}{4\pi v^2} \frac{}{r} \mathbf{a}_\theta \quad \text{and} \quad \mathbf{H} = -\frac{\omega^2 p_0 \cos\beta \sin\theta}{4\pi v} \frac{}{r} \mathbf{a}_\phi,$$

the Poynting vector at large distances is

$$\mathbf{S} = \mathbf{E} \times \mathbf{H} = \frac{\omega^4 p_0^2 \cos^2\beta \sin^2\theta}{(4\pi)^2 \epsilon v^3} \frac{}{r^2} \mathbf{a}_r$$

and the time average of the Poynting vector is

$$\bar{\mathbf{S}} = \frac{\omega^4 p_0^2}{2(4\pi)^2 \epsilon v^3} \frac{\sin^2 \theta}{r^2} \mathbf{a}_r. \qquad (12\text{–}23)$$

For many problems, it is more convenient to express Eq. (12–23) in terms of current and wavelength. The effective current is defined as $I = I_0/\sqrt{2}$, where I_0 is the maximum current and $p_0 = I_0 l/\omega$. Furthermore, $\mu \epsilon v^2 = 1$ and $\omega/v = 2\pi/\lambda$, where λ is the wavelength of the electromagnetic wave. Equation (12–23) can then be expressed as

$$\bar{\mathbf{S}} = \sqrt{\frac{\mu}{\epsilon}} \left(\frac{Il}{2\lambda} \right)^2 \frac{\sin^2 \theta}{r^2} \mathbf{a}_r, \qquad (12\text{–}24)$$

which shows that the radiation is proportional to the square of the current and inversely proportional to the square of the wavelength.

The average power radiated can be found by integrating Eq. (12–24) over a large sphere surrounding the dipole, which yields

$$\bar{P} = \int_s \bar{\mathbf{S}} \cdot d\mathbf{s}$$

$$= \sqrt{\frac{\mu}{\epsilon}} \left(\frac{Il}{2\lambda} \right)^2 \int_0^{2\pi} \int_0^\pi \frac{\sin^2 \theta}{r^2} r^2 \sin \theta \, d\phi \, d\theta$$

$$= \sqrt{\frac{\mu}{\epsilon}} \left(\frac{Il}{2\lambda} \right)^2 2\pi \int_0^\pi \sin^3 \theta \, d\theta = \frac{2\pi}{3} \sqrt{\frac{\mu}{\epsilon}} \left(\frac{Il}{\lambda} \right)^2,$$

which is the power radiated throughout space that must be supplied to the dipole by some power source if the current oscillations are to be maintained. In addition to this loss by radiation, the power lost in heating conductors is given by $\bar{P} = I^2 R_j$, where R_j may be greater than the dc resistance $\rho l/A$, because of the tendency for high-frequency currents to concentrate in the outer surface or "skin" of the wire. The total power lost in the dipole is given by

$$\bar{P}_t = \frac{2\pi}{3} \sqrt{\frac{\mu}{\epsilon}} \left(\frac{Il}{\lambda} \right)^2 + I^2 R_j$$

and the effective ac resistance of the dipole is then $R = \bar{P}_t/I^2$, in ohms.

General energy flow near an oscillating dipole. It has been shown that the only terms of the **E** and **H** fields that contribute significantly to the power radiated are those proportional to $1/r$. To determine what happens to the Poynting vector flux represented by the higher-order terms, every term of the Poynting vector must be investigated. The **E** and **H** fields should be written in their time-dependent form by taking the real parts of the complex expressions in Eqs. (12–21) and (12–22). Since there are five terms in the **E** field and two terms in the **H** field, the Poynting vector **E** × **H** contains ten terms. However, examination of the **E** terms shows that they can be paired off and written as three terms by reintroducing the characteristic dipole-field vector that is defined in Section 2–13 as

$$\mathbf{d}_f \equiv \mathbf{a}_r 2 \cos \theta + \mathbf{a}_\theta \sin \theta.$$

When this simplifying vector is introduced, the electric field of the oscillating dipole can be written as

$$\mathbf{E} = -(\mathbf{a}_\theta g_1 \sin \theta \cos \beta + \mathbf{d}_f g_2 \sin \beta - \mathbf{d}_f g_3 \cos \beta) \qquad (12\text{–}25)$$

and its magnetic field can be written as

$$\mathbf{H} = -\mathbf{a}_\phi \sin \theta (h_1 \cos \beta + h_2 \sin \beta), \qquad (12\text{–}26)$$

where the g and h constants are defined by

$$g_1 = \frac{p_0 \omega^2}{4\pi\epsilon v^2 r}, \qquad g_2 = \frac{p_0 \omega}{4\pi\epsilon v r^2}, \qquad g_3 = \frac{p_0}{4\pi\epsilon r^3},$$

and

$$h_1 = \frac{p_0 \omega^2}{4\pi v r}, \qquad h_2 = \frac{p_0 \omega}{4\pi r^2}.$$

In calculating Poynting vector terms the following cross products are needed: the product $\mathbf{a}_\theta \times \mathbf{a}_\phi = \mathbf{a}_r$ and the product $\mathbf{d}_f \times \mathbf{a}_\phi \equiv \mathbf{d}_s$, which is a newly defined vector that has the value

$$\mathbf{d}_s = \mathbf{a}_r \times \mathbf{a}_\phi 2 \cos \theta + \mathbf{a}_\theta \times \mathbf{a}_\phi \sin \theta$$

or

$$\mathbf{d}_s = -\mathbf{a}_\phi 2 \cos \theta + \mathbf{a}_r \sin \theta.$$

The Poynting vector $\mathbf{S} = \mathbf{E} \times \mathbf{H}$ for the oscillating dipole can now be written:

$$\begin{aligned} \mathbf{S} = {} & \mathbf{a}_r g_1 h_1 \sin^2 \theta \cos^2 \beta + \mathbf{a}_r g_1 h_2 \sin^2 \theta \sin \beta \cos \beta \\ & + \mathbf{d}_s g_2 h_1 \sin \theta \sin \beta \cos \beta + \mathbf{d}_s g_2 h_2 \sin \theta \sin^2 \beta \\ & - \mathbf{d}_s g_3 h_1 \sin \theta \cos^2 \beta - \mathbf{d}_s g_3 h_2 \sin \theta \cos \beta \sin \beta, \end{aligned}$$

where each of these six terms represents a rate of energy flow. To determine whether there is a net rate of energy flow outward, $\bar{\mathbf{S}}$, the time average of $\mathbf{S}$, must be calculated. If the period of oscillation is T,

$$\bar{\mathbf{S}} = \frac{1}{T} \int_0^T \mathbf{S} \, dt = \frac{1}{2\pi} \int_0^{2\pi} \mathbf{S} \, d\beta,$$

where the integrand consists of the six terms. Three of the six terms may be eliminated immediately by using the relation

$$\frac{1}{2\pi} \int_0^{2\pi} \cos \beta \sin \beta \, d\beta = 0.$$

Cancellation of two of the remaining terms results from

$$\frac{1}{2\pi} \int_0^{2\pi} \cos^2 \beta \, d\beta = \frac{1}{2\pi} \int_0^{2\pi} \sin^2 \beta \, d\beta = \tfrac{1}{2}$$

together with the relation $g_2 h_2 = g_3 h_1$, which is easily verified. The remaining term, containing $g_1 h_1$, is just the radiation term previously discussed, while the other five terms represent reversible conversions of energy between the electric and magnetic fields.

Some of this local-field energy flow will be absorbed if there are conductors, or dielectric or magnetic materials in the neighborhood of the dipole. Such materials modify the fields so as to cause energy flow to them from the dipole.

The magnetic and electric fields due to a coil of wire in which there is an oscillating current can be calculated with the relations used for the oscillating dipole. As in the oscillating dipole, there are reversible-energy, local-field terms as well as nonreversible-energy radiation terms. Simple examples calculated in Section 11–12 show how nearby conductors, or dielectric or magnetic materials can absorb energy from such oscillating circuits.

12–9 Practical antennas

The problem of the simplest of all antennas, the oscillating dipole, has now been solved. The solution applies rather well to any short-wire antenna in which current oscillates in a wire, alternately charging and discharging the ends. It also applies to a radio transmitting antenna consisting of a single wire rising vertically above the ground, provided the antenna is short compared with the wavelength of the radiation. The ground acts as a conducting surface, reflecting the radiation of the wire in such a way that the radiation appears to be coming from a dipole whose bottom end is just as far below the ground as the top of the antenna is above ground. It is known from image solutions that a charge above a conducting surface creates the same field above that surface as would be created by the charge and an oppositely charged image below the surface.

Antennas that are nearly as long as or longer than the wavelength of the transmitted radiation have more complicated solutions. The complication arises from the fact that the current is not uniform along the length of the antenna and the charge is not concentrated at the two ends. Once the current and charge distributions are determined along the length of the wire, the vector and scalar potentials may be found in the same manner as for the oscillating dipole.

A brief treatment of long antennas, loop antennas, and arrays of antennas is given by Harnwell.* For more detailed discussions, books on radio engineering should be consulted.

12–10 Transmission lines of negligible resistance

The power required to maintain oscillations in an antenna is usually transported to it by means of transmission lines, or waveguides. The variation of current along a wire as a function of time for high-frequency alternating current is illustrated in Fig. 12–10, with an ac source located at $x = 0$ supplying the current $I_0 \cos 2\pi f t$. The current maximum from the generator takes a time t_0 to travel the distance x_0, where $v = x_0/t_0$ is the velocity of propagation of the current. In most cases, this velocity is somewhat less than the velocity of light in a vacuum. Figure 12–10 shows how the current varies sinusoidally along the wire at any instant of time and how it varies sinusoidally with time at any position such as $x = x_0$ along the wire.

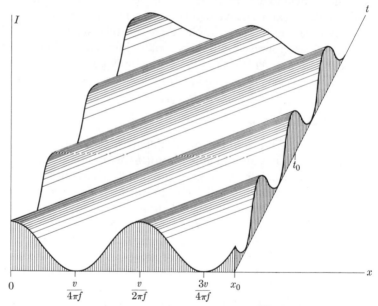

Fig. 12–10 The variation of current along a wire as a function of time for high-frequency alternating current.

The problems arising from such current variations were avoided in Chapter 11 by assuming that the wires in any circuit were much shorter than λ, where $\lambda = v/2\pi f$. Certainly the treatment of Chapter 11 can be used with

* G. P. Harnwell, *Principles of Electricity and Electromagnetism*, 2nd ed., pp. 599–623, McGraw-Hill, New York, 1949.

little error whenever the relation between l, the length of the wires, and f is given by $l \le v/2000\pi f$. For 60 Hz current, this is true for wires shorter than 800 m, a condition which is easily satisfied. However, for frequencies in the MHz range used by commercial radio and television stations, the wires in the circuit would have to be shorter than 5 cm.

The simple Kirchhoff's law relations that are based upon $\mathbf{\nabla} \cdot \mathbf{J} = 0$ and $\mathbf{\nabla} \times \mathbf{E} = 0$ do not hold for these short wavelengths because the current and charge vary along the length of the wire. At these high frequencies energy is carried from point to point by the electric and magnetic fields around the conductor, and may be lost from the system by radiation into space or through energy absorption by nearby materials. The lumped circuit parameter approximations of Chapter 11 are no longer valid. A general solution for circuits above this frequency requires a solution of the electromagnetic field relations for the boundary conditions of interest as is done for a rectangular wave guide in Section 12–11.

A valid solution up to limited frequencies exists for transmission lines carrying a current with wavelength *short* compared to the length of the line but *long* compared to the separation between the two conductors. For conductors separated by 1 cm, this approximation is good for frequencies below 100 MHz or for a wavelength of 1 m or so. In this case a length of line that is short compared to the wavelength of the current can be replaced by an equivalent circuit as is pictured in Fig. 12–11(b), because there is a well-defined homogeneous field between the conductors in the specified region. A long transmission line can be represented by a long series of such equivalent circuits; but, of course, the time delay between portions of the line must be considered.

Within the limits of this approximation, the velocity of propagation of electrical signals and the characteristic impedance for a parallel wire transmission line with *negligible resistance* will now be calculated. The form of the analysis actually applies equally well to coaxial cable or any two-conductor line in which the conductors preserve the same cross-sectional geometry for distances that are very long compared to their other dimensions.

Parallel-wire transmission line. When an ac generator is connected to a long transmission line consisting of two parallel wires, as shown in Fig. 12–11, energy is propagated parallel to the wires. From the Poynting vector relation, Eq. (10–2), it is evident that the fields $\mathbf{E}$ and $\mathbf{H}$ must both be perpendicular to the wires in order that $\mathbf{S}$ be parallel. Thus, the charges and currents have identical magnitudes but opposite signs on the two wires at any given x-coordinate. These can be represented by σ and i, respectively, both being functions of x and t.

The wave equation and propagation velocity will first be obtained using two fundamental relations: the law of charge conservation, $\mathbf{\nabla} \cdot \mathbf{J} + \partial \rho/\partial t = 0$, and Maxwell's third relation, $\mathbf{\nabla} \cdot \mathbf{E} = -\dot{\mathbf{B}}$. Integrating the first of these over the volume τ enclosed by the imaginary cylinder shown between a and b

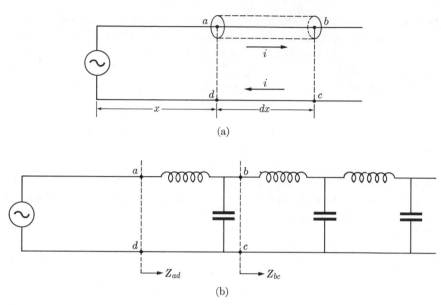

Fig. 12–11 (a) Parallel wires transmitting power from a high-frequency electric generator. (b) Equivalent circuit representation of the parallel wire transmission line.

in Fig. 12–11(a) and invoking Gauss's Theorem gives

$$\int_\tau \frac{\partial \rho}{\partial t}\, d\tau = -\int_\tau \nabla \cdot \mathbf{J}\, d\tau = -\int_s \mathbf{J} \cdot d\mathbf{s},$$

where s is the total surface of the cylinder. The left integral reduces to $(\partial \lambda / \partial t)\, dx$, since λ is the charge density per unit length along the wire, and the right integral reduces to $i - [i + (\partial i / \partial x)\, dx] = -(\partial i / \partial x)\, dx$, so long as the total change in λ or i is small over the distance dx, as is consistent with our assumptions. Therefore

$$\frac{\partial \lambda}{\partial t} = -\frac{\partial i}{\partial x}. \tag{12–27}$$

Integrating Maxwell's relation over the rectangular plane with corners at a, b, c, and d and taking the direction into the paper to be positive results in

$$-\int_{s'} \dot{\mathbf{B}} \cdot d\mathbf{s}' = \int_{s'} \nabla \times \mathbf{E} \cdot d\mathbf{s}' = \oint \mathbf{E} \cdot d\mathbf{l}.$$

The left integral is $-\partial \phi / \partial t$, which can be replaced by $-(\partial i / \partial t) l\, dx$, where l is the inductance per unit length and $l\, dx$ is the inductance of the two parallel lines over the length dx. With the resistivity of the lines assumed to be negligible, the right integral reduces to the potential difference between a

and d subtracted from that between b and c. That is,

$$\left(V + \frac{\partial V}{\partial x}\, dx\right) - V = \frac{\partial V}{\partial x}\, dx = \frac{1}{c}\frac{\partial \lambda}{\partial x}\, dx,$$

where the last term has been obtained from the fact that $V(x, t) = \lambda(x, t)/c$, and c is capacitance per unit length of the wire. From Maxwell's relation we thus obtain

$$\frac{1}{c}\frac{\partial \lambda}{\partial x} = -l\frac{\partial i}{\partial t}. \qquad (12\text{–}28)$$

Equation (12–28) simply states that in any arbitrary loop and at any given time, the sum of the capacitive voltage drops is equal to the inductive emf in the loop. By differentiating Eq. (12–27) with respect to x and Eq. (12–28) with respect to t, the two can be combined to give

$$\frac{\partial^2 i}{\partial x^2} = lc\,\frac{\partial^2 i}{\partial t^2}, \qquad (12\text{–}29)$$

which is a wave equation having the general solution

$$i = f(x \pm vt), \qquad (12\text{–}30)$$

where the velocity $v = 1/\sqrt{lc}$. The minus sign corresponds to wave propagation in the positive x-direction. A similar wave equation can be obtained for λ by reversing the differentiation variables for Eqs. (12–27) and (12–28) before combining. Thus i and λ are both propagated as waves along the wire as illustrated for sinusoidal current in Fig. 12–10.

By using the complex functions discussed in Section 4–9, the values of l and c can be determined with steady-state conditions by finding some function of the complex variable $z = x + jy$ such as $F(z) = G + jH$, where either H or G will fit the boundary conditions. It will be assumed that the medium surrounding the wire is a homogeneous one for which $\epsilon = K\epsilon_0$ and $\mu = K_m\mu_0$. According to Eq. (4–29), when the function $H(x, y)$ fits the boundary conditions for the potential, the capacitance per unit length is given by

$$c = \epsilon(G_2 - G_1)/(H_2 - H_1),$$

where the difference of potential between the wires is $H_2 - H_1$ and the charge per unit length is $\epsilon(G_2 - G_1)$. Under these steady-state conditions, the magnetic flux per unit length is given by $\mu(H_2 - H_1)$ and the current is given by $G_2 - G_1$, so that $l = \mu(H_2 - H_1)/(G_2 - G_1)$ and

$$lc = \mu\epsilon.$$

The velocity of the waves is $v = \sqrt{1/\mu\epsilon}$, which is the velocity of light in the medium.

Since the velocity v is independent of the frequency, it is possible to transmit several different frequencies over the line without a phase shift between the different frequencies (so long as the transmission distance is short enough that the resistance of the lines can be neglected). This is particularly important for digital computer applications, since it allows undistorted transmission of any shape of current pulse. Any shape of pulse can be synthesized with sinusoidal waves of many frequencies; e.g., pulses resembling the rectangular require many frequencies, some of which are quite high. With electronic computers or pulse-code modulated communications systems, the height, width, rise times, and fall times of signal pulses must be preserved over relatively long distances.

From the derivation of Eq. 12–28 it should be apparent that Fig. 12–11(b) is an equivalent circuit for the transmission line of Fig. 12–11(a). This equivalent circuit form is convenient for calculating the impedance of the line. The impedance of the line observed at point ad, Z_{ad}, is equal to the impedance of the inductance $l\,dx$ in series with the combination of $c\,dx$ in parallel with Z_{bc}, where Z_{bc} is the impedance of the remainder of the line. That is

$$Z_{ad} = j\omega l\,dx + \frac{Z_{bc}/j\omega c\,dx}{Z_{bc} + 1/j\omega c\,dx}.$$

Now, if the line is so long that the wave front fails to reach the end until after the period of observation, then the impedance observed must be *independent of the length* and in fact, independent of whether it is observed at point ad or point bc. We can therefore set $Z_0 = Z_{ad} = Z_{bc}$ in the previous equation; and after multiplying both sides by the denominator, we obtain

$$Z_0^2 = Z_0 j\omega l\,dx + l/c.$$

In the limit as $dx \to 0$ this yields

$$Z_0 = \sqrt{l/c}, \tag{12–31}$$

which is known as the *characteristic impedance* of the line.

Strip lines. In high-frequency electronic circuitry, transmission lines are frequently used which consist of two long strips of copper separated by a thin sheet of mylar or epoxy dielectric. Such lines have become known as *strip lines*. They are usually fabricated by photoetching the desired line patterns from an original sheet consisting of the dielectric laminated between two thin sheets of copper. In this manner relatively complicated strip line structures for interconnecting many circuit elements can be fabricated in one photoetching process.

A three-dimensional view of the end section of such a line with terminating resistor is shown in Fig. 12–12. Figure 12–11(a) and (b) can be made to apply to this geometry by assuming that the "wire" of Fig. 12–11a is actually

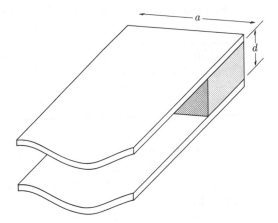

Fig. 12–12 Three-dimensional view of end section of a strip line with terminating resistor.

a two-dimensional view of the desired strip line. The cylindrical surface shown around the wire between a and b should be converted to a rectangular box shaped to properly enclose the strip line element. The derivation of Eq. (12–29) for this geometry can now be followed exactly as was done for the parallel line geometry. The velocity of propagation is again found to be $v = 1/\sqrt{lc}$ and the characteristic impedance is $Z_0 = \sqrt{l/c}$. The capacitance and inductance per unit length can be obtained simply, in the approximation where $a \gg d$, and are found to be $c = \epsilon a/d$ and $l = \mu\, d/a$. The first expression is the same as for a parallel plate capacitor with negligible fringing fields. The inductance per unit length is derived from $l = \phi/I = \mu H\, d/a$, where ϕ is flux per unit length. The last term is obtained from the fact that the integral of $\mathbf{H} \cdot d\mathbf{l}$ about one of the plates has a contribution only over the length a between the plates because the H field, due to the current in the upper and lower plates, cancels to zero everywhere else. This can readily be seen by sketching the lines of flux from the top line and then from the lower line. This calculation assumes the thickness of the conductors is small compared to their separation or that the frequency is high enough that the current exists only on the inner surfaces of the conducting strips.

The resultant basic equations for strip lines are:

$$v = 1/\sqrt{\epsilon\mu} \quad \text{and} \quad Z_0 = (d/a)\sqrt{\mu/\epsilon}. \tag{12–32}$$

Common transmission line geometries. In addition to the parallel wire and strip line geometries three other configurations are very common: a wire or flat strip parallel to a conducting plane and a coaxial cable. The first of these behaves very much as a transmission line consisting of two parallel wires. Using image solutions we know that the electric field-lines between a charged wire and a conducting plane are identical to those between two parallel wires placed twice as far apart. At high frequencies, magnetic lines of flux cannot penetrate the conducting plate due to eddy currents induced in the plate. Thus, it serves to create boundary conditions and a field configuration for lines of **H** and **B** which are identical above the conducting plane to those produced by two parallel current-carrying wires.

The capacitance per unit length for two hollow parallel wires and a hollow wire parallel to a conducting plane is given by Eqs. (4–35) and (4–34) while the inductance is given by Eq. (8–30). These are included in Table 12–2 along with the resultant propagation velocity and characteristic impedance. These results are also valid for solid wires at high frequencies where currents are confined to a surface region that is thin compared to the distance from the wire to the plane or between the wires.

The capacitance and inductance per unit length of a strip line parallel to a conducting plate are identical to those for two parallel strips of equal width so long as the separation is small enough that fringing fields may be neglected and, as for the wire over a plane, so long as the frequency is high enough that the magnetic fields do not penetrate the conducting plane.

The transmission line characteristics of a coaxial cable can be verified using the methods of this section. The inductance for a coaxial cable with hollow center wire is given by Eq. (8–29) and capacitance is given in Eq. (6–15).

These results together with propagation velocity and characteristic impedance are also listed in Table 12–2.

Termination of transmission lines. Because Z_0 is real, the transmission line can be terminated at any point with a resistance R of magnitude Z_0 without altering its transmission properties as observed anywhere else in the line. This is common practice in electronic circuitry where reflected pulses from the end of the line are to be avoided. Such lines are said to be *terminated in their characteristic impedance.* If $R = Z_0$ for the case illustrated in Fig. 12–13(a), then the current pulse, consisting of a positive charge density $\lambda(x, t)$ moving to the right on the top line and an equal pulse of negative charge density $-\lambda(x, t)$ moving to the right on the bottom line, vanishes as the positive and negative charges combine in the resistor.

A qualitative understanding of the effect of terminating in a higher or lower resistance can be obtained by considering the extreme cases of $R \to \infty$, Fig. 12–13(b), and $R = 0$, Fig. 12–13(c). In the first case the positive and negative charge pulses are reflected back from the end of their respective wires thus increasing the charge density and hence voltage near the end of the line. The oppositely flowing charges of same polarity reduce the current near the end of the line—obviously the current at the end must be exactly zero. For a shorted line (Fig. 12–13c), the oppositely charged pulses flow through each other and return on the opposite lines. The net charge on each line is

Transmission lines of negligible resistance

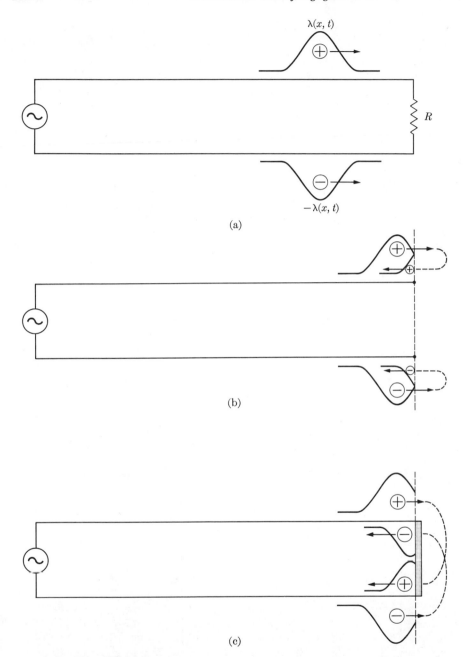

Fig. 12–13 (a) Parallel wire transmission line with electric pulse transmitted from the generator and propagating toward the termination resistor R. (b) Parallel wire transmission line open ended, $R \to \infty$. (c) Parallel wire transmission line shorted at end, $R = 0$.

Table 12–2

TRANSMISSION LINE PARAMETERS

	Parallel wires	Wire parallel to conducting plane	Strip line	Strip parallel to conducting plane	Coaxial cable
l	$\dfrac{\mu}{\pi}\ln\left[\dfrac{d}{a}+\sqrt{\left(\dfrac{d}{a}\right)^2-1}\right]$	$\dfrac{\mu}{2\pi}\ln\left[\dfrac{d}{a}+\sqrt{\left(\dfrac{d}{a}\right)^2-1}\right]$	$\dfrac{\mu d}{a}$	$\dfrac{\mu d}{a}$	$\dfrac{\mu}{2\pi}\ln\dfrac{b}{a}$
c	$\dfrac{\pi\epsilon}{\ln\left[\dfrac{d}{a}+\sqrt{\left(\dfrac{d}{a}\right)^2-1}\right]}$	$\dfrac{2\pi\epsilon}{\ln\left[\dfrac{d}{a}+\sqrt{\left(\dfrac{d}{a}\right)^2-1}\right]}$	$\dfrac{\epsilon a}{d}$	$\dfrac{\epsilon a}{d}$	$\dfrac{2\pi\epsilon}{\ln b/a}$
Z_0	$\dfrac{1}{\pi}\sqrt{\dfrac{\mu}{\epsilon}}\ln\left[\dfrac{d}{a}+\sqrt{\left(\dfrac{d}{a}\right)^2-1}\right]$	$\dfrac{1}{2\pi}\sqrt{\dfrac{\mu}{\epsilon}}\ln\left[\dfrac{d}{a}+\sqrt{\left(\dfrac{d}{a}\right)^2-1}\right]$	$\dfrac{d}{a}\sqrt{\dfrac{\mu}{\epsilon}}$	$\dfrac{d}{a}\sqrt{\dfrac{\mu}{\epsilon}}$	$\dfrac{1}{2\pi}\sqrt{\dfrac{\mu}{\epsilon}}\ln\dfrac{b}{a}$
v	$\dfrac{1}{\sqrt{\mu\epsilon}}$	$\dfrac{1}{\sqrt{\mu\epsilon}}$	$\dfrac{1}{\sqrt{\mu\epsilon}}$	$\dfrac{1}{\sqrt{\mu\epsilon}}$	$\dfrac{1}{\sqrt{\mu\epsilon}}$

Parallel wires diagram: two circles of radius a separated by $2d$.

Wire parallel to conducting plane diagram: circle of radius a at distance d from plane.

Strip line diagram: $a \gg d$.

Strip parallel to conducting plane diagram: $a \gg d$.

Coaxial cable diagram: inner radius a, outer radius b.

reduced near the end, becoming zero at the end as required by the condition that the voltage drop across the termination is zero. The oppositely flowing and oppositely charged pulses on each line exactly double the current *at the end* of the line with a lesser current increase *near the end.*

In actual transmission lines the finite resistance of the line itself causes the pulse to spread out and its peak height to diminish as it travels along the line. This attenuation in pulse height can be reduced near the end by terminating the line in $R < Z_0$ when maintaining the current level is important or in $R > Z_0$ when maintaining the voltage level is important. While this technique helps maintain the pulse height it does not prevent distortion of the pulse shape. The optimum value of R must be determined experimentally or by detailed analyses for the particular case of interest.*

12–11 Waveguides

Waveguides are constructed so as to confine all electromagnetic radiation within their walls, thus eliminating effects of inductive and capacitive coupling as well as preventing power loss due to radiation. Such a waveguide may consist of a hollow cylinder formed from a highly conducting material such as copper. Electric energy is carried through the tube by electromagnetic waves generated and received at the ends of the tube by antennalike devices. Because electric fields are zero within perfect conductors, the waves are unable to pass through the walls of the guide but are reflected repeatedly until they have traveled the length of the guide.

Even for copper walls which are not "perfect" conductors, the radiation escaping through them is negligible. The amplitude of the electric field penetrating the copper walls decreases with the distance of penetration according to an exponential law. For a plane wave polarized in the *x*-direction,

$$E_x = E_{0x} e^{-(z/\delta)} \cos \omega \left(t - \frac{z}{\delta \omega} \right),$$

where z is the distance of penetration and

$$\delta = (\mu \sigma \pi f)^{-1/2} = \lambda (\epsilon / \pi \sigma)^{1/2}$$

is the skin depth that was obtained in Section 12–5. For copper conductors this skin depth is approximately equal to $2.2 \times 10^{-10} \lambda$. Some power is dissipated by currents induced in the waveguide walls, but this is usually small and is proportional to the skin depth times the frequency. Hence-

---

* For a detailed treatment of transmission lines for modern high-frequency applications see R. E. Matick, *Transmission Lines for Digital and Communication Networks*, McGraw-Hill, New York, 1969.

forth, it will be assumed that the waveguide walls are perfectly conducting and that the medium is air or vacuum, where the velocity of light is c.

Reflection from an infinite sheet. To gain an understanding of the principles involved in waveguides, the reflection of electromagnetic waves from the surface of a large conducting sheet located in the xz-plane of Fig. 12–14 will be considered. Since the angle of incidence equals the angle of reflection, the angle made by $\mathbf{a}'$ with the z-axis is the negative of the angle made by $\mathbf{a}$, and both rays lie in the yz-plane. It is convenient to express the waves in an exponential form, where the incident and reflected x-components of $\mathbf{E}$ are the real parts of

$$E_x = E_{0x} \exp\{j\omega[t - (-y \sin \theta + z \cos \theta)/c]\}$$

and

$$E'_x = E'_{0x} \exp\{j\omega[t - (y \sin \theta + z \cos \theta)/c]\}.$$

Since the tangential components of $\mathbf{E}$ must be continuous across the boundary and since $\mathbf{E} = 0$ inside the plate, the boundary condition at $y = 0$ is $E_x + E'_x = 0$. This requires that $E_{0x} = -E'_{0x}$, which is equivalent to a 180°

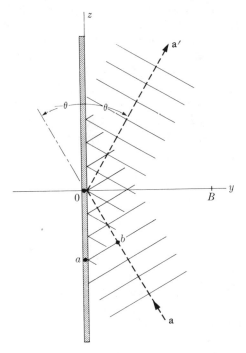

Fig. 12–14 Plane electromagnetic waves reflected from a conducting sheet. The lines perpendicular to the unit propagation vectors represent the positive peaks of the magnetic field.

phase shift upon reflection. The superposition of the incident and reflected waves must be done in such a way that the resulting function of y is sinusoidal; this can be accomplished by subtracting E'_x from E_x and dividing by $2j$. The resulting electric field is given by

$$E_{0x} \exp\left[j\omega\left(t - \frac{z}{c}\cos\theta \right) \right] \sin\left(\frac{\omega}{c} y \sin\theta \right), \qquad (12\text{–}33)$$

which represents an electric disturbance traveling along the z-axis parallel to the plates with a velocity $v_p = c/\cos\theta$. This velocity is known as *phase velocity* and clearly is greater than the velocity of light. No violation of relativity occurs here, however, since no signal can be transmitted along the guide with this velocity.

Phase and group velocity. The significance of phase velocity can be understood by studying Fig. 12–14, where an incident ray traveling in the direction **a** is reflected in the direction **a′** by the conducting sheet. The dotted line represents the path of the beam and the solid lines perpendicular to the path represent amplitude maxima of the magnetic field. The points of intersection between the incident maxima and the reflected maxima have twice the amplitude of either beam, and one of these points is indicated by a on the conducting sheet. As the incident ray progresses upward in the z-direction, this double-amplitude maximum reaches the origin in the same time it takes the wave to travel from b to the origin. The velocities of these double-amplitude maxima along the plate are therefore $c/\cos\theta$, which is the phase velocity, since phase velocity is defined as the velocity with which a disturbance in the electromagnetic wave is propagated along the waveguide.

The velocity with which the signal* is propagated along the guide is somewhat slower. It is just the component of the wave velocity parallel to the guide. This velocity, $v_g = c\cos\theta$, is known as group velocity. Group velocity and phase velocity are therefore related in a waveguide by

$$v_g v_p = c^2. \qquad (12\text{–}34)$$

Propagation between parallel plates. In addition to wave propagation in the z-direction, Eq. (12–33) also reveals a standing wave pattern along the y-axis, normal to the plate, having the wavelength $\lambda_n = 2\pi c/\omega \sin\theta = \lambda_c/\sin\theta$, where λ_c is the actual wavelength of the radiation. If a second plate is placed at $y = B$ in Fig. 12–14, then the boundary conditions require that $E_x + E'_x = 0$ at $y = B$. Equation (12–33) immediately gives the condition at $y = B$ that $\sin[(\omega/c)y \sin\theta] = 0$ or $(\omega/c)B\sin\theta = \pi m$, where m is an

---

* A good discussion of waveguides and the velocity with which signals are propagated is given in *Fundamentals of Electric Waves*, by Skilling, Chapter XIII, John Wiley & Sons, New York, 1942.

integer. This may be written as

$$B = \frac{m\lambda_c}{2 \sin \theta} = \frac{1}{2} m\lambda_n.$$

The variation of E_x in the y-direction cannot have arbitrary wavelengths, but only a discrete set.

If a wavelength along the z-axis is defined by the phase velocity as $\lambda_p = v_p/f$, the following relations between λ_p and λ_n may be obtained:

$$1 = \sin^2 \theta + \cos^2 \theta = \frac{\lambda_c^2}{\lambda_p^2} + \frac{\lambda_c^2}{\lambda_n^2}$$

or

$$\frac{1}{\lambda_c^2} = \frac{1}{\lambda_p^2} + \frac{1}{\lambda_n^2}. \qquad (12\text{--}35)$$

If λ_c is as large as λ_n, then λ_p will be infinite. Since $\lambda_p f = v_p$, this corresponds to an infinite phase velocity, which, according to Eq. (12–34), means that the group velocity is zero and no signal is propagated. The frequency corresponding to $\lambda_c = \lambda_n$ is known as the *cutoff frequency*. For frequencies below the cutoff frequency, λ_c is larger than λ_n and this required λ_p in Eq. (12–35) to be an imaginary number which may be represented by jR. Substitution of jR for $\lambda_p = 2\pi c/\omega \cos \theta$ in Eq. (12–33) yields

$$E_{0x} e^{j\omega t} e^{-(2\pi z/R)} \sin \left(\frac{\omega}{c} y \sin \theta\right)$$

and the E_x-components of all waves with frequencies below the cutoff frequency are attenuated exponentially in the z-direction.

Waves containing only an E_y-component can propagate between the parallel conductors in the z-direction and in fact there is no frequency limitation for such waves. This is the transmission line mode discussed in the previous section where **E** was perpendicular to the wires of the parallel wire transmission line. The parallel plate transmission line (strip line) for this mode has been discussed in Section 12–10 and the reader can easily verify that the geometry of a coaxial cable also permits a mode in which **E** is everywhere perpendicular to the conductors and **H** is parallel to the conductors but orthogonal to the direction of propagation.

If the center wire of the coaxial cable were removed, however, a waveform with **E** everywhere perpendicular to the conductor *could not be constructed.* Similarly if conducting sides were added in the yz-plane of Fig. 12–14 to form a hollow rectangular waveguide, then again orthogonality of **E** with the conducting sides at all points could not be achieved. Thus in waveguides, as opposed to transmission lines, the cutoff frequency applies to all possible waveforms which can be transmitted.

PROBLEMS

12–1 What is the frequency of electromagnetic waves (in free space) of the following wavelengths: 10^{-8} cm (x-rays), 5×10^{-5} cm (yellow light), 10 cm (microwaves), 300 m (broadcast band)?

12–2 What is the maximum magnetic intensity in a plane radiowave in which the maximum electric intensity is 100 μV m^{-1}?

12–3 The average power radiated by a broadcasting station is 10 kW. Assume the power to be radiated uniformly over the surface of any hemisphere with the station at its center.
a) What is the magnitude of the Poynting vector at points on the surface of a hemisphere 10 km in radius? At this distance, the waves can be considered plane.
b) Find the maximum electric and magnetic intensities at points on this hemispherical surface.

12–4 A radar transmitter emits a uniform conical beam of radiant energy. The solid angle of the cone is 0.01 steradian. The maximum electric intensity at a distance of 1 km from the transmitter is 10 V m^{-1}.
a) What is the maximum magnetic intensity?
b) What is the total power in the beam?

12–5 Assume that a 100-W incandescent lamp radiates all its energy in a sinusoidal electromagnetic wave, and that the energy is emitted uniformly in all directions. Compute the maximum electric and magnetic intensities at a distance of 2 m from the lamp.

12–6 Pure water has a dielectric constant of 80 in static electric fields but its index of refraction for visible light is 1.33. Calculate the ratio of the static to this high-frequency dielectric constant and account qualitatively for the discrepancy.

12–7 A certain broadcasting station radiates 50,000 W at 10 MHz. If its antenna can be considered to be a dipole of length $l = 4\lambda$ what effective current must it generate?

12–8 A plane-polarized beam of monochromatic light, directed along the positive y-axis, is represented by $E_z = G \cos \omega(t - y/c)$. An atom located at the origin can be polarized by displacing the center of a single electron through the distance **r** from the center of the very massive nucleus, located at the origin. The motion of the electron is given by

$$\frac{d^2\mathbf{r}}{dt^2} + k\frac{d\mathbf{r}}{dt} + \omega_0^2\mathbf{r} = -e\mathbf{E}/m.$$

a) Derive an expression for the motion of the electron due to the polarized light.
b) Write the expression for the oscillating dipole moment of the atom, assuming k to be negligible, and obtain an expression in spherical coordinates for the energy being re-radiated by the atom.
c) Find the value of k in the differential equation for **r**, assuming that the only energy loss is that due to the re-radiation (scattering) by the atom.
d) If ordinary light is scattered along the positive x-axis, does it appear polarized?

12–9 Sunlight is scattered by atoms and molecules in the upper atmosphere. Ordinary light can be considered as a collection of randomly timed pulses with random directions of polarization, perpendicular to the direction of propagation. With light traveling parallel to the y-axis, each pulse contains a long train of waves whose x- and z-components are given by $E_x = G_x \cos \omega(t - y/c)$ and $E_z = G_z \cos \omega(t - y/c)$, where $G^2 = G_x^2 + G_z^2$.

a) If sunlight parallel to the y-axis strikes the atom described in Problem 12–8, describe the state of polarization, if any, in the light scattered along the positive x-axis.

b) Determine the frequency-dependence of the scattered light and state which colors should predominate in this light if ω_0 corresponds to ultraviolet frequencies. Do your conclusions agree with observations on light scattered by the sky?

12–10 Prove the first two of Fresnel's equations.

12–11 Prove the last two of Fresnel's equations.

12–12 A beam of light is directed perpendicular to the polished surface of glass having the index of refraction n. Calculate the fractions of the light that are reflected and transmitted.

12–13 Show that when a plane electromagnetic wave is incident normally on a plane boundary between two different dielectrics, the rate at which energy leaves the surface in the refracted and reflected beams is equal to the rate at which energy is incident on the surface.

12–14 A beam of polarized light impinges normally on a plate glass window whose index of refraction is 1.5. Calculate the fraction of the original light energy that is reflected (a) from the first surface, and (b) from the second surface. [*Hint:* Sines and tangents of small angles are approximately equal to these angles.]

12–15 When ordinary light strikes a plane surface of glass at Brewster's angle, all of the light having its electric vector in the plane of incidence is refracted and none is reflected. The reflected light is then polarized perpendicular to the plane of incidence. However, not all of the light with this polarization in the original beam is reflected.

a) Calculate the fraction of the original light energy with this polarization that is reflected by glass having the index of refraction 1.5.

b) What fraction of the original light energy, with all polarizations, is reflected?

12–16 What fraction of the energy of a plane electromagnetic wave in air, incident normally on a plane water surface, is reflected if the frequency of the wave is such that the dielectric coefficient of the water is 80? Compare with the fraction reflected if the wave is of visible light, for which the index of refraction of water is 1.33.

12–17 When a beam of light strikes an interface between air and glass normally, determine any change of phase in the reflected light when the beam progresses (a) from air to glass, and (b) from glass to air.

12–18 A beam of light striking a metal surface exerts a radiation pressure. Calculate this pressure from the force-on-a-charge equation, assuming that plane-polarized and monochromatic light impinges normal to the surface of a conductor having very low resistivity ρ.

12–19 If the incident beam in Fig. 12–5 is in glass of index n and the space above contains air, Snell's law gives $\sin \theta'' = n \sin \theta$. When θ is large enough to make $n \sin \theta > 1$, $\sin \theta'' > 1$, then θ is imaginary and total reflection results. However, some electromagnetic fields in the air are required to satisfy boundary conditions.

a) Show that the boundary conditions can be satisfied for this situation by such fields in the air as that described by

$$E = M \exp\left[j\omega\left(t - \frac{x \sin \theta''}{c}\right)\right] \exp\left[-\omega_z \frac{\sqrt{n^2 \sin^2 \theta - 1}}{c}\right].$$

b) Show that the Poynting vector in this field is directed parallel to the surface.

12–20 At the origin of spherical coordinates a charge suddenly changes from $q = 0$ to $q = q_0$ at $t = 0$ and then remains constant.
a) Plot q versus t at the origin.
b) Plot V versus t at $r = c$ times 1 s.

12–21 An electric field pulse is described by the relation

$$E_z = \exp\left[-(vt - y)^{100}\right].$$

Calculate **H** and plot **E** and **H** versus y at $vt = 0$ and at $vt = 3$.

12–22 A certain oscillating dipole like that in Fig. 12–8 has such a low frequency that the second time derivatives in Eqs. (12–14) and (12–15) can be neglected and the potential equations like Eq. (8–25) are valid.

a) Calculate V and **A** for this dipole, using the simpler expressions.
b) Calculate **E** and **H** from these potentials and express them in the forms of Eqs. (12–25) and (12–26), even though there are fewer terms in these simpler expressions.
c) For both **E** and **H** calculate the amplitudes of the terms that are missing from the expressions in (b) to the largest amplitudes in these expressions, assuming $f = 10^4$ Hz and $r = 10$ m.

12–23 Calculate the time average of the total power radiated away from the low-frequency dipole, using the simpler field expressions calculated in Problem 12–22(b).

12–24 Calculate the vector potential **A** between the cylinders of the long coaxial line described in Problem 10–8(a) for direct currents, and (b) for low-frequency sinusoidal oscillations.

12–25 Show that for any two-conductor and two-dimensional transmission line the self-inductance per unit length is given by $l = \mu(G_1 - G_2)/(H_1 - H_2)$, where the G's and H's are obtained from the complex relation $f(Z) = G + jH$. It should be assumed that $G = G_1$ over the first conductor and $G = G_2$ over the second, while $\int_s dH = H_1 - H_2$ when integrated over that part of the surface of either conductor that is in contact with the electromagnetic waves.

12–26 Assuming that the conductors of the oscillating dipole have negligible resistance, calculate the effective resistance due to the power radiated. This is called the *radiation resistance*.

12–27 For a strip line consisting of two copper strips 0.50 mm wide and separated by a 0.025-mm-thick dielectric of $\mu/\mu_0 = 1$ and $\epsilon/\epsilon_0 = 2$, calculate the characteristic impedance and propagation velocity.

12–28 Show that the power input to a strip line $P = I^2 Z_0$ can be obtained from the Poynting vector using the same approximations in which the fringing fields are neglected. Let w be the width and s the separation of the two lines.

12–29 If higher performance semiconductor drivers and sense amplifiers were utilized on the 360/65 memory described in Section 11–13, a faster read-write cycle could be achieved. Good transmission line characteristics would then be desirable for the lines threading the ferrite cores in order to retain the current pulse forms over the length of the lines. A conducting ground plane placed parallel to each xy-plane will accomplish this.

a) Show qualitatively by graphical means that the boundary conditions are such that a line a distance d from a highly conducting plane has the same transmission line characteristics as two parallel lines separated by a distance $2d$.

b) To provide adequate clearance for the 0.35 mm × 0.53 mm cores the ground plane could be placed about 0.25 mm from the center of the wire. Calculate the velocity of propagation and the characteristic impedance for a 0.05 mm diameter wire so placed. [Note. Assume $\epsilon = \epsilon_0$ while the *average* permeability is $\mu = 3\mu_0$ due to the presence of ferrite cores along the lines whose relative permeability when saturated is about 20.]

c) Assume that coincident current selection of the magnetic core used in this memory requires the rectangular current pulse on the x-line to be coincident with that on the y-line at the core for at least 100 ns. If a given core is 2 m down the x-line but only 10 cm down the y-line, how much extra pulse length must be added to the y-pulse to assure adequate overlap?

Chapter 13 RELATIVITY

A treatment of electromagnetic theory hardly seems complete without some discussion of the key role it played in the development of the theory of relativity. It was, in fact, the inconsistencies between Newtonian mechanics and electromagnetic theory and experiments that led Einstein to develop his theory of special relativity. This chapter discusses some of these inconsistencies and shows how they are obviated by Einstein's theory. Furthermore, the Lorentz force-on-a-charge equation and magnetic fields themselves are shown to be a direct consequence of Coulomb's law and the coordinate transformations of special relativity.

13–1 Galilean Relativity

If you are sitting in a train or car intently watching another train or car, you may experience the sensation of motion when the car you are watching starts to move in the opposite direction. If you travel on a large ocean liner making good speed through a calm ocean, you find you can walk and play ping-pong or tennis as easily as you can on solid ground. Your motions relative to the liner follow the same laws as did your motions relative to the earth. The earth itself is moving rapidly with respect to the sun or the stars in space. Most of our experience indicates that it is our motions relative to our environment that count.

Nevertheless, Sir Isaac Newton based his laws of mechanics upon an absolute frame of reference. This absolute frame was assumed to be established by the so-called "fixed" stars in space. His laws, however, can be shown to describe motion relative to any frame that is moving at constant velocity, so long as the Galilean transformations of Table 13–1 are valid. The relation $t = t'$ was considered to be self-evident but is included here to permit comparison with special relativity. Such frames, moving with constant velocity, are called inertial frames.

Table 13–1

GALILEAN TRANSFORMATION
(FOR MOTION PARALLEL
TO THE x AXIS)

$$x' = x - vt$$
$$y' = y$$
$$z' = z$$
$$t' = t$$

The validity of Newtonian mechanics under Galilean transformations can be demonstrated by considering the equation of motion for the ith particle in an arbitrary assemblage of n particles in which the force between particles is describable by means of potential functions depending on their separation,

$$m_i\ddot{\mathbf{r}}_i = - \sum_{j \neq i}^{n} \mathbf{\nabla}_i[V(|\mathbf{r}_i - \mathbf{r}_j|)]. \tag{13–1}$$

Direct substitution of the transformation equations of Table 13–1 leaves this expression unaltered except for primes on the coordinates. This point can be made more simply by recognizing that the form of the equation $\mathbf{F} = m\mathbf{a}$ is invariant with respect to a Galilean transformation since the acceleration is the same in both coordinate systems.

By use of Newtonian mechanics, the rotation of the earth about its axis can be shown to be responsible for the fact that the earth attracts objects on its surface with less force at the equator than at either of the poles. Strictly speaking the earth's surface is not an inertial frame because it spins about its own axis and also because of its motion about the sun. However, the angular velocities are small and are largely compensated by using the observed gravitational attractions as real forces.

13–2 Low-velocity electromagnetic phenomena

Before discussing the limitations of Galilean relativity which occur at high velocities, it is well to consider low-velocity phenomena which it satisfactorily explains. Of all the basic equations, only the Lorentz force-on-a-charge equation is velocity-dependent and so our attention will be focused on it.

The simple linear generator of Fig. 7–5 has already been analyzed assuming the observer is in the reference frame of the voltmeter. It can as well be analyzed from a frame moving with the sliding rod. From this point of view no emf is generated in the rod because it does not move in the magnetic field, and the magnetic field is time-independent. However, the connections through the voltmeter do move in the opposite direction through the field. These produce the same emf and the same reading on the voltmeter as was found in the other frame.

The homopolar generator in Fig. 7–7 provides another informative example. As is shown in Section 7–6, the emf to be expected is obtained from the second term of Eq. (7–14), since $\dot{\mathbf{B}}$ is zero. This generator also can be analyzed from a frame rotating with the disk, since the rotation is too slow to require analysis based upon general relativity. From the rotating frame, all conductors but the disk in the diagram of Fig. 7–7 rotate in the opposite direction and thus move through the magnetic field. The emf generated in the wire shown above the resistor cancels some of that generated in the wire shown below the resistor. The difference can be shown to be just equal to the emf generated by the disk rotating in the stationary frame. The detailed mechanism of induction may appear to be different when viewed from the different frames but our five electromagnetic equations can be used in either frame to determine the observable quantities.

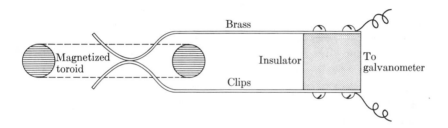

Fig. 13–1 Magnetized toroid and ballistic galvanometer circuit.

A puzzle involving only slow speeds. An experiment using the devices shown in Fig. 13–1 has brought forth so many wrong guesses from scientists with great competence in other fields, that we have called it a puzzle.* Consider a toroid made of homogeneous and isotropic ferromagnetic metal (Fig. 13–1). If this is permanently magnetized parallel to the circle passing through the centers of its circular cross sections, no magnetic field exists outside the metal. Provide two spring brass clips fastened to an insulating block as shown in the diagram. With the two clips connected to a ballistic galvanometer, the clips can be slid back and forth across the toroid without breaking the galvanometer circuit. In the position shown in the diagram all of the magnetic flux ϕ_m passing through the toroid passes through the galvanometer circuit. After this device is pulled to the right over the toroid

---

* The following material is taken from E. M. Pugh, "Electromagnetic Relations in a Single Coordinate System," *Am. J. Phys.*, **32**, 879, Nov. 1964.

no magnetic flux threads the circuit. The flux through the circuit has been changed from ϕ_m to zero without breaking the circuit. Does the galvanometer receive a ballistic impulse? Since no part of the circuit moves in a magnetic field, and since the **B** field is constant in time, Eq. (7–14) predicts a null result. This prediction is correct. The galvanometer does not deflect.

A perfect toroid is not readily available in most laboratories but a magnetron magnet whose gap is filled with a soft iron cylinder can be used instead. Since in this case the magnetic fields surrounding the iron cylinder are not quite zero, some deflection will be observed but it will be much smaller than when the brass clips are pulled through the gap with the iron cylinder removed.

Instead of moving the brass clips to the right over the toroid, the clips can be held still and the toroid moved to the left. In that case, the metallic toroid moves in its own magnetic field. The second term of Eq. (7–14) is not zero but neither is the first. These two terms are of opposite sign and exactly cancel so that $\mathcal{E} = 0$ in this case also.

The case of the moving clips could have been analyzed in a frame attached to the toroid, while the case of the moving toroid could have been analyzed in a frame attached to the clips. In both cases the analysis and the results would be the same.

Frequently scientists and engineers, who are asked to guess the result before the experiment is performed, expect a ballistic deflection greater than that obtained when the clips are pulled through the air gap between the poles of the magnetron magnet. This guess is based on the assumption that the soft iron cylinder between the poles will introduce more magnetic flux into the circuit of the ballistic galvanometer than the air gap. The experimental result is a surprise, but it should present no problem to the student who bases his analysis on Eq. (7–14).

In the above analysis use is made of the fact that, when a permanent magnet that is a conductor is moved, it constitutes a conductor moving in its self-produced field. Thus a type of monopolar generator can be produced by a permanent cylindrical magnet rotating about its own axis with no other magnetic field. Brush contacts are required on the cylindrical surface and on an axis about which the cylinder rotates. Experiments show that such a generator does work.

In all such problems it is well to follow systematically the following steps:

1. Establish the frame of reference in which you plan to solve the problem.

2. Determine whether $\partial \mathbf{B}/\partial t = 0$ in that frame.

3. Determine whether any part of the conducting circuit is moving with a component of motion orthogonal to a **B** field as viewed from the chosen reference frame.

If $\partial B/\partial t \neq 0$ in step 2 or if there is an orthogonal component of motion in step 3, then there will be an emf—except in the special case in which the two effects exactly cancel. If neither of these conditions is met, then there will be no induced emf.

The solutions of these problems are consistent with the Galilean transformations and utilize the Lorentz force equation or the associated equations for emf. While special relativity does not appear to be involved in these low-velocity problems, it actually is. For the Lorentz force equation, which was originally deduced empirically, can be shown to be derivable from Coulomb's law and the Lorentz transformations of special relativity. This is discussed further in Section 13–7.

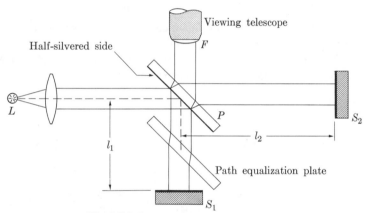

The Michelson-Morley experiment

Fig. 13–2 The Michelson-Morley experiment. (From Panofsky and Phillips, *Classical Electricity and Magnetism*, 2nd ed., Addison-Wesley Publishing Company, Inc., Reading, Mass., 1962.)

13–3 Electromagnetic inconsistencies with Galilean relativity

The velocity of electromagnetic radiation in vacuo is predicted by Maxwell's equations to be $c = 1/\sqrt{\mu_0 \epsilon_0}$, as was shown in Chapter 12. This result is independent, not only of the velocity of the frame of reference, but also of the velocity of the frame relative to the source of radiation. During the late 1800's considerable effort was devoted to attempting to resolve the inconsistencies between this prediction and Galilean relativity.

Michelson-Morley experiment. A number of experiments were devised to determine the absolute velocity of the earth through space by measuring the velocity of light relative to the earth's motion. The most famous of these was the experiment reported by Michelson and Morley in 1887. As shown in Fig. 13–2, light from a source L is split by a half-silvered mirror at P. The two beams then are reflected by S_1 and S_2, respectively, and pass back through the half silvered mirror to the telescope at F. Here the interference fringes are observed. Assuming the instrument is passing through space with a velocity v parallel to S_1P, the time required for the light to traverse the path PS_1P is, according to Galilean relativity, given by

$$t_1 = l_1\left(\frac{1}{c - v} + \frac{1}{c + v}\right) = \frac{2l_1}{c(1 - \beta^2)}, \tag{13–2}$$

where $\beta = v/c$. The time required for the light to traverse the path PS_2P is

$$t_2 = \frac{2}{c}\sqrt{l_2^2 + \delta^2} = \frac{2l_2}{c\sqrt{1 - \beta^2}}, \tag{13–3}$$

where δ is the distance the interferometer travels perpendicular to the line PS_2P during time $1/2t_2$. It has been eliminated from the equation using the relation $2\delta/v = t_2$, from which one obtains $\delta^2 = \beta^2 l_2^2/(1 - \beta^2)$. The difference in optical paths parallel and perpendicular to the velocity v is given by

$$\Delta_1 = c(t_1 - t_2) = \frac{2}{\sqrt{1 - \beta^2}}\left(\frac{l_1}{\sqrt{1 - \beta^2}} - l_2\right). \tag{13–4}$$

If the interferometer is rotated 90°, l_1 and l_2 exchange roles and the new difference in path lengths is

$$\Delta_2 = \frac{2}{\sqrt{1 - \beta^2}}\left(l_1 - \frac{l_2}{\sqrt{1 - \beta^2}}\right). \tag{13–5}$$

As a result of this 90° rotation, the interference fringes for light of wave length λ should shift by N fringes, where

$$N = \frac{\Delta_2 - \Delta_1}{\lambda} = \frac{2(l_1 + l_2)}{\lambda\sqrt{1 - \beta^2}}\left(1 - \frac{1}{\sqrt{1 - \beta^2}}\right); \tag{13–6}$$

and since v is small compared to c, this yields

$$N \cong -\frac{(l_1 + l_2)}{\lambda}\beta^2. \tag{13–7}$$

This predicted shift was not observed by Michelson and Morley even though the sensitivity of their equipment was three times greater than required to observe the 30 km s^{-1} velocity of the earth in its orbit about the sun. Other observers also failed to detect any velocity by this and related techniques.

Ether drag. It was widely felt that the noncorpuscular nature of light predicted by Maxwell's equations required a medium in which the wave motion could be propagated. This medium was called ether and was presumed to fill all of space. If one postulated that the ether was dragged through space by massive bodies such as the earth, then the negative results of Michelson and Morley could be explained.

The concept of an ether drag was, however, inconsistent with a number of other facts. Most evident of those was the aberration of fixed stars. Because of the orbital motion of the earth about the sun, the fixed stars appear from the earth to move in tiny orbits. This is because a telescope must be tipped toward the direction of the earth's motion in order to permit the light to pass down the center of the tube. The angle of tip is α, where classically $\tan \alpha = v/c$. Since the orbital velocity of the earth is about 30 km s^{-1}, $\tan \alpha = 10^{-4}$ or $\alpha \cong 10^{-4}$ rad. This is in good agreement with experimental observations and hence precluded the possibility that the ether, if it existed at all, was dragged with the earth.

The Michelson-Morley experiment thus showed that the velocity of light was independent of the velocity of the earth, relative to the source or to some postulated fixed frame of reference. The aberration of the stars showed that there was no "ether drag" since the angle at which light impinged on the earth did depend on the relative velocity of the source and receiver and the absolute velocity of light. These two results combined to provide a dilemma which was unresolvable in terms of Galilean relativity.

Force between charged particles. Electromagnetic theory presented problems other than those which were obviously associated with the fixed velocity of light. The force between two charged particles stationary in one frame of reference is given simply by Coulomb's Law. If these same particles are observed from another frame moving perpendicular to the line joining the particles, then each particle will produce a magnetic field by its relative motion. The force between the particles is then predicted to be the same identical Coulomb force plus the Biot-Savart term $(\mathbf{v} \times \mathbf{B})$ in which $\mathbf{v}$ is the velocity of the particle and $\mathbf{B}$ is the field generated by the motion of its neighbor. Yet Galilean relativity predicted that the force between particles should be independent of the inertial frame from which the observations are made.

13–4 The postulates of special relativity

The incompatibility of Newtonian mechanics and electrodynamics led to efforts to modify either or both of these otherwise very acceptable theoretical formulations. The unsuccessful efforts, while numerous and instructive in themselves, will not be discussed here.

In 1905 Einstein proposed the postulates of special relativity which were consistent with the experimental observations of the time:

1. All laws of electrodynamics, including the fixed velocity of light in free space, as well as all the laws of mechanics are the same in all inertial frames.

2. The absolute motion of an inertial frame cannot be determined experimentally and accordingly has no physical meaning. Only motion of reference frames relative to each other is significant.

An inertial frame is defined to be one in which objects are not accelerated in the absence of known forces. This is somewhat unsatisfactory as a definition of force. This same vagueness also existed in Galilean relativity. There is no way to avoid this vagueness outside of the broader concepts of general relativity which were developed over the next ten years. In 1911 Einstein published a paper entitled "On the Influence of Gravitation on the Propagation of Light" and in 1916 he published a paper called "The Foundation of the General Theory of Relativity." In these papers he considered systems rotating and accelerating with respect to each other. Only special relativity will be treated in this chapter.

Implicit in the postulates of special relativity is the concept that physical quantities of matter are the same in any frame of reference. For example, the length of a given spectral line, the ratio of the masses of two elementary particles, and the decay time for radioactive material are identical in any frame in which they are at rest. A second concept which should be considered an integral part of special relativity is the invariance of the sense of time. Events progress "forward" in time and never "backward." It is interesting to note that in formal physics only the concept of entropy is asymmetric in time. All other concepts are symmetrical. Entropy can be used to determine the sense of time and we shall consider the positive direction of time to be that of statistically increasing disorder. With these two additional concepts it is possible to retain an exact meaning for length and time when examining the consequences of the postulates of special relativity.

By the seemingly simple concepts of this section, Einstein was able to resolve the dilemmas presented in the previous section and to point the way to a consistent theory for all physical phenomena. In 1864 Maxwell certainly could not have appreciated that his electromagnetic formulation would remain valid under so drastic a change in point of view, any more than Newton could have anticipated the dramatic reformulation of his laws required to explain phenomena at high velocities.

13–5 Derivation of the Lorentz transformation

In order to satisfy his postulates, Einstein found that he needed the coordinate transformation equations published one year earlier by H. A. Lorentz.* These were derived in order to answer the question posed by Lorentz: namely, what kind of transformational relation would have to exist between two reference frames moving relative to each other in order to

---

* H. A. Lorentz, *Proc. Amsterdam Acad.* **6,** 809, 1904.

produce the invariance in Maxwell's equations from one coordinate system to another as suggested by the experimental work of Michelson and Morley? Specifically, he wished to determine what relationship must exist between $x, y, z, t, v, \mathbf{E}, \mathbf{D}, \mathbf{H}, \mathbf{B}$ and ρ in one reference frame and their prime counterparts in the other frame, so that Maxwell's equations would have the same form in both. That is, Maxwell's equations were to transform as follows:

$$\mathbf{\nabla} \cdot \mathbf{D} = \rho \qquad \text{to} \qquad \mathbf{\nabla}' \cdot \mathbf{D}' = \rho',$$
$$\mathbf{\nabla} \cdot \mathbf{B} = 0 \qquad \text{to} \qquad \mathbf{\nabla}' \cdot \mathbf{B}' = 0,$$
$$\mathbf{\nabla} \times \mathbf{E} = -\dot{\mathbf{B}} \qquad \text{to} \qquad \mathbf{\nabla}' \times \mathbf{E}' = -\dot{\mathbf{B}}', \qquad (13\text{–}8)$$
$$\mathbf{\nabla} \times \mathbf{H} = \mathbf{J} + \dot{\mathbf{D}} \qquad \text{to} \qquad \mathbf{\nabla}' \times \mathbf{H}' = \mathbf{J}' + \dot{\mathbf{D}}',$$

where $\mathbf{\nabla}' \times \mathbf{E}'$ is a vector with components $(\partial E_z'/\partial y' - \partial E_y'/\partial z')$, etc. This is a well-defined mathematical problem whose solution is, however, rather complicated.

We will take a somewhat simpler approach to derive these relations beginning with only two requirements, that the velocity of light be the same in two frames of reference moving relative to each other and that the transformation relation must reduce to the Galilean transformation for small velocities. Consider two inertial systems S and S' moving relative to each other along their common x-axis with a constant velocity v. For convenience we will define our axes and synchronize the clocks in the two systems so that at time $t = t' = 0$, the space coordinates are $x = x' = 0$, $y = y' = 0$, and $z = z' = 0$.

Now the origin of S' as observed in the S reference frame is observed to move according to the relation $x - vt = 0$, $y = 0$, and $z = 0$ where t is indicated by a clock fixed in S, located at point x, y, z and synchronized with all other clocks in the S frame. In the S' system, the motion of the origin of S' is given simply by $x' = 0$, $y' = 0$, and $z' = 0$. Since these two sets of equations describe the same thing, both sets must be true if one is true. They can therefore be related by the equations

$$x' = \gamma(x - vt),$$
$$y' = \alpha y, \qquad (13\text{–}9)$$
$$z' = \alpha z,$$

where γ and α are constants which are expected to be dependent on v. The possibility of a nonlinear relationship has been ruled out, since if the relation were nonlinear, then one event in S would correspond to two or more in S', and vice versa. If we now apply the same argument for the position of the origin of S as viewed from the S' and S systems, the resultant equations are

$$x = \gamma'(x' + vt'),$$
$$y = \alpha' y', \qquad (13\text{–}10)$$
$$z = \alpha' z'.$$

Galilean relativity would require $\alpha = \gamma = 1$ in Eqs. (13–9) and (13–10). We can no longer assume this; however, the second postulate of special relativity does require that the parameters be identical in the two frames; that is, $\gamma = \gamma'$ and $\alpha = \alpha'$. Clearly the only possible value for α is one.

Before solving for γ it is desirable to find the relation between t and t' and also between velocities in the two coordinate systems. Substitution for x' from Eq. (13–9) into Eq. (13–10) yields

$$t' = \gamma t + \left(\frac{1 - \gamma^2}{v\gamma}\right)x. \tag{13–11}$$

The differential of Eq. (13–9) is

$$dx' = \gamma(dx - v\,dt) \tag{13–12}$$

and of Eq. (13–11) is

$$dt' = \gamma\,dt + \left(\frac{1 - \gamma^2}{v\gamma}\right)dx. \tag{13–13}$$

Dividing Eq. (13–12) by Eq. (13–13) gives

$$\frac{dx'}{dt'} = \frac{dx - v\,dt}{dt + \left(\dfrac{1 - \gamma^2}{v\gamma}\right)dx} = \frac{\dfrac{dx}{dt} - v}{1 + \left(\dfrac{1 - \gamma^2}{v\gamma^2}\right)\dfrac{dx}{dt}}. \tag{13–14}$$

By definition dx'/dt' and dx/dt are the expressions for instantaneous velocity in S' and S, respectively. Equation (13–14) therefore, is the general relation between velocities as observed in the two coordinate systems.

To determine the form of γ we can take the special case of a photon traveling with velocity c. By the first postulate of special relativity this velocity must appear to have the same value in both reference frames. Therefore, for this case

$$dx'/dt' = dx/dt = c.$$

Substitution of this into Eq. 13–14 results in

$$c = \frac{c - v}{1 + \left(\dfrac{1 - \gamma^2}{v\gamma^2}\right)c},$$

which reduces to

$$\gamma = \frac{1}{\sqrt{1 - \beta^2}}, \tag{13–15}$$

where we have defined $\beta = v/c$. Substitution of this value for γ into Eq. 13–9 and Eq. 13–11 produces directly the Lorentz transformation of Table 13–2.

Table 13–2

LORENTZ TRANSFORMATION
(RELATIVE MOTION
PARALLEL TO THE x
AXES OF BOTH
COORDINATE SYSTEMS)

$$x' = \frac{x - vt}{\sqrt{1 - \beta^2}}$$

$$y' = y$$

$$z' = z$$

$$t' = \frac{t - xv/c^2}{\sqrt{1 - \beta^2}}$$

13–6 Implications of the Lorentz transformation

From the derivation of the Lorentz transformation (Table 13–2) given in the previous section, it is clear that it satisfies the postulate that the velocity of light be the same in all inertial frames. It is also evident that it reduces to the Galilean transformation for small velocities where $v/c \to 0$. Lorentz in his original derivation showed that Maxwell's equations were invariant under this transformation. It only remained to show that formulations could be achieved for all other physical phenomena such that they too were invariant under this transformation.

Demonstration of this has been the sole or partial purpose of much experimental and theoretical work since the postulates were first presented by Einstein in 1905. It must have been very exciting to find that each phenomena investigated did indeed obey the postulates of special relativity for which Lorentz had provided the basic mathematical relation. In this section we will summarize the implications of these relations to mechanics.

Length. In frames of reference moving at the velocity **v** past the observer's frame, lengths perpendicular to the motion appear unchanged ($l_\perp = l$) while lengths parallel to the motion appear shortened ($l_\parallel = l/\gamma$).

Time. When two clocks A and B are synchronized in one frame of reference and B is placed in a frame moving with the velocity v relative to the first frame, B will appear to slow down with respect to A. If the observer is transferred to the moving frame so that the original frame appears to be moving in the opposite direction at the velocity v, then A will appear to slow down with respect to B. This is often stated as "Moving clocks appear to slow down." The time intervals shown by the two clocks are related by $\Delta t' = \gamma \, \Delta t$, where $\Delta t'$ is the "proper" time interval in the observer's frame and Δt is the time interval in the moving frame.

The penetration of the atmosphere by cosmic ray mesons provides a striking example of the physical significance of time dilation. The life time of a μ-meson as measured when at rest is so short that, even with the velocity of light, it could only travel a few centimeters. How can it make the long journey through the atmosphere? Time dilation provides the answer since time for the meson in its frame of reference is slowed down relative to that of the observer on earth. On the other hand, from the meson's point of view, since it cannot exceed the velocity of light, it manages to make the trip through the atmosphere in a short time because the distance it must travel is relativistically shortened.

Simultaneity. The concept of a time that is not absolute, but depends on the motion and position of the observer relative to the event, requires a readjustment of our understanding of simultaneity. Consider for example two frames moving relative to each other with a velocity v parallel to the x-axis. Just as the origin of the two frames coincide a light pulse is emitted at the origin. According to special relativity the wave front as viewed by an observer in *each* frame must be spherical in shape with the center at the origin of *his reference frame*. This paradox results from the erroneous assumption that simultaneity is independent of the frame of the observer. Special relativity, however, requires a mechanism for determining simultaneity such that a measurement of the velocity of light gives c in any reference frame. A suitable mechanism is to define two different times t_1 and t_2 at locations x_1 and x_2 in a given reference frame to be simultaneous if a light wave front emitted at the midpoint between x_1 and x_2 arrives at these points at times t_1 and t_2, respectively.

The consistency of the Lorentz transformation can be checked by formally requiring the wave front to be spherical with radius ct and ct', respectively, in the two coordinate frames moving relative to each other:

$$x^2 + y^2 + z^2 - c^2t^2 = 0 \qquad (13\text{–}16a)$$

and

$$x'^2 + y'^2 + z'^2 - c^2t'^2 = 0. \qquad (13\text{–}16b)$$

Direct substitution of the Lorentz transformation into the second equation shows through simple algebraic manipulation that it does indeed reduce to the first equation as required by our concept of simultaneity.

Because simultaneity is defined in terms of the velocity of light, the constant c takes on greater significance than just the velocity of light. It states in fact that no information can travel faster than the velocity c; for if it could, it would open the possibility that a frame of reference could exist such that the cause of an event would occur after the event itself. In this manner c affects many physical concepts. For example, the "ideal rigid body" used in mechanics cannot exist. If it could, it would permit an impulse at one end to be observed instantaneously at the other and thus establish a time relation independent of location in violation of the simultaneity concept of special relativity.

Velocity. The velocity of an object in the unprimed coordinate system as observed from the primed coordinate system can be obtained by differentiating the equations in Table 13–2. This results in

$$u'_x = \frac{dx'}{dt'} = \frac{u_x - v}{1 - u_x v/c^2}, \tag{13-17a}$$

$$u'_y = \frac{dy'}{dt'} = \frac{\gamma u_y}{1 - u_x v/c^2}, \tag{13-17b}$$

$$u'_z = \frac{dz'}{dt'} = \frac{\gamma u_z}{1 - u_x v/c^2}, \tag{13-17c}$$

where elimination of dt/dt' in the expressions is achieved from $dt'/dt = \gamma(1 - \beta u_x/c)$, as obtained by differentiating the time transformation equation.

Mass and energy. The famous equation

$$U = mc^2 = \gamma m_0 c^2$$

states that the total energy of a particle is proportional to its mass. This concept of mass is different from the Newtonian concept since it consists of an invariant "rest mass" m_0 plus a velocity dependent term which is different for observers in different inertial frames. At low velocities it can be expanded to

$$U = m_0 c^2 + (1/2)m_0 v^2 + (3/8)m_0 v^4/c^2 + \cdots$$

This first-order increase in energy with the square of velocity is consistent with Newtonian mechanics.

Momentum and force. Since mass is velocity- and time-dependent, we must write Newton's laws as

$$\mathbf{F}' = \frac{d}{dt}\mathbf{P} \tag{13-18}$$

rather than in the more familiar form $\mathbf{F} = m\mathbf{a}$. In Eq. (13–18) $\mathbf{F}'$ is the force viewed by the observer while $\mathbf{P}$ is momentum in the reference frame. Interestingly enough, Newton originally expressed his laws as in Eq. (13–18). Performing the differentiation first for a parallel and then for a perpendicular component, we get

$$F'_x = \frac{d}{dt} P_x = \frac{d}{dt}\left(\frac{m_0 u_x}{\sqrt{1 - v_x^2/c^2}}\right)$$

$$= \frac{m_0 \dot{u}_x}{\sqrt{1 - v_x^2/c^2}}\left[\frac{v_x^2}{c^2}\left(1 - \frac{v_x^2}{c^2}\right)^{-1} + 1\right] = \frac{m_0 \dot{u}_x}{(1 - v_x^2/c^2)^{3/2}} \tag{13-19a}$$

and

$$F_y' = \frac{m_0}{(1 - v_x^2/c^2)^{1/2}} \frac{du_y}{dt} = \frac{m_0 \dot{u}_y}{(1 - v_x^2/c^2)^{1/2}}. \qquad (13\text{–}19b)$$

We find that the "force" required to increase the momentum parallel to the reference frame's motion is greater than that required for perpendicular motion. This is conceptually reasonable, since no matter how large the force, the object's velocity plus the reference frame velocity can never exceed c as viewed by the observer.

13–7 Derivation of magnetic field from Coulomb's law

It has been noted in Section 13–3 that the interaction between charged particles cannot be described within the framework of Galilean relativity. Specifically, the force between two charged particles at rest in one inertial frame is described by Coulomb's law, whereas in an inertial frame moving rapidly with respect to the particles, each particle creates a magnetic field by its relative motion and in turn is acted on by the induced magnetic field of its neighbor. Since the Coulomb force is expected to be identical in either frame of reference, the added force from induced magnetic fields indicates that the force between the particles is not independent of the frame of reference. We will now show how this dilemma for Galilean relativity is resolved by Einstein's special relativity.

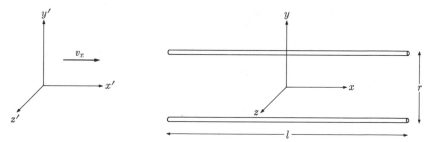

Fig. 13–3 Two positively charged wires moving with velocity $-v_x$ relative to the primed coordinate system.

In order to minimize mathematical complexities we will consider two parallel charged wires of length l large compared to the separation r and with charge per unit length λ as observed in an inertial frame in which the wires are at rest. The entire system will be assumed to be in a vacuum. From an inertial frame moving parallel to the wires with velocity $v = v_x$ (see Fig. 13–3) these parameters are observed to be l', r', and λ'.

From the frame stationary with respect to the wires we expect Newton's laws of motion to be obeyed, namely $F_y = m_y a$ and

$$\Delta r = \tfrac{1}{2}at^2 = \frac{1}{2} \frac{F_y}{m_y} t^2, \qquad (13\text{–}20)$$

where t is the time required for one of the wires to be displaced by a distance Δr which is small enough compared to r that the change in Coulomb force is negligible. In the other inertial frame a similar equation can be written

$$\Delta r' = \tfrac{1}{2}a't'^2 = \frac{1}{2}\frac{F'_y}{m'_y}t'^2. \tag{13–21}$$

From the Lorentz transformation of Section 13–5 the following relations may easily be deduced: $\Delta r'/\Delta r = 1$, since these directions are perpendicular to the relative motion v_x; $m'_y/m_y = \gamma$, since the force is perpendicular, as opposed to parallel, to the relative motion v_x; and $t'^2/t^2 = \gamma^2$. Substituting these into the above equation results in

$$\frac{F'}{F} = \frac{m'}{m}\frac{t^2}{t'^2} = \frac{1}{\gamma}. \tag{13–22}$$

It should be no more disconcerting to learn that the force as observed in the two different frames is different than to learn that length, mass, and time are different. The greater time observed for the motion Δr in the moving frame is in fact consistent with the larger observed mass and weaker force.

We will now express the force between the wires in terms of Coulomb's law for the stationary frame, and Coulomb's law plus an unknown force F'_i in the moving frame. This unknown force will be shown to be exactly the force predicted from induced magnetic fields of nonrelativistic electromagnetic theory. The force equations in the two frames are

$$F = l\lambda E = l\lambda^2/2\pi\epsilon_0 r \tag{13–23a}$$

and

$$F' = l'\lambda'E' = l'\lambda'^2/2\pi\epsilon_0 r' + F'_i, \tag{13–23b}$$

where E is the electric field on one wire due to the charges on the other. We can now solve for F'_i using the previously determined relation $F/F' = \gamma$ and $\lambda'/\lambda = \gamma$. This latter expression results from the shortening of the wires as observed in the moving frame combined with the necessity of conserving total charge on the system independent of the reference frame. The result is

$$F'_i = F/\gamma - l'\lambda'^2/2\pi\epsilon r'$$

$$= \frac{1}{\gamma}l'\gamma\frac{\lambda'^2/\gamma^2}{2\pi\epsilon r'} - l'\lambda'^2/2\pi\epsilon r'$$

$$= -(l'\lambda'^2/2\pi\epsilon r')[\mu\epsilon v^2] = -l'\lambda'\left(\frac{\lambda'\mu v^2}{2\pi r'}\right), \tag{13–24}$$

where by classical electrodynamic theory $l'\lambda'$ is the total charge $Q' = Q$, $-\lambda'v$ equals the current I', and $H' = I'/2\pi r'$. With these substitutions we obtain

$$F'_i = Q\mu H'v = QB'v. \tag{13–25}$$

Thus beginning with the Coulomb force and the Lorentz transformation we have been able to deduce the necessity for a magnetic field term. Our geometry was selected to minimize mathematical complexity; however, the reader can verify by inspection that the total force between the wires, as observed from the relatively moving inertial frame,

$$F' = Q(E' + vB') \tag{13-26}$$

is consistent with the more general form for the Lorentz force-on-a-charge equation

$$\mathbf{F}' = q(\mathbf{E}' + \mathbf{v} \times \mathbf{B}'). \tag{13-27}$$

PROBLEMS

13-1 Show that direct substitution of the Galilean transformation equations into Eq. (13-1) leaves the expression unaltered except for primes on the coordinates.

13-2 A permanent magnet shaped like a large oval has a 1000-turn solenoid of insulated copper wire wrapped around one leg as illustrated in Fig. 13-4. The insulation has been removed from the side of the solenoid so that contact can be made by the brush b anywhere along the length of the solenoid. The brush contact is now caused to slide back and forth over the solenoid at 60 cps so that the galvanometer G is in a closed electrical circuit in which the number of turns of wire about the magnet varies according to the relation $N = N_0 (1 - 0.5 \sin \omega t)$, where $N_0 = 1000$ and $\omega = 2\pi 60$. Calculate the galvanometer voltage assuming the flux in the magnet is $\phi = 10^{-3}$ Wb, that there is no stray flux outside the magnet itself, and that the resistance of the meter is large compared to that of the solenoid.

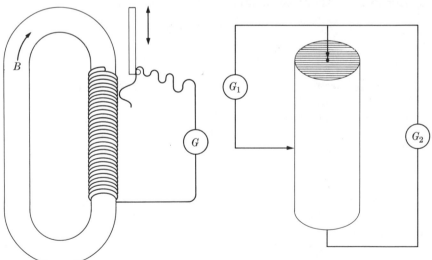

Fig. 13-4 A proposed ac generator consisting of a coil wrapped around an oval-shaped permanent magnet with sliding contact.

Fig. 13-5 A proposed dc generator consisting of a rotating alnico magnet with electrical brush contact to external circuit as shown.

13–3 A cylindrical permanent magnet of Alnico 3 (an electrically conducting material composed of Fe, Ni, and Al) is caused to rotate about its axis at 60 cps. Its length is 0.2 m, cross-sectional area is 0.001 m², and remanence magnetization parallel to its axis is $B_r = 0.7$ Wb. Two galvanometers of very high internal impedance are connected to the rotating cylindrical magnet as shown in Fig. 13–5. Galvanometer G_1 is connected by brush contact between the center of one end and the outside surface while G_2 is contacted to the centers of the two end surfaces.

a) Calculate the voltage measured by G_1 and G_2 in their frame of reference.

b) What is the voltage measured by G_1 and G_2 as observed from the reference frame of the rotating magnet?

13–4 Show that substitution of the Lorentz transformation into Eq. (13–8b) produces Eq. (13–8a).

13–5 In the upper atmosphere cosmic ray particles create π-mesons which in turn decay to μ-mesons and neutrinos. The μ-mesons, which have a rest mass 207 times that of an electron and a mean life of 2.5×10^{-8} s in the rest frame, can be observed at the surface of the earth after making the long journey through the atmosphere. For the purpose of the following calculations assume they start 10 miles or 16 km above the earth and travel at a constant velocity.

a) What is the minimum possible velocity of the meson as observed from the earth?

b) From the meson's point of view, at what minimum velocity is the earth coming up to meet it?

c) How much larger is the meson's energy in the earth's frame of reference than in its own?

d) What is the total distance the meson travels to the earth as measured by the meson?

13–6 An intriguing gedanken experiment consists of placing a spotlight inside a long box at one end so that the light impinges on perfectly absorbing material at the other end. The box is placed on a frictionless surface and is initially at rest. The momentum of the light beam when turned on is $\mathbf{P} = ALS/c^2$ where A is the area of the beam, L is the length of the box, and $\mathbf{S}$ is the Poynting vector. The box of mass M must move in the opposite direction to conserve momentum and will move a distance x in time t. Since the center of mass of the system cannot move without external forces, the light beam must have carried some mass with it.

a) Find the total light energy U absorbed at the far end of the box.

b) Find the velocity of the box when the light is on.

c) How far does the box move in time t?

d) In terms of U (the light energy transmitted), find the mass m transmitted by the light beam to the far end of the box.

13–7 Two flat metallic plates, large compared to the separation d between them, are placed perpendicular to the z-axis. A voltage $V = Ed$ is placed across them so that they each acquire a charge density per square meter of σ. A particle having a positive charge q and mass m is permitted to move freely in the influence of the field from a rest position on the positive plate to the negative plate.

a) Using the Lorentz transformation, calculate the ratio of forces F'/F acting on the particle as observed from a reference frame, moving with velocity $v = v_x$ relative to the plates, divided by the force observed from the rest frame.

b) Using Coulomb's law and the Lorentz transformation, calculate the force F' acting on the particle as observed from the moving frame.

c) Show that the force F' consists of a Coulomb force plus a term which can be expressed as qvB' where B' is the magnetic field generated by the relative motion v of the charged plates.

13–8 As the charged particle in Problem 13–7 begins to gain velocity in the z-direction, there will be an x-component of the Lorentz force as observed from the moving frame. Since the particle must arrive at the same point on the negative plate, independent of the motion of the reference frame, we must conclude that this force does not alter the component of velocity in the x-direction. How is this possible? Give a qualitative argument.

13–9 Using Eq. (13–10) to provide a general solution for Problem 13–7, show that a consistent solution for F_x and F_y can be found by requiring that the velocity of the particle in the x-direction remains constant. Use $F_y' = (1 - \beta^2)q\sigma'/\epsilon_0$ as found in Problem 13–7 and $F_x' = qv_z'v_x'\sigma'\mu_0$ as inferred from Problem 13–8. Note that γ is time dependent for both the F_x and F_y equations.

ANSWERS TO ODD NUMBERED PROBLEMS

To eliminate ambiguity, and at the same time to save space, the following commonly used conventions have been adopted in these answers:

$$a/b + c \equiv \frac{a}{b} + c, \qquad a/(b + c) \equiv \frac{a}{b + c},$$

and

$$\cos \theta/4\pi r^2 \sin \beta \equiv \frac{\cos \theta}{4\pi r^2 \sin \beta}.$$

Chapter 1

1-1 a) 1.52×10^{11} N   b) 17,100 tons   c) 2.02×10^{-16}
1-3 0.119 m
1-5 0.66×10^{16} rev/s
1-7 $a/\sqrt{2}$
1-9 24 and 2 statcoul
1-11 900 dynes, 8.854×10^{-12}
1-13 If 1 statvolt $= \alpha$ abvolts, $\alpha = 2.998 \times 10^{10}$

Chapter 2

2-3 a) $\partial a_r/\partial \phi = -\mathbf{i} \sin \phi + \mathbf{j} \cos \phi$
b) $\partial a_R/\partial \theta = +\mathbf{i} \cos \theta \cos \phi + \mathbf{j} \cos \theta \sin \phi - \mathbf{k} \sin \theta$
c) $\partial a_R/\partial \phi = -\mathbf{i} \sin \theta \sin \phi + \mathbf{j} \sin \theta \cos \phi$
d) $\partial a_\phi/\partial \phi = -\mathbf{i} \cos \phi - \mathbf{j} \sin \phi$
a') a_ϕ, (b') a_θ, (c') $a_\phi \sin \theta$, (d') $-a_r$ or $-a_R \sin \theta - a_\theta \cos \theta$
2-5 a) $\mathbf{i} B_x(A_y C_y + A_z C_z) - \mathbf{i} C_x(A_y B_y + A_z B_z)$
$\qquad + \mathbf{j} B_y(A_x C_x + A_z C_z) - \mathbf{j} C_y(A_x B_x + A_z B_z)$
$\qquad + \mathbf{k} B_z(A_x C_x + A_y C_y) - \mathbf{k} C_z(A_x B_x + A_y B_y)$
b) Add $A_x B_x C_x$ in the plus $\mathbf{i}$ and subtract it in the minus term. Treat the $\mathbf{j}$ and $\mathbf{k}$ terms similarly.
2-7 -25×10^{-9} C

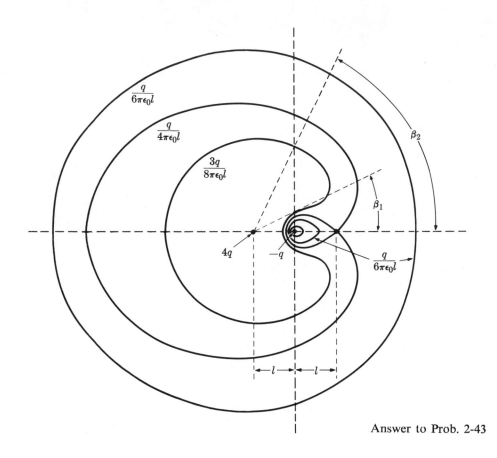

$$\frac{q}{6\pi\epsilon_0 l}$$

$$\frac{q}{4\pi\epsilon_0 l}$$

$$\frac{3q}{8\pi\epsilon_0 l}$$

β_2

β_1

$4q$ $-q$

$$\frac{q}{6\pi\epsilon_0 l}$$

$\leftarrow l \rightarrow \leftarrow l \rightarrow$

Answer to Prob. 2-43

2-9 a) $10^4(-\mathbf{j}\,1.55 + \mathbf{k}\,4.23)$ b) $V_A = -3600\ \text{V},\ V_B = +3600\ \text{V}$
c) $21.6 \times 10^{-6}\ \text{J}$

2-11 a) $k\sigma(1 - z/\sqrt{b^2 + z^2})/2\epsilon_0$ b) $k\sigma/2\epsilon_0$
c) $k\sigma b^2/4\epsilon_0 z^2$, since $(b^2 + z^2)^{-1/2} \cong (1 - b^2/2z^2)/z$

2-13 a) $4\ \text{NC}^{-1}$ upward   b) $1.28 \times 10^{-18}\ \text{N}$ upward

2-15 a) $\mathbf{j}0.48\ \text{N} + \mathbf{k}1.31\ \text{N}$ and $\mathbf{j}0.48\ \text{N} - \mathbf{k}1.31\ \text{N}$ b) $4\ \text{V}$

2-17 $87.0 \times 10^6\ \text{m s}^{-1}$

2-19 a) σ/ϵ_0,   b) zero

2-21 a) $k(1 - 3d/2z + \cdots)\sigma a^2 d/2\epsilon_0 z^3$ b) $kp/2\pi\epsilon_0 z^3$, where $p = \sigma\pi a^2 d$

2-25 c) zero   d) $\mathbf{i}39 + \mathbf{j}156 - \mathbf{k}52$ e) $90°$

2-27 $\cos\theta = l_1 l_2 + m_1 m_2 + n_1 n_2$

2-29 $b(\sqrt{b^2 + a^2} - b)$

2-31 a) $q/4\pi\epsilon_0 a$ b) $q/4\pi\epsilon_0\sqrt{a^2 + z^2}$
c) $(qe/4\pi\epsilon_0 am)^{1/2}$, max. vel. $(qe/2\pi\epsilon_0 am)^{1/2}$

2-33 a) $2\pi A$ b) $2\pi A$ c) zero

2-35 a) $\mathbf{i}qx/2\pi\epsilon_0(b^2 + x^2)^{3/2}$ b) $0.657qq'/\pi\epsilon_0 b$

2-39 a) $\{a_r Z - a_z(\sqrt{Z^2 + r^2} - r)\}\lambda/4\pi\epsilon_0 r\sqrt{Z^2 + r^2}$

b) $\{\ln (Z + \sqrt{Z^2 + r^2}) - \ln r\}\lambda/4\pi\epsilon_0$

d) The point P is sufficiently general for the infinite wire, but not for the half-infinite wire.

2-41 $r = \sqrt{bd}$, a sphere

2-43 a) Distant l from $-q$ on the extended centerline   b) $q/4\pi\epsilon_0 l$

c) Crosses axis at l, $-0.235l$, and $-4.235l$. Angles $\beta_1 = 19°55'$ and $\beta_2 = 70°5'$ (see figure on p. 482).

Chapter 3

3-3 9440 V

3-5 5760, 7200, 7200, 7200, and 7200; all in V

3-9 a) 1.2×10^{-3} C, 2.4×10^{-3} C, and 1200 V

b) 0.4×10^{-3} C, 0.8×10^{-3} C, and 400 V

3-11 a) $Q/4\pi\epsilon_0 r^2$ b) $Q(b - a)/4\pi\epsilon_0 ab$ c) $4\pi\epsilon_0 ab/(b - a)$

3-13 a) $Q/2\pi\epsilon_0 lr$ and $Q \ln (r/a)/2\pi\epsilon_0 l$ b) $2\pi\epsilon_0 l/\ln (b/a)$

3-15 60°

3-17 $jG(y^2/2 - a^2/4)$, $jGa^2/4$, $jGa^2/4$

3-19 $a_r \rho r/2\epsilon_0$, $a_r \rho p^2/2\epsilon_0 r$, $a_r \rho(p^2 + b^2 - a^2)/2\epsilon_0 r$

3-21 a) $kq_a/r + kq_b/(r^2 + c^2 - 2rc \cos \theta)^{1/2}$

b) $kq_a/a + kq_b/b$

c) $kq_a/a + kq_b/(r^2 + c^2 - 2rc \cos \theta)^{1/2}$ where $k = 1/4\pi\epsilon_0$

3-23 6°58'

3-25 $(e^{\alpha r^3} - e^{\alpha a^3})\rho_0/3\alpha\epsilon_0 r^2$

3-27 a) $-\epsilon_0/r^3$ b) zero   c) zero   d) $-2\epsilon_0$ e) zero

f) $-\epsilon_0/r^2$ g) $\epsilon_0 n r^{n-2} (\sin n\theta \, \text{ctn} \, \theta - \cos n\theta)$ h) zero

3-29 $\nabla \times \mathbf{A} = 0$ for (a) and (c) but not for (b) or (d). $\nabla^2 V = 0$ for (c) but not for (a)

3-31 $\epsilon_0 xy$ at $z = 0$, $-\epsilon_0 xy$ at $z = l$

3-33 $\rho = -B/r^2$, $\rho = 0$

3-35 $(4M/3) \cos \theta$

3-37 18

3-39 a) zero   b) No, since $\nabla \times \mathbf{F} = -\mathbf{k}4y \neq 0$.

3-41 $\mathbf{H}$ is not conservative.

Chapter 4

4-1 a) $q[h/(h^2 + r^2)^{1/2} - 1]$ b) 60°   c) 84°16'

4-3 b) and c) $-kqh/2\pi\epsilon_0(h^2 + r^2)^{3/2}$

4-5 $q(d^2 - a^2)/a(a^2 + d^2 - 2ad \cos \theta)^{3/2}$

4-7 $\nabla^2 V \neq 0$, where $\mathbf{r}_i = 0$

4-9 a) $\sigma = -(q_1 + q_2)a/2\pi(a^2 + r^2)^{3/2}$

b) $f = q_1^2/16\pi\epsilon_0 a^2$

4-11 a) $q/8\pi\epsilon_0 a^2$ b) $\rho r/3\epsilon_0 + q/4\pi\epsilon_0(4a - r)^2$

4-13 a) $kq/a + kq/[r^2 + (z - b)^2]^{1/2}$

b) $kq/[r^2 + (z - b)^2]^{1/2} + kq/[r^2 + (z + b)^2]^{1/2}$ where $k = 1/4\pi\epsilon_0$

4-15 a) $2kq/r$

b) $2kq(1/a - 1/b) + kq/(r^2 + d^2 - 2rd\cos\theta)^{1/2} + kq/(r^2 + d^2 + 2rd\cos\theta)^{1/2}$

4-17 $p = a^2/d$

4-23 Taking the center of the small sphere as the origin of spherical coordinates,

a) $\sigma a^2(hl - br)/\epsilon_0 hlr$ between spheres and $\sigma a(hl - ba)/\epsilon_0 hl$ inside the small sphere

b) $\rho a^3(hl - br)/3\epsilon_0 hlr$ between spheres and $\rho a(3hla^2 - 2ba^2 - hlr^2)/6\epsilon_0 hl$ inside the small sphere. In each case $l^2 = w^2 + r^2 - 2wr\cos\theta$, where $w = (b^2 - h^2)/h$

4-25 a) $q(1/l - \lambda/l_1 + \lambda/r)/4\pi\epsilon_0$,   where $l^2 = p^2 + r^2 - 2pr\cos\theta$ and $l_1^2 = \lambda^4 p^2 + r^2 - 2\lambda^2 pr\cos\theta$

b) $q\{l_a^3 - p^3(1 - \lambda^2)\}/4\pi a p l_a^3$,   where $l_a^2 = p^2 + a^2 - 2pa\cos\theta$ and $l_{a1} = \lambda l_a$

c) $-q^2\lambda^3(2 - \lambda^2)/(1 - \lambda^2)^2\, 4\pi\epsilon_0 p^2$ d) No   e) $q/4\pi\epsilon_0 p$

4-27 $V = (1 - a/r)bD/(b - a) - Ca^2(r - b^3/r^2)\cos\theta/(b^3 - a^3)$

4-29 For $r > a$, $V = kq(1/r - P_2 a^2/r^3 + 3P_4 a^4/8r^5 \cdots)$

and for $r < a$, $V = kq(1/a - P_2 r^2/2a^3 - 3P_4 r^4/8a^5 \cdots)$. Here $k = 1/4\pi\epsilon_0$.

4-31 $2\pi\epsilon_0/\ln 2$

4-35 a) $\dfrac{4\pi\epsilon_0 V}{p} = \dfrac{z - h}{\{(h - z)^2 + r^2\}^{3/2}} + \dfrac{z + h}{\{(h + z)^2 + r^2\}^{3/2}}$,

$\sigma = -p(r^2 - 2h^2)/2\pi(r^2 + h^2)^{5/2}$

b) $\dfrac{4\pi\epsilon_0 V}{pr\cos\phi} = \dfrac{1}{\{(h - z)^2 + r^2\}^{3/2}} - \dfrac{1}{\{(h + z)^2 + r^2\}^{3/2}}$

$\sigma = -3phr\cos\phi/2\pi(h^2 + r^2)^{5/2}$

where r, ϕ, and z are the cylindrical coordinates

4-37 a) $V = (1/l^3 - a^3/d^3 l_1^3)pr\sin\theta\cos\phi/4\pi\epsilon_0$,

where $l^2 = r^2 + d^2 - 2rd\cos\theta$, $l_1^2 = u^2 + r^2 - 2ru\cos\theta$,

and $u = a^2/d$

$\sigma = -3p(a^2 - d^2)\sin\theta\cos\phi/4\pi l_a^5$,

where $l_a^2 = a^2 + d^2 - 2ad\cos\theta$

b) $V = \dfrac{p}{4\pi\epsilon_0}\left[\dfrac{r\cos\theta - d}{l^3} - \dfrac{a^3(a^2/d - r\cos\theta)}{d^3 l_1^3} + \dfrac{a}{d^2 l_1}\right]$

$\sigma = \dfrac{p}{4\pi a l_a^5}[(5a^2 - d^2)d - (3a^2 + d^2)a\cos\theta]$

4-39 $-\lambda^2 h/2\pi\epsilon_0(a^2 - h^2)$

4-41 a) $(1/l^3 - a^3/d^3 l_1^3)pr\sin\theta\cos\phi/4\pi\epsilon_0$,   where $l^2 = r^2 + d^2 - 2rd\cos\theta$ and $l_1^2 = r^2 + a^4/d^2 - 2a^2 r\cos\theta/d$

b) $\dfrac{-p\sin\theta\cos\phi}{\epsilon_0\tau_s}\left[\dfrac{\gamma^3(1 - \gamma^2)}{(1 + \gamma^2 - 2\gamma\cos\theta)^{5/2}}\right]$

is the general expression for E at the surface and the general charge density is obtained from (σ in the plane) $\cos\phi$.

4-43 a) $2\epsilon_0(3X^2Y - Y^3)/A^3$　　b) $3\epsilon_0 V_e/A$

c) $-4\epsilon_0 V_e x^2/A^3$　and　$-16\epsilon_0 V_e x^3/3\sqrt{3}\,A^3$

4-45 a) $2\pi\epsilon_0/\ln 4/3$　　b) $a/s = 8/15$, $b/s = 3/4$, $d_e/s = 7/60$

4-47 a) $\pi\epsilon_0/\ln 3$　　b) $4.428\epsilon_0/\ln 3$　　c) $y/b = 8/3$

4-49 $C_e/C_0 = 1 + \delta^2/2(B^2 - 1)\ln B$, where $B = b/a$

4-51 a) $2\,V$　and　$-2\,V$　　b) zero　　c) $-0.04\pi\epsilon_0$

4-53 At $\phi = \pi/2$　or　$y = s$

4-55 a) $+1$, $-0.5 + 0.866j$, $-0.5 - 0.866j$

Chapter 5

5-1 a) $26.56\ \mu\,C\,m^{-2}$　　b) $-17.71\ \mu\,C\,m^{-2}$

5-3 a) $0.38\ \mu C$　　b) $7600\,V$　　c) $1.44 \times 10^{-3}\,J$　　d) $1.37 \times 10^{-3}\,J$

5-5 a) $0.020\,C$　　b) $800\,V$　　c) $8\,J$　　d) $2\,J$

5-7 $\epsilon_0 h^2/\pi m e^2$

5-9 a) No　　b) $D_{n2} - D_{n1} = \sigma$

5-11 a) $-(K - 1)q/K4\pi a^2$ and $(K - 1)q/4\pi b^2$　　b) $q/K\epsilon_0 4\pi r^2$

5-13 $D = q/4\pi r^2$, $E = q/4\pi\epsilon_0 Kr^2$, $P = q(K - 1)/K4\pi r^2$, $P_p = (q\,\partial K/\partial r)/K^2 4\pi r^2$

5-15 a) $2\pi K\epsilon_0 l/\ln (b/a)$　　b) $4\pi K\epsilon_0 l/(K + 1)\ln (b/a)$

5-17 $P_{1t} - P_{2t}$

5-19 a) $\epsilon_0 E_0 \cos\theta 3(K - 1)/(K + 2)$

b) $\epsilon_0 E_0 \cos\phi 2(K - 1)/(K + 1)$

5-21 b) $q'/q = (1 - K)/(1 + K)$, $q''/q = 2K(K + 1)$

c) $q^2(K - 1)/(K + 1)16\pi\epsilon_0 h^2$

d) $qh(K - 1)/(K + 1)2\pi(h^2 + r^2)^{3/2}$

e) $q(K - 1)/(K + 1)$

5-23 a) $\lambda^2(K - 1)/(K + 1)4\pi\epsilon_0 h$　　b) $-\lambda h/2\pi(K + 1)(x^2 + h^2)$

5-25 $C = 2P/3$, $F = Pa^3/3$

5-27 $k3KE_0/(2K + 1)$

5-29 $E_r = cr(r - a)/4\epsilon_0$, where $a < r < 2a$, and $E_r = 2ca^4/\epsilon_0 r^2$ where $r > 2a$

5-31 a) $D = q/4\pi r^2$　　b) $q/4\pi a^2$ at $r = a$ and $q/4\pi b^2$ at $r = b$

c) $\sigma_p = -q(b - a)/4\pi a^2 b$ at $r = a$ and $\sigma_p = $ zero at $r = b$

d) $\rho_p = q/4\pi r^2 b$

5-33 a) $\rho_1 = -B\epsilon_1(2\theta + \text{ctn}\ \theta)/r$　　b) $\rho_2 = (-Ba^2\ \text{ctn}\ \theta)/r^3$

c) $\sigma = B\theta(\epsilon_2 + \epsilon_1)$

5-35 $10^{-9}\,J$

5-37 a) $-i25 + j48 - k36$ in N-m　　b) $-i4 + k3$ in N

5-39 a) $a_\theta(kp^2 \sin\theta)/r^3$　　b) $-(2kp^2/r^4)(3a_r \cos\theta + a_\theta \sin\theta)$

5-41 $V = \dfrac{k}{r}\displaystyle\int_\tau \rho_i\, d\tau + \dfrac{k}{r}\int_\tau \dfrac{\rho_i(x_i x + y_i y + z_i z)\, d\tau}{r^2}.$

The second integrand is small and can be made zero by proper choice of co-ordinates. Whence $\mathbf{D} = \mathbf{a}_r q/4\pi r^2$.

5-43 a) $q^2 d/2\epsilon_0 a^2$　　b) $q^2/2\pi\epsilon_0 a^2$　　c) $-q^2\ \delta d/2\pi\epsilon_0 a^2$. The work done charging the battery is $q^2\ \delta d/\pi\epsilon_0 a^2$.

5-45 a) $0.766 \times 10^{-18}\,m$　　b) $0.383Z \times 10^{-18}\,m$　　c) $1.40 \times 10^{-15}\,m$

Chapter 6

6–1 37.5×10^{18}

6–3 a) 425 W   b) generator 18.1 W, battery 36.1 W, and rheostat 90.3 W
c) 3.03×10^6 J

6–5 2 A, 53 V and 17 V

6–7 a) 6 μA,   b) 2.4 μA   c) 4.5×10^{12} Ω-m

6–9 165 Ω, 1.31×10^{-8} A mm^{-1}

6–11 a) 12 A   b) 4.8 A   c) 57.6 V and 28.8 V

6–13 8.75 A

6–15 1.34

6–17 Maximum allowable resistance is 0.1 Ω, for which 0.306 cm^2 is the minimum cross section (B & S gauge #2). However, the underwriters will allow only 325 A through the largest listed gauge #0000, which has 1.072 cm^2 cross section.

6–19 a) $\{Plk(b - a)/(b + a)\pi\rho\}^{1/2}$ b) independent of d

6–21 $J_i = 3J_0/7$

6–23 a) $-\mathbf{a}_z Ih/2\pi(r^2 + h^2)^{3/2}$ b) $0.0276I$

6–25 $2\pi\epsilon_0 Vt/\ln (b/a)$, $20\pi\epsilon_0 Vt/\ln (b/a)$

6–27 a) $\lambda = 2I\rho\epsilon_0/b$ b) $J = 4I(1 - 1/3 + 1/5 - 1/7 + 1/9 - \cdots)/\pi bd$

6–29 $\rho(1 - a/l - a^4/l^4 \cdots)/4\pi a$

6–31 $\rho(1 - 1.29a/l + \cdots)/4\pi a$

6–33 a) $2\pi\rho_{Cu}/t \ln (b/a)$ b) $2\pi N^2\rho_{Cu}/lk \ln (b/a)$ c) 1.015

6–35 Balanced when closing of the key does not change the galvanometer reading. 107.7 Ω

6–37 a) On 220 V use 0.518 Ω and 0.0706 Ω. On 110 V use 0.129 Ω and 0.0176 Ω. Batteries in series on 220 V and in parallel on 110 V
b) 0.00349   c) 0.00349

6–39 a) $\epsilon_0 J(\rho_2 - \rho_1)/d$ b) $Jd(\rho_1 + \rho_2)/2$

Chapter 7

7–1 a) 1.98 N, 1.98 N, and zero
b) 0.36 A m^2 or 0.45×10^{-6} Wb·m, L $= 0.396$ N·m parallel to *ab* with counterclockwise current

7–2 b) $C = B\beta \cosh \delta$, $D = (\tanh \delta)/\beta$

7–3 a) $l^2V/V_a 4s$; no   b) $V^2/V_a 2B^2 s^2$; no
c) $V \gtrless sBc/10$, $V_a \gtrless (m/2e)(c/10)^2$ d) $v = V/sB$

7–5 a) $y^2/z = x_0^2 B^2 e/2mE$ b) No error with sufficiently high velocities

7–7 a) $m \, dv/dt = e(\mathbf{E} + \mathbf{v} \times \mathbf{B})$ b) $v_z = eEt/m$
c) $v_x = v_0 \sin (eB/m)t$ and $v_y = v_0 \cos (eB/m)t$

7–9 a) 30.7×10^6 cps   b) 2.15×10^8 m s^{-1} c) 19.2×10^6 cps

7–11 a) +1.5 A and 54 W   b) zero A and zero W   c) -1.5 A and -54 W

7–13 a) 54 m s^{-2}, 26.7 m s^{-1} b) 15.7 m s^{-1}
c) 33.3 m s^{-1} d) 4.05 N   e) 135 W

7–15 a) $v_f = \mathcal{E}_B Bl/(B^2 l^2 + RG)$, $p_m = Gv_f^2$ b) $\mathcal{E}_B(\mathcal{E}_B - Blv_f)/R$

7–17 a) $\pi a^2 B_0 \alpha e^{-\alpha t}/R$, $\pi a^2 B_0 e^{-\alpha t}/R$

7–19 a) $B_0 \omega r \sin \omega t/2\rho$ b) $(\omega^2 B_0^2 \pi a^4 h/8\rho) \sin^2 \omega t$

7-21 a) Opposite direction   b) 1000 rad s^{-1} c) 667 rad s^{-1}
7-29 a) $\theta = \theta_p[1 - 0.8e^{-\alpha t} \sin(\beta t + 0.93)]$, $\alpha = 3\omega_0/5$, and $\beta = 4\omega_0/5$
7-31 $\theta_i/q = \omega_0\theta R_x/I(R_x + R_g)e$
7-33 a) Fluxmeter treatment in Section 7–8 is valid since $\omega_0 \ll N^2\phi_m/2G(R_g + R_s)$

Chapter 8

8-1 a) Zero   b) 1.41×10^{-6} Wb m^{-2} c) 3×10^{-6} Wb m^{-2} in the x-direction
8-3 2.4×10^{-20} N
8-5 3.42×10^{-3} Wb m^{-2}
8-7 10 μC
8-9 a) $\sin \alpha = \sin \beta \simeq L/2a$
8-11 $\mathbf{B} = (\mathbf{i} + \mathbf{j})2I\mu_0/3\pi a$
8-13 $Ia^2\{1/(a^2 + x^2)^{3/2} - 4/(4a^2 + x^2)^{3/2}\}/2$
8-15 $B = \mu_0 Ir/2\pi a^2$
8-17 a) $(V_e 2\pi t/\rho_c) \ln b/a$ b) 0, $(V_e/r) \ln b/a$, 0, 0
c) $Ir/2\pi a^2$, $I/2\pi r$, $I(c^2 - r^2)/(c^2 - b^2)2\pi r$, 0
8-19 a) If $2d$ = distance between wire centers, $s^2 = d^2 - a^2$ and the results for Problem 8–18 are correct.
b) Flux lines follow circles outside the wires but are distorted where they pass through the wires.
8-21 $B_x = m(y - x) + c_1$, $B_z = my + c_2$
8-23 a) $\mathbf{A} = k(\mu_0 I/4\pi)\{\ln 2Z + \ln(\sqrt{r^2 + z^2} + z) - 2\ln r\}$

b) $\mathbf{B} = \mathbf{a}_\phi \dfrac{\mu_0 I}{4\pi r}\left\{2 - \dfrac{r^2}{r^2 + z^2 + z\sqrt{r^2 + z^2}}\right\}$

8-25 a) $\mathbf{J} = \mathbf{a}_\phi(3G \sin \theta)/2$ in A m^{-1} b) $\mathbf{B}_0 = \mathbf{d}_f\mu_0 Ga^3/2r^3$ in Wb m^{-2}
8-27 yes
8-29 a) 121 mH   b) 102 mH
8-31 a) $L_1 + L_2 + 2M$ b) $L_1 + L_2 - 2M$
8-33 a) 23×10^{-4} n·m   b) the diameter of the coil
8-35 a) $NI/\pi(a + b)$ b) max $NI/2\pi a$, min $NI/2\pi b$
8-37 a) zero   b) $F = 0$, $L = \mu_0 I_1 I_2 b/\pi$

Chapter 9

9-5 $\mathbf{B}_i = \mathbf{k}1.0$ Wb m^{-2}, $\mu_0\mathbf{H}_i = -\mathbf{k}0.5$ Wb m^{-2}
9-7 $\mathbf{B}_i = \mathbf{k}2\mathcal{M}/3$, $\mu_0\mathbf{H}_i = -\mathbf{k}\mathcal{M}/3$, $\mathbf{B}_e = \mathbf{d}_f\mathcal{M}a^3/3r^3$
9-9 a) 1430δ J   b) 1430 N or 320 lb
9-11 a) 94×10^{-6} Wb   b) 9.4×10^{-6} Wb·m
c) 7.5×10^{-3} N·m   d) 1.8% of H_c
9-15 a) $\mathbf{a}_z(1 - a^2/2l^2)\mathcal{M}/2$, $\mathbf{H} = \mathbf{B}/\mu_0$
b) $\mathbf{a}_z(1 - a^2/2l^2)\mathcal{M}/2$, $-\mathbf{a}_z(1 + a^2/2l^2)\mathcal{M}/2\mu_0$
c) $\mathbf{a}_z(1 - 2a^2/l^2)\mathcal{M}$, $-\mathbf{a}_z 2\mathcal{M}a^2/\mu_0 l^2$
d) $-\mathbf{a}_z 2\mathcal{M}a^2/l^2$, $\mathbf{H} = \mathbf{B}/\mu_0$
9-17 $-\mathbf{a}_r I\mathcal{M}la^2/2r^2$

9–19 $\Omega = p_m \cos \theta / 4\pi\mu_0 K_m r^2$

$H = d_f p_m / 4\pi\mu r^3$

$B = d_f p_m / 4\pi r^3$

9–21 a) $\mu_0 H + \mathcal{M}/3$ b) $\mu_0 H$ c) $\mu_0 H + \mathcal{M}$ d) $\mu_0 H + \mathcal{M}/2$

9–23 a) $B = \mu_0 H = 0.293\,\mathcal{M}_i$ b) $B = \mu_0 H = 0.552\,\mathcal{M}_i$

c) $\mu_0 H = -0.448\,\mathcal{M}_i$, $B = 0.552\,\mathcal{M}_i$

d) $\mu_0 H = -\mathcal{M}_i a^3 / 16 r_c^3$, $B = \mathcal{M}_i(1 - a^3 / 16 r_c^3)$

9–25 $B_i/B_0 = 4b^2 K_m / [b^2(K_m + 1)^2 - a^2(K_m - 1)^2]$

9–27 a) $L_a = \mu_0 K_m N_a^2 a^2 / 2 r_c$, $M = \mu_0 K_m N_a N_b a^2 / 2 r_c$

$L_b = \mu_0 N_b^2 (K_m a^2 - a^2 + b^2)/2 r_c$

b) $L_b = \mu_0 K_m N_b^2 b^2 / 2 r_c$, and L_a and M are the same as in (a).

Chapter 10

10–1 a) ρ_m is unique   b) **S** is not unique

10–3 a) $\mathcal{E}_1(\mathcal{E}_1 - \mathcal{E}_2)\theta^2 / R\theta_0^2$

b) Flux builds from zero at $\theta = 0$ to $\mathcal{E}_1(\mathcal{E}_1 - \mathcal{E}_2)/R$ at $\theta = \theta_0$.

10–5 a) Angular velocity increases from zero to ω_0 and then remains steady.

b) Initial $\mathbf{S} = \mathbf{a}_z IB\omega_0 / 2\pi + \mathbf{a}_\phi B^2 \omega_0 r / \mu_0$, and final $\mathbf{S} = \mathbf{a}_\phi B^2 \omega_0 r / \mu_0$, since I then equals zero

c) $\mathcal{E}_0 I_i$, zero

10–7 a) $(\mu_0 I^2 l / 4\pi) \ln b/a$ b) $\epsilon_0 V_m^2 \pi l / \ln (b/a)$

10–9 a) $\mathbf{j} 2G \sin \omega t \sin (\omega z/v)$

b) $\mathbf{k}(4G^2/\mu r) \sin \omega t \cos \omega t \sin (\omega z/v) \cos (\omega z/v)$ c) zero

Chapter 11

11–5 a) 70.7 V   b) 86.6 V

11–7 $X_L = X_0 f/f_0$, $X_C = X_0 f_0/f$, and $X = X_L - X_C = X_0(f/f_0 - f_0/f)$.

11–9 a) 150 μF   b) 917 W

11–11 a) 0.987   b) 6.9×10^{-4} J

c) 8×10^{-4} J   d) 5320 Ω

11–13 b) $\delta = 1 + 8|\mathfrak{Z}_m|3/\mathfrak{V}$ c) $\delta - 1 = 320 \times 10^{-6} e^{j\phi}$, a circle

11–15 $\mathfrak{Z}_n \mathfrak{Z}_n = \mathfrak{V}_t$

11–17 a) $CVB[1 - e^{-\alpha t} \sec \delta \cos (\beta t - \delta)]$, where $\text{ctn}^2 \delta = (4L/R^2 C - 1)$

b) $CVB[1 - e^{-\alpha t} \text{sech} \delta \cosh (\gamma t + \delta)]$, where

$\coth^2 \delta = (1 - 4L/R^2 C)$ and $\gamma^2 = R^2/4L^2 - 1/LC$

11–19 a) Assuming the generator voltage to be $v = V_m \cos \omega t$, the transient current is $i_c = K\beta e^{-\alpha t} \cos (\beta t + \delta)$, where $K = V_m \sin \theta / \omega Z \sin \delta$ and $\text{ctn } \delta = (\alpha - \omega \text{ ctn } \theta)/\beta$. The symbols Z and θ are defined in Fig. 11–3 and the values of α and β are given in Section 11–2.

b) Assuming the generator voltage to be $v = V_m \sin \omega t$, the transient current is $i_c = -K_1 \beta e^{-\alpha t} \sin (\beta t + \delta_1)$, where $K_1 = V_m \cos \theta / \omega Z \cos \delta_1$, $\tan \delta_1 = (\alpha + \omega \tan \theta)/\beta$, and the symbols α, β, Z, and θ have the same values as given in (a).

c) $f_0 = 3.13 \times 10^3$ cps

11-21 a) $\theta_2 - \theta_1 = 158°8'$, $\theta_3 - \theta_1 = 228°11'$
b) $i_1 = 3.3$ A, $i_2 = -4.6$ A, and $i_3 = 1.3$ A
11-23 b) $\mathcal{V}_J = (\mathcal{V}_1 \mathfrak{z}_2 \mathfrak{z}_3 + \mathcal{V}_2 \mathfrak{z}_1 \mathfrak{z}_3 + \mathcal{V}_3 \mathfrak{z}_1 \mathfrak{z}_2)/(\mathfrak{z}_1 \mathfrak{z}_2 + \mathfrak{z}_1 \mathfrak{z}_3 + \mathfrak{z}_2 \mathfrak{z}_3)$
c) $\mathcal{V}_J = -4.85 - j3.18$ V
11-25 a) 0.679 W   b) 0.751 W   c) 0.093 W
11-27 a) $R_{ac} = 1\ \Omega$, $X = 40.0\ \Omega$
b) $R_{ac} = 9\ \Omega$, $X = 4.36\ \Omega$
11-29 a) 2.40 A, 2.078 A, 2.29 A   b) 14.59 Ω must be added
c) 0.202 μF   d) 19.9×10^{-6} A or 10^{-5} of the steady 2.00 A
11-31 a) 286 turns   b) 0.0248 Ω c) 26.46 W and 75.9 Ω d) 75.93 Ω
11-33 a) 4.7×10^{-19} Wb   b) Output signal is 4.7 V·ns and peak voltage is
47 mV.   c) 3.0×10^{-19} Wb
11-35 a) 2.81×10^{-9} A Wb or W·s   b) 1.2×10^{-4}
11-37 a) 20×10^{-6} V   b) The large area enclosed by the hysteresis loop results in large losses per cycle.

Chapter 12

12-1 3×10^{18}, 0.6×10^{15}, 3×10^9, 10^6; all in Hz
12-3 a) 16×10^{-6} W m^{-2} b) 0.11 V m^{-1}, 2.9×10^{-4} A m^{-1}
12-5 38.6 V m^{-1}, 0.103 A m^{-1}
12-7 63 A
12-9 a) Parallel to z-axis
b) $\omega^4/(\omega_0^2 - \omega^2)^2$ becomes large at high frequencies so that blue light is strongly
scattered by the sky.
12-15 a) 0.15   b) 0.074
12-17 a) 180° phase change in **E**   b) No phase change
12-21 $H_y = -\epsilon_0 c E_z$. The pulse is rectangular.
12-23 Zero everywhere.
12-27 $v = c/\sqrt{2}$, $z_0 = (\sqrt{\mu_0/2\epsilon_0})/20$
12-29 b) $z_0 = (\sqrt{3\mu_0/\epsilon_0} \ln 19.95)/2\pi$ c) 11 ns

Chapter 13

13-3 a) $G_1 = 0.042$ V and $G_2 = 0$ b) same as (a)
13-5 a) $v = c(1 - 0.11 \times 10^{-7})$ b) same as (a)   c) 3,000 times larger
d) 7.5 m
13-7 a) $F'/F = 1/\gamma$ b) $F = q\sigma/\epsilon_0$ and $F' = q\sigma'/\epsilon_0 + F_i'$. Therefore $F_i' = F/\gamma - q\sigma'/\epsilon_0 = (q\sigma'/\epsilon_0)(1 - \beta^2 - 1)$ since $\sigma' = \sigma\gamma$. Finally, $F' = (1 - \beta^2)q\sigma'/\epsilon_0$.
c) $F' = q\sigma'/\epsilon_0 - \beta^2 q\sigma'/\epsilon_0$ from (b). The first term is the Coulomb term while the
second term can be transformed as follows:

$$F_i' = -v^2 \mu_0 \epsilon_0 q\sigma'/\epsilon_0 = qvB'$$

where $-\sigma'v\mu_0 = B'$.

13–9 By requiring $\dot{v}_x = 0$ we get for F'_x and F'_z, respectively:

$$q\sigma' \mu_0 v'_z v'_x = (m_0\gamma^3/c^2)\dot{v}'_z v'_z v'_x$$

and

$$(1 - \beta_0^2)q\sigma'/\epsilon_0 = (m_0\gamma^3/c^2)(v'^2_z + c^2/\gamma^2)\dot{u}'_z$$

Consistency of the solutions can be shown by reducing both equations to the same form, i.e.,

$$q\sigma' \mu = (m_0\gamma^3/c^2)\dot{v}'_z.$$

INDEX